BIOENGINEERING INNOVATIVE SOLUTIONS FOR CANCER

BIOENGINEERING INNOVATIVE SOLUTIONS FOR CANCER

Edited by

SYLVAIN LADAME

Department of Bioengineering, Imperial College London, London, United Kingdom

JASON Y.H. CHANG

Koch Institute for Integrative Cancer Research, Massachusetts Institute of Technology, Cambridge, MA, United States

ELSEVIER

ACADEMIC PRESS

An imprint of Elsevier

Academic Press is an imprint of Elsevier
125 London Wall, London EC2Y 5AS, United Kingdom
525 B Street, Suite 1650, San Diego, CA 92101, United States
50 Hampshire Street, 5th Floor, Cambridge, MA 02139, United States
The Boulevard, Langford Lane, Kidlington, Oxford OX5 1GB, United Kingdom

Library of Congress Cataloging-in-Publication Data
A catalog record for this book is available from the Library of Congress

British Library Cataloguing-in-Publication Data
A catalogue record for this book is available from the British Library

ISBN : 978-0-12-813886-1

For information on all Academic Press publications
visit our website at https://www.elsevier.com/books-and-journals

Publisher: Mara Conner
Acquisition Editor: Fiona Geraghty
Editorial Project Manager: Emma Hayes
Production Project Manager: Selvaraj Raviraj
Cover Designer: Christian J. Bilbow

Typeset by SPi Global, India

Contents

Contributors

Dana Al Sulaiman
Department of Bioengineering, Imperial College London, London, United Kingdom

Neal K. Bangerter
Department of Bioengineering, Imperial College London, London, United Kingdom

Alexandra Boussommier-Calleja
Department of Mechanical Engineering, Massachusetts Institute of Technology, Cambridge, MA, United States

Rona Chandrawati
School of Chemical Engineering and Australian Centre for Nanomedicine (ACN), The University of New South Wales (UNSW Sydney), Sydney, NSW, Australia

Jason Y.H. Chang
Koch Institute for Integrative Cancer Research, Massachusetts Institute of Technology, Cambridge, MA, United States

Robert B. Channon
Department of Bioengineering, Imperial College London, London, United Kingdom

Armando del Rio Hernandez
Cellular and Molecular Biomechanics Laboratory, Department of Bioengineering, Imperial College London, London, United Kingdom

Derfogail Delcassian
Koch Institute for Integrative Cancer Research, Massachusetts Institute of Technology, Cambridge, MA, United States; Division of Regenerative Medicine and Cellular Therapies, University of Nottingham, Nottingham, United Kingdom

Daniel S. Elson
Hamlyn Centre for Robotic Surgery, Institute of Global Health Innovation and Department of Surgery and Cancer, Imperial College London, London, United Kingdom

Matthew Grech-Sollars
Department of Surgery & Cancer, Imperial College London, London, United Kingdom

Md. Nazmul Islam
Department of Bioengineering, Imperial College London, London, United Kingdom

Sylvain Ladame
Department of Bioengineering, Imperial College London, London, United Kingdom

Leyuan Ma
Koch Institute for Integrative Cancer Research, Massachusetts Institute of Technology, Cambridge, MA, United States

Glen Morrell
Department of Radiology and Imaging Sciences, University of Utah, Salt Lake City, UT, United States

George P. Mylonas
Department of Surgery and Cancer, Imperial College London, London, United Kingdom

Timo Joric Corman Oude Vrielink
Department of Surgery and Cancer, Imperial College London, London, United Kingdom

Asha K. Patel
National Heart & Lung Institute, Imperial College London, London, United Kingdom

Suraj Pavagada
Department of Bioengineering, Imperial College London, London, United Kingdom

Alistair Rice
Cellular and Molecular Biomechanics Laboratory, Department of Bioengineering, Imperial College London, London, United Kingdom

Charles A. Sennoga
Auto entrepreneur & Visiting Researcher UMR Inserm U930 - Imagerie et cerveau Université François Rabelais de Tours France

Murillo Silva
Koch Institute for Integrative Cancer Research, Massachusetts Institute of Technology, Cambridge, MA, United States

Isobel Steer
Department of Bioengineering, Imperial College London, London, United Kingdom

Angie Davina Tjandra
School of Chemical Engineering and Australian Centre for Nanomedicine (ACN), The University of New South Wales (UNSW Sydney), Sydney, NSW, Australia

Valentina Vitiello
School of Biomedical Engineering & Imaging Sciences, Kings College London, London, United Kingdom

Chensu Wang
Koch Institute for Integrative Cancer Research, Massachusetts Institute of Technology, Cambridge, MA, United States

Acknowledgments

Contributions toward the book cover:

Top panel: "Microwell array for massively parallel combinatorial screening"

Credit: Jared Kehe

Affiliation:

Department of Biological Engineering, Massachusetts Institute of Technology, Cambridge, MA, United States

The Broad Institute of Massachusetts Institute of Technology and Harvard, Cambridge, MA, United States

Middle panel: "Microneedle-based sampling of miRNAs from interstitial fluid"

Credit: Jason Y.H. Chang[a], Dana Al Sulaiman[b], and Sylvain Ladame[b]

Affiliation:

[a]Koch Institute for Integrative Cancer Research, Massachusetts Institute of Technology, Cambridge, MA, United States

[b]Department of Bioengineering, Imperial College London, London, United Kingdom

Bottom panel: "Cellular uptake of aluminum hydroxide particulates"

Credit: Yash Agarwal[a,b] and Jason Y.H. Chang[a]

Affiliation:

[a]Koch Institute for Integrative Cancer Research, Massachusetts Institute of Technology, Cambridge, MA, United States

[b]Department of Biological Engineering, Massachusetts Institute of Technology, Cambridge, MA, United States

Introduction

Cancer is the second leading cause of death globally and is responsible for an estimated 9.6 million deaths in 2018, according to the World Health Organization. Globally, about one in six deaths is due to cancer. Thirty to fifty percent of cancers can currently be prevented by avoiding risk factors and implementing existing evidence-based prevention strategies. But what about those 50%–70% of other cancer types that have no clearly identified risk factors and cannot be rationally prevented? Early detection and better management of cancer patients hold the greatest promise for reducing the burden associated with cancer. Bioengineers play a critical role in both diagnostics and treatment intervention thanks to their ability to design and develop devices and technologies that can be implemented into the clinic to ensure cancers are diagnosed at their early onset and that cancer patients are offered the best suited and most effective treatment. This textbook illustrates how bioengineers have contributed, in recent years, to the emergence of new sensing and imaging technologies that enable specific cancer types to be detected earlier with increased accuracy, allowing cancer patients to be treated sooner to maximize survival rates. It also reports on ways molecular bioengineers have also created new ways of treating cancers by developing new drugs (chemotherapies and more recently immunotherapies), improving their site-specific delivery and specificity to minimize the side effects commonly associated with the current treatments. Cancer surgery can also be optimized through technological advances brought up by bioengineers. In addition to providing clinicians with new tools and technologies, bioengineering innovative solutions for cancer does also involve helping researchers understand how cancer cells form and proliferate, how they respond to new treatments, and what specific features make them different from healthy cells. This fundamental knowledge is mainly acquired through in vitro or ex vivo experiments that make use of accurate cancer models designed and developed by bioengineers.

What is cancer? Cancer, as defined by the National Cancer Institute at the National Institutes of Health, is a general term for diseases in which abnormal cells divide without control and can invade nearby tissues. Other terms used include malignancy (or malignant tumor) and neoplasm. In a typical life cycle, healthy human cells divide to form new cells when required and then ultimately die after growing old or becoming damaged, at which point they are replaced by new, healthy cells. When cancer forms and develops, cells start behaving in an abnormal way, surviving, and duplicating despite growing old, damaged, and dysfunctional. This abnormal and uncontrolled cell growth may ultimately lead to the formation of tumors. Cancer can start almost anywhere in the human body but can also spread to nearby or more distant tissues/organs, for example, through the blood or the lymphatics, thus forming new malignancies far from the original tumor. The ability to spread is what distinguishes malignant tumors from benign tumors. When benign tumors remain localized and can typically be easily removed from almost anywhere in the body (unless when present in the brain), malignant tumors are significantly more aggressive and difficult to treat.

Cancer types are usually named after the organs or tissues the cancer originated form. For example, prostate cancer starts in cells of the prostate, and breast cancer starts in cells of the breast. They can also be broadly categorized as follows: *carcinoma* for cancers originating in the skin or in tissues that line or cover internal organs; *sarcoma* for cancers originating in bone, cartilage, fat, muscle, blood vessels, or other connective or supportive tissue; *leukemia* for cancers originating in blood-forming tissue, such as the bone marrow; and *lymphoma* for cancers originating in cells from the immune system.

How does cancer arise? Cancer is a genetic disease, meaning it is the result of modifications affecting genes that control the way healthy cells grow, divide, and function. These abnormal modifications can either be inherited from our parents or be acquired during our lifetime as a result of our lifestyle or specific environmental exposure. While inherited factors can help clinicians predict the likelihood of developing a certain cancer in your lifetime, environmental factors (e.g., smoking, alcohol consumption, unbalanced diet, and sun exposure) suggest that certain cancer types can be avoided by adjusting our lifestyle. Genetic modifications (e.g., changes in DNA sequences or DNA mutations) are typically more prevalent in cancer cells than in healthy cells and are often called cancer "drivers." Each person's cancer has a unique combination of genetic changes that keeps evolving while cancer grows. Genes most commonly affected by these changes in cancer cells are proto-oncogenes, tumor suppressor genes, and DNA repair genes. *Proto-oncogenes* are genes naturally involved in controlling cell growth and division, which, when altered in specific ways, become cancer-causing genes (or oncogenes) by allowing cells to grow erratically and survive when they should not. *Tumor suppressor genes* are genes normally acting to inhibit cell proliferation and tumor development. Modifications to such genes may remove negative regulators of cell proliferation and contribute to the abnormal proliferation of tumor cells. *DNA repair genes* are genes responsible for maintaining the integrity of our genes by fixing/correcting any damaged DNA. Modifications to this DNA repair machinery make cells more susceptible to cancer-causing irreversible modifications.

How is cancer diagnosed? Symptoms vary greatly with cancer types, which can be very difficult to identify, and in some cases only appear once the cancer is already at a fairly advanced stage. Cancer is most commonly diagnosed when patients approach their doctor after appearance of the first persistent physical symptoms. Further tests including scans or invasive tissue biopsies will then typically be performed to confirm the diagnosis. If confirmed, further sets of tests will typically be ordered to assess the aggressive and advanced nature of the cancer. Detecting cancers early, even before the appearance of the first symptoms, is possible by widespread checking of broad populations via a process called screening, that is, looking for cancer-associated abnormalities in people who have no apparent symptoms. Early detection through public screening of at risk-individuals is the most effective way toward early detection, when the cancer is more localized and easier to treat. Only very few cancer screening programs are available (for very specific cancer types), with large geographical disparities. Despite their proven ability to detect cancers early and increase patient survival rate, screening tests also have some potential pitfalls. False-positive results, for instance, (test responding positive even though the patient is healthy) can not only cause unnecessary anxiety to patient who may even undergo potentially harmful procedures unnecessarily but also lead to overdiagnosis of slow growing cancers and overtreatment of patients who could have lived happily without any treatment due to the low aggressiveness of the cancer.

How is cancer treated? Following diagnosis, patients will be offered different treatment options that depend hugely on the cancer type and how advanced the cancer is. Most patients will undergo a combination of treatments, such as surgery to remove the primary tumor followed by chemotherapy and/or radiation therapy. New therapies are also becoming available that have been developed, at least in part, by molecular bioengineers, which include immunotherapy, targeted therapy, or hormone therapy. Surgery is the procedure by which a surgeon removes a malignant tumor from your body; it requires extreme precision, which has been improved hugely by the introduction of *robotic surgery*, as discussed later in this textbook. *Radiation therapy* also remains widely used to shrink tumors through high doses of radiations, while *chemotherapy* refers to the use of chemical drugs to treat cancers. Although typically very effective at killing cancer cells, chemotherapies are also commonly associated with a number of undesirable side effects (e.g., hair loss, weakened immune system, nausea). While *hormone therapy* can help treat cancers that rely on hormones to grow (e.g., prostate and breast cancer), *immunotherapy* is a more recent type of treatment that helps your immune system fight cancer.

Because every cancer patient requiring medical attention is unique, the optimal way to deal with their condition will inevitably differ between individuals. This is the concept of *precision medicine*, which proposes a customized approach for treatment and prevention compared to the current "one size fits all" model of patient management. By getting a better genetic understanding of a person's cancer, clinicians will be able to select the best suited treatment.

This textbook presents how bioengineers are tackling the many challenges our society faces to reduce the burden of cancer. Finding innovative solutions to cancer can take many forms that are presented and discussed in this book: developing new sensing and imaging technologies to detect cancer earlier and with increased precision; engineering new therapies and new nanotechnologies to improve their delivery; to developing new robotic tools to improve surgery's precision; and developing new models to better understand the various ways cancer cells form, propagate, and respond to treatment. Every branch of the very diverse field that is bioengineering has a role to play in addressing the many challenges caused by cancer, helping cancer patients live longer, healthier, and ultimately cancer-free.

Sylvain Ladame
Department of Bioengineering, Imperial College London,
London, United Kingdom

PART 1

Sensing and monitoring

CHAPTER 1.1

Diagnostic, prognostic, and predictive biomarkers for cancer

Jason Y.H. Chang[a], **Sylvain Ladame**[b]
[a]Koch Institute for Integrative Cancer Research, Massachusetts Institute of Technology, Cambridge, MA, United States
[b]Department of Bioengineering, Imperial College London, London, United Kingdom

A biomarker, as defined by Naylor [1], is a "biological characteristic that is objectively measured and evaluated as an indicator of normal biological processes, pathogenic processes, or pharmacological response to a therapeutic intervention." In the context of cancer specific biomarkers, these can be subclassed into either biological molecules (molecular biomarkers) or biological processes. They can be produced or induced either by the tumor itself or by other noncancerous tissues in response to the tumor. The clinical performance of a cancer biomarker is defined by its sensitivity and specificity.

Sensitivity or true positive rate (TPR) is the probability that the biomarker will be detected positive in the cancer patient. Specificity, also known as true negative rate (TNR), is the probability that the biomarker will be absent or below a given threshold in a "noncancer" control. An ideal biomarker would have 100% sensitivity (all individuals with confirmed cancer test positive for the biomarker) and 100% specificity (all individuals without cancer test negative for the biomarker). The relationship between sensitivity and specificity can be represented graphically as a receiver operating characteristic (ROC) curve (Fig. 1) [2]. It is a plot of the true positive rate (sensitivity) against the false positive rate (1 − specificity) and is commonly used to assess the efficacy of a biomarker for every possible cutoff point (i.e., for every chosen threshold value that distinguishes healthy from sick). A change in cutoff value that leads to an increase in sensitivity will typically be at the expense of lowering specificity, and vice versa.

The area under the ROC curve (AUROC) for a given biomarker (or test) measures its ability to discriminate between "healthy" and "sick" states [2]. An AUROC of 1 indicates 100% sensitivity and 100% specificity, whereas a biomarker with an AUROC of 0.5 is equivalent to flipping a coin to discriminate between a healthy and sick subject and is considered useless. The closer the ROC curve is to the upper left corner, the larger the AUROC and the more effective it is at being a discriminative biomarker.

Biomarkers can be diagnostic (determine the presence and type of cancer), prognostic (give information on the patient's overall cancer outcome with or without standard treatment), or predictive (help to identify which treatment the patient is most likely to

Bioengineering Innovative Solutions for Cancer
https://doi.org/10.1016/B978-0-12-813886-1.00001-2

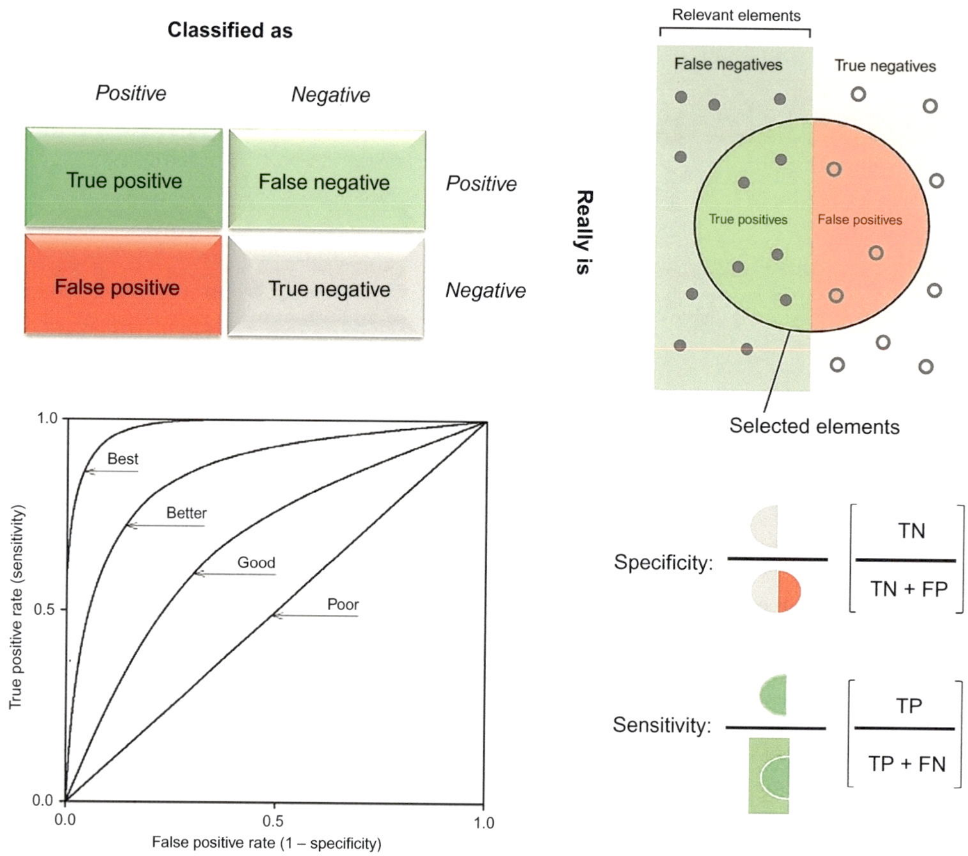

Fig. 1 Sensitivity, specificity and receiver operating characteristic (ROC): assessing the performance of a diagnostic, prognostic or predictive cancer test.

respond to or benefit from) [3]. Certain biomarkers can be diagnostic, prognostic, and predictive at the same time.

Detecting cancer at an early and potentially treatable stage is one of the greatest challenges in medicine today, with the identification of biomarkers for this purpose and the development of technologies to detect them currently being one of the top priorities in cancer research. Early detection of cancer can be achieved via identification of individuals at increased risk of developing a cancer or via widespread public screening and is commonly associated with significantly higher survival rates (Fig. 2).

Many genes predisposing to hereditary cancers have already been identified, enabling genetic testing for cancer susceptibility to be carried out within high-risk families [4]. Typically, a positive test indicates that the individual has an increased risk of developing a malignancy, while a negative result will indicate that the individual has the same probability of developing cancer as a similar sex and age-matched person from a family

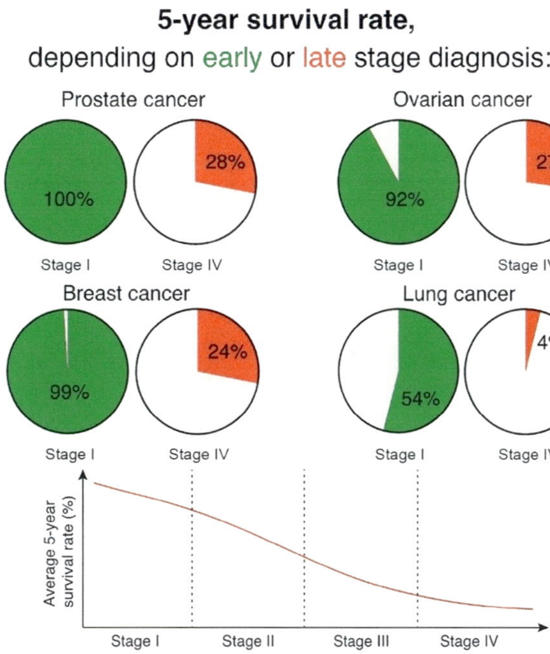

Fig. 2 Early intervention leads to higher probability of cure and survival. *(Modified from SEER Cancer Statistics, National Cancer Institute 2013.)*

not at risk. While full genome or targeted sequencing may help identify individuals at risk of developing cancer, rapid and low-cost qualitative and quantitative detection of specific cancer biomarkers is essential for early cancer diagnosis, monitoring of disease progression, and ability to assess individual patients' response to treatment, providing a snapshot of the state of the disease.

Cancer biomarkers can be nucleic acids, proteins, metabolites, cells, extracellular vesicles, or processes such as apoptosis, angiogenesis, or proliferation [5]. Depending on where they are found in the body (e.g., in bodily fluids such as blood, urine, or saliva; in tissues; or in cells), these biomarkers will be classified as either noninvasive, minimally invasive or invasive. Noninvasive biomarkers are typically found in exhaled breath or bodily fluids such as urine or saliva and can be sampled without causing any harm to the patient. Minimally invasive biomarkers are typically found in fluids such as blood that need to be sampled via procedures such as finger pricking or blood draw. Finally, invasive biomarkers are typically found and detected within tissue biopsies.

1 Nucleic acid biomarkers

In cancer, nucleic acid-based biomarkers refer to pathogenic changes in deoxyribonucleic acid (DNA) or ribonucleic acid (RNA), molecules which can be used for cancer screening, classification, diagnosis, prognosis, and therapy monitoring. The transformation of a normal (somatic) cell into a malignant cancer cell requires both genetic and epigenetic modifications [6]. These alterations include genome rearrangements (incorrectly placed or translocated genes), amplifications (duplicated genes), and deletions (missing genes), as well as minor changes in the gene's primary sequence (point mutation or single-nucleotide polymorphism). Although these genomic alterations are not all necessarily tumor specific, they can all be used as cancer biomarkers. In this section, we will highlight how specific modifications of the sequence, length, and level of expression of both intracellular and cell-free nucleic acids (cfNAs) can be used as cancer biomarkers.

1.1 Gene translocation

Translocation occurs when a segment from one chromosome is transferred to either a different chromosome or a different site of the same chromosome. It can result in the disruption or deregulation of a normal gene's function and is often considered to be the primary cause of various cancers such as leukemia, lymphoma, and sarcoma. As shown in Fig. 3, translocations can lead to the fusion of the coding sequences of two genes, thereby generating a new gene, so-called oncogene, which carries the ability to induce cancer. Hundreds of translocations have now been linked to cancer, and this information has been reviewed elsewhere [7]. The database continues to grow as investigators uncover new links between genes, chromosomes, and cancer.

1.2 Gene amplification

Gene amplification refers to an increase in the number of copies of a given gene that can also be associated with an increase in the amount of RNA and protein produced from that gene. Gene amplification is common in cancer cells, and some amplified genes may cause cancer cells to grow or become resistant to anticancer drugs. Gene amplification is commonly associated with increased gene expression, with many statistical analyses showing a correlation between the copy number of a gene and its expression level. For example, amplification of the HER2 (for human epidermal growth factor receptor 2) gene is considered a major mechanism for its overexpression in breast cancer [8]. Breast cancers can have up to 25–50 copies of the HER2 gene, and up to 40–100-fold increase in HER2 protein resulting in 2 million receptors expressed at the surface of the tumor cell [9]. HER2 is found overexpressed in 15%–30% of breast tumors (where it is associated with higher rates of recurrence and mortality) but also in 10%–30% of gastric cancers. HER2 overexpression can serve as both predictive and prognostic biomarker and has been observed in other cancer types, including bladder, colon, lung, and ovary.

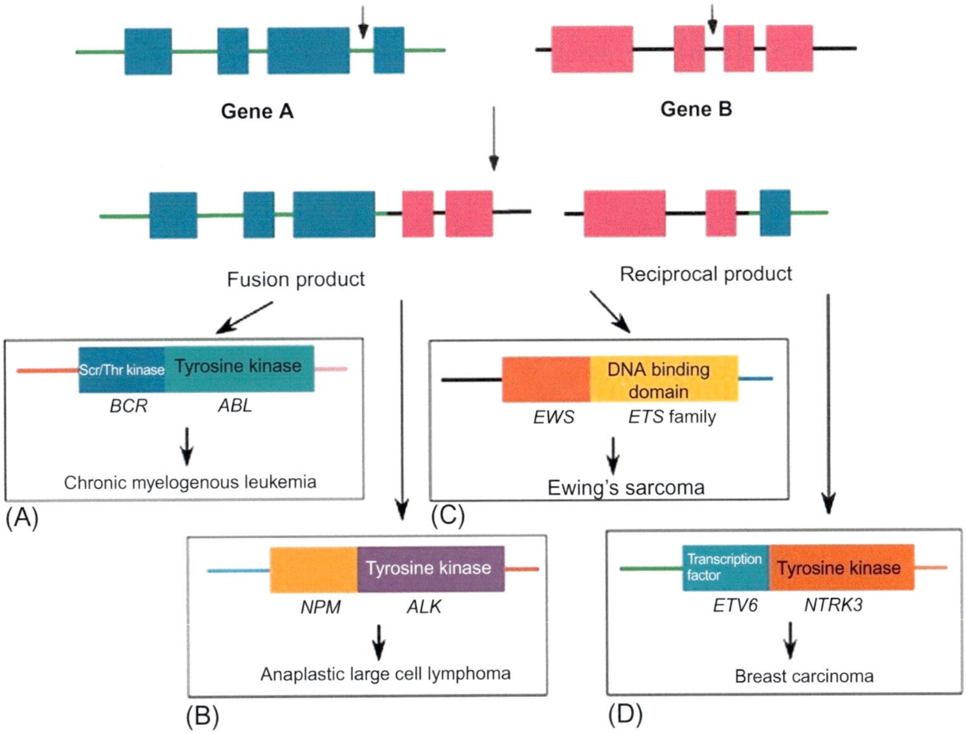

Fig. 3 Chromosomal translocation results in chimeric proteins. In many cancers, translocations result in the formation of a fusion/chimeric protein, which may have a new or enhanced function. For example, upon translocation, gene A breaks and joins with already-broken gene B to result into a fusion product. Joining of other two ends leads to the generation of a reciprocal product, which generally would not result in a functional protein. (A) Chronic myelogenous leukemia is characteristically associated with the presence of the chromosomal translocation between chromosomes 9 and 22. The breaks at chromosome 22 occur in the BCR gene, whereas breaks on chromosome 9 occur in the ABL gene. Upon translocation, the derivative chromosome 22 contains the fused BCR-ABL gene, which is commonly known as the Philadelphia chromosome. This chimeric protein is larger than the ABL protein and has an enhanced tyrosine kinase activity. (B) Anaplastic large-cell lymphoma is a non-Hodgkin's lymphoma, which contains a fusion between the NPM gene on chromosome 5 and ALK gene on chromosome 2. The fusion results in the formation of an activated tyrosine kinase. (C) Ewing's sarcoma is an osteosarcoma, which is characterized by the presence of the fusions between EWS and ETS family of genes. The resultant chimeric proteins act as aberrant transcription factors. (D) One of the common translocations in breast cancers results in the fusion between the ETV6 and NTRK3 genes. ETV6 is a transcription factor, which fuses to the NTRK3 gene, which has tyrosine kinase activity. The chimeric protein acts as an aberrant transcription factor and plays a role in the cancer development. The *lines in bold* represent the double-stranded DNA and the *boxes* represent the coding regions of the genes. *(Data from N. Nambiar, et al., BBA 1786 (2) (2008) 139–152.)*

1.3 Gene deletion

Gene deletion refers to the loss of all or a part of a given gene. In a typical cancer sample, about 25% of the genome is affected by chromosome arm-level deletions and 10% by focal deletions, with about 2% overlap. It is also well established that deletions of tumor suppressor genes could confer selective growth advantage to the cancer cells. For example, phosphatase and tensin homolog (PTEN) is a tumor suppressor gene on chromosome 10 found inactivated by either deletion or mutation in approximately 20% of primary prostate tumor samples and in as many as 50% of castration-resistant tumors. PTEN genomic deletion in prostate cancer was first identified almost 2 decades ago, with PTEN being the most commonly lost tumor suppressor gene in primary disease [10].

1.4 Point mutations and single-nucleotide polymorphism

A point mutation refers to a genetic alteration caused by the substitution of a single nucleotide for another nucleotide. When point mutations are selectively maintained in populations, they are also called single-nucleotide polymorphism (SNP) [11]. The total number of SNPs in the human genome is estimated to be more than 10 million. They are widespread in the human genome with an estimated overall frequency of at least one in every 1000 base pairs (bp), but with marked regional differences. SNPs found in genes that regulate DNA mismatch repair, cell cycle regulation, metabolism, and immunity are associated with genetic susceptibility to cancer although their effect is strongly dependent on their location (Fig. 4). SNPs found in gene promoter affect gene expression by altering promoter activity, transcription-factor binding, DNA methylation, and histone modifications. SNPs in exons affect cancer susceptibility by interfering with gene transcription and RNA translation. SNPs in introns interfere with gene splicing and promote or disrupt binding and function of long noncoding RNAs (lncRNAs). SNPs in the 5′-UTR (untranslated region) affect RNA translation, whereas SNPs in the 3′-UTR affect micro-RNA (miRNA) binding. SNPs in regions that are located far from the actual genes reduce or enhance gene transcription through long-range cis effects.

1.5 Epigenetic modifications

Epigenetic modifications in cancer generally refer to DNA alterations and histone modifications that lead to the silencing of tumor suppressor genes or the activation of oncogenic genes [12]. Epigenetics is formally defined as a heritable change in gene expression or chromosomal stability as a result of chemical modifications of DNA (e.g., methylation) or covalent modification of histones and of noncoding RNAs without a change in DNA sequence [13].

Epigenetic modification of chromatin plays an important role in the regulation of gene expression. DNA is methylated postsynthetically on cytosine residues

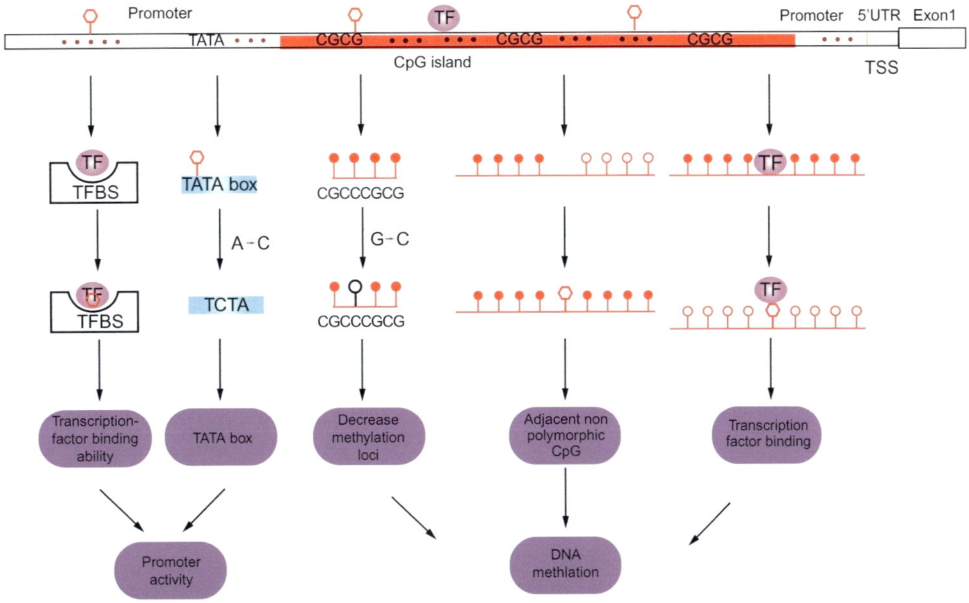

Fig. 4 Schematic representation of mechanisms associated with promoter SNPs and cancer susceptibility. SNPs in transcription factor binding sites affect transcription factor binding to the gene promoter. SNPs in the TATA box affect promoter activity with A to C substitutions decreasing the number of the TATA boxes. SNPs in the CpG islands decrease methylation, affecting adjacent nonpolymorphic CpG and transcription factor binding. The *red hexagon* represents SNP; *red arrows* show substitutions of SNPs; *red hollow circle* represents unmethylated loci; *red solid circle* represents methylated loci. *TF*, transcription factor. *(Modified from N. Deng, H. Zhou, H. Fan, Y. Yuan, Single nucleotide polymorphisms and cancer susceptibility, Oncotarget 8 (66) (2017) 110635–110649.)*

(predominantly in the sequence CpG) and DNA methylation patterns (as well as histone modifications) have been found to be different between normal tissues and tumor-derived tissues. More recently, it has been shown that 5-methylcytosine can be oxidized to 5-hydroxymethylcytosine, resulting in the subsequent removal of the repressive methyl group by a mechanism that appears to include base excision repair processes. Other DNA modifications are also described such as methylation at sites other than CpG and the generation of formyl and carboxyl derivatives of DNA. Expression of every gene is ultimately controlled by their epigenetic status. Therefore, it is not surprising that epigenetic changes play a key role in the process of tumorigenesis.

1.6 Noncoding RNA biomarkers

Work from the past decade has altered our perception of noncoding RNA (ncRNA) from "junk" transcriptional product to functional regulatory molecules involved in key cellular processes (chromatin remodeling, transcription, posttranscriptional modifications, and signal transduction). ncRNAs have been identified as oncogenic drivers and tumor suppressors in every major cancer type and are commonly abnormally expressed (or deregulated) in tumors when compared to normal tissues [14]. Advances in sequencing technologies have led to the discovery of a multitude of ncRNA species. While microRNAs (miRNAs), and circular RNAs (circRNAs) are highly conserved across species, others such as long ncRNAs (lncRNAs) are not. Cellular lncRNAs are primarily found in the nucleus and cytoplasm and adopt a range of secondary structures, such as stem loops and hairpins, through which they can interact with proteins and chromatin [15]. Cellular miRNAs have also recently emerged as a new type of cancer-specific biomarker [16]. miRNAs are small (~22–25 nucleotides) highly conserved ncRNAs endogenously expressed in every cell type. They function as key regulators of gene expression by silencing target transcripts through sequence-specific Watson-Crick base pairing. miRNA deregulation in cancer was first reported in 2002, when a cluster of two miRNAs, miR-15 and miR-16, was identified at a chromosomal region frequently deleted in chronic lymphocytic leukemia (CLL) [17]. This miRNA deletion was shown to act at least in part through allowing higher expression of the miR-15/16 antiapoptotic target B-cell lymphoma 2 (BCL2). Since then, it has been documented that miRNAs have roles in all of the cancer hallmarks defined by Hanahan and Weinberg [18], and are implicated in the clinical management of cancers at every stage. Ubiquitously aberrant expression profiles of miRNAs have been identified in a broad array of human cancers, showing great potential as novel diagnostic and prognostic biomarkers of cancer with high specificity and sensitivity. miRNAs overexpressed in cancer tissues are often regarded as oncogenes or "oncomiRs," which accelerate tumor occurrence, development, and metastasis. Meanwhile, those found at abnormally low levels in cancer patients in relation to healthy individuals were considered tumor suppressors.

2 Circulating, cfNA biomarkers

cfNAs are freely circulating molecules of DNA and RNA in biofluids (including blood, saliva, and urine). CfNAs were first reported in 1948 with Mandel and Métais describing their presence in human blood [19]. However, this attracted little interest from the scientific community until the early 1990s, when researchers detected mutated Rat sarcoma (RAS) gene fragments (associated with multiple cell signaling pathways in cell growth and cell death) in blood samples from cancer patients. Since this time, cfNAs have received a growing amount of interest, with studies suggesting their value as biomarkers of a broad range of pathologies, including cancer [20]. The concentration of cfNAs in healthy individuals is relatively low (ng/mL), due to most nonliving cells being efficiently removed from circulation by phagocytes. However, in cancer patients, the amount of cfNAs present in blood is reportedly up to 40 times higher, thus making cfNA a promising biomarker. The ways cfNAs typically end up in bodily fluids are not yet fully understood but are believed to involve either cell leakage via apoptotic or necrotic mechanisms, or active secretion (e.g., via extracellular vesicles such as microvesicles, exosomes and transport proteins) from both healthy and tumor cells. A broad structural diversity of nucleic acids is found in bodily fluids that is very comparable to that found within cells, including short and long noncoding RNA and double-stranded DNA fragments. When compared to tissue biopsies, liquid biopsies can overcome limitations such as sampling bias, tumor heterogeneity, and difficulties in repeated extraction for longitudinal monitoring. Although the total amount of cfNA per unit of volume of bodily fluid can provide information on tumor progression or response to treatment, the quantitative detection of specific NA biomarkers based on their primary sequence, epigenetic modification, or fragment length typically provides more clinically useful information.

2.1 Circulating miRNAs

miRNAs secreted from cells are found in circulation, making then ideal candidates as minimally invasive biomarkers for diagnosis, prognosis and prediction of cancer. Because single-stranded RNAs are highly sensitive to chemical and enzymatic degradation, circulating miRNAs are assumed to be predominantly bound to either RNA-binding proteins (e.g., Argonaute or Ago) or to be encapsulated within either high-density lipoproteins (HDL) or extracellular vesicles such as exosomes. Unlike initially suspected, circulating miRNAs originating from tumor cells may not simply act as junk in the blood but could in fact be involved in mechanisms of tumor growth and metastasis (traveling to distant organ via transport in exosomes). Mitchell et al. first proved that tumor-derived miRNAs could enter the blood stream and could be detected at high enough levels in bodily fluids to be exploited as a biomarker [21]. Since then, circulating miRNAs have received a lot of attention as either diagnostic or prognostic cancer biomarkers with a large number of distinct miRNAs being deregulated at early stage, during progression or at metastatic stage of specific cancer types (Fig. 5) [22].

Circulating miRNAs also proved to be clinically useful biomarkers to discriminate between different cancer subtypes. For instance, miR-16-5p, miR-21-5p, and miR-199a-5p were all found to be downregulated in plasma of triple negative breast cancer (TNBC) patients compared with both non-TNBC and healthy individuals, while both miR-92a-3p and miR-342-3p were found to be overexpressed in TNBC patients [23]. This can have a major impact on the way cancer patients are treated, with TNBC being typically unresponsive to hormone therapies and preferentially treated by the removal of primary tumors followed by either radiotherapy or chemotherapy.

Despite the huge promise of circulating miRNA for cancer screening, cancer grading, and monitoring response to treatment, they are by no means considered ideal biomarkers [24]. Because miRNAs in circulation originate from both healthy and tumor cells, differences in their expression levels between cancer patients and healthy individuals are usually small. Endogenous concentrations of specific miRNAs in bodily fluids are also highly variable between individuals and strongly dependent on the way samples are collected and

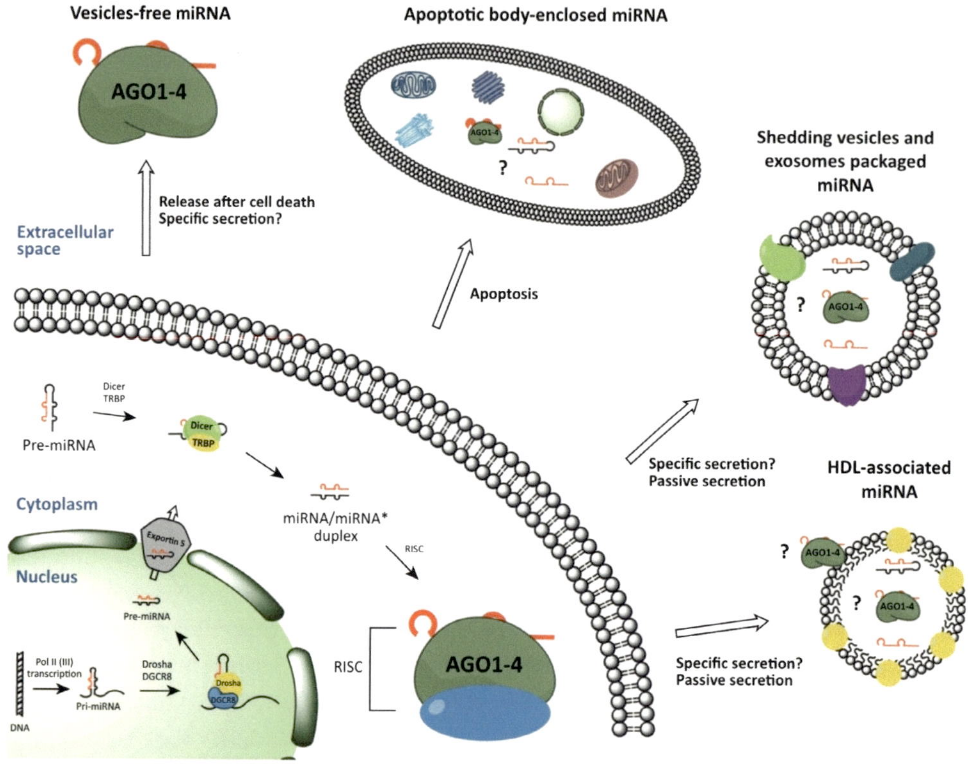

T/BS

Fig. 5 Biogenesis of miRNAs in cells and possible modes of extracellular miRNA packaging. Nearly all miRNAs are synthesized by RNA polymerase II as >200 nucleotide primary miRNA (pri-miRNA) transcripts. Pri-miRNAs must then be matured, starting in the nucleus with cleavage by the endonuclease complex Drosha-DGCR8 into 60–70-nucleotide hairpin intermediates known as miRNA precursors (pre-miRNA). The next step is active transport from the nucleus into the cytoplasm by exportin-5; there, they are further cut by the endonuclease Dicer into ~22-nucleotide miRNA/miRNA* duplexes. Finally, the guide miRNA strand becomes incorporated into one of the four proteins of the Argonaute family (AGO1, AGO2, AGO3, or AGO4), which are the key components of RNA-induced silencing complex (RISC), while the miRNA* strand is degraded. The mature miRNA strand eventually serves as the guide for RISC-mediated mRNA targeting resulting in either mRNA cleavage or translational interference. The three types of membranous vesicles that contain extracellular miRNA are apoptotic bodies, shedding vesicles, and exosomes. Besides these, extracellular miRNA can also be vesicle free and associated with either AGO proteins alone or be incorporated into HDL particles. Apoptotic bodies can contain various cellular organelles, including mitochondria and nuclei. Shedding both vesicles and exosomes belong to the class of microvesicles and are limited by a lipid bilayer; however, they differ in size, content, and mechanism of generation. It remains unknown whether microvesicles can also contain pre-miRNA or protein-free mature miRNA *(question marks)*. It is also not clear whether miRNA is incorporated inside high-density lipoprotein particles or adsorbed on their surface *(question marks)*. *(Data from A. Turchinovich, et al., Trends Biochem. Sci. 37 (11) (2012) 460–465.)*

processed. However, some of these pitfalls can be avoided by measuring multiple miRNAs simultaneously (also called multiplexed analysis or miRNA profiling).

2.2 Circulating lncRNAs

Circulating lncRNAs share many common properties with circulating miRNAs, from the way they are secreted and transported to the extracellular matrix to their surprisingly high stability in bodily fluids. Similar to miRNAs, lncRNA deregulation in cancer patients compared to those of healthy individuals mirrors the deregulation observed in primary tumor tissues. Although characterized by a very low natural abundance in these fluids making them challenging to extract and measure, lncRNAs have already been approved by the US Food and Drug Administration (FDA) for use as diagnostic or prognostic biomarkers in clinical practice. This is the case of PCA3, a lncRNA overexpressed in the urine of prostate cancer patients and outperforming the well-known Prostate-Specific Antigen (PSA) blood test in terms of sensitivity (62%) and specificity (75%) [25]. Several other lncRNAs such as HOTAIR (Ovarian), MALAT1 (lung), and UCA1 (bladder) are also aberrantly upregulated in both malignant tumors and bodily fluids with changes of expression level reaching up to 40-fold [26].

2.3 Circulating tumor DNA

Apoptosis is typically characterized by a pattern of DNA fragmentation, also known as the "apoptotic ladder," which results from internucleosome cleavage of genomic DNA. In most cases, cfDNAs typically display apoptotic ladder profiles due to fragments being 185–200 base pairs (bp) in length as a result of programmed enzymatic cleavage. It is also common to detect larger-sized, nonuniform, NA fragments that are believed to be the result of random and incomplete NA digestion observed in necrotic cells. Circulating tumor DNA (ctDNA) refers to cfDNA that originates from cancerous cells and solid tumors [27]. As a tumor grows, cells die and are replaced by new ones. The dead cells get broken down and their genomic content gets released into the bloodstream. ctDNA are therefore fragments of DNA of globally comparable size (although often slightly shorter on average) to that of cfDNA but with specific genetic differences (e.g., point mutations or epigenetic modifications) [28]. The quantity of ctDNA varies among individuals and depends on the type of tumor, its location, and the cancer stage (e.g., more advanced cancers and typically characterized by larger amounts of ctDNA). Detection of ctDNA can be used for early cancer diagnosis without the need for invasive tumor biopsies, which are particularly challenging to perform when the tumor is located in areas difficult to access, such as in the brain or lung [29]. Detection and genomic characterization of ctDNA can also help select the best suited treatment for personalized medicine by providing an insight into the tumor cell's genome. CtDNAs are also good indicators for monitoring how a given patient responds to treatment (e.g., a decrease in the total amount to ctDNA suggests a reduction in tumor size and successful treatment) and for longitudinal monitoring of remission periods (the detection of ctDNA after apparently curative treatment has been shown to be associated with a high risk of recurrence while the absence of ctDNA suggests the cancer has not returned).

Unfortunately, ctDNAs in bodily fluids are usually vastly outnumbered by DNA fragments originating from noncancer cells making them particularly difficult to detect. Recent studies suggest that differences in fragment lengths of circulating DNA could be exploited to enhance sensitivity for detecting the presence of ctDNA and for noninvasive genomic analysis of cancer [30].

3 Protein biomarkers

Protein-based cancer biomarkers are probably the type of molecular biomarkers most commonly used in clinic. They can be detected either directly in the tumor tissues (via either minimally invasive advanced imaging technologies or ex vivo following highly invasive tissue biopsies) or in liquid biopsies (e.g., blood or urine samples) where they can be detected less invasively [31]. The Bence-Jones protein, immunoglobulin light chains found at abnormally high concentrations in the urine of patients with multiple myeloma, is commonly considered as being the first protein biomarker for cancer discovered more than 160 years ago. Since then, many tumor-associated proteins were identified as either diagnostic or prognostic biomarkers for a broad range of cancer types. Despite tremendous efforts from the scientific community to identify, in recent years, clinically useful protein cancer biomarkers, only a very small proportion of them have received approval from the US FDA. These include, to name only a few, protein CA-125 (also known as mucin 16) for ovarian cancer, CA19-9 for pancreatic cancer, CA15-3 for breast cancer, or prostate-specific antigen (or PSA) for prostate cancer. Although clinically useful for longitudinal monitoring of disease progression, these biomarkers, taken individually, often lack specificity for early stage diagnosis and at best can discriminate between cancer patients and healthy controls. Looking at protein signatures (via multiplexed analysis) instead of individual protein biomarkers has recently emerged as a highly promising way to improve diagnosis accuracy (i.e., increasing rate of true positive and reducing rate of false negative) [32].

The OVA-1 and ROMA diagnostic tests for ovarian cancer illustrate the benefits of multiplexed protein analysis for improved diagnostic accuracy [32]. Despite CA-125 being a very good protein biomarker for ovarian cancer, recent advances showed that combining CA-125 with other protein biomarkers could significantly increase test accuracy. Combining CA-125 and HE4 (or human epididymal secretory protein 4) led to a new scoring model termed Risk of Ovarian Malignancy Algorithm (ROMA) for assessing malignancy risk to women presenting with a pelvic mass. Another variation is the so-called OVA-1 test (Vermillion, Inc.) that is based on a combination of five protein biomarkers (CA-125, TTR, ApoA1, β-2 microglobulin, TF). FDA-approved in 2009, it is offered to patients with a pelvic mass, for use as an adjunct to physical examination and imaging.

4 Exosome biomarkers

Exosomes are extracellular vesicles released by cells that carry a broad diversity of biomolecules such as proteins, nucleic acids, and lipids. Exosomes are endocytic and heterogeneous membrane-derived vesicles, ~50–100 nm in diameter, actively secreted by cells [33]. Because of their ability to transport biologically active cargo from cell to cell (intercellular communication), they play very important roles in immune regulation as well as in cancer initiation and growth [34].

The process of exosome formation is illustrated in Fig. 6. Briefly, the formation of intraluminal vesicles (or ILVs) in multivesicular endosomes (or MVEs) is followed by the transport of MVEs to the plasma membrane. Fusion of MVEs with the plasma membrane then releases the vesicles it contains as exosomes [33].

The potential of exosomes as cancer biomarkers is of two types: (1) Recent studies have reported higher levels of exosomes in cancer patients compared to healthy controls; (2) Exosomes represent a valuable source of molecular biomarkers similar to those described in Sections 2 and 3 and can therefore help in detection and monitoring of cancers. Fig. 6D summarizes the main types of biomarkers associated with exosomes and includes proteins or nucleic acids entrapped within exosomes as well as lipids and proteins associated with the exosomal membrane [33].

4.1 Membrane associated proteins

Tetraspanins are membrane proteins present on almost any exosome from any cell type. Some of the major tetraspanins which are associated with exosomes and promote cancer cell motility, invasion, metastasis, tumor initiation, promotion, progression, and angiogenesis include CD9, CD81, or CD82, to only name a few [35]. They are currently used as markers for oral, colorectal, and colon cancers as well as for glioblastoma. CD9 suppresses motility and promotes adherence to the surrounding matrix subsequently leading to the suppression of tumor progression. Its downregulation has been correlated with tumor progression in several types of cancer, including prostate and bladder. CD82 has been frequently downregulated in advanced stages of cancer and low levels of CD82 are strongly associated with a poor prognosis in patients with prostate, gastric, colon, cervical, ovarian, breast, skin, bladder, endometrial, lung, pancreatic, liver and thyroid cancer, melanoma, and myeloma [36].

GTPases members of the Rab family (e.g., Rab5, Rab11, Rab27, and Rab35) are also commonly found attached to the inner surface of exosome membranes. Because of the

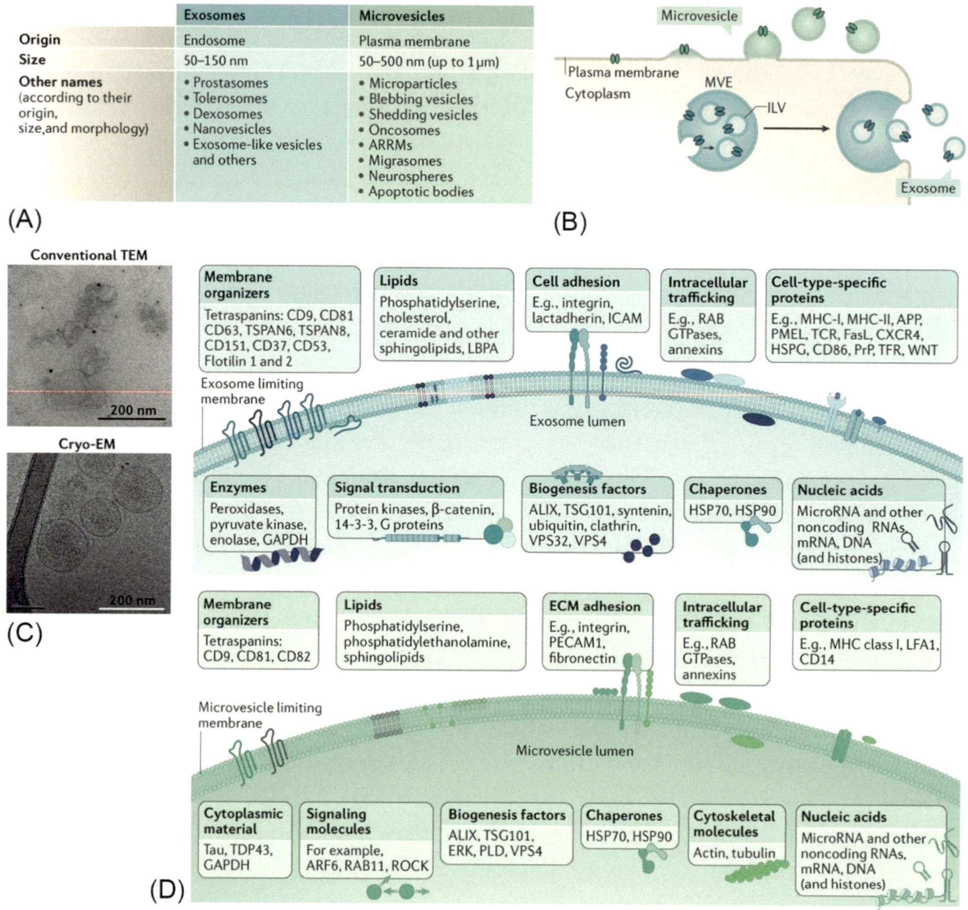

Fig. 6 Main features of extracellular vesicles. (A) Extracellular vesicles comprise a heterogeneous population of membrane vesicles of various origins. Their size may vary (typically between 50 nm and 500 nm, but they can be even larger, measuring 1–10 μm). Over the past two decades, extracellular vesicles have been named based on their origin (cell type), size, morphology, and cargo content but can now be classified into two distinct classes: exosomes and microvesicles. (B) Extracellular vesicles are formed either by budding of the plasma membrane, in which case they are referred to as microvesicles, or as intraluminal vesicles (ILVs) within the lumen of multivesicular endosomes (MVEs). MVEs fuse with the plasma membrane to release ILVs that are then called exosomes. (C) Processing of extracellular vesicles for observation by conventional transmission electron microscopy (TEM) causes their shrinking, leading to an artefactual cup-shaped morphology (top panel). However, when observed in a close-to-native state by cryo-electron microscopy (cryo-EM), they appear as round structures enclosed by double-leaflet membranes (bottom panel). (D) Study of extracellular vesicle composition revealed that they can carry various cargoes, including proteins, lipids, and nucleic acids, and this content can vary widely between cells and conditions. The particular composition will directly affect the fate and function of extracellular vesicles, strengthening the importance of selective cargo-sorting mechanisms. Of note, depending on the cell type, extracellular vesicles will display a set of cell-type-specific proteins that account for their specific fates and functions. Despite a different mode of biogenesis, exosomes and microvesicles display a similar appearance, overlapping size and often common composition that make it difficult to ascertain their origin once isolated from the extracellular medium or from biological fluids. *(Modified from G. van Niel et al., Nat. Rev. Mol. Cell Biol. 19 (2018) 213–228.)*

role they play in exosome release, dysregulation in their expression patterns has been linked with cancer progression [37].

Glypican-1 (GPC1), a cell surface proteoglycan, was recently identified to be enriched on exosomes originating from cancer cells. Such GPC1-enriched exosomes were detected in the serum of patients with pancreatic cancer with absolute (100%) specificity and sensitivity, distinguishing healthy subjects and patients with a benign pancreatic disease from patients with early- and late-stage pancreatic cancer. Levels of these exosomes were also shown to correlate with tumor burden and patient survival, pre- or posttumor removal [38]. However, it is noteworthy that many surface markers are shared by exosomes originating from both tumor and healthy tissues, significantly limiting their application as highly specific and clinically useful cancer biomarkers.

4.2 Nucleic acid exosomal biomarkers

The nucleic acid content of exosomes is extremely diverse, including double-stranded DNA [39], messenger RNA as well as noncoding RNA molecules (e.g., miRNA, lncRNA, etc.) [40]. This content reflects the mutational status of the cells from which exosomes originate and has therefore great potential for diagnosis and/or prognosis of cancer. The miRNA content of exosomes has recently been investigated [41]. However, contradictory studies suggest either that each exosome contains one copy of each miRNA or that miRNAs are concentrated in a few exosomes while the others remain empty. This can be explained by the current lack of standardized protocols to isolate and then analyze exosomes. Most commonly used strategies to isolate exosomes include (i) ultracentrifugation, (ii) pull-down using antibody-coated magnetic beads and (iii) -microfluidic-based platforms for on-chip separation.

5 Circulating tumor cells

The idea of circulating tumor cells (CTCs) dates back to 1869, following observations made by Ashworth and has since been strongly linked with tumor metastases, which account for approximately 90% of cancer-related deaths. Although the metastatic process is not yet fully understood, it involves the detachment of malignant cells from the primary tumor followed by extravasation of cancer cells into the bloodstream [42]. CTCs that survive during the migration in the vascular system may extravasate to colonize and proliferate within distant organs and/or tissues to create a metastatic site (Fig. 7). CTCs can be observed in the peripheral blood mainly as single cells, although clusters ranging from 2 cells up to 100 cells can also form that represent 2%–5% of the total number of CTCs. Although the lifetime of CTC clusters in the bloodstream is significantly shorter than that of single CTCs, they are highly metastatic. Between 1 and 10 CTCs can be typically detected per milliliter of peripheral blood of patients with metastatic stage cancer, with <1 CTC for 100 peripheral blood mononuclear cells. It is noteworthy that the number of

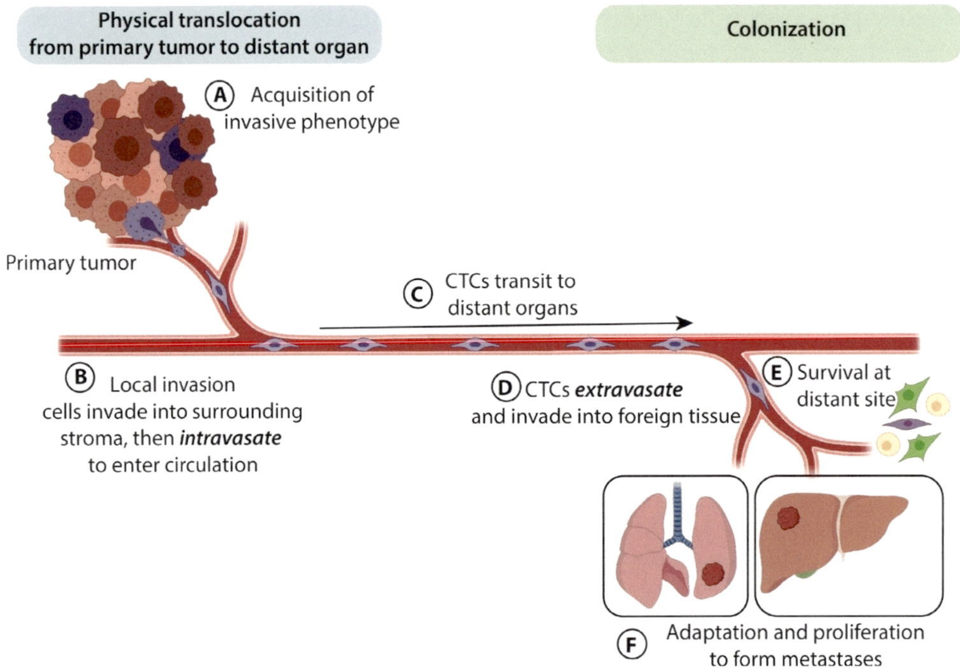

Fig. 7 From circulating tumor cell (CTC) to metastasis.

CTC far exceeds the number of metastatic lesions that will actually develop: in other words, not every CTC will create a metastasis. Cancer cells that survive after infiltrating distant organs (so-called disseminated tumor cells or DTCs) can be present in the bone marrow of cancer patients for years without developing a metastasis. This is because most cancer cells die while only a minority of the surviving cells end up forming macrometastases [43].

Detecting metastasis is an important step in cancer diagnosis and cancer grading and current imaging technologies lack sensitivity or resolution to detect small clusters of tumor cells and early metastases. DTCs are more easily detectable in bone marrow although bone marrow biopsies are highly invasive and cannot be done routinely or repeatedly. Detecting CTCs directly from liquid biopsies (e.g., blood test) offers the advantage of being minimally invasive and suitable for both public screening and longitudinal monitoring.

The main challenge in CTC analysis from whole blood is their low abundance against the high background of red and white blood cells. Extensive research has been conducted to discern CTC from blood cells based on unique biological (e.g., protein signature) and/or physical (size, deformability, density, or dielectric properties) properties of CTCs [44].

6 Metabolomics and cancer biomarkers

The presence or absence of certain metabolites and the changes in their concentrations in selected matrices (blood, breath, urine, tissues) may be indicative of a particular disease. Metabolic reprogramming is a key feature of most cancers with characteristic changes, including mitochondrial malfunction, adaptation to hypoxic tumor microenvironment, enhanced glycolysis, or being essential to promote cancer cell growth [45]. Metabolomics refers to the study, identification, and quantification of low-molecular-weight small molecules (<1500 Da) so-called metabolites that are the product of metabolic reactions. It can therefore play a key role in better understanding the biological changes occurring during tumorigenesis while identifying molecular signatures characteristic of specific cancer types. Metabolomics enables the systematic interrogation of small molecules in cells, tissues, bodily fluids, or organisms. Metabolites whose abnormal accumulation in any of these matrices induces malignancy are commonly named oncometabolites and are most commonly the result of cancer-associated mutations in genes encoding enzymes with significant roles in cellular metabolism [46]. For instance, loss-of-function mutations in genes encoding the Krebs cycle enzymes fumarate hydratase (FH) and succinate dehydrogenase (SDH) cause the accumulation of fumarate and succinate, respectively. In contrary, gain-of-function isocitrate dehydrogenase (IDH) mutations increase levels of D-2-hydroxyglutarate (D-2HG). Increasing evidence supports the hypothesis that fumarate, succinate, and D-2HG play a key role in cellular transformation and oncogenesis and have great potential as cancer biomarkers.

7 Conclusion

This chapter summarizes the main types of molecular cancer biomarkers, highlighting for each category, where they can be found and their potential clinical use. Cancer biomarkers cover a broad range of sizes (from small molecule metabolite to cells), are found in many different matrices (tissues, cells, and bodily fluids) and can provide a broad range of clinically useful information (early diagnosis, cancer type, tumor grading, or tumor progression). Depending on their nature and natural abundance, these molecular biomarkers can either be detected in vivo, ex vivo or in vitro, in centralized laboratories or at the point-of-care using technologies developed by bioengineers. Examples of such technologies are discussed in more detail in the following chapters and are sorted based on the nature of the readout being either optical (Chapter 1.2) or electrochemical (Chapter 1.3).

References

[1] S. Naylor, Biomarkers: current perspectives and future prospects, Expert. Rev. Mol. Diagn. 3 (2003) 525–529.

[2] S.G. Baker, The central role of receiver operating characteristic (ROC) curves in evaluating tests for the early detection of cancer, J. Natl. Cancer Inst. 95 (2003) 511–515.

[3] F.M. Buonaguro, D. Pauza, M.L. Tornesello, P. Hainaut, R. Franco, F.M. Marincola, Cancer diagnostic and predictive biomarkers, Biomed. Res. Int. (2014) 980163.
[4] J. Lin, L. Ma, D. Zhang, J. Gao, Y. Jin, Z. Han, D. Lin, Tumour biomarkers-tracing the molecular function and clinical implication, Cell Prolif. 52 (2019).
[5] S.A. Frank, Genetic predisposition to cancer—insights from population genetics, Nat. Rev. Genet. 5 (2004) 764–772.
[6] J.S. You, P.A. Jones, Cancer genetics and epigenetics: two sides of the same coin? Cancer Cell 22 (2012) 9–20.
[7] M.R. Lieber, Mechanisms of human lymphoid chromosomal translocations, Nat. Rev. Cancer 16 (2016) 387–398.
[8] H.J. Burstein, The distinctive nature of HER2-positive breast cancers, N. Engl. J. Med. 353 (2005) 1652–1654.
[9] O.-P. Kallioniemi, A. Kallioniemi, W. Kurisu, et al., ERBB2 amplification in breast cancer analyzed by fluorescence in situ hybridization, Proc. Natl. Acad. Sci. U. S. A. 89 (1992) 5321–5325.
[10] T. Jamaspishvili, D.M. Berman, A.E. Ross, H.I. Scher, A.M. De Marzo, J.A. Squire, T.L. Lotan, Clinical implications of PTEN loss in prostate cancer, Nat. Rev. Urol. 15 (2018) 222–234.
[11] N. Deng, H. Zhou, H. Fan, Y. Yuan, Single nucleotide polymorphisms and cancer susceptibility, Oncotarget 8 (2017) 110635–110649.
[12] E. Baxter, K. Windloch, F. Gannon, J.S. Lee, Epigenetic regulation in cancer progression, Cell Biosci. 4 (2014) 45.
[13] C.D. Allis, T. Jenuwein, The molecular hallmarks of epigenetic control, Nat. Rev. Genet. 17 (2016) 487–500.
[14] E. Anastasiadou, L.S. Jacob, F.J. Slack, Non-coding RNA networks in cancer, Nat. Rev. Cancer 18 (2018) 5–18.
[15] A. Sanchez Calle, Y. Kawamura, Y. Yamamoto, F. Takeshita, T. Ochiya, Emerging roles of long non-coding RNA in cancer, Cancer Sci. 109 (2018) 2093–2100.
[16] J. Hayes, P.P. Peruzzi, S. Lawler, MicroRNAs in cancer: biomarkers, functions and therapy, Trends Mol. Med. 20 (2014) 460–469.
[17] G.A. Calin, et al., Frequent deletions and down-regulation of micro-RNA genes miR15 and miR16 at 13q14 in chronic lymphocytic leukemia, Proc. Natl. Acad. Sci. U. S. A. 99 (2002) 15524–15529.
[18] D. Hanahan, R.A. Weinberg, Hallmarks of cancer: the next generation, Cell 144 (2011) 646–674.
[19] P. Mandel, P. Metais, Les acides nucléiques du plasma sanguin chez l'homme, C. R. Seances Soc. Biol. Fil. 142 (1948) 241–243.
[20] M. Fleischhacker, B. Schmidt, Circulating nucleic acids (CNAs) and cancer—a survey, Biochim. Biophys. Acta 1775 (2007) 181–232.
[21] P.S. Mitchell, et al., Circulating microRNAs as stable blood-based markers for cancer detection, Proc. Natl. Acad. Sci. U. S. A. 105 (2008) 10513–10518.
[22] V. Armand-Labit, A. Pradines, Circulating cell-free microRNAs as clinical cancer biomarkers, Biomol. Concepts 8 (2017) 61–81.
[23] V.Y. Shin, J.M. Siu, I. Cheuk, E.K.O. Ng, A. Kwong, Circulating cell-free miRNAs as biomarker for triple-negative breast cancer, Br. J. Cancer 112 (2015) 1751–1759.
[24] H. Wang, R. Peng, J. Wang, Z. Qin, L. Xue, Circulating microRNAs as potential cancer biomarkers: the advantage and disadvantage, Clin. Epigenetics 10 (2018) 59.
[25] S. Loeb, A.W. Partin, PCA3 urinary biomarker for prostate cancer, Rev. Urol. 12 (2010) e205–e206.
[26] L. Bolha, M. Ravnik-Glavac, D. Glavac, Long noncoding RNAs as biomarkers in cancer, Dis. Markers (2017) 7243968.
[27] J. Donaldson, B.H. Park, Circulating tumor DNA: measurement and clinical utility, Annu. Rev. Med. 69 (2018) 223–234.
[28] F. Cheng, L. Su, C. Qian, Circulating tumor DNA: a promising biomarker in the liquid biopsy of cancer, Oncotarget 7 (2016) 48832–48841.
[29] C. Fiala, E.P. Diamandis, Utility of circulating tumor DNA in cancer diagnostics with emphasis on early detection, BMC Med. 16 (2018) 166.
[30] H.R. Underhill, et al., Fragment length of circulating tumor DNA, PLoS Genet. 12 (2016).

[31] C.A.K. Borrebaek, Precision diagnostics: moving towards protein biomarker signatures of clinical utility in cancer, Nat. Rev. Cancer 17 (2017) 199–204.
[32] B.M. Nolen, A.E. Lokshin, Biomarker testing for ovarian cancer: clinical utility of multiplex assays, Mol. Diagn. Ther. 17 (2013) 139–146.
[33] M. Colombo, G. Raposo, C. Théry, Biogenesis, secretion, and intercellular interactions of exosomes and other extracellular vesicles, Annu. Rev. Cell Dev. Biol. 30 (2014) 255–289.
[34] S.H. Jalalian, M. Ramezani, S.A. Jalalian, K. Abnous, S.M. Taghdisi, Exosomes, new biomarkers in early cancer detection, Anal. Biochem. 571 (2019) 1–13.
[35] R.R. Malla, S. Pandrangi, S. Kumari, M.M. Gavara, A.K. Badana, Exosomal tetraspanins as regulators of cancer progression and metastasis and novel diagnostic markers, Asia Pac. J. Clin. Oncol. 14 (2018) 383–391.
[36] X. Wang, W. Zhong, J. Bu, Y. Li, R. Li, R. Nie, C. Xiao, K. Ma, X. Huang, Y. Li, Exosomal protein CD82 as a diagnostic biomarker for precision medicine for breast cancer, Mol. Carcinog. 58 (2019) 674–685.
[37] H. Stenmark, Rab GTPases as coordinators of vesicle traffic, Nat. Rev. Mol. Cell Biol. 10 (2009) 513–525.
[38] S.A. Melo, L.B. Luecke, C. Kahlert, A.F. Fernandez, S.T. Gammon, J. Kaye, V.S. LeBleu, E.A. Mittendorf, J. Weitz, N. Rahbari, C. Reissfelder, C. Pilarsky, M.F. Fraga, D. Piwnica-Worms, R. Kalluri, Glypican-1 identifies cancer exosomes and detects early pancreatic cancer, Nature 523 (2015) 177–182.
[39] B.K. Thakur, H. Zhang, A. Becker, I. Matei, Y. Huang, B. Costa-Silva, Y. Zheng, A. Hoshino, H. Brazier, J. Xiang, C. Williams, R. Rodriguez-Barrueco, J.M. Silva, W. Zhang, S. Hearn, O. Elemento, N. Paknejad, K. Manova-Todorova, K. Welte, J. Bromberg, H. Peinado, D. Lyden, Double-stranded DNA in exosomes: a novel biomarker in cancer detection, Cell Res. 24 (2014) 766–769.
[40] Y. Xie, W. Dang, S. Zhang, W. Yue, L. Yang, X. Zhai, Q. Yan, L. Lu, The role of exosomal noncoding RNAs in cancer, Mol. Cancer 18 (2019) 37.
[41] M. Wang, F. Yu, H. Ding, Y. Wang, P. Li, K. Wang, Emerging function and clinical values of exosomal microRNAs in cancer, Mol. Ther. Nucleic Acids 16 (2019) 791–804.
[42] D. Woo, M. Yu, Circulating tumor cells as "liquid biopsies" to understand cancer metastasis, Transl. Res. 201 (2018) 128–135.
[43] S. Riethdorf, H. Wikman, K. Pantel, Review: biological relevance of disseminated tumor cells in cancer patients, Int. J. Cancer 123 (2008) 1991–2006.
[44] S. Sharma, R. Zhuang, M. Long, M. Pavlovic, Y. Kang, A. Ilyas, W. Asghar, Circulating tumor cell isolation, culture, and downstream molecular analysis, Biotechnol. Adv. 36 (2018) 1063–1078.
[45] A. Kumar, B.B. Misra, Challenges and opportunities in cancer metabolomics, Proteomics (2019), in press. https://doi.org/10.1002/pmic.201900042.
[46] K. Sajnani, F. Islam, R.A. Smith, V. Gopalan, A.K. Lam, Genetic alterations in Krebs cycle and its impact on cancer pathogenesis, Biochimie 135 (2017) 164–172.

CHAPTER 1.2

Optical sensors

Angie Davina Tjandra[a], Jason Y.H. Chang[b], Sylvain Ladame[c], Rona Chandrawati[a]
[a]School of Chemical Engineering and Australian Centre for Nanomedicine (ACN), The University of New South Wales (UNSW Sydney), Sydney, NSW, Australia
[b]Koch Institute for Integrative Cancer Research, Massachusetts Institute of Technology, Cambridge, MA, United States
[c]Department of Bioengineering, Imperial College London, London, United Kingdom

1 Introduction

Until recently, cancer diagnosis and longitudinal monitoring primarily relied on highly invasive tissue biopsy procedures and/or imaging tests (e.g., computed tomography, ultrasound, magnetic resonance imaging, positron-emission tomography, and X-ray) [1]. While they are an essential part of clinical management of cancer with the ability to distinguish between benign and malignant tissues and will be discussed in detail in the following chapters, these techniques are typically expensive, require trained professionals, have generally low sensitivity, and are often not effective at detecting cancer at early stages.

The detection of cancer biomarkers in vitro (e.g., from liquid biopsies), ex vivo (e.g., from tissue biopsies) and in vivo can all provide highly valuable information and guide clinicians in their decision making of when and how to treat. Cancer diagnostic and prognostic tests require sensing technologies that can detect biomarkers such as those presented in Chapter 1.1 with both high affinity and high specificity, sometimes directly in complex biological fluids (e.g., blood), especially when used at the point of care. Detection technologies that can be used in low-resource settings without the need for highly trained medical staff have the potential to revolutionize medical diagnostics. Fluorescence-based sensors are among the most common and most powerful analytical platforms offering a large dynamic range, high sensitivity, and multiplexing capabilities.

Biosensors are analytical tools that measure biological or chemical reactions and generate signals proportional to the concentration of a target analyte in the reaction [2]. Biosensors are typically made of three main components: (1) a recognition element (receptor) that specifically interacts with a target analyte; (2) a signal transducer to transform an analyte-receptor recognition event into a measurable signal; and (3) a reader that interprets the emitted signal and outputs a quantifiable result [3]. Outputs can be quantitative or qualitative depending on the functionality of the sensor and/or the biomedical need. Fluorescent and plasmonic sensors (small molecules and nanoparticles) are the most common types of optical sensors because they exhibit characteristic responses upon excitation with incident light that can be affected in an often predictable and quantitative

Bioengineering Innovative Solutions for Cancer
https://doi.org/10.1016/B978-0-12-813886-1.00003-6

manner upon binding to specific target analytes. For these reasons, they represent excellent signal transducers [4,5].

In this chapter, we first introduce the general principles behind fluorescence emission and how they can be exploited for sensing purposes. A general overview of the most common technologies engineered for the detection of cancer biomarkers that are based on a fluorescence or absorbance readout using small molecules, nanoparticles, or a combination of both is then presented. Particular attention is given to promising technological developments validated using clinically relevant samples.

2 Brief introduction to fluorescence

2.1 Jablonski diagram: Absorption and emission

Fluorescence describes the phenomenon of light emission by a molecule within nanoseconds after the absorption of higher energy photons. The exact wavelengths of absorption (also called excitation) and emission are determined by the outermost electron orbitals of a fluorophore. The processes of fluorescence excitation and emission can be represented by the Jablonski diagram, first introduced by Alexander Jablonski in the 1930s (Fig. 1). The left side of the diagram displays the singlet states, which outer electron pairs can occupy, with each electron having an opposite spin of +½ or −½. The right side displays the triplet states occupied by outer electrons, which have undergone a boost into a new orbital followed by a reversal in spin. When a molecule is not being excited by light, it occupies the lowest vibrational states of the ground state, S_0. The principle of absorption is that light of wavelength (λ) or frequency (f) can be absorbed by a fluorophore if the incident photon energy matches the energy difference between two electronic states (e.g., S_0 and S_1). The energy (E) of the photons is defined by Eqs. (1) and (2), where h is Planck's constant and c is the speed of light in vacuum. The process of absorption will only occur if the energy of the incident photon E is similar or slightly more than what is necessary for an electronic transition. The minimum energy necessary for a fluorophore to absorb a photon is that required to promote an electron from S_0 to the lowest energy level of the first excited state S_1. If the photon energy is slightly greater than this transition, excess energy is usually converted into vibrational or rotational energy (in the same electronic state) or used to move the molecule into an even higher electronic orbital singlet state (i.e., S_2). If the photon energy E is too high, excess energy cannot be dissipated through vibrational-level transitions to the same electronic state; thus, absorption does not occur. Consequently, there is a specific range of photon energies or wavelengths which may excite a fluorophore, as represented by the excitation or absorption spectrum.

$$E = hf \tag{1}$$

$$f = c/\lambda \tag{2}$$

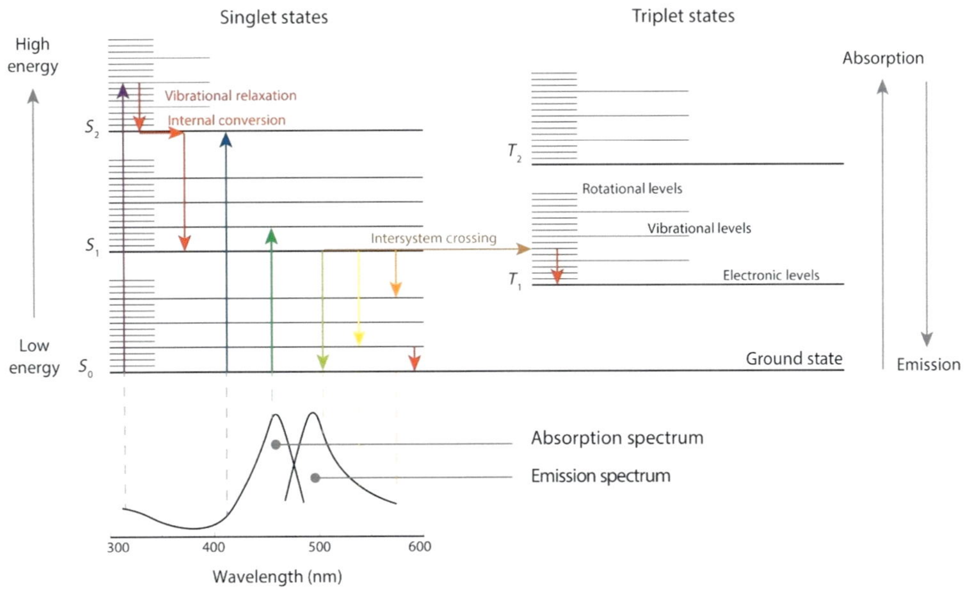

Fig. 1 Jablonski diagram illustrating the energy states of a typical fluorophore. Colors of the *arrows* represent the wavelength of the absorbed or emitted photon. For example, *blue arrow* represents the energy of a blue photon, which can transition a molecule from the ground state to the lowest energy level of the second excited state S_2. Vertical lines trace the energy absorbed *(arrow up)* or emitted *(arrow down)* to the corresponding wavelength on the absorption or emission spectra shown below. *(Adapted from J.W. Lichtman, J.-A. Conchello, Fluorescence microscopy, Nat. Methods 2 (2005) 910–919.)*

Following absorption of a photon, a molecule will remain in a so-called excited state for a very short period of time before losing the absorbed energy and eventually returning to the ground state via several competing pathways. Internal conversion is an isoenergetic electronic transition from low vibrational energies of one electronic state to high vibrational states of a lower electronic state (Fig. 1, horizontal red arrow). No energy is lost in this transition; however, extra vibrational energy is eventually lost by direct transfer to nearby molecules (e.g., water molecules in aqueous media) through the process of vibrational relaxation (Fig. 1, vertical red arrow). These two nonradioactive decay pathways typically bring the molecule down to the lowest energy level of S_1 (wherein the vibrational energy of the excited molecule is similar to the ground state, but the outer orbital electrons still have extra energy).

Sometimes, internal conversion may transition the molecule all the way down to the ground state, but this pathway is not preferred in most fluorophores. Instead, the preferred energy pathway is a radiative decay pathway from the lowest energy state of S_1 to the ground state S_0 involving the emission of a photon. The energy of the emitted photon covers the gap between the lowest vibrational state of S_1 to any vibrational or rotational state of S_0. Thus, a range of emission wavelengths can exist as represented by the emission spectrum in Fig. 1. Because emission always starts from the lowest energy level of S_1, the emitted photon is typically of lower energy than the absorbed photon. This results in the emission spectrum being shifted to higher wavelengths relative to the excitation spectrum by an amount called Stokes' shift. A larger shift results in easier separation between the exciting and emitting light. It is also possible to detect multiple fluorophores from the same sample as long as their characteristic fluorescence emission spectra are well separated. Too much overlap can result in bleed through or crosstalk between different fluorophores and thus loss or misperception of the signal information.

Fluorescence emission is not the only transition pathway from the excited state S_1 to the ground state. An excited fluorophore may also lose energy by intersystem crossing, which involves a forbidden transition to a triplet state (Fig. 1, brown arrow). When a molecule is in this long-lived triplet state, an excited electron may interact with neighboring molecules such as oxygen. A triplet state fluorophore can transfer its energy to a neighboring oxygen, thus exciting it (from a triplet in the ground state to an excited singlet state). In this state, oxygen is a reactive molecule, which may in turn inactivate the fluorophore's ability to fluoresce. This is one of the mechanisms that leads to the phenomenon of photobleaching. While, in theory, fluorophores can undergo an unlimited number of excitation and emission cycles, there is in fact a limited number of cycles (about 10,000–40,000) before permanent photobleaching of the fluorophore occurs. Photobleaching thus refers to all processes which lead fluorescence to fade permanently. Highly photostable dyes (e.g., Alexa dyes) have been engineered with enhanced photostability and much slower bleaching rates than conventional dyes [6]. Quantum dots in comparison, due to their composition and structure, have typically much greater photostability than molecular fluorophores.

2.2 Fluorescence quantum yield and fluorescence lifetime

Fluorescence quantum yield defines the capacity of a fluorophore (small molecule or nanoparticle) to convert every absorbed photon into an emitted photon (i.e., fluorescence emission). It corresponds to the ratio between the number of photons emitted and the number of photons absorbed. A quantum yield of 100% ($\Phi = 1$) means that every absorbed photon leads to an emitted photon while a quantum yield of 0% ($\Phi = 0$) means that the molecule is nonfluorescent.

Fluorescence lifetime (τ) is the characteristic time that a molecule remains in an excited state prior to returning to the ground state and is an indicator of the time available for information to be gathered from the emission profile. It is given by Eq. (3) and corresponds to the time (t) at which the fluorescence intensity (I) emitted by the fluorophore has decayed to $1/e$ of its original value (I_0) after excitation.

$$I(t) = I_0 \cdot e^{(-t/\tau)} \tag{3}$$

2.3 Förster resonance energy transfer (FRET)

FRET is a mechanism responsible for an energy transfer between two chromophores (Fig. 2). When a donor molecule absorbs a photon, and if there is an acceptor molecule close to the donor molecule, then nonradiative energy transfer can occur from the donor to the acceptor. FRET results in a decrease of the fluorescence intensity and lifetime of the donor probe and enhances the fluorescence of the acceptor probe when the acceptor is also a fluorophore. Conditions for FRET to occur are: (1) that there is a significant overlap between the emission spectrum of the donor and the absorption spectrum of the acceptor, and (2) that both the donor and the acceptor molecules are in close proximity to each other.

2.4 Fluorescent labels and probes

Fluorescence-based sensors can rely on three different types of fluorophores: fluorescent labels, fluorescent probes, and enzyme substrates. Fluorescent labels are fluorescent molecules (or nanoparticles) attached to receptors directed against a specific analyte or biomarker. Upon binding of the analyte to the receptor molecule a complex is formed which is fluorescent and the amount of fluorescence is directly proportional to the amount of analyte present (quantitative detection). In that case, it is essential that the fluorophore has highly stable optical properties that do not vary significantly

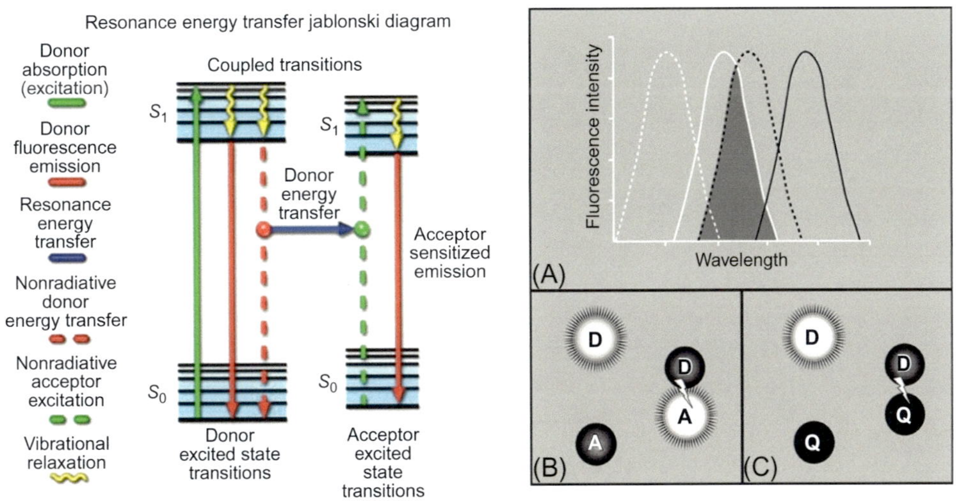

Fig. 2 Mechanism of Förster resonance energy transfer (FRET). (Left) FRET explained through the Jablonski diagram; (right) conditions required for FRET to occur between a fluorescence donor (D) and fluorescence acceptor (A) (panel B) or between a fluorescence donor (D) and quencher (Q) (panel C) are (i) an overlap between the emission spectrum of the donor and the absorption spectrum of the acceptor (or quencher) (panel A) and (ii) spatial proximity between the donor and the acceptor (or quencher) (panels B and C).

as a function of the local environment they are placed in. Responsive probes measure a change in fluorescence properties (intensity, excitation or emission wavelength, fluorescence lifetime) as a response to a change in environment upon binding to a specific analyte or biomarker. Fluorescent probes can be either ratiometric or nonratiometric (Fig. 3).

Nonratiometric probes are also called single indicator probes and use the sole change in fluorescence intensity (either up or down) to detect the presence of a specific biomarker. Although very easy to measure experimentally, this is also often unreliable as changes in fluorescence intensity can be caused by various (unpredictable) factors such as probe degradation and irregular depth of the tissue/sample being analyzed. Ratiometric probes in comparison are called dual indicator probes and are characterized by both a change in fluorescence intensity and a change in either fluorescence excitation or emission wavelength upon binding to the biomarker. In other words, the complex formed between the probe and the analyte has optical features distinct to those of the unbound probe. It is therefore possible to monitor binding by looking simultaneously at the fluorescence of the bound and unbound probes, making the measurement significantly more accurate (although also slightly more complex). Indeed, measurement becomes independent of the probe concentration and of the optical path length and both degradation of the probe (e.g., due to prolonged exposure to the excitation source) and variations in the excitation intensity are compensated by the ratio technique.

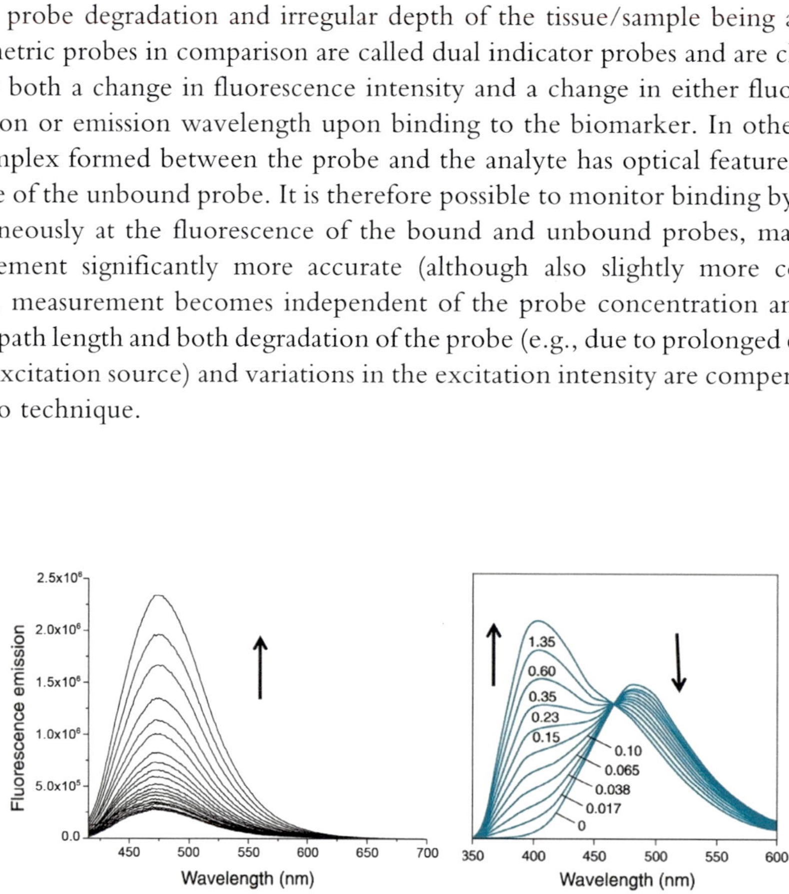

Fig. 3 Examples of fluorescent probes responsive to specific analytes. (Left) Example of nonratiometric (or single indicator) probe. Upon addition of increasing concentrations of analyte *(black arrow)*, the intensity of the emitted fluorescence increases but the fluorescence emission wavelength remains unchanged. (Right) Example of ratiometric (or dual indicator) probe. Upon addition of increasing concentrations of analyte *(black arrow)*, the fluorescence of the unbound probe (here centered at 490 nm) decreases and a new fluorescence signal corresponding to the probe bound to the analyte appears and increases with increasing analyte concentration that is centered at a different wavelength (here 400 nm).

2.5 Background fluorescence and autofluorescence

Fluorescence detection sensitivity can be severely compromised by background signals, which may originate from endogenous sample constituents (referred to as autofluorescence) or from unbound or nonspecifically bound probes (referred to as background fluorescence). Autofluorescence can be a major problem when trying to detect biomarkers from cells, tissues or biofluids in which naturally fluorescent molecules are present. Signal distortion caused by autofluorescence of these matrices is most readily minimized by using probes that can be excited at longer wavelengths (>500 nm). Otherwise it is often required (and preferable) to extract the biomarkers from their original environment prior to analysis. This extra sample processing step however can be a source of significant error and variability, the outcome of the analysis being strongly dependent on the extraction efficiency, and the purity of the extracted biomarker.

2.6 Small molecules versus quantum dots

Fluorescent biosensors are typically based on either engineered fluorescent small molecules or engineered nanoparticles. Although naturally occurring small molecules or proteins can exhibit fluorescence properties, they typically suffer from low fluorescence quantum yield and low excitation/emission wavelengths making them unsuitable for most sensing applications. Fluorescent labels and probes have therefore been engineered and tailored for specific applications and can be categorized as either small molecules or nanoparticles. Small molecules can either be designed to bind preferentially to cancer biomarkers or be attached to receptors designed to target them. In contrast, fluorescent nanoparticles require their surface to be functionalized with receptors to confer them the required biomarker specificity. Both small molecules and nanoparticles will absorb light at a characteristic wavelength and re-emit photons of lower energy (i.e., higher wavelength). The intensity and stability of fluorescence emission play important roles in determining the sensitivity and dynamic range of the assay. Nanoparticles are typically more photostable (i.e., resistant to photobleaching) and have longer fluorescence lifetime (i.e., longer time to record emitted photons after excitation) than small molecules. They also have broader absorption spectra meaning that it is possible to use a single light source to excite multiple nanoparticles simultaneously that will all emit at different (but characteristic) wavelengths (Fig. 4). While small-molecule fluorescence properties can be finely tuned by chemically modifying the structure of the chromophore, the fluorescence emission profile of nanoparticles is controlled by their size, smaller nanoparticles typically emitting at shorter wavelengths. To date, fluorescent nanoparticle probes that are most widely used in cancer diagnostic applications include quantum dots (QDs) [7], upconversion nanoparticles [8], fluorescent dye-doped silica nanoparticles [9], and gold nanoclusters (AuNCs) [10]. They are typically functionalized with receptors, including antibodies, peptides, and aptamers to enable highly specific interaction between the nanoparticle and the target analyte.

Fig. 4 Ten distinguishable emission colors of ZnS-capped CdSe QDs excited with a near-UV lamp. QD tunable emission wavelengths allow multicolor labeling to achieve the simultaneous detection of multiple target analytes (multiplexing), where a mixed population of QDs can be excited with a single excitation source. *(Reproduced with permission from M. Han, X. Gao, J.Z. Su, S. Nie, Quantum-dot-tagged microbeads for multiplexed optical coding of biomolecules, Nat. Biotechnol. 19 (2001) 631.)*

In the following section, selected examples of optical sensors engineered to detect cancer biomarkers belonging to the families of nucleic acids and proteins will be presented. This will include sensors based on either small molecules or nanoparticles (or a combination of both).

3 Optical sensing of nucleic acid biomarkers

3.1 Small-molecule sensing

3.1.1 Molecular beacons

Molecular beacons (or MBs) are probably the most commonly used probes for sequence-specific detection of nucleic acid biomarkers. They are typically single-stranded oligonucleotide hybridization probes that can adopt a stable stem-and-loop structure in solution. The loop contains a probe sequence that is complementary to that of the nucleic acid biomarker target, and the stem is formed by the annealing of two complementary "arm" sequences that are located on either side of the probe sequence (i.e., at the 3′ and 5′ end of the oligonucleotide). A fluorophore is covalently linked to the end of one arm and a quencher is covalently linked to the end of the other arm. Molecular beacons do not fluoresce when they are free in solution. However, when they hybridize to an oligonucleotide containing a target sequence they undergo a conformational change that enables them to fluoresce brightly (Fig. 5).

This strategy can be applied to the detection of specific DNA and RNA biomarkers both in cells and ex vivo (e.g., in liquid biopsies) [11]. Multiplexed analysis is also possible by using a combination of MBs functionalized with nonoverlapping fluorophores. For instance, this sensing technology has proven highly promising for miRNA profiling (i.e., simultaneous analysis of multiple miRNAs from the same sample).

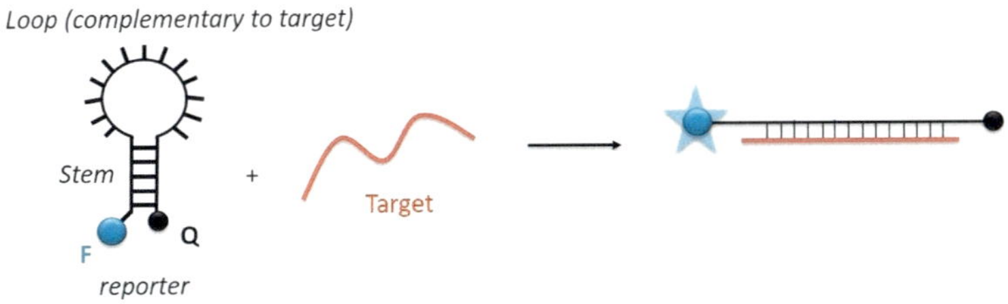

Fig. 5 Molecular beacons (MBs) and applications for nucleic acid sensing. Schematic representation of the recognition mechanism of single-stranded DNA (or RNA) by a MB.

This strategy was also successfully applied to the detection of point mutations or single-nucleotide polymorphism (SNP), the probe sequence contained in the MB's loop being highly sequence specific.

3.1.2 Fluorogenic intercalating dyes

Intercalating dyes, as their name suggests, are small molecules that can intercalate between Watson-Crick base pairs of double-stranded DNA and emit fluorescence of intensity orders of magnitude greater than when the dye is free in solution. The fluorescence intensity emitted by the intercalated dye correlates with the amount of double-stranded DNA present in solution [12]. The most commonly used intercalating dye is SYBR Green, but many other dyes have been developed that often belong to the family of molecular rotors. Briefly, molecular rotors are probes known to form twisted intramolecular charge transfer (TICT) complexes in the excited state producing a fluorescence quantum yield that is dependent on the surrounding environment. Common to the structure of all molecular rotors is a motif that consists of an electron donor group in π-conjugation with an electron acceptor group. Following photoexcitation, this motif has the unique ability to relax either via fluorescence emission or via an internal nonradiative process that involves molecular rotation between the donor and the acceptor. When this rotation is hindered (e.g., because the dye is intercalated between base pairs), the relaxation occurs via an increased fluorescence emission. In contrast, when the dye is in solution, relaxation proceeds mainly via a nonradiative pathway, meaning no (or lower) emission of fluorescence.

3.1.3 Sensing through oligonucleotide templated reaction (OTR)

Sensing strategies based on bio-orthogonal oligonucleotide-templated chemistries have recently emerged as highly promising for the detection of endogenous nucleic acids (NAs) both in vitro and in vivo [13]. Widespread in Nature, OTRs use a DNA or RNA strand as a template to catalyze an unfavorable bimolecular chemical reaction by significantly increasing the effective molarity of two monomers otherwise present in solution at too low concentrations to crossreact. The general mechanism relies on

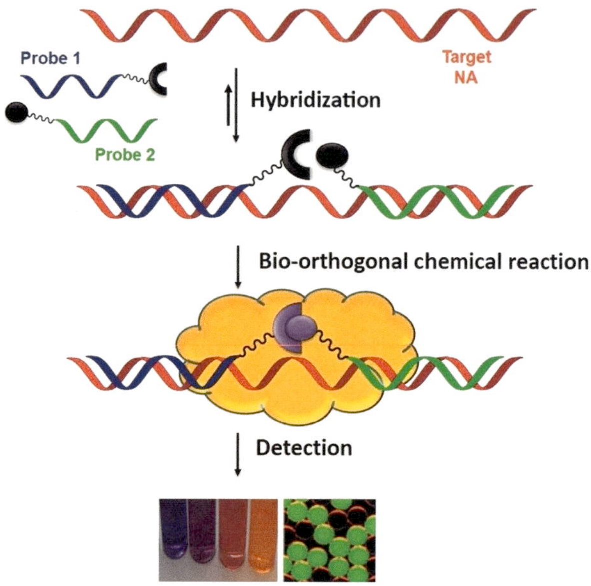

Fig. 6 Nucleic acid sensing through oligonucleotide-templated reaction (OTR). *(Reproduced with permission from S. Pavagada, S. Ladame, Platforms for bioorthogonal oligonucleotide-templated reactions: translating concepts into devices, Chimia (Aarau) 72 (2018) 809–814.)*

sequence-specific Watson-Crick base pairing between the template strand and two engineered oligonucleotide (or oligonucleotide analog) probes (functionalized with carefully designed probe heads), to facilitate a proximity-induced bio-orthogonal chemical reaction and unleash a detectable signal, most commonly fluorescence (Fig. 6). Overall, OTR-based technologies are very well suited for in vitro sensing of nucleic acid biomarkers, and in recent years they have been successfully used for sensing single-stranded, double-stranded, and four-stranded (e.g., G-quadruplexes) nucleic acid structures. In 2016, OTR probes were successfully engineered that could detect endogenous concentrations of circulating miRNAs extracted from human blood without the need for any amplification step and could distinguish between prostate cancer patients and healthy controls [14].

3.2 Nanoparticle sensing

Similar strategies to those presented here make use of nanoparticles instead of small molecules either as a fluorescence emitter or fluorescence quencher. Most common strategies rely on the functionalization of nanoparticles with hybridization probes complementary to the nucleic acid biomarker of interest.

3.2.1 Gold nanoparticle (AuNP)-based sensors

AuNP-based fluorescent sensors typically take advantage of the excellent fluorescence quenching capability of AuNPs, which takes place when the emission of a fluorescent molecule overlaps with the plasmon band of the nanoparticles. Interestingly, AuNPs can quench the fluorescence of chromophores 100-fold better than molecular quenchers and present higher quenching efficiencies even for dyes emitting in the near-infrared region. In addition to their exceptional quenching properties, the ability of fluorescent oligonucleotide probes to distinguish single-base mutations is about 8-fold greater when using AuNPs as quenchers, when compared to the use of conventional molecular quenchers. Therefore, replacing a molecular quencher with AuNPs results in a more sensitive oligonucleotide probe and further improves its selectivity. As a result, MB as described in Section 3.1.1 can be engineered for sequence specific sensing of nucleic acid biomarkers by replacing the molecular quencher with an AuNP [15,16].

Nanoflares have also been engineered that represent an alternative to MBs (Fig. 7). They consist of AuNPs functionalized with a fragment of double-stranded DNA, in which one strand is complementary to the target sequence and the other is a short fluorescently tagged complementary reporter sequence. In the absence of nucleic acid biomarker, the fluorophore is held in close proximity to the AuNP quencher and no fluorescence is emitted. In the presence of the only biomarker of interest, the reporter sequence is displaced from the AuNP and can start emitting a characteristic fluorescence signal.

It is also noteworthy that many AuNPs functionalized with oligonucleotides can be easily internalized into cells, without the need for transfection agents, making them valuable sensors for the detection of intracellular nucleic acid cancer biomarkers.

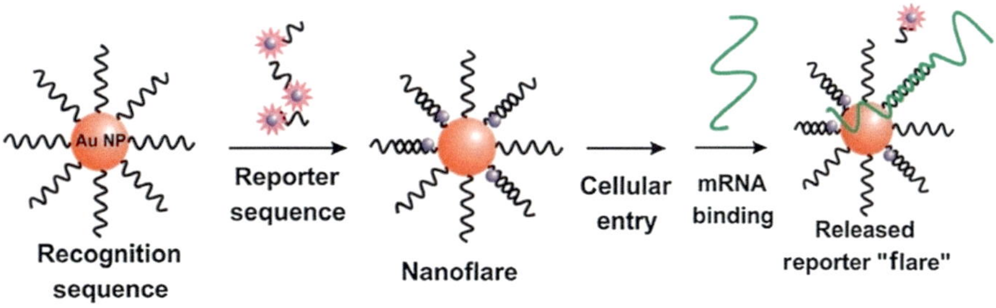

Fig. 7 Schematic illustration of nanoflares capable of entering cells and detecting a specific mRNA target. Upon mRNA binding, the flare strand is released through competitive hybridization leading to a detectable increase in fluorescence. *(Reproduced with permission from A. Heuer-Jungemann, P.K. Harimech, T. Brown, A.G. Kanaras, Gold nanoparticles and fluorescently-labelled DNA as a platform for biological sensing, Nanoscale 5 (2013) 9503–9510.)*

AuNP-based oligonucleotide sensors can also be engineered that enable colorimetric detection. In most cases, detection relies on a color change induced by aggregation of AuNPs upon binding to a nucleic acid biomarker of interest with resolution at the single nucleotide level (hence suitable for SNP detection).

3.2.2 Quantum dots (QD)-based sensors

The overall design of QD-based sensors for nucleic acid detection is often very similar to that of AuNP-based sensors although exploiting the characteristic fluorescent properties of QD instead of the quenching properties of AuNP. In the case of QD-based sensors, the fluorescence readout relies on biomarker-induced changes in FRET or fluorescence quenching between the QD and an oligonucleotide probe at its surface functionalized with either an acceptor fluorophore or a quencher, respectively. Upon binding of the biomarker of interest, a displacement of the oligonucleotide probe restores the intrinsic fluorescent properties of the QD.

4 Optical sensing of protein biomarkers

4.1 Small-molecule sensing

Although protein biomarkers may contain some naturally occurring fluorophores (e.g., tryptophan or tyrosine amino acid), the intrinsic optical properties of these fluorophores (very low quantum yield, short excitation and emission wavelength, sensitivity to environment) make them unsuitable for sensing applications. It is therefore necessary to engineer fluorescent probes that can detect protein biomarkers with both high affinity and high specificity. As for nucleic acid sensing, this often relies on combining a protein-specific recognition element and a reporter dye (for visualization). The vast majority of recognition elements for protein biomarkers are either protein-based (e.g., growth factors or engineered antibodies) or DNA-based (e.g., engineered DNA aptamers).

4.1.1 Growth factor-based sensors

Epidermal growth factor receptor (EGFR) is one of a family of receptor tyrosine kinases found on the surface of epithelial cells, with abnormally high EGFR levels on the surface of many types of cancer cells. Human epidermal growth factor (EGF) is a small protein (6 kDa, 53 amino acids) that binds EGFR with high affinity and specificity. Chemical modification of EGF with molecular fluorophores has proven effective for in vivo monitoring of specific solid tumors overexpressing EGFR [17].

4.1.2 Antibody-based sensors

A typical Antibody (Ab) is made up of four peptide chains, including two identical γ heavy chains and two identical light chains, and adopt a tetrameric Y-like shaped quaternary structure. The two heavy chains are linked to each other and to a light chain by disulfide bonds. Each end of the fork contains an identical antigen-binding site that provides the Ab with exceptional specificity. Interestingly, antibodies can be engineered that bind to specific proteins or to other antibodies. Antibodies can also be chemically modified with fluorophores for easy visualization of the targeted protein biomarker. In practice, two or more antibodies are typically used for each protein biomarker: a primary, nonmodified antibody enables the specific detection of the target while a secondary antibody (labeled with fluorophores) directed against the primary antibody is used for visualization purposes. Providing a detail description of how antibodies are designed and engineered is beyond the scope of this chapter, but further information on this topic can be found in reference [18].

The main applications of antibodies for the detection of cancer biomarkers are either in immunohistochemistry (IHC) or in immunoassays.

The aim of IHC is the detection of proteins and their cellular localization in tissue sections that have been fixed and embedded in a form of medium (typically paraffin, agarose, or optimal cutting temperature (OCT) compound) and is commonly used for the diagnosis and monitoring of cancers. Particularly useful (and broadly used) for cancer diagnostics are antibodies directed against specific types of keratins, key constituents of the cytoskeleton of endothelial cells. Interestingly, keratins have long and extensively been used as immunohistochemical markers in diagnostic tumor pathology and examples of keratins commonly used in the diagnosis of human epithelial malignancies are shown in Fig. 8. Antibodies directed against specific keratins have been engineered and are now commercially available.

In a standard immunoassay, a first capture antibody is immobilized onto a surface and used to selectively isolate the protein biomarker of interest. A second antibody, labeled with molecular fluorophore(s) and specific of the same protein biomarker, is then added for visualization purposes.

Skin	K1, K4-6, K10, K13-19		**Lung**	K7, K8, K18, K19
Liver	K8, K18, K19		**Mesothelioma**	K5, K7, K8, K18, K19
Biliary Tract	K7, K8, K18-20		**Breast**	K5, K6-8, K14, K17-19
Colon	K7, K8, K18-20		**Stomach**	K7, K8, K18-20
Ovary	K7, K8, K18, K19		**Kidney**	K7, K8, K18, K19
Prostate	K8, K18, K19		**Pancreas**	K7, K8, K18-20
Uterus	K5, K7, K8, K18, K19		**Cervix**	K4-8, K10, K13-19

Fig. 8 Keratin expression in human cancer. Keratins are normally expressed in a cell type-, differentiation- and functional status-dependent manner, and epithelial cancers largely maintain the characteristics of keratin expression associated with their respective cell type of origin, so keratins have long been recognized as diagnostic markers in tumor pathology. Examples of keratins commonly used in the diagnosis of human epithelial malignancies are presented in this table.

4.1.3 Aptamer-based sensors

Nucleic acid aptamers are short DNA or RNA molecules able to bind with high affinity and specificity to a wide range of targets when folded into their unique three-dimensional architecture [19]. Specific aptamer-protein interactions may include hydrophobic and electrostatic interactions, hydrogen bonding, van der Waals interactions, and π-π stacking, often combined with overall shape complementarity. When compared to antibodies, aptamers offer a number of advantages, including lower cost, ease of synthesis, and greater tolerance to chemical modifications. The aptamer-based fluorescent biosensors can be broadly divided into fluorescently labeled aptamers and label-free aptamers. Within each category, there are "signal-on" and "signal-off" sensor-based strategies, typically employing fluorescence resonance energy transfer (FRET). The signal change, either an increase (i.e., the "signal-on" mode) or a decrease (i.e., the "signal-off" mode), reflects the extent of the binding process, thereby allowing for quantitative measurements of the target concentration (Fig. 9).

4.2 Nanoparticle sensing

4.2.1 Immunosandwich assays

Similar strategies as described in Section 4.1 can be adapted by substituting molecular fluorophores with fluorescent nanoparticles. For instance, antibodies can be labeled with fluorescent nanoparticles for use in immunoassays (Fig. 10, top panel). In such cases, binding events are characterized by a fluorescence turn-on and the amount of fluorescence emitted correlates with the amount of biomarker present.

4.2.2 Probing protein activity

Nanoparticle optical sensors have also been developed that can detect protein activities, which can be exploited for diagnosis or monitoring of cancer. Proteases for example are an important class of enzymes that play a role in every hallmark of cancer. As a result, their specific activities represent highly valuable cancer biomarkers. Proteases are categorized by their catalytic type (e.g., metalloprotease, serine protease, etc.) and are often dysregulated in cancer. Approaches for measuring protease activity focus most commonly on measuring the cleavage of defined peptide substrates. Successful strategies were reported that use natural peptide sequences recognized by the protease to bring together either a fluorophore and a quencher or a donor and an acceptor. Protease-induced cleavage of the peptide releases the quencher from the fluorophore (or the acceptor from the donor) to restore its intrinsic fluorescence properties. A representative example is given in Fig. 11 where a fluorescent nanoparticle is functionalized with a fluorescently labeled peptide so that FRET can occur from the nanoparticle to the molecular fluorophore. Upon cleavage of the peptide by the protease, the fluorophore is released, and the intrinsic fluorescence properties of the nanoparticle are restored.

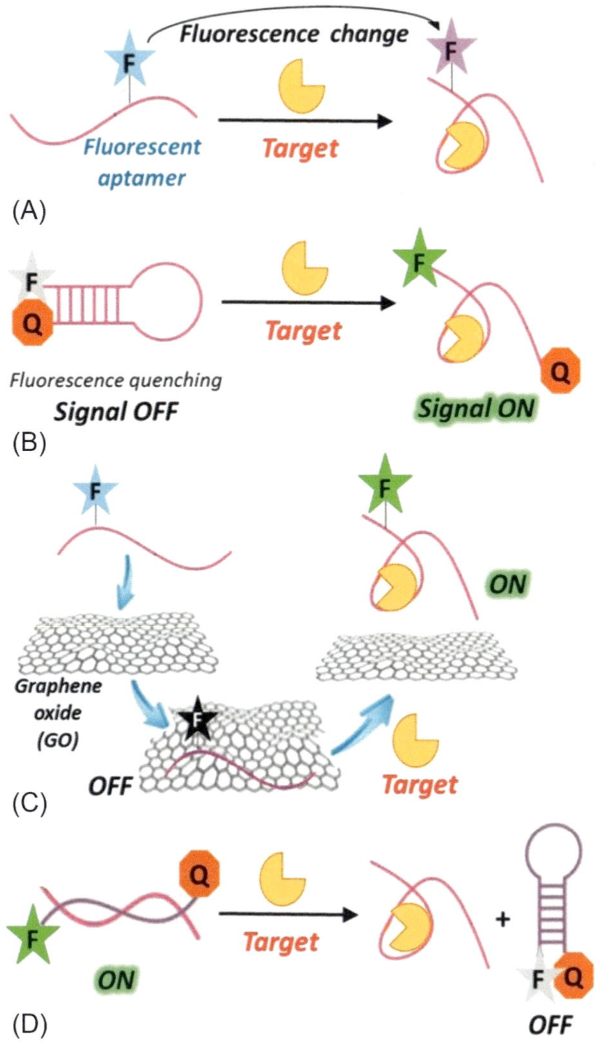

Fig. 9 Aptamer-based fluorescent sensors. Schematic representation of four general sensing strategies based on aptamers and fluorescence readout (A)–(D). *F*, fluorophore; *Q*, quencher. *(Reproduced with permission from D. Musumeci, C. Platella, C. Riccardi, F. Moccia, D. Montesarchio, Fluorescence sensing using DNA aptamers in cancer research and clinical diagnostics, Cancers (Basel) 9 (2017), 174.)*

A variation of the same strategy has been developed for colorimetric detection of protease activity. Here, AuNPs are covalently linked via specific peptide sequences and held as clusters. Upon protease-induced cleavage of the peptide linkages, the AuNPs become free to diffuse away from each other, which can be monitored optically with the naked eye (Fig. 12).

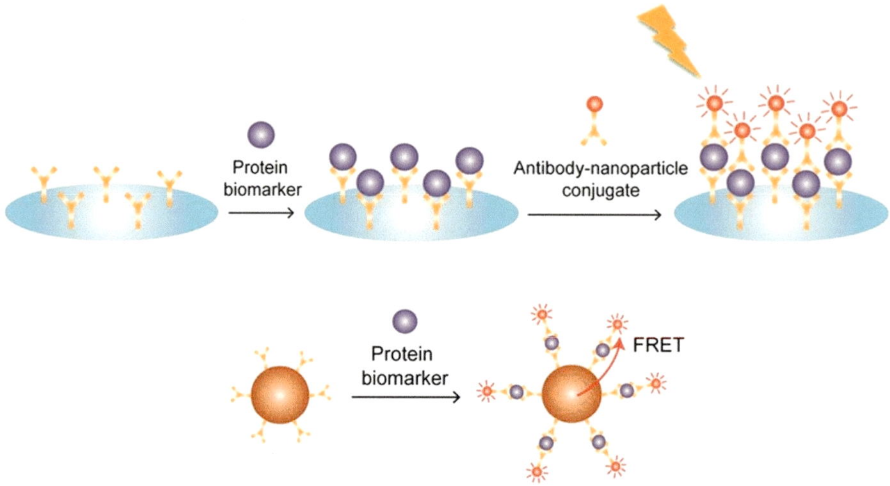

Fig. 10 Schematic illustration of fluorescence turn-on sensors for the detection of protein biomarkers of cancer. (Top) Capture antibodies are immobilized onto a substrate. In the presence of target proteins, the analytes bind to the capture antibodies. Detection antibodies with fluorescent nanoparticle probes sandwich the analytes and produce fluorescence emission; (bottom) fluorescent nanoparticles are functionalized with capture antibodies. In the presence of target proteins, the analytes bind to the capture antibodies. Detection antibodies (labeled with molecular fluorophores) sandwich the analytes and produce characteristic fluorescence emission through FRET between the nanoparticle and the molecular fluorophore.

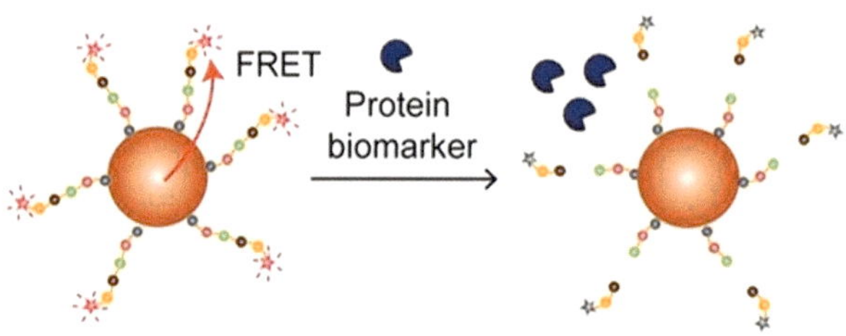

Fig. 11 Schematic illustration of a FRET-based sensor for monitoring protease activity.

5 Applications to in vivo tissue imaging and point-of-care testing

5.1 Engineered AuNP for in vivo cancer tissue imaging

Although the main optical imaging technologies for diagnosis and/or monitoring of cancer will be discussed in more detail in Chapter 2.1, a specific example is described here. Perrault and Chan reported the use of biotinylated 30-nm-diameter AuNPs coupled with fluorescently labeled streptavidin for in vivo cancer tissue imaging in a xenograft breast cancer (MDA-MD-435) mouse model (Fig. 13) [20]. This approach combines passive

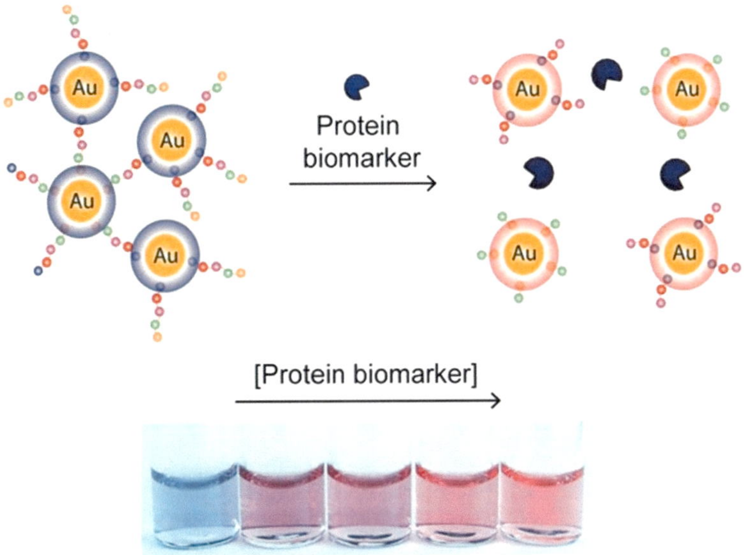

Fig. 12 Schematic illustration of a colorimetric-based sensor for the detection of protease activity. *(Adapted with permission from P.D. Howes, R. Chandrawati, M.M. Stevens, Colloidal nanoparticles as advanced biological sensors, Science 346 (2014) 1247390.)*

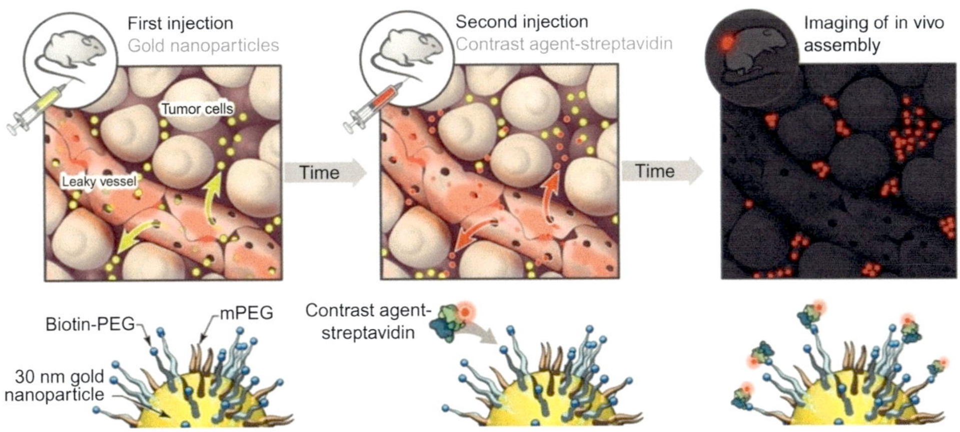

Fig. 13 Schematic of the working mechanism of biotin-AuNPs/fluorescently labeled streptavidin for in vivo cancer tissue imaging. Biotin-AuNPs enter tumors through leaky vasculature and passively accumulate in the extracellular matrix. Fluorescently labeled streptavidin is then injected and interacts with biotin on the AuNP surface. *(Reproduced with permission from S.D. Perrault, W.C.W. Chan, In vivo assembly of nanoparticle components to improve targeted cancer imaging, PNAS 107 (2010) 11194–11199.)*

targeting of the AuNPs to tumor tissue and active targeting of the fluorophores to label the AuNPs in vivo. The biotin-AuNPs were first injected to enter tumors through leaky vasculature, where particles passively accumulated in the extracellular matrix over 24 h. Fluorescently labeled streptavidin was then injected and reacted with the biotin on the AuNP surface, producing fluorescent signal. The authors showed that this method resulted in probe accumulation in tumor tissue at a rate 200 times faster compared to fluorophore-labeled AuNPs of the same size.

5.2 Point-of-care tests based on optical sensing of cancer biomarker

Although more detailed case studies will be provided in Chapter 1.4, the following section presents selected examples of platform technologies for the colorimetric detection of cancer biomarker at the point of care. The development of a highly sensitive sensor with eye-detectable cancer indicator is of utmost importance for a point-of-care (POC) device. Over the past few years, the concurrent advancement in bioengineering and optical biosensors or sensing modalities has shown a promising future in the development of POC diagnostic devices. Liquid biopsies such as urine, blood, and saliva, or even a simple test using exhaled breath can be used to diagnose a complicated disease like cancer.

Breath test as a means to diagnose lung cancer has been in the clinical phase for many years. The base concept of this assay is owing to the fact that lung cancer increases the concentration of several volatile organic compounds (VOCs) in the human breath, from 1–20 ppb to 10–100 ppb [21]. Mazzone et al. designed a colorimetric sensor array based on chemically responsive dyes printed on a disposable cartridge to identify the altered VOC composition in lung cancer patients (Fig. 14) [22]. Three classes of dyes are used: dyes containing metal ions responsive to Lewis basicity, pH indicator dyes responsive to Brønsted acidity/basicity, and solvatochromic dyes responsive to local polarity. The colorimetric sensor arrays generated red, green, and blue color changes for each dye with 72 dimensional vectors, which allowed for facile discrimination among complex mixtures. Exhaled breath of study subjects consisting of 92 patients with lung cancer and 137 healthy controls was drawn across the sensor array, and the sensor was moderately accurate at distinguishing lung cancer from control subjects. The study suggested that the use of breath biosignatures could predict cancer stage and survival rate. However, challenges with inability to detect and quantify all potentially relevant trace compounds still remain.

Another suitable candidate for POC cancer diagnosis is polydiacetylene (PDA), which is a class of conjugated polymer that has gained significant interest in sensing technologies due to its unique optical properties, versatility, stability, and simple synthesis method [23]. A PDA-based POC device to diagnose early stage ovarian cancer in ascites and blood plasma was recently developed by Wang et al. [24] In this work, a facile "lock-key" strategy was employed based on the synergistic electrostatic and hydrophobic

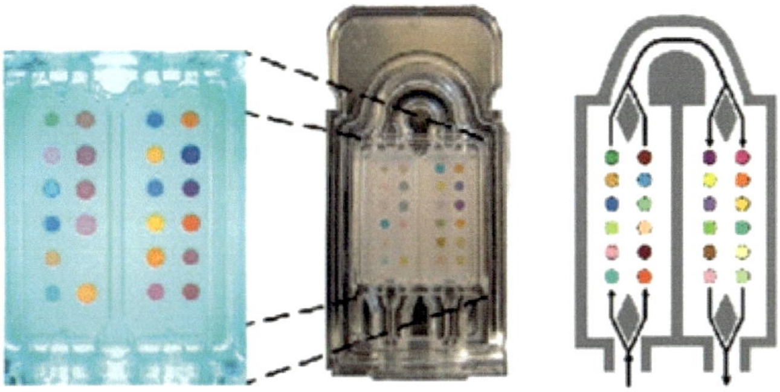

H2TPP+TsOH	CoTPP	Malachite green	Methyl red + NaOH
Zn2,6-F2PP	ZnTPP	Chlorophenol red	Bromophenol red + NaOH
Reichardt's dye RD1	ZnTPP	Nile red	Chlorophenol red + NaOH
Malachite green	F258 ZnTMP	Bromophenol blue	Chlorophenol red + NaOH
Reichardt's dye RD3	Lead acetate	Bromocresol green	Cresol red + NaOH
Copper neodecanoate	Bromophone Red +TsOH	Thymol blue	Cresol red + NaOH

Fig. 14 Colorimetric sensor array for identification and characterization of lung cancer using chemically responsive dyes printed on a disposable cartridge. Exhaled breath drawn moves across the cartridge in the direction indicated by the *black arrows*. The material in each colorant is listed in the table. *(Reproduced with permission from P.J. Mazzone, X.-F. Wang, Y. Xu, T. Mekhail, M.C. Beukemann, J. Na, et al., Exhaled breath analysis with a colorimetric sensor array for the identification and characterization of lung cancer, J. Thoracic Oncol. 7 (2012) 137–142.)*

reaction between cancer biomarker lysophosphatic (LPA) "key," which causes the blue-to-red shift by opening the PDA "lock" (Fig. 15). The color change was directly visible by the naked eye and was attributed to the reorganization of the PDA backbone in response to the stimuli [25,26]. Additionally, the blue color was nonfluorescent, whereas the red color showed strong fluorescence. Using this concept, a paper-based lateral flow assay (LFA) was fabricated and the deposition of LPA-containing blood plasma changed the blue line to red in 5 min, indicating the presence of cancer. A limit of detection (LOD) of 0.5 μM was reported, which is well below plasma LPA concentrations typically observed in patients with ovarian cancer.

Chapman et al. reported a liposome-based POC sensor for the detection of phospholipase A_2 (PLA_2) (Fig. 16) [27]. PLA_2 is known to be upregulated in prostate cancer [28] and lung cancer [29]. In their system, biotinylated polymer linkers were encapsulated in liposomes. Phospholipids are natural PLA_2 substrates. In the presence of biomarkers, PLA_2 cleaved the liposomes and released the encapsulated linkers, which caused aggregation of streptavidin-labeled AuNPs that could be readily detected by the naked eye due to the surface plasmon resonance effect of AuNPs. A LOD of 1 nM PLA_2 in human serum was achieved and the assay could be completed within 20 min without the need for an expensive instrument.

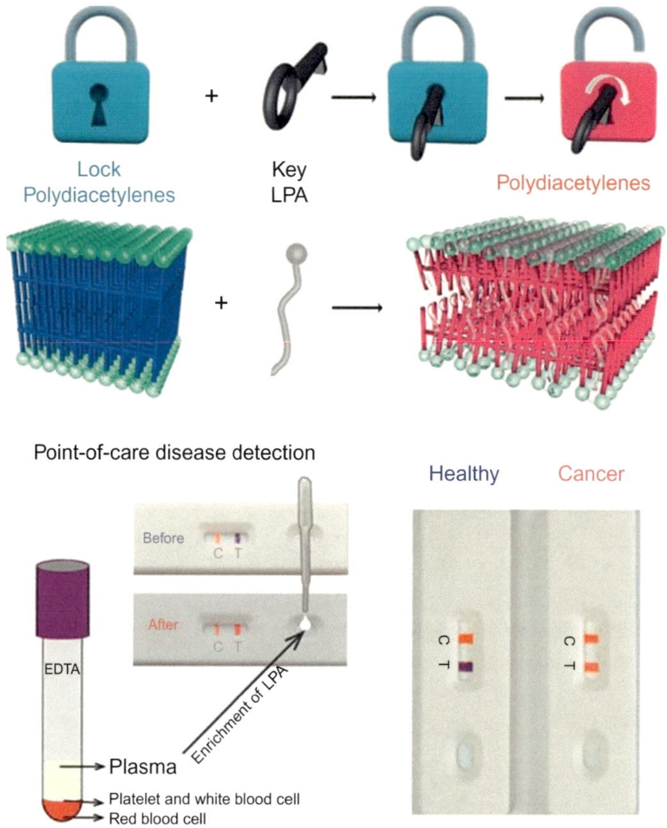

Fig. 15 Paper-based lateral flow assay (LFA) as a point-of-care ovarian cancer diagnosis. Device was developed using a lock (polydiacetylene, PDA) and key (cancer biomarker lysophosphatic, LPA) mechanism. The changes in PDA backbone conformation in the presence of LPA cause a dramatic color change from *blue* to *red* that is visible to the naked eye. *(Reproduced with permission from Y. Wang, H. Pei, J. Liu, Z. Li, K. Ai, Z. Lu, et al., Synergistic tailoring of electrostatic and hydrophobic interactions for rapid and specific recognition of lysophosphatidic acid, an early-stage ovarian cancer biomarker, J. Am. Chem. Soc. 139 (2017) 11616–11621.)*

6 Conclusions

In summary, we have illustrated a number of promising examples of optical sensors for the detection of cancer biomarkers that are based on either small-molecule fluorophores or nanoparticles. Nanotechnology-based assays offer enhanced sensitivity and selectivity compared to traditional cancer imaging techniques. In some cases, they offer entirely new and unique capabilities that are not attainable with conventional methods. While a variety of probes have been designed for the detection of cancer biomarkers, no specific assay has been selected as the "gold standard" for clinical use and there are many interesting challenges ahead before their full potential can be realized. Ultimately, translation of

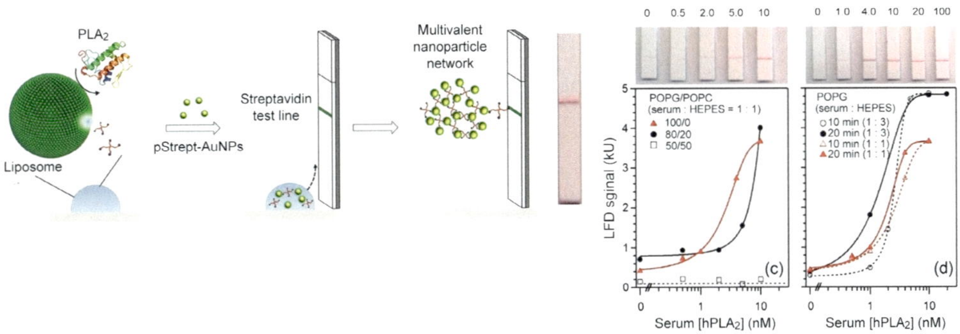

Fig. 16 Liposome sensors for the detection of phospholipase A_2 (PLA_2). Liposomes encapsulating biotinylated polymer linkers are mixed with a sample solution and streptavidin-coated AuNPs. A lateral flow strip is introduced and the solution mixture runs up the nitrocellulose membrane by capillary forces. In the presence of PLA_2, the enzyme cleaves the liposome and releases the polymer linkers. The linkers adhere streptavidin-coated AuNPs to the test line and multivalent nanoparticle networks are formed, which can be read with the naked eye due to the localized surface plasmon resonance (LSPR) effect of AuNPs. *(Reproduced with permission from R. Chapman, Y. Lin, M. Burnapp, A. Bentham, D. Hillier, A. Zabron, et al., Multivalent nanoparticle networks enable point-of-care detection of human phospholipase-A2 in serum, ACS Nano 9 (2015) 2565–2573.)*

optical probes to clinical use will require further knowledge of the correlation between levels of cancer biomarkers present in a patient and the stage of their disease to enhance prognostic and diagnostic capabilities.

Optical detection can also be coupled with smartphone technologies for better translation. Recently, the use of smartphone-enabled strategies to facilitate cancer detection and monitoring has been largely explored. The use of microfluidic barcode platform is particularly interesting due to its accuracy and possibility to provide quantitative results from minute sample volume [30]. Biochips [31,32] and microfluidic methods integrated with smartphones [33,34] to detect tumor-derived circulating tumor cells [35,36] and exosomes [37,38], as well as technologies to facilitate cancer monitoring in POC device, have been reported. Although problems with eliminating the influence of background light and inconsistent smartphone camera quality to affirm the accuracy of the results still exist [39], the advancement of this technology could revolutionize liquid biopsy screening to diagnose early stage cancers and for enabling personalized medicine in an affordable and easily accessible way.

Acknowledgments

R.C. acknowledges the support from the Australian Research Council Discovery Early Career Researcher Award (ARC DECRA DE170100068) and the UNSW Scientia Fellowship.

References

[1] L. Fass, Imaging and cancer: a review, Mol. Oncol. 2 (2008) 115–152.

[2] A.P.F. Turner, Biosensors: sense and sensibility, Chem. Soc. Rev. 42 (2013) 3184–3196.

[3] P.D. Howes, R. Chandrawati, M.M. Stevens, Colloidal nanoparticles as advanced biological sensors, Science 346 (2014).

[4] D. Vilela, M.C. González, A. Escarpa, Sensing colorimetric approaches based on gold and silver nanoparticles aggregation: chemical creativity behind the assay. A review, Anal. Chim. Acta 751 (2012) 24–43.

[5] R. Freeman, I. Willner, Optical molecular sensing with semiconductor quantum dots (QDs), Chem. Soc. Rev. 41 (2012) 4067–4085.

[6] N. Panchuk-Voloshina, R.P. Haugland, J. Bishop-Stewart, M.K. Bhalgat, P.J. Millard, F. Mao, W.-Y. Leung, R.P. Haugland, Alexa dyes, a series of new fluorescent dyes that yield exceptionally bright, photostable conjugates, J. Histochem. Cytochem. 47 (1999) 1179–1188.

[7] K.D. Wegner, N. Hildebrandt, Quantum dots: bright and versatile in vitro and in vivo fluorescence imaging biosensors, Chem. Soc. Rev. 44 (2015) 4792–4834.

[8] X. Wu, G. Chen, J. Shen, Z. Li, Y. Zhang, G. Han, Upconversion nanoparticles: a versatile solution to multiscale biological imaging, Bioconjug. Chem. 26 (2015) 166–175.

[9] M. Montalti, L. Prodi, E. Rampazzo, N. Zaccheroni, Dye-doped silica nanoparticles as luminescent organized systems for nanomedicine, Chem. Soc. Rev. 43 (2014) 4243–4268.

[10] L.-Y. Chen, C.-W. Wang, Z. Yuan, H.-T. Chang, Fluorescent gold nanoclusters: recent advances in sensing and imaging, Anal. Chem. 87 (2015) 216–229.

[11] J. Zheng, R. Yang, M. Shi, C. Wu, X. Fang, Y. Li, J. Li, W. Tan, Rationally designed molecular beacons for bioanalytical and biomedical applications, Chem. Soc. Rev. 44 (2015) 3036–3055.

[12] A.N. Glazer, H.S. Rye, Stable dye-DNA intercalation complexes as reagents for high-sensitivity fluorescence detection, Nature 359 (1992) 859–861.

[13] A. Shibata, H. Abe, Y. Ito, Oligonucleotide-templated reactions for sensing nucleic acids, Molecules 17 (2012) 2446–2463.

[14] G.A. Metcalf, A. Shibakawa, H. Patel, A. Sita-Lumsden, A. Zivi, N. Rama, C.L. Bevan, S. Ladame, Amplification-free detection of circulating microRNA biomarkers from body fluids based on fluorogenic oligonucleotide-templated reaction between engineered peptide nucleic acid probes: application to prostate cancer diagnosis, Anal. Chem. 88 (2016) 8091–8098.

[15] C. Coutinho, A. Somoza, MicroRNA sensors based on gold nanoparticles, Anal. Bioanal. Chem. 411 (2019) 1807–1824.

[16] A. Heuer-Jungemann, P.K. Harimech, T. Brown, A.G. Kanaras, Gold nanoparticles and fluorescently-labelled DNA as a platform for biological sensing, Nanoscale 5 (2013) 9503–9510.

[17] M. Bai, D.J. Bornshop, Recent advances in receptor-targeted fluorescent probes for in vivo cancer imaging, Curr. Med. Chem. 19 (2012) 4742–4758.

[18] Y. Hillman, D. Lustiger, Y. Wine, Antibody-based nanotechnology, Nanotechnology 30 (2019).

[19] T. Wang, C. Chen, L.M. Larcher, R.A. Barrero, R.N. Veedu, Three decades of nucleic acid aptamer technologies: lessons learned, progress and opportunities on aptamer development, Biotechnol. Adv. 37 (2019) 28–50.

[20] S.D. Perrault, W.C.W. Chan, In vivo assembly of nanoparticle components to improve targeted cancer imaging, PNAS 107 (2010) 11194–11199.

[21] G. Peng, U. Tisch, O. Adams, M. Hakim, N. Shehada, Y.V. Broza, S. Billan, R. Abdah-Bortnyak, A. Kuten, Diagnosing lung cancer in exhaled breath using gold nanoparticles, Nat. Nanotechnol. 4 (2009) 669–673.

[22] P.J. Mazzone, X.-F. Wang, Y. Xu, T. Mekhail, M.C. Beukemann, J. Na, J.W. Kemling, K.S. Suslick, M. Sashidar, Exhaled breath analysis with a colorimetric sensor array for the identification and characterization of lung cancer, J. Thorac. Oncol. 7 (2012) 137–142.

[23] F. Mazur, M. Bally, B. Städler, R. Chandrawati, Liposomes and lipid bilayers in biosensors, Adv. Colloid Interf. Sci. 249 (2017) 88–99.

[24] Y. Wang, H. Pei, J. Liu, Z. Li, K. Ai, Z. Lu, L. Lu, Synergistic tailoring of electrostatic and hydrophobic interactions for rapid and specific recognition of lysophosphatidic acid, an early-stage ovarian cancer biomarker, J. Am. Chem. Soc. 139 (2017) 11616–11621.

[25] Y.K. Jung, H.G. Park, Colorimetric detection of clinical DNA samples using an intercalator-conjugated polydiacetylene sensor, Biosens. Bioelectron. 72 (2015) 127–132.
[26] X. Qian, B. Städler, Recent developments in polydiacetylene-based sensors, Chem. Mater. 31 (2019) 1196–1222.
[27] R. Chapman, Y. Lin, M. Burnapp, A. Bentham, D. Hillier, A. Zabron, S. Khan, M. Tyreman, M. M. Stevens, Multivalent nanoparticle networks enable point-of-care detection of human phospholipase-A2 in serum, ACS Nano 9 (2015) 2565–2573.
[28] M. Menschikowski, A. Hagelgans, S. Fuessel, O.A. Mareninova, V. Neumeister, M.P. Wirth, G. Siegert, Serum levels of secreted group IIA phospholipase A2 in benign prostatic hyperplasia and prostate cancer: a biomarker for inflammation or neoplasia? Inflammation 35 (2012) 1113–1118.
[29] E. Kupert, M. Anderson, Y. Liu, P. Succop, L. Levin, J. Wang, K. Wikenheiser-brokamp, P. Chen, S. M. Pinney, T. Macdonald, Z. Dong, S. Starnes, S. Lu, Plasma secretory phospholipase A2-IIa as a potential biomarker for lung cancer in patients with solitary pulmonary nodules, BMC Cancer 11 (2011) 513.
[30] B. Hayes, C. Murphy, A. Crawley, R. O'Kennedy, Developments in point-of-care diagnostic technology for cancer detection, Diagnostics 8 (2018) 1–18.
[31] H. Shao, J. Chung, K. Lee, L. Balaj, C. Min, B.S. Carter, F.H. Hochberg, X.O. Breakefield, H. Lee, R. Weissleder, Chip-based analysis of exosomal mRNA mediating drug resistance in glioblastoma, Nat. Commun. 6 (2015) 6999.
[32] M. He, J. Crow, M. Roth, Y. Zeng, A.K. Godwin, Integrated immunoisolation and protein analysis of circulating exosomes using microfluidic technology, Lab Chip 14 (2014) 3773–3780.
[33] D. Xu, X. Huang, J. Guo, X. Ma, Automatic smartphone-based microfluidic biosensor system at the point of care, Biosens. Bioelectron. 110 (2018) 78–88.
[34] S.K. Vashist, P.B. Luppa, L.Y. Yeo, A. Ozcan, J.H.T. Luong, Review: emerging technologies for next-generation point-of-care testing, Trends Biotechnol. 33 (2015) 692–705.
[35] K.H. Neoh, A.A. Hassan, A. Chen, Y. Sun, P. Liu, K.-F. Xu, A.S.T. Wong, R.P.S. Han, Rethinking liquid biopsy: microfluidic assays for mobile tumor cells in human body fluids, Biomaterials 150 (2018) 112–124.
[36] J.M. Jackson, M.A. Witek, J.W. Kamande, S.A. Soper, Materials and microfluidics: enabling the efficient isolation and analysis of circulating tumour cells, Chem. Soc. Rev. 46 (2017) 4245–4280.
[37] M. He, Y. Zeng, Microfluidic exosome analysis toward liquid biopsy for cancer, SLAS Technol. 21 (2016) 599–608.
[38] J.C. Contreras-Naranjo, H.-J. Wu, V.M. Ugaz, Microfluidics for exosome isolation and analysis: enabling liquid biopsy for personalized medicine, Lab Chip 17 (2017) 3558–3577.
[39] Z. Geng, X. Zhang, Z. Fan, X. Lv, Y. Su, H. Chen, Recent progress in optical biosensors based on smartphone platforms, Sensors. 17 (2017) 2449.

Further reading

[40] J.W. Lichtman, J.-A. Conchello, Fluorescence microscopy, Nat. Methods 2 (2005) 910–919.
[41] S. Feng, Y. Fang, W. Feng, Q. Xia, G. Feng, A colorimetric and ratiometric fluorescent probe with enhanced near-infrared fluorescence for selective detection of cysteine and its application in living cells, Dyes Pigments 146 (2017) 103–111.
[42] M. Han, X. Gao, J.Z. Su, S. Nie, Quantum-dot-tagged microbeads for multiplexed optical coding of biomolecules, Nat. Biotechnol. 19 (2001) 631.
[43] S. Pavagada, S. Ladame, Platforms for bioorthogonal oligonucleotide-templated reactions: translating concepts into devices, Chimia (Aarau) 72 (2018) 809–814.
[44] V. Karantza, Keratins in health and cancer: more than mere epithelial cell markers, Oncogene 30 (2011) 127–138.
[45] D. Musumeci, C. Platella, C. Riccardi, F. Moccia, D. Montesarchio, Fluorescence sensing using DNA aptamers in cancer research and clinical diagnostics, Cancers 9 (2017) 174.

CHAPTER 1.3

Electrochemical sensors

Md. Nazmul Islam, Robert B. Channon
Department of Bioengineering, Imperial College London, London, United Kingdom

1 Introduction to electrochemical analysis

Electrochemistry is a 200-year-old discipline that is integral to a wide range of fields, including battery technologies, solar cells, glucose monitors for diabetes, fuel cells, supercapacitors, and pH sensors to name a few [1]. As an analytical platform, electrochemical sensors are typically fast (detection in less than a minute), inexpensive (<£1 per measurement), capable of high sensitivity and low detection limits (~nM/1000 targets per mL) and are ideally suited for miniaturization and for point-of-care analysis.

At its most basic, electrochemistry investigates charge transfer at interfaces. Consider a one electron redox reaction of a species *r* in its reduced (r^n) and oxidized forms (o^{n+1}),

$$r^n \rightleftharpoons o^{n+1} + e^-, \tag{1}$$

If we perturb this equilibrium by applying a potential (E), we can drive the reaction in either direction to oxidize *r* into *o*, or reduce *o* into *r*. This results in a flow of electrons in the form of a current (i). This current is directly proportional to the concentration of the species in solution, facilitating quantification.

A typical electrochemical cell is shown in Fig. 1. The solution contains the species of interest, a solvent (typically water), and a background electrolyte (e.g., 0.1 M KNO_3) to increase the conductivity of the solution. If the working electrode is held at a positive potential, this transfers an electron from the species *r* to the electrode surface, inducing a current flow and the oxidation of *r* into *o*. Conversely, a negative working electrode potential induces reduction of the target species (*o* to *r*). This convention is known as OIL RIG (oxidation is loss of electrons, reduction is gain of electrons).

The potentiostat serves as the power source and applies a voltage to the working electrode while measuring the current (amperometry) or applies a current and measures the voltage (potentiometry). Potentiostat can be homemade or sourced commercially, ranging in price based on their versatility and sensitivity (£500–£10,000). The reference electrode serves to maintain a constant potential, from which the working electrode potential can be applied (e.g., 1.0 V vs an Ag/AgCl reference). While several conductive materials can be used as reference electrodes, model reference electrodes are designed to maintain a constant potential irrespective of solution conditions (e.g., pH, temperature). A typical Ag/AgCl reference electrode (Fig. 1) consists of a silver wire coated with an

Bioengineering Innovative Solutions for Cancer
https://doi.org/10.1016/B978-0-12-813886-1.00004-8

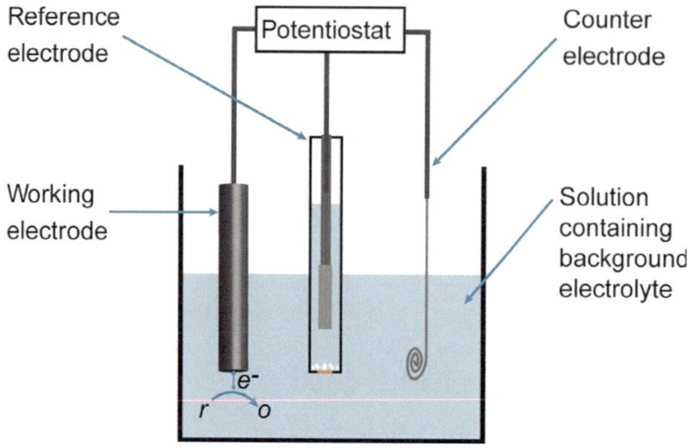

Fig. 1 Illustration of a typical three-electrode electrolytic cell.

AgCl surface layer, immersed in a 3.0 M KCl solution, with a frit in between this solution and the target analyte solution. Here, the potential is linked only to the chloride concentration in the reference solution and thus will maintain a constant potential irrespective of the target solution identity or pH. The potentiostat also measures the current at the working electrode and applies an equal but opposite current to the counter electrode in order to complete the electrical circuit. The counter electrode is typically a conductive metal with a high surface area such as a platinum gauze.

1.1 Theoretical considerations

The current generated from the oxidation or reduction of a target analyte is related to several experimental variables and constants, but a general expression can be written as,

$$i = nAFj, \tag{2}$$

where n is the number of electrons transferred, A is the surface area of the working electrode (cm^2), F is a Faraday constant (96,485 C mol^{-1}), and j is the flux of the target analyte toward the electrode (expressed in mol cm^{-2} s^{-1}). The flux term is critical as, in order to undergo a redox reaction, a species must reach the electrode surface from the bulk solution (mass transport) and exchange an electron with the surface through electron tunneling (electron transfer). These processes are outlined in more detail in Fig. 2.

In many electrochemical systems, the electron transfer is significantly faster than the mass transport, such that any species which reaches the electrode is turned over and the mass transport is the limiting step (mass transport limited). The three main forms of mass transport are diffusion (movement of a species down a concentration gradient), migration (movement of charged species under an electric field), and convection (movement of a species by mechanical forces or temperature gradients). Therefore, the total flux can be

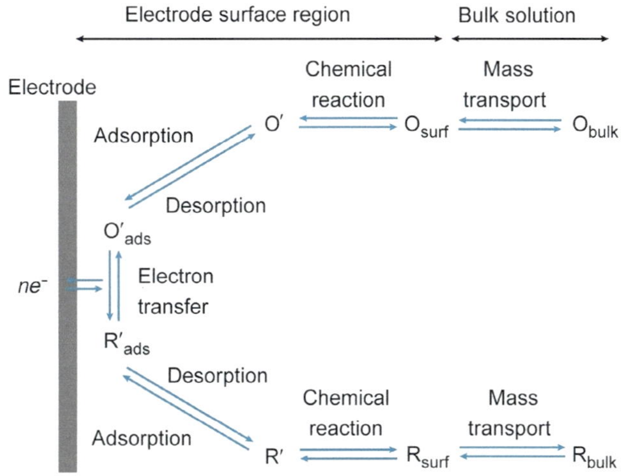

Fig. 2 Illustration of heterogeneous electron transfer, where all of the processes within the electron surface region are encompassed under electron transfer [2].

described by the Nernst-Planck equation, which describes the one-dimensional mass transport to an electrode as [2],

$$j_x = D\frac{\partial C}{\partial x} + \frac{zF}{RT}DC\frac{\partial \varphi}{\partial x} - v_x C^*, \tag{3}$$

where the three terms represent the diffusive, migratory, and convective contributions to mass transport, respectively. D is the diffusion coefficient (expressed in $\text{cm}^2\,\text{s}^{-1}$), C is the concentration (expressed in $\text{mol}\,\text{cm}^{-2}$), z is the charge of the species, R is the molar gas constant ($8.314\,\text{J}\,\text{K}^{-1}\,\text{mol}^{-1}$), T is the temperature (expressed in K), $\delta C/\delta x$ and $\delta\varphi/\delta x$ are the concentration and potential gradients at point x and v_x is the velocity. While diffusion, migration, and convection can all be used to increase the flux and enhance the analytical signals, often migration is negated by adding high concentration of background electrolyte (which shields the solution from the electric fields at the electrode surface) and convection is negated by avoiding stirring the solution and maintain solutions at room temperature (to avoid generation of convective flows). This way, diffusion is the only form of mass transport, the analytical expressions to describe the current are simpler and the results are simpler to interpret.

1.2 Experimental variables

Many experimental variables can affect electrochemical measurements and thus should be optimized to enhance sensitivity, specificity, reliability, and accuracy of the sensing.

- **Electrode material:** Common electrode materials include metals (Pt, Au) carbon-based materials (glassy carbon, graphene, highly ordered pyrolytic graphite) and composites (boron doped diamond, thermoplastic carbon electrodes, screen-printed carbon). Electrode choice is driven by the application requirements, such as: does the electrode need to be disposable (screen printed carbon is cheap) or used for multiple experiments (boron doped diamond is robust and inert to fouling)? Is the sensing system simple (commercial Pt or Au electrodes) or complex (home-made inlaid band electrodes for flow systems)? Are small (nanopipette electrodes for skin patches) or large (stripping voltammetry of heavy metals) electrodes needed?

- **Solution conditions:** While in vivo measurements are conducted under physiological conditions, for systems where the target analyte is extracted/purified from the body, conditions can be chosen to promote and optimize the sensing. For example, redox reactions involving proton transfer are significantly affected by the pH (in terms of mechanism, onset potential, and current). Similarly, background electrolyte can be added to extracted/purified samples to negate migratory effects.
- **Direct/indirect sensing:** Many biological analytes can be oxidized or reduced electrochemically for direct detection, such as glucose, oxygen, and most neurotransmitters. However, most cancer biomarkers require indirect detection, where the presence of a target molecule generates electrochemical species for detection (electrochemical enzyme-linked immunosorbent assays or ELISAs) or, capturing of the target alters the electrode's ability to facilitate redox reactions (on/off sensing). For indirect sensing, specific recognition elements are required for every target of interest, which is often achieved through electrode modification.

1.3 Modified electrodes

Many strategies have been employed for electrode modification, including π-π stacking, silanization, and nanoparticle deposition, though the two most common approaches are thiol [3] and diazonium [4] coupling. Gold-thiol modification is driven thermodynamically by the strength of the Au-S bond, such that exposure of Au to a thiol (R-SH) (Scheme 1A) or a disulfide (R-S-S-R) [5] (Scheme 1B) results in the formation of a self-assembled monolayer on the Au surface. Diazonium coupling proceeds via reduction of an aryldiazonium salt (electrochemically or chemically), followed by radical addition to the electrode surface (Scheme 1C, commonly carbon electrodes). EDC/NHS coupling (1-ethyl-3-(-3-dimethylaminopropyl) carbodiimide hydrochloride/*N*-hydroxysuccinimide) employs carbodiimide conjugation to link a primary amine ($R-NH_2$) to a carboxylic acid ($R'-COOH$) (Scheme 1D). This is a reliable tool for linking modified electrodes (e.g., Au-Thiol-COOH) with biorecognition elements. Finally, the coupling reaction between biotin (vitamin B7) and streptavidin (a protein) is one of the strongest noncovalent bonds in nature and can also be used to link electrodes to biorecognition elements.

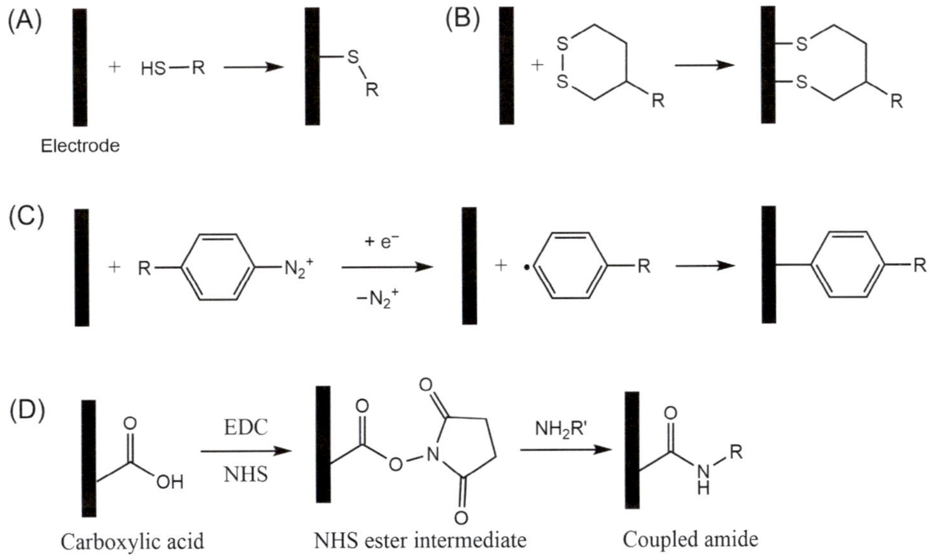

Scheme 1 (A) Au-thiol, (B) Au-dithiol, (C) diazonium and (D) EDC/NHS coupling reactions.

2 Sensing techniques

Electrochemical analysis encompasses a range of techniques, each with their own advantages and drawbacks. The following approaches are not an exhaustive list but serve as a guide to the most common electrochemical analysis methods for sensing of cancer biomarkers.

2.1 Chronoamperometry

Chronoamperometry is the simplest amperometry technique whereby a constant potential is applied, and the resulting current measured over time (e.g., 10 s). This is usually accomplished through a single potential step: from a potential where no redox reactions are occurring to a high positive/negative potential to drive the turnover of the target species (e.g., 0 V to +0.5 V). A spike in current is observed over short times ($t = 10$ ms) followed by a decaying current to a constant value (Fig. 3), which is proportional to the analyte concentration through the Cottrell equation,

$$i = \frac{nAFD^{1/2}C}{(\pi t)^{1/2}}, \tag{4}$$

Chronoamperometry is a fast (10 s) and simple technique to apply (cheap instrumentation), but the signal-to-noise ratios are often poor resulting in modest detection limits. Chronoamperometry is also used in combination with convective fluid transport for flow

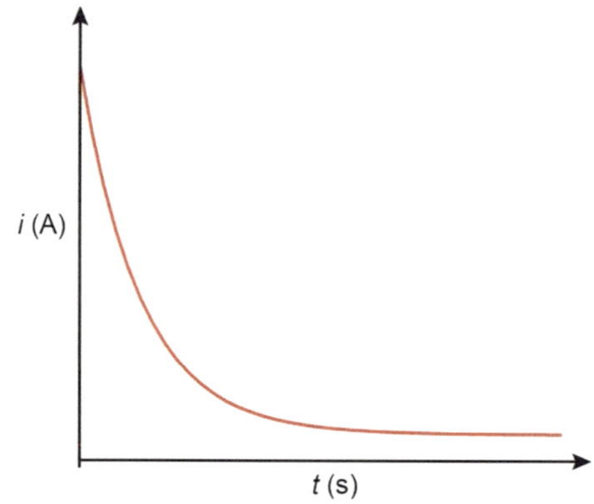

Fig. 3 Typical current-time plot from a chronoamperometry experiment.

injection analysis. Here, a small volume of sample analyte (5–50 μL) is added to a flowing carrier stream (1 mL min^{-1}), which transports the plug to the electrode surface for chronoamperometric detection. As a result of the extra convective contribution to mass transport (Eq. 3), flow injection analysis exhibits marked improvements over conventional chronoamperometry in terms of both sensitivity (better/lower detection limit) and sample throughput (~6 samples per minute).

2.2 Cyclic voltammetry

In voltammetric techniques, the potential is changed with time and the resulting current recorded. Cyclic voltammetry is the most basic of these techniques, whereby the potential is scanned between two potentials (E_s and E_f) in a staircase fashion as shown in Fig. 4A. As the potential becomes more positive, this causes the target analyte to be oxidized by the working electrode, resulting in an increase in current. This depletes the analyte from the electrode surface generating a linear concentration gradient for a macrometer-sized electrode (1-mm disk) and a hemispherical concentration gradient for a micrometer-sized electrode (10-μm-diameter disk) as shown in Fig. 5.

The shape of the current voltage plot (known generally as a cyclic voltammogram) depends on the working electrode dimensions. For a microelectrode, at high oxidizing potentials, the current reaches a steady state where the mass transport is fast and the current is limited by how quickly the electrode surface can turn over the analyte (Fig. 4B). Conversely, for a macroelectrode, the mass transport is slower (1 dimensional) and limiting, such that the current reaches a maximum before decaying to a steady-state value where the current is limited by how quickly the analyte is transported to the electrode surface (Fig. 4C). Note that if the waveform is stopped after the first sweep, this is known as linear sweep voltammetry.

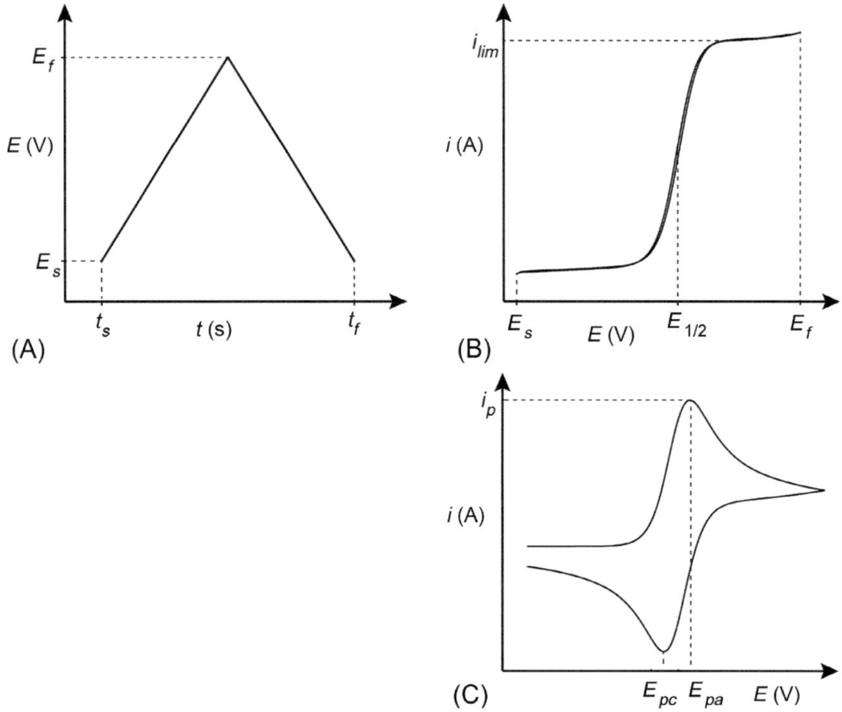

Fig. 4 (A) Applied potential waveform with time and the resulting cyclic voltammogram (current vs potential) at a (B) micrometer and (C) macrometer-sized electrode, for an oxidation process. E_{pa} and E_{pc} are the anodic and cathodic peak potentials, respectively, i_{lim} is the limiting current, and i_p is the peak current.

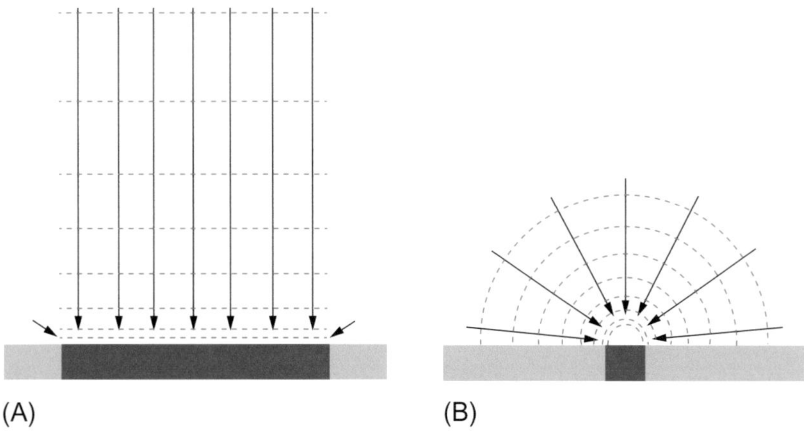

Fig. 5 Diffusion-generated concentration gradient for (A) a macroelectrode or (B) a microelectode, where the concentration is zero at the electrode surface, and bulk concentration far from the surface for both cases.

For the microelectrode, the steady-state current (also sometimes referred to as the limiting current, i_{lim}) is given by,

$$i_{lim} = 4nFDCr, \tag{5}$$

where r is the radii of the microelectrode (cm).

For a macroelectrode, the peak current of the forward wave is given by,

$$i_p = 2.69 \times 10^5 n^{3/2} D^{1/2} CA\nu^{1/2}, \tag{6}$$

where ν is the scan rate ($V\,s^{-1}$) and D and C use units of $cm^2\,s^{-1}$ and $mol\,cm^{-3}$, respectively.

Cyclic voltammograms are typically quantified in terms of the i_p or i_{lim} as well as the peak-to-peak potential (ΔE_p), which is $E_{pa} - E_{pc}$ for the macrocase and $E_{3/4} - E_{1/4}$ for the microcase ($E_{3/4}$ and $E_{1/4}$ represent the potentials at 75% and 25% of the i_{lim}). For a 1-electron reversible redox process under standard conditions, ΔE_p should be 59.1 mV, although slow kinetics of the redox process, poor reversibility ($r \leftrightarrow o$ vs $r \rightarrow o$), or resistances within the electrochemical cell can lead to larger ΔE_p and smaller currents.

2.3 Pulsed voltammetry

In pulsed voltammetry, a series of pulses are superimposed on the cyclic voltammetry staircase waveform as shown in Fig. 6A. Instead of constant current sampling as used in cyclic voltammetry, the current in differential pulse voltammetry is sampled just before the start (τ) and at the end (τ') of a potential step, i.e.,

$$\Delta i = i_{\tau'} - i_{\tau}, \tag{7}$$

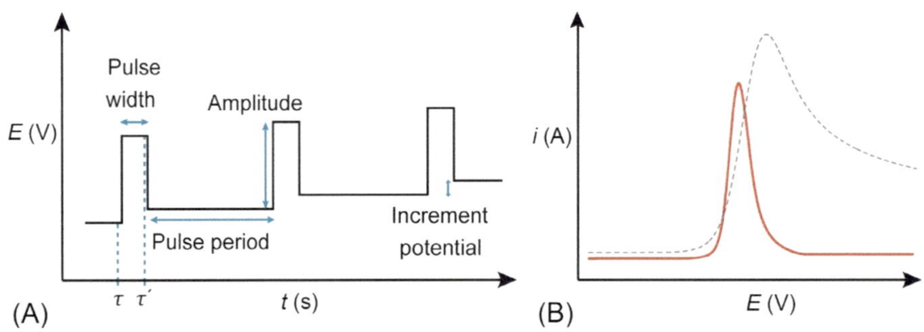

Fig. 6 (A) Potential staircase and (B) current-potential plot for an oxidation via differential pulse voltammetry (—) at a macroelectrode overlaid with the corresponding linear sweep voltammogram (- - -).

where Δi is the differential current, $i_{\tau'}$ is the current at time τ', and i_{τ} is the current at time τ. The peak current is given by the following equation [6],

$$\Delta i = \frac{nFAD^{1/2}C}{\pi^{1/2}(\tau' - \tau)^{1/2}}\left(\frac{1-\sigma}{1+\sigma}\right), \tag{8}$$

where σ is given by

$$\sigma = exp\left(\frac{nF\Delta E}{2RT}\right), \tag{9}$$

and ΔE represents the incremental potential.

The main advantage of differential pulsed voltammetry over other electrochemical techniques is the significantly better detection limits resulting from the high signal-to-noise ratio. After the potential pulse, a background capacitive current is generated as a result of the changing potential of the electrode versus the solution potential. This background charging current decays faster with time ($i \propto e^{-t}$) than the faradaic current from the analyte of interest ($i \propto 1/t^{1/2}$, Eq. 4). Therefore, the differential current (Δi, Eq. 8) generates a smaller current than cyclic voltammetry, but with a significantly smaller background current as shown in Fig. 6B. Note that sharper peaks (smaller width) are observed in pulsed voltammetry due to the differential current, which can be used to measure mixtures of species with similar onset potentials. Another commonly used pulsed voltammetry technique is square wave voltammetry. This employs a similar potential waveform to differential pulse voltammetry, but with a faster scanning speed resulting in a quicker experiment (30 s vs 3 min) and only marginal losses in sensitivity (higher detection limit).

2.4 Electrochemical impedance spectroscopy

Electrochemical impedance spectroscopy (EIS) is one of the most sensitive electrochemical techniques, capable of delivering measurable signal changes resulting from small changes in biomarker concentration [6]. Here, an alternating potential is applied at a range of frequencies to the working electrode. This generates an alternating current from the reduction and oxidation (charge transfer) of the electrochemical species at the electrode surface (Eq. 1). This technique probes the electrode-solution interface and the redox couple to measure the following variables:

- R_{sol} (the solution resistance includes resistances in the electrode material, the electrode contacts, and in the solution between the working and reference electrodes)
- R_{ct} (the resistance to charge transfer for the redox couple)
- C_d (the double-layer capacitance)
- Z (the impedance, which is the effective resistance of the electrochemical cell to the alternating current, including reactance and ohmic resistance)

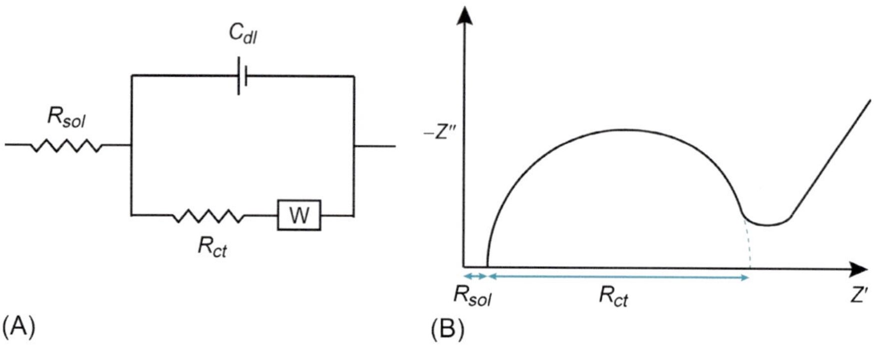

Fig. 7 (A) Circuit diagram (Randel's circuit) and (B) Nyquist plot (real vs imaginary forms of the impedance, *Z*) for a simple redox reaction.

These variables can be combined to generate an equivalent electrochemical circuit diagram for the electrochemical cell as shown in Fig. 7A. Typically, a Nyquist plot is generated (Fig. 7B) by plotting the real (Z') and imaginary (Z'') components of the impedance. The diameter of the semicircle corresponds to the resistance to charge transfer. For electrochemical biosensors, species captured on modified electrodes alter both the ability of the analyte to diffuse to the surface (the Walberg diffusive element, *W*) and the kinetics of the charge transfer process, typically resulting in a large increase in R_{ct} (semicircle diameter), which can be correlated with analyte concentration.

3 Electrochemical sensing of cancer biomarkers

Cancer biomarkers comprise a variety of molecular entities of cellular and subcellular origin such as nucleic acids (DNA and RNA), functional subgroups of proteins (e.g., enzymes, glycoproteins, tumor antigens, and receptors), circulating tumor cells (CTCs) as well as cells with overexpressed tumor antigens, and extracellular vesicles (e.g., tumor-derived exosomes) [7]. (For more detailed discussion on cancer biomarkers, see Chapter 1.1.) Of these markers, CTCs, extracellular vesicles, cell-free nucleic acids (cfNAs), and circulating tumor DNA (ctDNA) are known to circulate in bodily fluids (including urine, blood, and saliva) and are becoming increasingly popular as minimally- or non-invasive biomarkers from liquid biopsies [8].

Recently, a wide range of sensing platforms have been developed to analyze these biomarkers. Among these, electrochemical biosensors represent a simple and inexpensive solution to achieve clinically relevant sensitivities and specificities in rapid measurement times [9]. Due to the ease of miniaturization and automation, electrochemical biosensors can easily be integrated into point-of-care platforms for in-field testing. Herein, we will discuss simple example approaches for the aforementioned cancer biomarkers. For further examples or a more detailed discussion, we recommend the following reviews [9–11].

3.1 Electrochemical sensors for nucleic acid biomarkers

The main approaches for detection of nucleic acid biomarkers to date are generally confined to several classic molecular biology-based methods, such as polymerase chain reaction (PCR) and next-generation sequencing (NGS). Despite their reliability, the scope of these approaches is limited outside of centralized laboratories, for example in resource-limited settings where sophisticated and expensive instruments may not be available [9].

Conversely, electrochemical sensors for nucleic acid biomarkers are suitable for analysis either in centralized labs or at the point of care. Due to their high sensitivity, many electrochemical sensors can detect endogenous concentrations of nucleic acids without the need for amplification steps, which reduces assay times and simplifies workflows. These methods are typically based on the hybridization of target sequences to complementary receptor probes (mostly DNA oligonucleotides) on an electrode surface. The hybridization event is then measured using intrinsic electrochemical properties of nucleobases, electroactive indicators, redox labels, or reporter enzymes to obtain an electrochemical signal, which is read via voltammetric, amperometric, or impedimetric approaches. The remainder of this chapter considers electrochemical approaches to different biomarkers and their inherent challenges.

3.1.1 Electrochemical DNA sensors

cfDNA and ctDNA

Circulating cell-free DNA (cfDNA) are short (~130–180 base pairs) double-stranded DNA fragments that are detectable in body fluids. Increased cfDNA levels are often associated with tumor progression and thus their detection can aid the diagnosis and prognosis of cancer [12]. Circulating tumor DNA (ctDNA) denote only a fraction of the cfDNA that specifically originates from tumors and may carry the same genetic signatures as the primary tumor. While overall levels of cfDNA can be quantified to track cancer progression, ctDNA analysis typically requires specific detection of its surrogate markers such as mutations, epigenetic alteration, and other molecular aberrations [13]. Despite huge potentials in precision medicine, routine use of cfDNA and ctDNA as biomarkers in clinical setups is yet to be demonstrated. One major problem is the difficulty of specifically analyzing ctDNA in an overwhelming excess of nontumor cfDNA. Existing methodologies are also limited by several biological (e.g., high fragmentation, low stability, and rare abundance of cfDNA), technical (e.g., ineffective extraction procedures), and clinical challenges. Recently, microfabricated devices based on electrochemical analysis of ctDNA have gained considerable attention through addressing these challenges.

The Kelley group have reported some of the important developments relating to chip-based electrochemical ctDNA sensors [14]. For example, nanostructured microelectrodes modified with peptide nucleic acid (PNA)-clutch probes were able to detect mutated ctDNA with high sensitivity. Following target hybridization, in the presence of an electrocatalytic reporter system comprising ruthenium hexaamine(III) chloride ($[Ru(NH_3)_6]^{3+}$), and potassium ferricyanide ($[Fe(CN)_6]^{3-}$), differential pulse voltammetry (DPV) was used to obtain an electrochemical signal. On the electrode surface, positively charged $[Ru(NH_3)_6]^{3+}$ was electrostatically attracted to the negatively charged phosphate backbone of ctDNA-PNA duplex. After further coupling with $[Fe(CN)_6]^{3-}$, a sharp amplification in the current was observed through $[Fe(CN)_6]^{3-}$ oxidizing $[Ru(NH_3)_6]^{2+}$ to regenerate $[Ru(NH_3)_6]^{3+}$ and initiate multiple turnovers of $Ru(NH_3)_6^{3+}$. Current intensity was directly correlated to the increased concentration of target ctDNA at the electrode surface and the assay could detect as low as 1 fg μL^{-1} of a DNA mutation in the presence of 100 pg μL^{-1} of wild-type DNA.

In another example, a glassy carbon electrode was modified with nanomaterials (molybdenum disulfide and graphene) to enhance its electrocatalytic activity [15]. Single-stranded DNA probes directed against specific ctDNA were then immobilized on the modified electrode and the hybridization of the ctDNA biomarker was monitored via DPV in the presence of a $K_3[Fe(CN)_6]$ electroactive indicator. For a more comprehensive review of recent technological advances in ctDNA analysis, refer to Gorgannezhad et al. [13].

DNA mutation

The analysis of disease-specific gene mutations in genomic DNA (e.g., mismatched base pairs, single-nucleotide polymorphisms or SNPs) can play a crucial role in understanding mechanisms of carcinogenesis and drug resistance. During the last two decades, several electrochemical sensors for mutation analysis have been developed [9,10]. Barton et al. have developed a technology for detection of point mutations that is based on DNA charge transport mechanisms [10,16]. DNA duplexes were deposited on polycrystalline Au electrodes using thiol-terminated linkers. As shown in Fig. 8, in the case of fully complementary DNA duplexes (i.e., no mutation), electrons could easily flow from the electrode surface through the DNA toward the molecules of methylene blue (MB^+) intercalated between base pairs on the other end of the DNA duplex. This resulted in MB^+ reduction to leucomethylene blue (LB), which in turns could reduce $[Fe(CN)_6]^{3-}$ in solution, thereby electrocatalytically regenerating MB^+ and causing an amplified chronocoulometric (CC) response. However, if the DNA contained as little as one single mismatch (i.e., one point mutation), current flow through the DNA duplex was hindered preventing the bound MB^+ from being reduced and leading to a loss in electrochemical signal (left panel, Fig. 8). This can be explained by the fact that long-range charge transport via π-stacking of DNA base pairs is highly sensitive to base stacking perturbations. This assay was successfully applied to the detection of all the possible single-base mismatches in a preassembled DNA sequence.

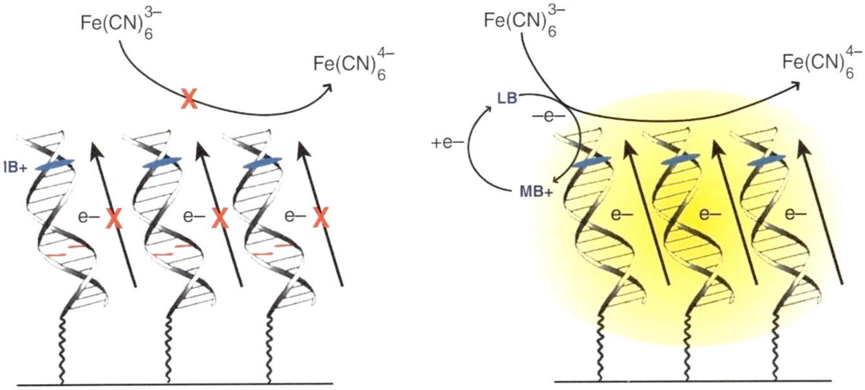

Fig. 8 Electrochemical assay for detecting DNA point mutation. *(Reproduced with permission from T.G. Drummond, M.G. Hill, J.K. Barton, Electrochemical DNA sensors, Nat. Biotechnol. 21 (10) (2003) 1192–1199. Copyright (2003) Nature Publishing Group.)*

DNA methylation

DNA methylation is a natural epigenetic (i.e., it does not alter the DNA sequence) phenomenon that occurs during development and involving addition of a methyl group to cytosines to produce 5-methylcytosine (or 5mC). In contrast, aberrant DNA methylation, one of the crucial players that inactivates tumor suppressor and DNA repair genes via inhibiting transcription, results in the onset of several diseases, including various cancers. Since this event takes place during the developmental stages of cells, aberrant DNA methylation is regarded as highly promising surrogate biomarker for the early diagnosis of cancer. The aberrancy is attributed to both global DNA hypomethylation (i.e., loss of the 5mC across the entire genome) and regional or gene-specific hypermethylation (i.e., an increase in locus-specific 5mC at regulatory regions, mostly in promoter regions) [17].

Bisulfite conversion of DNA is the most common approach for the analysis of both gene-specific and global DNA methylation, where sodium bisulfite mediates the conversion of cytosine (C) to uracil (U) in single-stranded DNA, while keeping 5mC intact. For precise quantification of the methylation levels, further suitable amplification or quantification of bisulfite-converted samples are required such as methylation-specific polymerase chain reaction. Most bisulfite conversion-based methods are limited by the uncertainty related to conversion efficiency, an inability to detect 5-methylcytosine oxidation products, DNA degradation, mispriming, and amplification-associated errors (i.e., PCR bias).

In recent years, a number of relatively simple but robust biosensing strategies coupled with optical, electrical, and electrochemical readouts have been developed, with bisulfite-free electrochemical sensors attracting specific interest. One of the most straightforward strategies is the direct electrochemical oxidation of 5mC, as each base has a defined oxidation potential reflecting the number of bases present in a target DNA sequence [18]. However, direct electrochemical oxidation of 5mC is challenging because conventional unmodified electrodes (i.e., carbon, gold, or indium tin oxide) are unstable at the high oxidation potential of 5mC. Moreover, the oxidative peak potentials of C, T, and 5mC are closely aligned due to their similar molecular structure and electrochemical properties, making it difficult to discriminate 5mC from the C and T bases. To address these issues, new electrode materials with wider potential window and higher electrode activity have been developed. Einaga and colleagues reported on a boron-doped diamond (BDD) thin-film electrode for 5mC detection, which showed a wider potential window than conventional materials at low pH [19]. Kato et al. developed an assay where electron cyclotron resonance (ECR) was used to deposit a sp^2-sp^3 carbon mixture on a sputtered nanocarbon film electrode to produce a high electrode stability and a significant potential difference (130–150 mV) between 5mC and C oxidation via square wave voltammetry (Fig. 9I) [20]. One disadvantage of this assay is false background peaks of the bases that are not exposed in the helical structure of longer DNA molecules. In a follow-up assay, the same group was able to address this issue through digestion of long DNA sequences into smaller identical mononucleotides prior to electrochemical oxidation [21].

More recently, an electrochemical sensor has been developed based on the hydrophobic properties of 5mC, where the solvation properties of DNA in aqueous solution changes with varying level of methylation [22]. Normal cells have a higher aggregation tendency in water compared to that of cancer cells due to different levels of DNA methylation, acting as a universal biomarker for cancer. Therefore, an electrochemical assay was engineered to quantitatively assess the amount of DNA adsorbed on an unmodified Au surface, exploiting different levels of adsorption between cytosine and 5mC. As shown in Fig. 9II, purified DNA samples were directly added on a disposable gold electrode and allowed to physically adsorb for 10 min. A $[Fe(CN)_6]^{3-/4-}$ redox system was used to quantify the amount of adsorbed DNA via DPV, whereby adsorbed DNA prevents $[Fe(CN)_6]^{3-/4-}$ from reaching the electrode surface resulting in a significant current reduction.

3.1.2 Electrochemical RNA sensors

Ribonucleic acids (RNAs) comprising different coding and noncoding transcripts such as messenger RNA (mRNA), microRNA (miRNA), and long noncoding RNA (lncRNA) have shown great potential as diagnostic and prognostic biomarkers for cancer [23]. Protein coding mRNAs denote a small portion of the transcriptome, while the majority are noncoding RNAs (e.g., miRNA and lncRNA). Cell-free circulating miRNAs are particularly attractive for minimally invasive testing as they are found in bodily fluids such as blood and urine. Reverse transcriptase quantitative PCR (RT-qPCR) is the gold standard method for miRNA analysis. However, RT-qPCR is hampered by amplification bias and may suffer from high and variable background

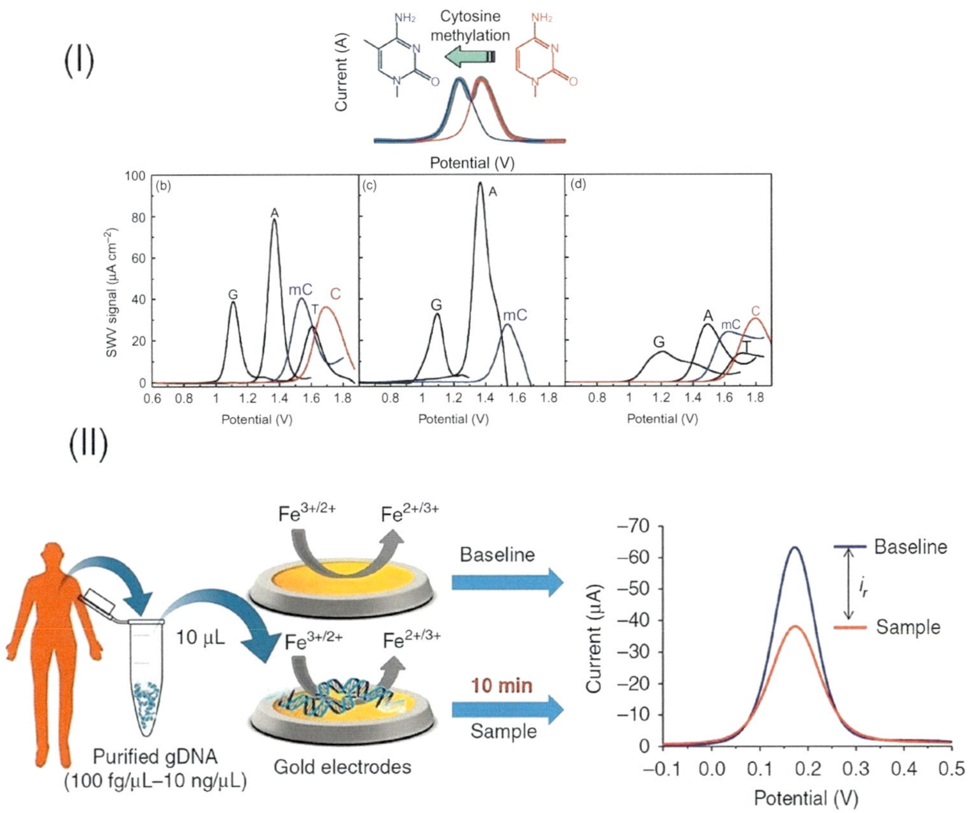

Fig. 9 (I) Direct electrochemical oxidation-based detection of 5-methylcytosine (5mC). (II) Electrochemical detection of global DNA methylation using different solvation properties of methylated and unmethylated DNA. *((I) Reproduced with permission from D. Kato, N. Sekioka, A. Ueda, R. Kurita, S. Hirono, K. Suzuki, O. Niwa, A nanocarbon film electrode as a platform for exploring DNA methylation, J. Am. Chem. Soc. 130 (12) (2008) 3716–3717. Copyright (2008) American Chemical Society. (II) Reproduced from A.A.I. Sina, L.G. Carrascosa, Z. Liang, Y.S. Grewal, A. Wardiana, M.J.A. Shiddiky, R.A. Gardiner, H. Samaratunga, M.K. Gandhi, R.J. Scott, D. Korbie, M. Trau, Epigenetically reprogrammed methylation landscape drives the DNA self-assembly and serves as a universal cancer biomarker, Nat. Commun. 9 (1) (2018) 4915. Copyright (2018): http://creativecommons.org/licenses/by/4.0/.)*

noise. Moreover, RT-qPCR is generally a laboratory-based method, making it less suitable for on-field screening than cheaper and faster approaches based on electrochemical detection. Electrochemical methods are particularly attractive for miRNA analysis because of their high sensitivity, scalability, and ease of automation.

Since the discovery of miRNA, several electrochemical sensors have been developed [24] and the majority of them rely on often complex isolation and purification steps (i.e., isolating target miRNAs from complex biological fluid) prior to sequence-specific hybridization to at least one sensing probe. Sensors for circulating miRNAs from liquid biopsies require high sequence specificity (i.e., ability to distinguish one miRNA from other family members and other nucleic acids) as well as high sensitivity (due to the low natural abundance of miRNA in bodily samples ~ng mL^{-1}).

Amplification-free miRNA sensors have been engineered that typically require labeling miRNAs with redox probes (i.e., direct labeling, electroactive intercalator), enzymes (i.e., conjugation of a predictive enzyme on the detection probe), and DNAzymes (i.e., single-stranded DNA aptamers with catalytic activity in the sensing probe). Dynabeads-based magnetic separation of RNA through multiple binding and washing steps is also used to reduce matrix effects and enhance assay specificity. To meet sensitivity requirements, electrochemical signal amplification is commonly required that can be achieved through a range of approaches. Representative example include surface modification with nanoparticles, conjugation with enzymatic and electrocatalytic reactions, or redox cycling [25]. Multifunctional nanomaterials such as tracers, catalysts, and electronic conductors are especially common. These nanomaterials have the inherent advantages of having high catalytic properties, large surface area, biomimetic activity, biocompatibility, good conductivity, and high sample loading capacity. Fig. 10 depicts an overall scheme of nanomaterials-based approaches for miRNA isolation and detection. MiRNAs are extracted and purified from various biological samples using magnetic nanomaterials, which can then also act as signal amplification labels in the miRNA electrochemical quantification step. Several enzymatic amplification-based miRNA electrochemical sensors were also reported such as one-step rolling circle amplification (RCA), which works isothermally. For more information on this specific topic, please refer to the following reviews [23,25–27].

3.2 Electrochemical sensors for protein biomarkers

The intrinsic challenges when engineering sensors for the detection of protein-based cancer biomarkers include: (i) unlike nucleic acids, proteins cannot be amplified; (ii) proteins have stability issues, such that optimal environments (e.g., temperature or pH) are needed; (iii) nonspecific interferences often occur that are caused by other proteins present in crude biological samples. Some progress has been made in addressing these obstacles toward the development of electrochemical protein sensors. A proof-of-concept impedimetric sensor was developed for the detection of vascular endothelial growth factor (VEGF), a well-characterized protein biomarker for breast cancer [28]. The recognition layer on the gold electrode was made up from a self-assembled monolayer of VEGF receptor-1 (VEGFR1). Interaction between VEGF and its receptor VEGFR1 on the electrode's surface altered both the mass transfer and kinetics of redox molecules, typically resulting in a large increase in charge transfer resistance (R_{ct}) in EIS, which could be correlated with VEGF concentration.

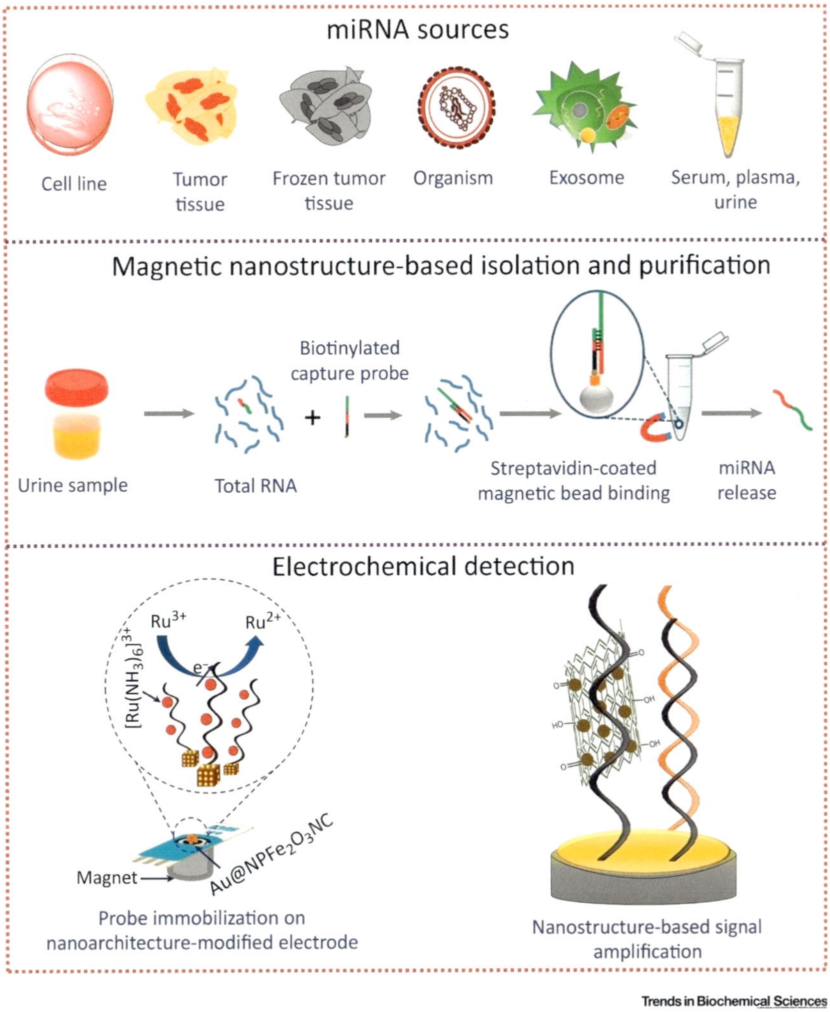

Fig. 10 Overall scheme of nanomaterial-based electrochemical sensing of miRNA. *(Reproduced with permission from M.K. Masud, M. Umer, M.S.A. Hossain, Y. Yamauchi, N.-T. Nguyen, M.J.A. Shiddiky, Nanoarchitecture frameworks for electrochemical miRNA detection, Trends Biochem. Sci. 2019.)*

A special class of engineered small proteins referred to as affibodies that mimic monoclonal antibodies can also be used as biomarker specific recognition elements in electrochemical sensors [29]. In an example of impedimetric assay based on detection of HER2 protein biomarkers, a gold nanoparticle (AuNP)-modified screen-printed graphite electrode was functionalized with anti-HER2 affibodies (Fig. 11I), then backfilled with blocking agents to avoid nonspecific adsorption. The quantity of bound target HER2 biomarker was evaluated from the changes in R_{ct} (via EIS) and the sensor offered an acceptable LOD of 6.0 μg L^{-1} of HER2 in spiked samples.

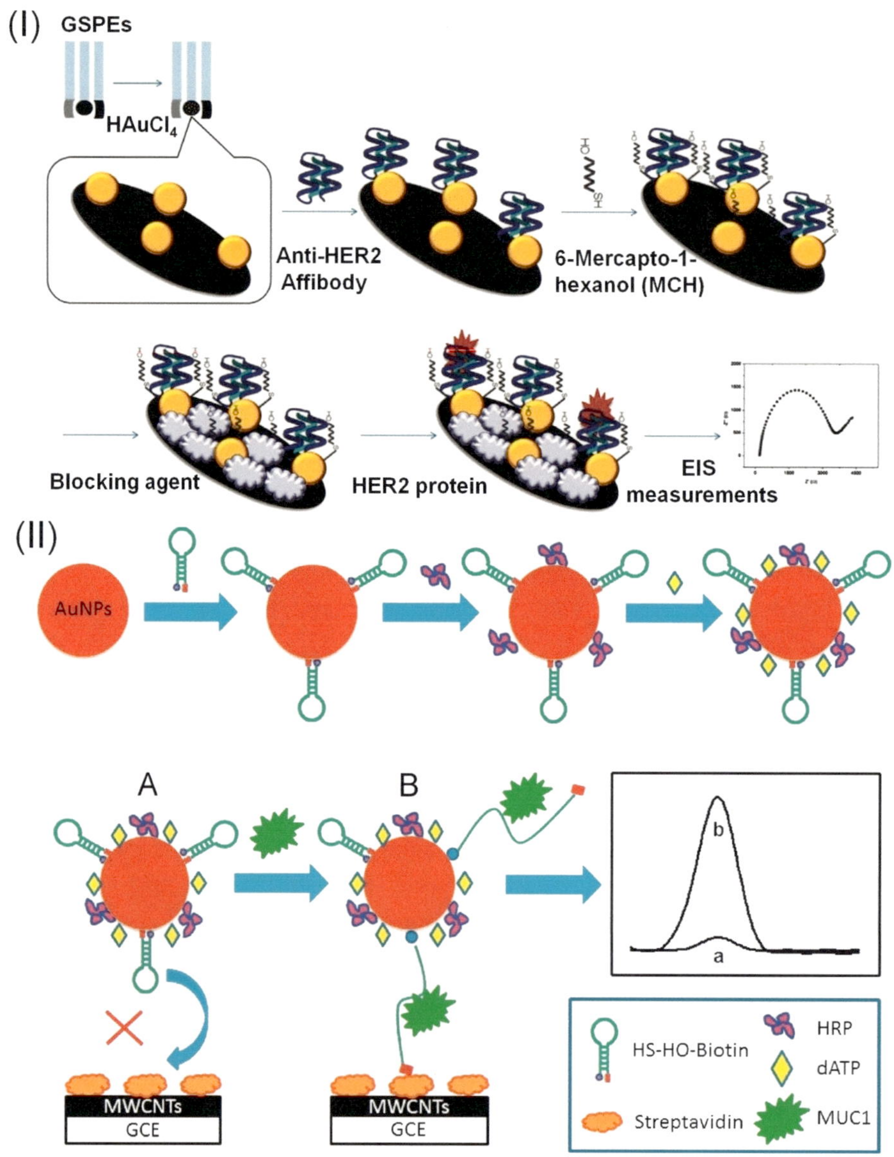

Fig. 11 (I) Impedimetric detection of HER2 protein biomarker. (II) Aptamer-based electrochemical sensing of MUC1 tumor protein marker. *((I) Reproduced with permission from A. Ravalli, C.G. da Rocha, H. Yamanaka, G. Marrazza, A label-free electrochemical affisensor for cancer marker detection: the case of HER2, Bioelectrochemistry 106 (2015) 268–275. (II) Reproduced with permission from R. Hu, W. Wen, Q. Wang, H. Xiong, X. Zhang, H. Gu, S. Wang, Novel electrochemical aptamer biosensor based on an enzyme-gold nanoparticle dual label for the ultrasensitive detection of epithelial tumour marker MUC1, Biosens. Bioelectron. 53 (2014) 384–9.)*

In an aptamer-based sensor, a mucin 1 protein (MUC1) cancer biomarker was detected using DPV (Fig. 11II) [30]. In this assay, horseradish peroxidase (HRP)-labeled AuNPs were modified with an MUC-specific biotinylated aptamer (via Au-SH immobilization). These AuNPs were then immobilized on a glassy carbon electrode (GCE), which was prefunctionalized with streptavidin-modified multiwall carbon nanotubes. In the presence of MUC1, the aptamer undergoes a conformational change that makes its biotinylated end accessible for capturing by the streptavidin-modified electrode. In the presence of H_2O_2, HRP catalyzes the oxidation of the working electrolyte (*o*-phenylenediamine into 2,3-diaminophenazine), which was measured electrochemically (via DPV). However, in the absence of target biomarker, the aptamer remained intact, resulting in no biotin binding and no DPV current peak.

3.3 Electrochemical sensors for circulating tumor cells

The unique phenotypic and molecular characteristics of circulating tumor cells (CTCs), which account for 1–10 cells out of 10^5–10^6 in peripheral blood, have an influential role in cancer metastasis [8]. Quantification of CTCs in bodily fluids is therefore of great interest for evaluating cancer dissemination and patient prognosis. A major limitation of current standard methods for the quantification of CTCs (e.g., the FDA-approved Cell Search methodology) is the high level of "biological noise" associated with low sensitivity, poor specificity, and long analysis times. In response to this challenge a number of integrated platforms (e.g., microfluidic devices with electrochemical sensors) have been developed in an attempt to effectively isolate, capture, and detect rare CTCs [11]. Generally, CTCs can be enriched through immunoaffinity as well as using commercial kits, prior to electrochemical detection. Recent progress in microfabrication techniques has enabled the development of several integrated electrochemical sensors for CTC capture and detection [11].

Costa et al. [31] engineered an electrochemical sensor that could detect a CTC model target known as human colon adenocarcinoma cell line (Caco2 cell) using AuNP labels and superparamagnetic beads modified with an antiepithelial cell adhesion molecule (or anti-EpCAM) antibody. As shown in Fig. 12, these two nanomaterials were dispersed into the sample, where the target cells were selectively labeled with anti-EpCAM modified AuNPs and magnetic beads. The target complex was then magnetically separated and quantified electrochemically.

A chronoamperometric current was obtained by leveraging the electrocatalytic properties of AuNPs to catalyze hydrogen (H_2) formation from hydrogen ions (hydrogen evolution reaction, HER) in the presence of HCl. This current was directly proportional to the amount of AuNPs present on the sensor, which in turn is directly related to the extent of AuNP-bound target cells. In another study, an array of nanochannel-ion channel hybrid coupled with an electrochemical sensor was shown to capture and then detect

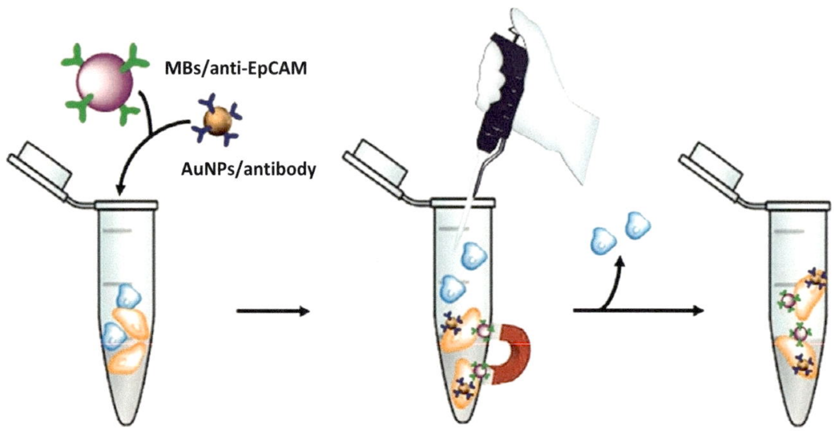

Fig. 12 Magnetic capture and amperometric quantification of CTCs. *(Reproduced with permission from M. Maltez-da Costa, A. de la Escosura-Muñiz, C. Nogués, L. Barrios, E. Ibáñez, A. Merkoçi, Simple monitoring of cancer cells using nanoparticles, Nano Lett. 12 (8) (2012) 4164–4171. Copyright (2012) American Chemical Society.)*

CTCs. The ion channel surface of the sensor was modified with aptamer probes to selectively bind surface protein markers of CTCs. When the sample was passed through the probe-functionalized channel, target CTCs, which overexpress the specific surface marker, were captured. This resulted in an alteration of the ionic flow inside the hybrid channel due to the variation of the mass-transfer properties, which was then measured via linear sweep voltammetry. This platform was reported to capture and detect as low as 100 cells mL^{-1} of leukemia CCRF-CEM CTCs. A considerable number of articles have been published on the biology and sensing aspects of CTC [32,33]. Of note are two excellent perspectives, which identify the grand challenges and solutions for CTC analysis [34,35].

3.4 Electrochemical sensors for extracellular vesicles

Exosomes are nanovesicles, which are actively secreted by cells and contain a specific combination of molecular biomarkers, including DNAs, miRNAs, and proteins. Due to the important roles of exosomes in uniquely representing parental tumor cells and transporting tumor-associated biomolecules throughout the body, they have emerged as a useful choice of stable, specific, and minimally invasive cancer biomarkers [36]. This increasing interest has led to the engineering of several isolation and detection technologies.

Recently, a portable electrochemical sensor was developed for exosome capture and detection (Fig. 13I) [37]. In this platform, four single-stranded DNA sequences were immobilized onto a gold electrode (via Au-thiol bonds) that could form a nanotetrahedron structure. One of the DNA strands also contained a DNA aptamer sequence

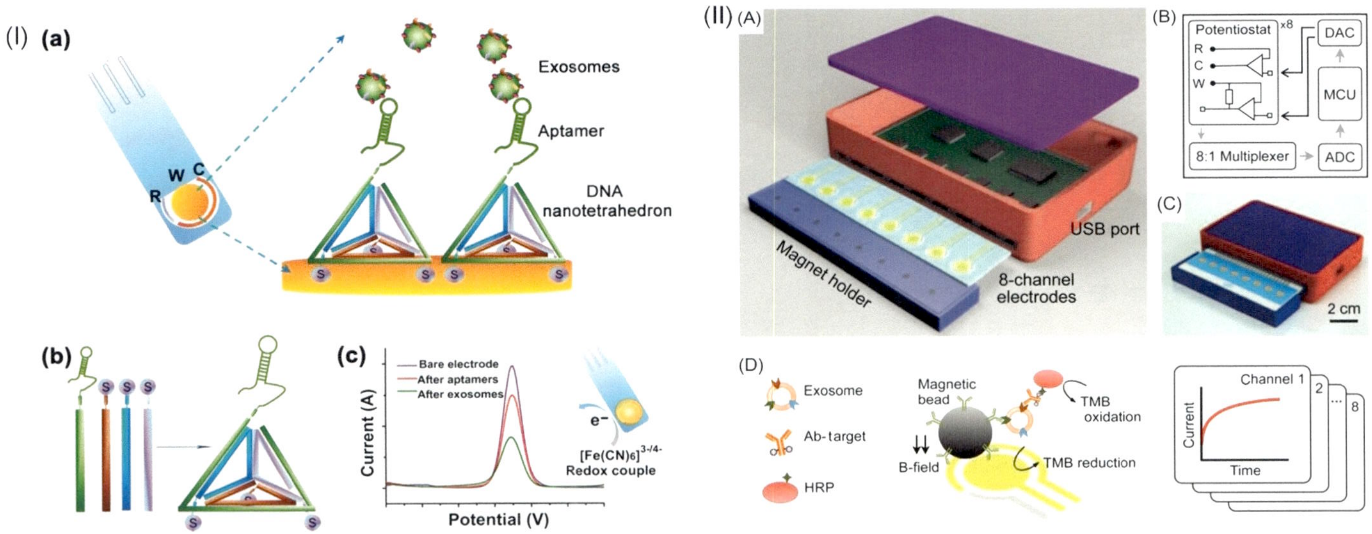

Fig. 13 (I) A nanotetrahedron-assisted aptasensor for direct capture and detection of hepatocellular exosomes. (II) A portable eight-channel electrochemical sensor for the isolation and quantification of cancer specific exosomes. *((I) Reproduced with permission from S. Wang, L. Zhang, S. Wan, S. Cansiz, C. Cui, Y. Liu, R. Cai, C. Hong, I.T. Teng, M. Shi, Y. Wu, Y. Dong, W. Tan, Aptasensor with expanded nucleotide using DNA nanotetrahedra for electrochemical detection of cancerous exosomes, ACS Nano 11 (4) (2017) 3943–3949. Copyright (2017) American Chemical Society. (II) Reproduced with permission from S. Jeong, J. Park, D. Pathania, C.M. Castro, R. Weissleder, H. Lee, Integrated magneto–electrochemical sensor for exosome analysis, ACS Nano 10 (2) (2016) 1802–1809. Copyright (2016) American Chemical Society.)*

directed against cancer-specific exosomes. The level of captured exosomes was then quantified using voltammetric readout in the presence of $[Fe(CN)_6]^{3-/4-}$ redox couple.

Another electrochemical sensor referred to as iMEX was recently engineered, which showed high potential for point-of-care testing [38]. iMEX reported detection down to $<10^5$ vesicles in biological samples, in rapid assay times (<1.0 h) and requiring small sample volumes (10 μL). The sensor featured eight channels, each comprising a potentiostat connected to complex circuits for different functionalities (i.e., potential control, signal digitization, selection, and system operation) (Fig. 13II). Eight cylindrical magnets were placed under these potentiostats to facilitate immunomagnetic capture of magnetic beads-bound target vesicles.

On the sensor, plasma samples were incubated with magnetic beads labeled with a generic antibody (anti-CD63), facilitating the magnetic capture of bulk exosomes (i.e., both cancer-specific and nonspecific exosomes). HRP labeled with a tumor-specific antibody was then dispersed on the sensor to selectively bind cancer-specific exosomes. After several washing and purification steps, the HRP-catalyzed oxidation of TMB was measured via amperometry to enable quantification of the HRP-bound target exosomes. Note that Boriachek et al. [36] organized a comprehensive review on the recent advances of both conventional and electrochemical strategies for isolation and detection of exosomes, as well as summarizing the major technical and biological challenges associated with each strategy.

4 Conclusions and perspectives

Electrochemistry is a powerful tool for the quantitative and specific analysis of cancer biomarkers. The sensing is often fast and very low detection limits are possible depending on the technique used, and the size, identity, and natural abundance of the biomarker. A wide range of cancer biomarkers, including RNA, DNA, proteins, circulating tumor cells, and extracellular vesicles, can be sensed indirectly through electrochemistry, where capturing of the target triggers an electrochemical response. The exact sensing approach required depends on the target of interest, the matrix, the required sensitivity and specificity as well as the feasible time and cost of the measurement.

When choosing between electrochemical methods, the following general guidelines can be applied: (i) Cyclic voltammetry is simple to run and analyze, inexpensive to implement, e.g., in point-of-care devices, but can suffer from moderate sensitivities; (ii) For more sensitive analyses, pulse voltammetry approaches such as DPV or SWV are preferred ($\sim 10\times$ improvement in detection limits); (iii) Compared to voltammetric approaches, amperometry generally offers more selectivity because a fixed oxidation or reduction potential is employed rather than a potential scan, though it is generally less sensitive than pulsed methods or impedance spectroscopy; (iv) Impedance spectroscopy is particularly suitable when highly sensitive and label-free analysis of biomarkers is needed such as analysis of trace concentrations of miRNA or immunosensing; however, impedance is a slower method (3 min per scan vs 30 s for voltammetry) and requires more complex and expensive potentiostats to run.

Compared to optical techniques, electrochemistry requires more complex and expensive instrumentation than colorimetry, and similar instrumentation size and cost to fluorescence detection. Electrochemical detection typically (but not always) offers an improvement over these optical techniques in terms of both sensitivity and specificity, depending on the target and sample matrix. Electrochemical devices are also scalable for clinical point-of-care sensing, for example the widely used glucose monitor. Nonspecific or background signals can be quenched or overcome through electrochemical tricks such as changing the technique, electrode material or pH, whereas background signals can be more difficult to circumvent via optical approaches (e.g., background fluorescence signals). The choice of sensing approach is therefore driven by (i) the abundance of the target (required sensitivity) and its matrix (required specificity), (ii) the time and cost requirements for the sensing and (iii) the sensing location (centralized lab vs in situ measurements).

References

[1] E.T. Whittaker, A History of the Theories of Aether and Electricity From the Age of Descartes to the Close of the Nineteenth Century, Longmans, Green and co., London, New York, 1910.

[2] A.J. Bard, L.R. Faulkner, Electrochemical Methods: Fundamentals and Applications, second ed., John Wiley & Sons, New York, 1980.

[3] L.H. Dubois, R.G. Nuzzo, Synthesis, structure, and properties of model organic surfaces, Annu. Rev. Phys. Chem. 43 (1) (1992) 437–463.

[4] P. Allongue, M. Delamar, B. Desbat, O. Fagebaume, R. Hitmi, J. Pinson, J.-M. Savéant, Covalent modification of carbon surfaces by aryl radicals generated from the electrochemical reduction of diazonium salts, J. Am. Chem. Soc. 119 (1) (1997) 201–207.

[5] R.G. Nuzzo, D.L. Allara, Adsorption of bifunctional organic disulfides on gold surfaces, J. Am. Chem. Soc. 105 (13) (1983) 4481–4483.

[6] C.M.A. Brett, A.M.O. Brett, Electrochemistry: Principles, Methods, and Applications, Oxford University Press, Oxford, 1993.

[7] N.L. Henry, D.F. Hayes, Cancer biomarkers, Mol. Oncol. 6 (2) (2012) 140–146.

[8] R. Vaidyanathan, R.H. Soon, P. Zhang, K. Jiang, C.T. Lim, Cancer diagnosis: from tumor to liquid biopsy and beyond, Lab Chip 19 (1) (2019) 11–34.

[9] M. Labib, E.H. Sargent, S.O. Kelley, Electrochemical methods for the analysis of clinically relevant biomolecules, Chem. Rev. 116 (16) (2016) 9001–9090.

[10] T.G. Drummond, M.G. Hill, J.K. Barton, Electrochemical DNA sensors, Nat. Biotechnol. 21 (10) (2003) 1192–1199.

[11] L. Wu, X. Qu, Cancer biomarker detection: recent achievements and challenges, Chem. Soc. Rev. 44 (10) (2015) 2963–2997.

[12] S. Valpione, G. Gremel, P. Mundra, P. Middlehurst, E. Galvani, M.R. Girotti, R.J. Lee, G. Garner, N. Dhomen, P.C. Lorigan, R. Marais, Plasma total cell-free DNA (cfDNA) is a surrogate biomarker for tumour burden and a prognostic biomarker for survival in metastatic melanoma patients, Eur. J. Cancer 88 (2018) 1–9.

[13] L. Gorgannezhad, M. Umer, M.N. Islam, N.-T. Nguyen, M.J.A. Shiddiky, Circulating tumor DNA and liquid biopsy: opportunities, challenges, and recent advances in detection technologies, Lab Chip 18 (8) (2018) 1174–1196.
[14] J. Das, I. Ivanov, E.H. Sargent, S.O. Kelley, DNA clutch probes for circulating tumor DNA analysis, J. Am. Chem. Soc. 138 (34) (2016) 11009–11016.
[15] Y. Chu, B. Cai, Y. Ma, M. Zhao, Z. Ye, J. Huang, Highly sensitive electrochemical detection of circulating tumor DNA based on thin-layer MoS2/graphene composites, RSC Adv. 6 (27) (2016) 22673–22678.
[16] E.M. Boon, D.M. Ceres, T.G. Drummond, M.G. Hill, J.K. Barton, Mutation detection by electrocatalysis at DNA-modified electrodes, Nat. Biotechnol. 18 (10) (2000) 1096–1100.
[17] T. Mikeska, J.M. Craig, DNA methylation biomarkers: cancer and beyond, Genes 5 (3) (2014) 821–864.
[18] T. Hossain, G. Mahmudunnabi, M.K. Masud, M.N. Islam, L. Ooi, K. Konstantinov, M.S.A. Hossain, B. Martinac, G. Alici, N.-T. Nguyen, M.J.A. Shiddiky, Electrochemical biosensing strategies for DNA methylation analysis, Biosens. Bioelectron. 94 (2017) 63–73.
[19] T.A. Ivandini, K. Honda, T.N. Rao, A. Fujishima, Y. Einaga, Simultaneous detection of purine and pyrimidine at highly boron-doped diamond electrodes by using liquid chromatography, Talanta 71 (2) (2007) 648–655.
[20] D. Kato, N. Sekioka, A. Ueda, R. Kurita, S. Hirono, K. Suzuki, O. Niwa, A nanocarbon film electrode as a platform for exploring DNA methylation, J. Am. Chem. Soc. 130 (12) (2008) 3716–3717.
[21] D. Kato, K. Goto, S.-i. Fujii, A. Takatsu, S. Hirono, O. Niwa, Electrochemical DNA methylation detection for enzymatically digested CpG oligonucleotides, Anal. Chem. 83 (20) (2011) 7595–7599.
[22] A.A.I. Sina, L.G. Carrascosa, Z. Liang, Y.S. Grewal, A. Wardiana, M.J.A. Shiddiky, R.A. Gardiner, H. Samaratunga, M.K. Gandhi, R.J. Scott, D. Korbie, M. Trau, Epigenetically reprogrammed methylation landscape drives the DNA self-assembly and serves as a universal cancer biomarker, Nat. Commun. 9 (1) (2018) 4915.
[23] M.N. Islam, M.K. Masud, M.H. Haque, M.S.A. Hossain, Y. Yamauchi, N.-T. Nguyen, M.J.A. Shiddiky, RNA biomarkers: diagnostic and prognostic potentials and recent developments of electrochemical biosensors, Small Methods 1 (2017).
[24] M. Labib, M.V. Berezovski, Electrochemical sensing of microRNAs: avenues and paradigms, Biosens. Bioelectron. 68 (2015) 83–94.
[25] M.K. Masud, M. Umer, M.S.A. Hossain, Y. Yamauchi, N.-T. Nguyen, M.J.A. Shiddiky, Nanoarchitecture frameworks for electrochemical miRNA detection, Trends Biochem. Sci. 44 (5) (2019) 433–452.
[26] P. Gillespie, S. Ladame, D. O'Hare, Molecular methods in electrochemical microRNA detection, Analyst 144 (1) (2019) 114–129.
[27] R. Tavallaie, S.R. De Almeida, J.J. Gooding, Toward biosensors for the detection of circulating microRNA as a cancer biomarker: an overview of the challenges and successes, Wiley Interdiscip. Rev. Nanomed. Nanobiotechnol. 7 (4) (2015) 580–592.
[28] M.K. Sezgintürk, A new impedimetric biosensor utilizing VEGF receptor-1 (Flt-1): early diagnosis of vascular endothelial growth factor in breast cancer, Biosens. Bioelectron. 26 (10) (2011) 4032–4039.
[29] A. Ravalli, C.G. da Rocha, H. Yamanaka, G. Marrazza, A label-free electrochemical affisensor for cancer marker detection: the case of HER2, Bioelectrochemistry 106 (2015) 268–275.
[30] R. Hu, W. Wen, Q. Wang, H. Xiong, X. Zhang, H. Gu, S. Wang, Novel electrochemical aptamer biosensor based on an enzyme-gold nanoparticle dual label for the ultrasensitive detection of epithelial tumour marker MUC1, Biosens. Bioelectron. 53 (2014) 384–389.
[31] M. Maltez-da Costa, A. de la Escosura-Muñiz, C. Nogués, L. Barrios, E. Ibáñez, A. Merkoçi, Simple monitoring of cancer cells using nanoparticles, Nano Lett. 12 (8) (2012) 4164–4171.
[32] C. Costa, M. Abal, R. Lopez-Lopez, L. Muinelo-Romay, Biosensors for the detection of circulating tumour cells, Sensors (Basel) 14 (3) (2014) 4856–4875.
[33] M. Yu, S. Stott, M. Toner, S. Maheswaran, D.A. Haber, Circulating tumor cells: approaches to isolation and characterization, J. Cell Biol. 192 (3) (2011) 373–382.
[34] J. den Toonder, Circulating tumor cells: the Grand Challenge, Lab Chip 11 (3) (2011) 375–377.

[35] C. Alix-Panabieres, K. Pantel, Challenges in circulating tumour cell research, Nat. Rev. Cancer 14 (9) (2014) 623–631.

[36] K. Boriachek, M.N. Islam, A. Möller, C. Salomon, N.-T. Nguyen, M.S.A. Hossain, Y. Yamauchi, M.J.A. Shiddiky, Biological functions and current advances in isolation and detection strategies for exosome nanovesicles, Small (2018).

[37] S. Wang, L. Zhang, S. Wan, S. Cansiz, C. Cui, Y. Liu, R. Cai, C. Hong, I.T. Teng, M. Shi, Y. Wu, Y. Dong, W. Tan, Aptasensor with expanded nucleotide using DNA nanotetrahedra for electrochemical detection of cancerous exosomes, ACS Nano 11 (4) (2017) 3943–3949.

[38] S. Jeong, J. Park, D. Pathania, C.M. Castro, R. Weissleder, H. Lee, Integrated magneto–electrochemical sensor for exosome analysis, ACS Nano 10 (2) (2016) 1802–1809.

CHAPTER 1.4

Device engineering

Dana Al Sulaiman, Isobel Steer, Suraj Pavagada, Sylvain Ladame
Department of Bioengineering, Imperial College London, London, United Kingdom

1 Introduction to designing a cancer biosensor

Biosensors are analytical devices, that can detect and provide information about a biological analyte from a biological sample. A typical biosensor is composed of three main elements: (i) a recognition or sensing element (i.e., bioreceptor molecule), (ii) a transducing element, which converts this detection into a measurable signal (typically electrical, optical, or electrochemical), and often (iii) a signal processor, which displays the signal correlating to the analyte concentration. A schematic representation of a standard biosensor is shown in Fig. 1.

1.1 Recognition element

Based on the nature of the cancer biomarker being targeted, recognition elements can be peptides, proteins, nucleic acids (or nucleic acid analogs), carbohydrates, or even whole cells. Ideally, this element should bind the target biomarker with both high affinity and high specificity, two parameters that will dictate the level of sensitivity and specificity of the sensing device (as defined in Chapter 1.1). Depending on the intended use, fast binding kinetics and long shelf-life may also be required. Antibodies are most commonly used for sensing biomarkers belonging to the families of proteins, viruses, or cells. Oligonucleotides such as DNA or RNA strands or oligonucleotide analogs such as peptide nucleic acids (PNA) or locked nucleic acids (LNA) are typically used for sensing of nucleic acid biomarkers. Recognition occurs via sequence-specific hybridization between the probe and the target. Short single-stranded nucleic acids that fold into a well-defined three-dimensional structure called aptamers can also be engineered that can bind to a broad range of analytes, including small-molecule metabolites and proteins. When compared to antibodies, DNA aptamers offer the advantage of being easily manufactured and chemically modified, therefore facilitating their immobilization or incorporation into devices. In addition, synthetic polymers such as molecularly imprinted polymers (MIPs) or engineered proteins such as affibodies have also been used as alternative recognition elements.

Bioengineering Innovative Solutions for Cancer
https://doi.org/10.1016/B978-0-12-813886-1.00005-X

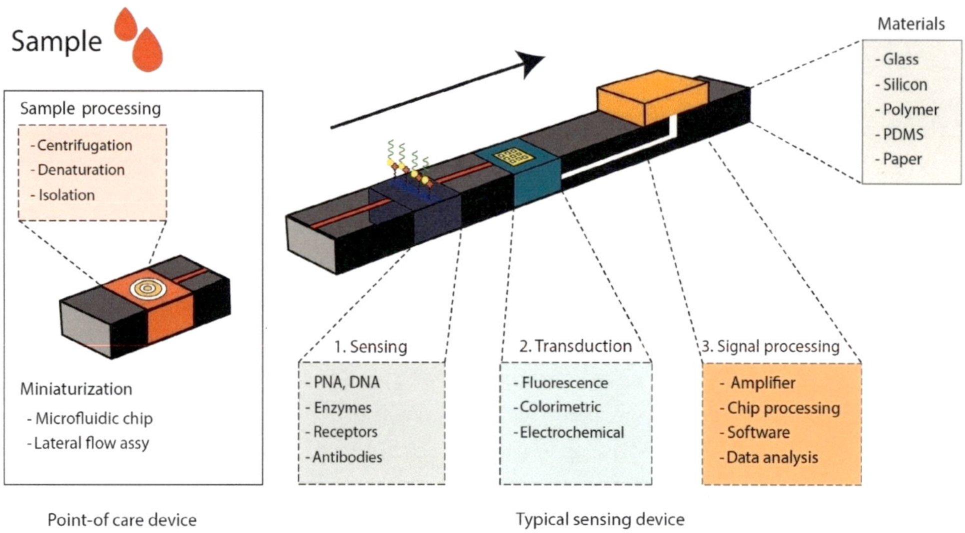

Fig. 1 Schematic representation of a standard biosensor. Typically composed of three main elements for sensing, transduction, and signal processing, a sensor can be adapted into a point of care (POC) device by miniaturization of all three elements onto a portable platform. It is noteworthy that in most cases of POC devices, a sample processing element is required that enables the prepurification or separation of the analyte of interest from the biological matrix it is in.

1.2 Transducing element: Converting binding into measurable signal

For the presence of the biomarker to be detectable, it is essential that the binding event between the target analyte and the recognition element is converted into a measurable signal. The signal produced can be either optical (see Chapter 1.2) or electrochemical (see Chapter 1.3), although other outputs can also be used (e.g., piezoelectric signals resulting from changes in mass). The transducer is the key element that helps convert the measured signal and dictates the type of detector that will be required based on the nature of the output signal generated. For each type of output signal, various detection platforms can be selected depending on the sensor's intended use and purpose.

1.3 Amplification steps

For a sensing device to meet its requirements in terms of sensitivity, amplification steps are often required, more specifically when trying to detect biomarkers with naturally low abundance (which is true for the majority of them). Two types of amplifications are most commonly used:

(i) Target amplification is mostly applicable to nucleic acid biomarkers and involves generating multiple copies of the biomarker of interest to facilitate its detection.

This is typically achieved by polymerase chain reaction (PCR) (see section 2 for more detailed description) and can either be used to preamplify the target number prior to detection or be monitored in real time (real time PCR).

(ii) Signal amplification requires each recognition event to produce multiple copies of the same signal output. This is most commonly found in enzyme-linked immunosorbent assays (ELISAs), a common strategy for detecting peptide and protein biomarkers (see section 3 for more detailed description).

1.4 Platforms

The nature of the sensing platform (material, size, and format) will be highly dependent on the sensor's intended use. For example, microarray-based analyses are typically performed on functionalized glass slides, microfluidic chips are commonly made from polydimethylsiloxane (PDMS) while other polymeric materials are commonly used for high-throughput screening tests. With the recent emergence of POC devices for public screening in a GP's clinic or in the field, low-cost materials such as paper have also recently gained in popularity.

1.5 Research tools versus point-of-care devices

Devices intended for research use only and those intended for use at the point of care (POC) are typically composed of the same elements. However, they may differ significantly with regards to device specifications. In an effort to maximize sensing sensitivity and specificity, devices engineered for research applications are usually bulky and expensive. They often accommodate liquid handling elements (for multistep/multireagents sensing processes) as well as temperature controllers (e.g., for thermocycling applications). In most cases, however, a trained professional is required to (i) safely and efficiently operate such large and costly devices and (ii) process, analyze, and interpret the generated data. Point-of-care devices are typically composed of the same three biosensor elements; however, they are all miniaturized and integrated into a smaller, portable platform (e.g., microfluidic chip) that also contains an additional sample processing element (when required). The biosensor components can either be self-contained in a single device or be individually connected to carry out sample processing and biomarker sensing.

Platform technologies have recently been developed by bioengineers to facilitate our understanding of the complex mechanisms involved in the initiation and progression of cancers. However, there is still a pressing need for newly engineered analytical devices that can rapidly and quantitatively monitor biomarker levels in clinical samples. The following chapter covers the basic principles governing the design and engineering of sensors and describes selected examples of sensing devices and technologies that have had or are likely to have a significant impact on the way cancer is diagnosed, monitored, and treated.

2 PCR-based detection methods

2.1 Introduction to PCR

Polymerase chain reaction (PCR) is an elegant enzymatic assay that enables the amplification of a specific DNA fragment among a pool of DNA. Since its introduction by Kary Mullis in 1985 [1], PCR has drastically revolutionized biological sciences, including disease diagnosis and the study and sequencing of genes [2]. There are four main components for a PCR assay: the DNA template, which is to be amplified, the four nucleotide building blocks of DNA (deoxynucleoside triphosphates: dATP, dCTP, dGTP, and dTTP), synthetic oligonucleotide primers, and a DNA polymerase enzyme. Primers are short (18–30 bases long) specific fragments of DNA that are complementary to the DNA sequence to be amplified. They act as extension points for DNA polymerase to build on. These four components are mixed and inserted into a PCR machine, which (acting as a thermal cycler) varies the block temperature in discrete steps. A single cycle of PCR is depicted in Fig. 2 and involves the following three steps:

(1) *Thermal denaturation*: The reaction solution is raised above the melting point of DNA (~94°C), separating the two complementary strands.

(2) *Primer annealing/hybridization*: The temperature is lowered (~55°C), to allow the primers to hybridize to the target DNA.

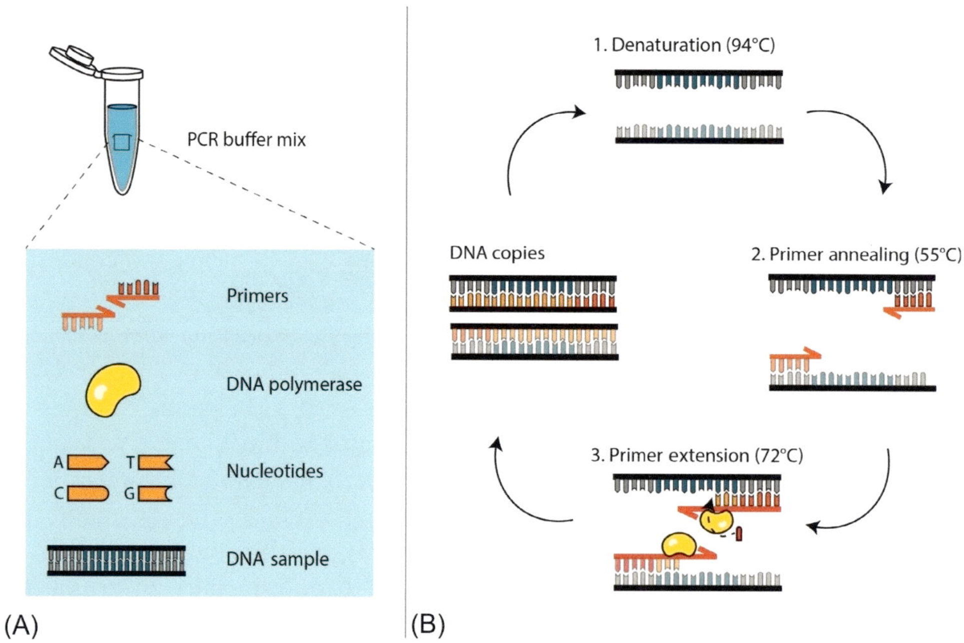

Fig. 2 Schematic representation of a conventional PCR assay, showing the major components of a PCR buffer mix (A) and the major steps involved in one PCR cycle (B).

(3) *Primer extension*: The temperature is raised once again (~72°C) so that the DNA polymerase can extend the primers and initiate the synthesis of a new DNA strand complementary to the template.

This cycle is repeated many times, each time doubling the copy number of DNA strands. Note that PCR can also be applied to RNA amplification by introducing a preliminary step of reverse transcription, whereby a target RNA is converted into complementary DNA (cDNA), which will subsequently be amplified exponentially as described earlier.

2.2 Visualization of PCR products

PCR-amplified DNA products can be visualized by either (i) unspecific staining of the amplified product with an intercalator dye (e.g., SYBR-green) or (ii) hybridization of sequence-specific oligonucleotide probes labeled with reporter fluorophores to the amplified DNA product. A common technique is to use a so-called Taqman probe, a short oligonucleotide complementary to specific regions of the target DNA and functionalized with a quencher and a reporter fluorophore at its 3′ and 5′ ends. During amplification, the 5′-3′ exonuclease activity of *Taq* DNA polymerase cleaves one end, which releases the reporter dye, generating a fluorescent signal, which increases at every cycle.

2.3 Limitations to PCR

Depending on the intended application, one of two PCR types is conducted. Qualitative PCR only determines the presence or absence of a particular DNA product, while real-time quantitative PCR (qPCR) can additionally determine how much of that DNA is present (i.e., detection and quantification). PCR is a very powerful technique whereby billions of copies of DNA can be produced from trace amounts of analyte. For sensing applications, this means that it can facilitate the detection of even the least abundant nucleic acid biomarkers. However, PCR also has several limitations. Any contamination occurring during the sample preparation or the PCR preparation may result in errors, which are then amplified. Furthermore, specificity may be compromised by unspecific binding of primers to DNA fragments sharing high sequence homology with the DNA target of interest. Errors may also arise from DNA polymerase incorporating incorrect nucleotides, although high enzyme fidelity makes this much less likely.

2.4 Toward miniaturization

The incorporation of PCR into POC devices requires that the PCR steps be automated and performed on smaller, self-contained platforms such as microfluidic chips. Miniaturizing amplification reactions presents several advantages, such as fewer manual steps, minimal contamination, and a significant reduction in the volume of expensive reagents. One strategy to achieve this is to design the space domain of a chip such that, as the sample flows through a microfluidic channel, its temperature is controlled by integrated heating blocks in a location-dependent manner. Various designs have been explored, including serpentine

[3] radial [4] channels, which exploit symmetry to reduce the number of heaters required. Another approach is to keep the sample solution stationary while controlling temperature by active heating and cooling. For example, heating can be achieved by using a tungsten lamp [5] or a laser, while cooling can be achieved by a fan [6]. Microfluidics is also a suitable platform to perform digital PCR, an advancement on conventional PCR whereby absolute quantification does not require the measurement of a standard curve. Digital PCR requires that the sample solution is divided into subsamples (or droplets), each containing an average of less than one molecule of DNA (i.e., containing either one or zero molecule of DNA) that can then be amplified independently. This compartmentalization can be easily performed by the creation of droplets or microwells in microfluidic chips. As an alternative to traditional PCR, isothermal amplification reactions can be designed that do not require any thermal cycling. As such, isothermal microsystems can be designed that are simple and low-energy consuming. These may outperform PCR in portable, battery-operated detection systems in the future [6a].

2.5 Commercial PCR platforms

Substantial advancements to basic PCR have led to the development of real-time (quantitative) PCR technology, which offers simultaneous amplification and quantification of specific nucleic acids. A wide range of high-throughput real-time PCR platforms are commercially available, all of which quantify PCR product optically (via fluorescence readout).

One of the first real-time PCR platforms on the market was the TaqMan OpenArray (Applied Biosystems). The plate design is based on a through-hole array, composed of 3072 holes of 33-nL reaction volume split into 48 groups of 64 holes each. The exterior surface of the plate is hydrophobic while the interior is hydrophilic, allowing fluid to efficiently move through. Today, the OpenArray technology works with the Applied Biosystems QuantStudio 12K Flex Real-Time PCR system. Probes can also be provided preloaded, meaning that the user simply needs to mix the sample with the PCR mix and load it onto the plate. The user can choose from the millions of predesigned assays on the company database or order custom assays. There also exist commercially available plates loaded with predesigned assays for specific applications such as detecting cancer-related genes.

With respect to cancer-specific PCR-based tests, there are a number of FDA-approved assays on the market that are intended to guide cancer patient therapy and monitor disease progression/response to treatment. One example is the Cobas EGFR Mutation Test v2 (Roche), a real-time PCR test that qualitatively detects mutations in exons 18, 19, 20, and 21 of the epidermal growth factor receptor (EGFR) gene in DNA either extracted from serum or Formalin-fixed, paraffin-embedded (FFPE) tissue specimens. The test is designed to help clinicians select patients with nonsmall-cell lung carcinoma (NSCLC) for therapy with an EGFR tyrosine kinase inhibitor (TKI).

Although these technologies are not yet ideally suited for POC testing, increasing efforts are being dedicated to achieving this goal.

2.6 Next-generation sequencing

Since Sanger sequencing, genome sequencing technologies have advanced in speed, through parallel analysis and high-throughput technology, as well as improved accuracy [7]. This in turn has led to reduced costs and increased automation. Next-generation sequencing (NGS) [8] represents one example of massively parallel sequencing technology that was made possible by the engineering of fluorescently labeled ddNTPs. One major platform company using NGS technology in their products for sequencing is Illumina. For example, Illumina's MiSeq Sequencer is a benchtop instrument using NGS technology for many applications, including clonal amplification as well as genomic DNA and micro-RNA sequencing. Other commercial NGS instruments include the Ion Torrent platforms (Thermo Fisher Scientific).

3 Sensing protein biomarkers by ELISA

Enzyme-linked immunosorbent assays (ELISAs) [9] are perhaps the most widely used sensing strategy for detecting peptide and protein biomarkers in a quantitative and highly specific manner and have been routinely used in cancer research and clinical diagnostic labs for biomarker/drug discovery, cancer screening, and early detection. In a typical ELISA, recognition elements (typically antibodies) are first immobilized onto a surface that can then capture the target biomarker (protein or antigen) present in the clinical sample. This complex is then labeled directly or indirectly with an enzyme, which eventually catalyzes a reaction resulting in a measurable output signal, most commonly optical (that is colorimetric, fluorescent, or luminescent). ELISAs are typically performed in 96-well (or 384-well) polystyrene plates, which passively bind antibodies and proteins. Having probes immobilized makes it easy to separate bound from nonbound material through extensive washing steps, making ELISA a powerful tool for measuring specific biomarkers from a crude or minimally processed biological sample. ELISAs can be performed in a number of different formats, as depicted in Fig. 3.

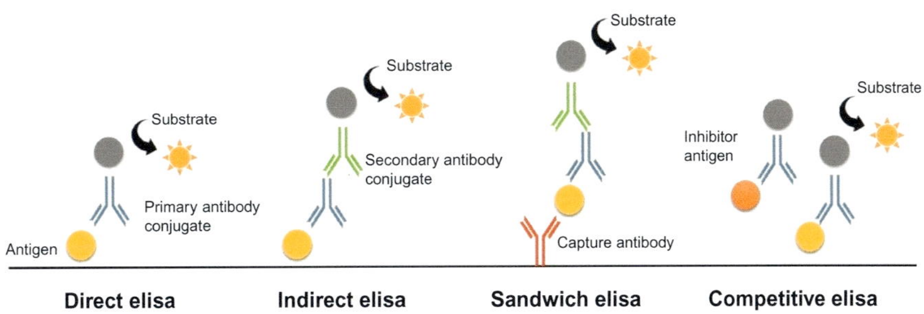

Fig. 3 Simplified illustration showing the most common ELISA formats.

Direct ELISA involves the immobilization of antigen biomarkers to a platform followed by the addition of an enzyme-labeled antibody.

Indirect ELISA also involves immobilization of antigens, but in this case, the primary antibody is not labeled. An enzyme-conjugated secondary antibody, directed against the first antibody, is then added. This format is often used to detect specific antibodies in serum samples.

Sandwich ELISA involves the attachment of a capture antibody to the platform surface. Samples containing known or unknown antigens are then added in a matrix or buffer that will minimize nonspecific attachment to the platform. Then, an enzyme-labeled antibody is added for detection. Sandwich ELISA is further divided into two main systems: (i) *Direct sandwich*, which uses a detecting antibody that is conjugated to an enzyme or fluorescence tag and (ii) *indirect sandwich,* which uses either an unconjugated or biotinylated detecting antibody to bind to the antibody-antigen complex on the well. Once bound, it is subsequently detected either by a secondary antibody or streptavidin that is conjugated to an enzyme or fluorescent tag.

Competitive ELISA relies on a competitive binding process between the original antigen (i.e., biomarkers from the sample analyte) and add-in antigens. Briefly, a primary unlabeled antibody is incubated with the sample antigen. The mixture is then added to a platform precoated with the same antigen and unbound antibodies are removed through extensive washing (Note: the more antigen biomarkers are present in your sample, the less antibodies will be able to bind the immobilized antigen, hence competition). A secondary antibody that is specific to the primary antibody and conjugated with an enzyme is then added, which produces a characteristic signal (color change or fluorescence) when the enzyme substrate is added. In that particular case, the more antigen is present in the sample analyte, the weaker the signal.

Several modifications have been made to ELISAs to incorporate them into POC diagnostic tests. Microfluidic technologies have been used to create portable lab-on-a-chip (LoC) ELISA devices, with added functionalities such as the ability to work with low sample volumes, multiplexing capabilities, and reduced costs. However, at this current time, conventional ELISAs are mostly confined to laboratory use.

Another emerging technology is immuno-PCR, which merges ELISA with PCR. Although ELISA can detect any protein (assuming the corresponding antibodies are available), it may not always be sensitive enough to detect proteins of naturally low abundance. While real-time PCR provides exponential signal amplification, it cannot be used directly for antigen detection since it does not utilize antibodies. Immuno-PCR thus provides the best of both worlds by combining these two assay formats via an antibody-oligonucleotide conjugate [10]. However, since traditional conjugation methods require large quantities of antibodies, generating antibody-oligonucleotide conjugates can be very costly. Immuno-PCR is however becoming more widely accessible, since user-friendly antibody-oligonucleotide conjugation kits are becoming increasingly more available.

4 Point-of-care devices

Point-of-care testing (POCT), also known as near-patient testing, is defined as the ability to provide testing outside of a central laboratory. This not only includes the patient's hospital bedside, but also the operating theater, critical care unit, maternity unit, and in the field, with little to no resources. The underlying goal of POCT is to make critical laboratory tests accessible to patients in a timely manner. The development of these tests has empowered patients with a more convenient, rapid, reliable, and cost-effective diagnostic tool for better management of their health.

Classical research techniques such as ELISA (presented in the earlier section) are confined to clinical laboratories, rely on kits often nonreusable and costly, and require multiple preparation steps, expensive labeling reagents (fluorescent/chemiluminescent), and lengthy processes, all of which make these research techniques unsuitable for real-time measurements. These prohibitively long assay times matter to those in need of rapid medical attention, while cost and portability matter to those in resource-limited or remote parts of the world. Thus, there is a strong need for the development of POC devices, a class of simple biosensor devices, which are small (ideally portable), simple to operate, and thus can be used in resource-limited areas or clinical settings.

4.1 End-user considerations

For a biosensor device to be classified as POC rather than research-based, it must satisfy a number of additional requirements, which are based on the needs of the end user. This could either be the patients themselves or clinical workers like nurses and physicians or volunteers for field testing. The end user determines the environment where the test will take place (i.e., at home, at a clinic, in the field, etc.), the level of expertise required to carry out the test and the time and the resources available to run the test. Therefore, designing POCT begins with a discussion about the requirements of the end user. These requirements will shape many other aspects of the sensor, such as the chosen output signal, the type of signal processing, and the device material. Important end-user considerations for POC applications are outlined below.

(i) Time: Ideally, a physician must be able to assess the need of the patient for the right test, source the test, get the sample, run the test, receive the test results, and be able to communicate any findings back to the patient, all within the timeframe of a typical consultation. This presents a significant technical challenge in the design of rapid tests with quick reaction times, while maintaining the required level of sensitivity.

(ii) Patient experience: One critical factor in determining patient experience is the selection of the right sample type. For example, a test that requires a large volume of blood sample may put off a significant proportion of potential users. Clinical samples should ideally be low volume and noninvasively obtained (e.g., urine or saliva is preferable over blood; low blood volumes such as finger prick tests are preferred over intravenous procedures).

(iii) Expertise: The level of expert supervision required by the test determines the level of complexity acceptable. For example, a nurse would not be equipped to carry out complex washing steps in a microfluidic assay. The requirement for trained professionals to carry out a test further increases its total cost burden.

(iv) Operation and storage: In an ideal scenario, the introduction of a POC test should have minimum impact on the current processes of the primary care center (e.g., infrastructures for storage or waste disposal). POCT should require as little space as possible since storage in a primary-care setting or pharmacy is highly valuable. The need for refrigeration is even more challenging as any temperature-controlled storage space is typically occupied by drugs, which are temperature sensitive.

(v) Relevant information: POC testing, like pathology lab testing, should provide healthcare professionals with the confidence required to make evidence-based clinical recommendations and take action. Conversely, it should not provide excessive information—e.g., in the case of a test, which is supposed to help healthcare professionals decide whether an antibiotic is the right course of action, it is of little value to provide the full microbiome map of a patient.

4.2 Miniaturization and sample processing

There are two key things a bioengineer can implement to fulfill these end-user requirements: miniaturization and sample processing capabilities.

Miniaturization: The ideal material for a miniaturized POC sensor should be cheap, robust and easy to manufacture. Research-based microfluidic devices often involve polydimethylsiloxane (PDMS), which is neither. However, advances in 3D printing of polymers offer great opportunities for cost-effective mass production. For example, Dolomite has recently launched the first microfluidic-targeted 3D printer. In the future, 3D printing of biocompatible hydrogels may also become common practice. Even without 3D printing, cheap materials such as hydrogels and paper offer advantages as low cost and easily tuneable materials for POC sensors.

Miniaturization can also involve utilizing preexisting signal-processing systems, thereby avoiding the extra bulk of incorporating in-device software. Smartphones can be directly used for colorimetric and electrochemical detection. For example, photos can be taken of assay plates or paper strips and uploaded to an app for quantification. Going further, functionalized electrodes can be physically integrated into smartphones, co-opting the existing electronics for power, signal processing, and readout.

Sample processing: An automated sample processing element is typically required to extract the biomarker of interest from its biological matrix prior to detection without the requirement for trained technicians. Sample processing may be done by centrifugation, material-based filtering, or nanoparticle-based extraction. Additionally, more complex steps may be required such as protein denaturation or lysis that are often complex and a major source of error. The processing techniques used to extract or enrich target analytes determine the sensors' accuracy and precision and are therefore crucial considerations in the design of effective biosensors. Sample amplification steps such as PCR for nucleic acid biomarkers may also be required, which can produce enough copies of analyte for on-chip detection.

5 Microfluidics-based technologies

Microfluidics refers to miniature systems that deal with microscale quantities of fluid flow, typically just a few microliters (μL) [11]. It is a key branch of micro electromechanical systems (MEMS)-based devices, which has recently gained immense popularity in biomedical research. Microfluidics offers a way to improve many sensors in terms of cost and portability. Global cost reduction per analysis is mostly due to low reagent consumption. In addition, there is faster analysis due to the shorter reactions/separation times as well as high-throughput and multiplexing capabilities.

5.1 Microchip design and fabrication

Microfluidic chips (microchips) are small "Lab on a chip" (LoC) platforms containing channels designed to handle tiny volumes of liquid. Depending on the design, they can filter, concentrate, mix, heat, separate, and detect. There are many ways to fabricate a microchip: early versions were made of silicon and glass, but optically clear polymers that are chemically stable are currently mainly favored (see Table 1).

The fluid flow through a microfluidic channel can be characterized by the Reynolds number (Re). Due to the small dimensions of the channels, the flow is completely laminar (turbulence occurs at $Re \sim 2000$, while Re in microfluidic channels is <100). This predominantly laminar flow provides an increased level of control to manipulate fluid flow and transport molecules in a relatively predictable manner. There are two common methods by which fluid flow through microchannels can be achieved: Pressure driven and Electrokinetic flow. New methods using surface acoustic waves (SAW) and bulk acoustic waves (BAW) have also been developed to drive fluids effectively. With these developments underway, the North American microfluidic market is expected to reach over 10 billion dollars by 2025, with the healthcare sector projected to lead over half of the market share [12].

Table 1 Comparison of commonly used substrate materials for microfluidic applications.

Substrate	Fabrication technique	Desired properties	Limitations	Most suited application
Glass	• Photo lithography	• Impermeable to gas • Low nonspecific adsorption • High thermoconductivity	• Expensive and complex fabrication process	Research, preliminary studies (requiring superior optics)
PDMS	• Soft lithography	• Easy, rapid, and inexpensive fabrication process • Good optical transparency and elastomeric properties	• Channel deformation, sample absorption, and leaching	Prototyping and research (low volume)
Thermoplastics (PMMA/PC/PS/PVC)	• Rapid prototyping (e.g., CNC milling, laser ablation) • Replication processes (e.g., hot embossing/imprinting, roller imprinting, and injection molding)	• Good mechanical stability • Low water absorption • Organic solvent/acid and base resistivity	• Unsuitable for long-term cell cultures due to low gas permeation • Difficult to make valves due to rigidity	Industry and mass production (high volume)

5.2 Lab-on-a-disc platforms

Lab-on-a-disc (LoaD) platforms, also known as Compact Disk microfluidics, are advantageous over pump-driven systems due to their simple operation via centrifugal force and the capability for complex sample handling, integration, and automation [13]. The system consists of microfluidic CDs that house the fluids and a CD player type device that controls the rotation rate (rpm). The sample is introduced to the center of the CD and carried axially outward by centrifugal force. The channels are mostly made of transparent plastics, which enable optical detection (the most commonly used detection mechanism for LoaD systems).

Optical disc drive technology was developed in the 1980s and CD-type cartridge manufacture was perfected by the music industry. With this preexisting technology, the commercialization is thus relatively easy compared to other platforms. One of the most well-known LoaD platforms is Abaxis, founded in 1989. Abaxis' Piccolo Xpress system provides panel assays for metabolic biomarkers, which are marketed to oncology wards, although the biomarkers are not yet specifically for cancer. Gyros is another company whose LoaD platforms are capable of biomarker detection, including cancer biomarkers such as tumor necrosis factor (TNF) Alpha. Their immunoassays have excellent correlation with a conventional ELISA. However, these products are currently exclusively for laboratory use and not compatible with POCT.

6 Paper-based diagnostic tests

Paper-based devices for disease diagnostics have already revolutionized POC approaches to health monitoring, by providing low-cost, disposable tools that can be utilized even in remote settings. Paper-based devices rely on capillary flow to achieve delivery and mixing of liquids [14]. These devices typically consist of chemical and biological components implemented on paper substrates. Paper presents several advantages that make them ideally suited for POC development:

(i) Paper is available ubiquitously and is inexpensive;
(ii) Paper wicks aqueous fluids, thus enabling passive transport of fluids without the need for active pumping;
(iii) Paper can be chemically modified to incorporate a wide variety of functional groups that can be covalently bound to proteins, DNA or RNA oligonucleotides, or small molecules;
(iv) Paper-based devices can work with very low sample volumes (usually below 100 μL)

Paper-based devices can be categorized in two typical formats: Lateral-flow assays (LFAs) and paper-microfluidic devices.

6.1 Lateral flow assays

Popularized by their early use in home pregnancy test kits, lateral flow assays are now used for a variety of devices designed for POC settings. Their simplest design is perhaps the "dipstick test", with a sample loading pad at one end, a reading zone where the signal is generated (test line), and an absorbent material at the other end to facilitate the wicking of fluid. LFAs may also be incorporated into housings/cassettes that support higher functionality, branding, and bespoke designs based on the application (Fig. 4) [15].

LFAs may be designed to be qualitative (i.e., indicating the presence or absence of an analyte), semiquantitative, or fully quantitative. Quantitative assays are required in situations where clinicians need the concentration of an analyte to make clinical decisions (e.g., concentrations of mutant DNA/cancer-specific proteins, etc.). However, for home testing or sometimes for screening tests, a qualitative result indicating the presence or absence of biomarker is considered adequate. Although several studies have reported

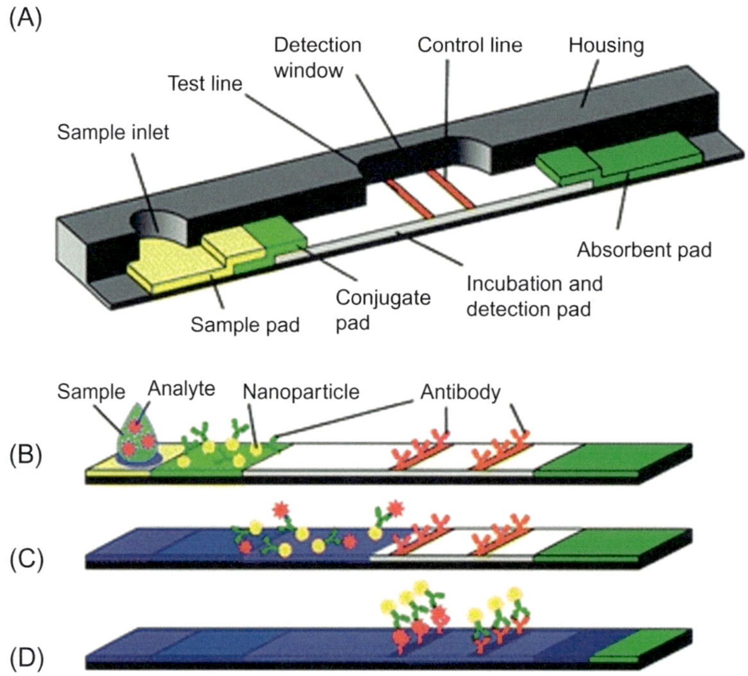

Fig. 4 Schematic representation of an immunolateral flow assay (LFA). (A) LFA strip incorporated into its bespoke cassette; (B–D) schematic representation of the three main steps of the assay: briefly, the analyte-containing sample is loaded onto the sample pad (B); the analyte can then interact with specific antibodies (conjugated to nanoparticles for visualization) preembedded onto the conjugate pad (C); the analyte antibody conjugates are then captured by analyte-specific antibodies immobilized onto the test lines, while unbound antibodies are captured by secondary antibodies immobilized onto the control line (D). *(Reproduced with permission from J. Kanter, Point-of-Care Testing in Sickle Cell Disease, Intechopen, 2016, https://doi.org/10.5772/64862.)*

Table 2 Summary of readout methods in paper-based devices and their preferred results.

Readout strategy	Common methods employed	Detection platform	Preferred result
Colorimetric	• Colloidal gold, carbon • Colored latex beads, colored dyes	• Naked eye • Smartphone • Colorimetric sensor	Qualitative
Fluorescence	• Fluorescent labels/dyes (e.g., FITC, Alexa Fluor, etc.)	• Fluorescent scanner • FRET	Semi quantitative to quantitative (depending on scanner used)
Electrochemical	• Screen-printed gold/carbon electrodes • Conductive polymers (PEDOT, Polyaniline, etc.)	• Voltammetry, amperometry, etc.	Quantitative

examples of LFAs with an electrochemical readout (using screen-printed Au/C electrodes and conductive polymers), optical readout strategies are still the preferred system for most applications. Common options for optical readout systems include use of colloidal gold nanoparticles, fluorescent or luminescent materials, or colored latex beads (Table 2).

6.2 Paper-microfluidics

Microfluidic paper-based analytical devices (μPADs), also known as micro total analysis systems (μTAS), are similar to LFAs but do not rely on two test lines to report results, thus allowing for a wider range of applications [16]. μPADs have garnered significant interest in recent years for their ability to perform complex reactions and fluid flow like those seen in traditional microfluidics. The ability to embed complex patterns onto μPADs reduces the number of steps required to run the test and allows for facile operation requiring little to no training prior to use. Since its introduction by the Whitesides group at Harvard University, 3-dimensional μPADs have gained popularity among researchers. 3D-μPADs use several layers of paper, each functionalized with reagents to perform a function, that would otherwise require a traditional lab-based protocol (such as sample processing, preconcentration of reagents, antibody-conjugation, etc.).

Initially, fabricating a 3D-μPAD involved cutting layers of paper, functionalizing each layer with reagents, and attaching them using adhesive tape between each layer. Recently, the use of origami (paper folding) and kirigami (paper cutting) techniques has provided simpler alternatives that has given researchers new opportunities for creating intricate 3D designs to facilitate complex (multistep) chemical reactions on paper [17].

μPADs are good prospects for manufacturing, since they are easy to produce and do not require complex machinery to analyze their outputs, which makes them ideal candidates for clinical use. Fabrication of μPADs is based on patterning hydrophilic channels (paper) bounded by hydrophobic barriers (e.g., wax) on sheets of paper. Several methods are now available to accomplish such patterning. One of the most commonly used methods of patterning is wax printing, which is rapid and utilizes commercially available printers. In this technique, a printer is used to print wax ink onto the surface of paper. The deposited ink is then melted through the paper on a hot plate to create a hydrophobic barrier along the thickness of the paper. Another common method of patterning on paper is photolithography, which is widely used to create channels in traditional microfluidic devices. However, photolithography exposes the hydrophilic regions to polymers and solvents that could lead to crossreactivity and affect the sensitivity of the assay.

6.3 Paper-based diagnostic device for cancer

Paper-based devices are now starting to attain the level of complexity required to address the main challenges in cancer diagnostics. A representative example of paper-based technology developed for the detection of colon cancer is presented here.

Several research groups from MIT have produced effective LFA-based devices to test for clinically relevant biomarkers from blood and urine. One such study was reported by Bhatia and colleagues that described a dipstick device that can be used to test for colon cancer [18]. They engineered nanoparticles to amplify the signal generated by tumor proteins called matrix metalloproteinases (MMPs) that would otherwise be hard to detect. Briefly, nanoparticles were engineered that were coated with specific peptides recognized as substrates for different MMPs. The nanoparticles proved capable of gathering at tumor sites, where MMPs would cleave hundreds of peptides, which would eventually move to the kidneys and be excreted in the urine. The peptides secreted in urine could then be captured by specific antibodies adsorbed in spatially multiplexed lines on nitrocellulose membrane and ultimately be imaged using streptavidin-gold detection nanoparticles. If one of these lines becomes visible, it means the target peptide is present in the sample. Further, the changes in color on the test lines are quantified using a smartphone app to provide an additional layer of information. The researchers claim that this technology can also easily be modified to detect multiple types of peptides released by different types or stages of cancer.

7 Smartphone-based clinical diagnostics

The advancements made in smartphone-based imaging and sensing platforms have produced several effective technologies for POC testing. They are contributing to the decentralization of diagnostic tests by offering practical features such as portability, cost effectiveness, and connectivity. External biosensors and lab-on-a-chip technologies have been engineered that are interfaced with smartphones, exploiting their computational power to produce high precision and sensitivity. Three useful features of smartphones have helped them revolutionize remote and primary healthcare:

(i) The built-in components (e.g., camera, LEDs, and microphone) can be used for the measurement of several different types of readout.

(ii) The powerful processors and memories for the analysis and storage of diagnostic results are particularly beneficial for a POC device.

(iii) Most smartphones are equipped with powerful data transmission capabilities, such as Global System for Mobile Communication (GSM), wireless fidelity (Wi-Fi), Bluetooth, and universal serial bus (USB), providing an effective platform for telemedicine.

The use of smartphones in cancer diagnostics can be classified into two main applications: **in vivo testing**, where signals from the body are measured directly with external sampling and **in vitro testing**, where signals are measured from an external sample (e.g., tissue, blood, saliva, etc.), usually for molecular testing. Two case studies of diagnostic tests are presented that use smartphone for either capturing or analyzing data.

7.1 Case study 1: Pancreatic cancer screening using a selfie

Pancreatic cancer is often diagnosed late once it has already spread, at which point any therapeutic/surgical options are drastically limited. A new smartphone application was developed that scans the white part of the eye for one of the disease's early tell-tale signs: jaundice [19]. Jaundice is characterized by the yellowing of the skin and eyes as a result of a substance called bilirubin in the blood. The application, dubbed "BiliScreen," uses a smartphone camera along with computer vision algorithms and machine learning to detect elevated bilirubin levels in the white sclera of the eye. To control the ambient light exposure and calibrate the color, the BiliScreen smartphone app was used in combination with either a special box or paper glasses. The team found it was possible to correctly identify causes for concern 89.7% of the times compared to the conventional blood test. Although further testing in a larger group of patients is required, this technology shows great promise as an effective tool for early diagnosis of pancreatic cancer.

7.2 Case study 2: Stanford CNN for skin cancer

A team of scientists from Stanford trained an artificial intelligence program to classify images of skin lesions as either benign or malignant skin cancer [20]. A single deep convolutional neural network (CNN) was trained end to end from images directly, using only pixels and disease labels as inputs. The program was then finetuned on a dataset of ~130,000 skin lesion images comprising over 2000 diseases. The CNN achieved performance on par with board-certified dermatologists in the critical classification of keratinocyte carcinomas versus benign seborrheic keratoses. This program can be deployed onto smartphones to analyze high-resolution images of biopsy specimen, and even those obtained from the phone's camera, which can potentially extend the reach of dermatologists outside of the clinic. This platform could also potentially be used for self-examination through pictures taken of the skin. Although multiple applications have already entered the market claiming to do just that, they lack extensive testing required on large datasets. These applications that claim to diagnose cancer present a significant risk of not recognizing rare/unusual cancers or fail to pick up on all red flag symptoms and as a result of this give patients false reassurances. In such cases, patients risk not seeking professional advice in the early stages and therefore miss out on early diagnosis. Therefore, application developers and regulators have a great deal of responsibility in ensuring that extensive testing has been carried out, before deploying it in the market for use.

8 Summary

From next-generation sequencing to paper-based devices with smartphone detection, the most recent technological developments in the field of cancer biosensor are the contribution of bioengineers who have the capacity to design and develop devices that can fulfill a broad range of requirements. Depending on their intended end use, devices will be bulky or handheld, automated or run by trained professional, low cost or expensive, qualitative or quantitative. Although traditional laboratory-based sensors have previously held the upper hand (gold standard) because of their high-throughput capabilities and ultra-low detection limits, the constraints of cost and portability and the move away from a "one size fits all" approach toward more personalized medicine have in recent years driven bioengineers' focus toward POC devices. This chapter however does not intend to suggest that POC devices will ever substitute research-based technologies. Both molecular information originating from research benches and from patient's bedside are transforming the future of healthcare in their own way and bioengineers have a key role to play in all these technological advances.

References

[1] K. Mullis, F. Faloona, S. Scharf, R. Saiki, G. Horn, H. Erlich, Specific enzymatic amplification of DNA in vitro: the polymerase chain reaction, Cold Spring Harb. Symp. Quant. Biol. 51 (1986) 263–273.

[2] I. Tournier, B.B. Paillerets, H. Sobol, et al., Significant contribution of germline BRCA2 rearrangements in male breast cancer families, Cancer Res. 64 (2004) 8143–8147.

[3] D. Moschou, N. Vourdas, G. Kokkoris, G. Papadakis, J. Parthenios, S. Chatzandroulis, A. Tserepi, All-plastic, low-power, disposable, continuous-flow PCR chip with integrated microheaters for rapid DNA amplification, Sensors Actuators B Chem. 199 (2014) 470–478.

[4] (a) Y. Schaerli, R.C. Wootton, T. Robinson, V. Stein, C. Dunsby, M.A.A. Neil, P.M.W. French, A.J. deMello, C. Abell, F. Hollfelder, Continuous-flow polymerase chain reaction of single-copy DNA in microfluidic microdroplets, Anal. Chem. 81 (2009) 302–306. (b) L. Jiang, M. Mancuso, Z. Lu, G. Akar, E. Cesarman, D. Erickson, Solar thermal polymerase chain reaction for smartphone-assisted molecular diagnostics, Sci. Rep. 4 (2014) 4137. (c) R. Snodgrass, A. Gardner, L. Jiang, C. Fu, E. Cesarman, D. Erickson, KS-detect—validation of solar thermal PCR for the diagnosis of Kaposi's sarcoma using pseudo-biopsy samples, PLoS ONE 11 (2016).

[5] Y. Ouyang, G.R.M. Duarte, B.L. Poe, P.S. Riehl, F.M. dos Santos, C.C.G. Martin-Didonet, E. Carrilho, J.P. Landers, A disposable laser print-cut-laminate polyester microchip for multiplexed PCR via infra-red-mediated thermal control, Anal. Chim. Acta 901 (2015) 59–67.

[6] S.O. Sundberg, C.T. Wittwer, R.M. Howell, J. Huuskonen, R.J. Pryor, J.S. Farrar, H.M. Stiles, R.A. Palais, I.T. Knight, Microfluidic genotyping by rapid serial PCR and high-speed melting analysis, Clin. Chem. 60 (2014) 1306–1313.

[6a] P.J. Asiello, A.J. Baeumner, Miniaturized isothermal nucleic acid amplification, a review, Lab Chip 11 (8) (2011) 1420–1430.

[7] J. Shendure, S. Balasubramanian, G.M. Church, W. Gilbert, J. Rogers, J.A. Schloss, R.H. Waterston, DNA sequencing at 40: past, present and future, Nature 550 (2017) 345–353.

[8] H.P. Buermans, J.T. den Dunnen, Next generation sequencing technology: advances and applications, Biochim. Biophys. Acta 1842 (2014) 1932–1941.

[9] JoVE Science Education Database, Basic Methods in Cellular and Molecular Biology. The ELISA Method, JoVE, Cambridge, MA, 2019.

[10] K.P.F. Janssen, K. Knez, D. Spasic, J. Lammertyn, Nucleic acids for ultra-sensitive protein detection, Sensors (Basel) 13 (2013) 1353–1384.

[11] P.N. Nge, C.I. Rogers, A.T. Woolley, Advances in microfluidic materials, functions, integration, and applications, Chem. Rev. 113 (2013) 2550–2583.

[12] https://www.businesswire.com/news/home/20180423005691/en/North-America-Microfluidics-Market-Analysis-Forecast-2025.

[13] H.V. Nguyen, V.D. Nguyen, H.Q. Nguyen, T.H.T. Chau, E.Y. Lee, T.S. Seo, Nucleic acid diagnostics on the total integrated lab-on-a-disc for point-of-care testing, Biosens. Bioelectron. 141 (2019) 111466.

[14] M. Sher, R. Zhuang, U. Demirci, W. Asghar, Paper-based analytical devices for clinical diagnosis: recent advances in the fabrication techniques and sensing mechanisms, Expert. Rev. Mol. Diagn. 17 (2017) 351–366.

[15] K.M. Koczula, A. Gallotta, Lateral flow assays, Essays Biochem. 60 (2016) 111–120.

[16] A.K. Yetisen, M.S. Akram, C.R. Lowe, Paper-based microfluidic point-of-care diagnostic devices, Lab Chip 13 (2013) 2210–2251.

[17] A.T. Singh, D. Lantigua, A. Meka, S. Taing, M. Pandher, G. Camci-Unal, Paper-based sensors: emerging themes and applications, Sensors (Basel) 18 (2018). pii: E2838.

[18] A.D. Warren, G.A. Kwong, D.K. Wood, K.Y. Lin, S.N. Bhatia, Point-of-care diagnostics for noncommunicable diseases using synthetic urinary biomarkers and paper microfluidics, Proc. Natl. Acad. Sci. U. S. A. 111 (2014) 3671–3676.

[19] A. Mariakakis, M.A. Banks, L. Phillipi, L. Yu, J. Taylor, S.N. Patel, BiliScreen: smartphone-based scleral jaundice monitoring for liver and pancreatic disorders, Proc. ACM Interact. Mob. Wearable Ubiquitous Technol. 1 (2017), https://doi.org/10.1145/3090085.

[20] A. Esteva, B. Kuprel, R.A. Novoa, J. Ko, S.M. Swetter, H.M. Blau, S. Thrun, Dermatologist-level classification of skin cancer with deep neural networks, Nature 542 (2017) 115–118.

Further reading

[21] J. Kanter, Point-of-Care Testing in Sickle Cell Disease, Intechopen, 2016. https://doi.org/10.5772/64862.

PART 2

Imaging

CHAPTER 2.1

Optical imaging

Daniel S. Elson
Hamlyn Centre for Robotic Surgery, Institute of Global Health Innovation and Department of Surgery and Cancer, Imperial College London, London, United Kingdom

1 Introduction

Since the beginning of medicine, optical imaging has formed a central pillar for the diagnosis and treatment of disease. The 20th century saw the development of many other types of diagnostic imaging methods—CT, MRI, nuclear methods, etc.—but optical imaging remains of paramount importance. Within the last few decades, optical imaging has started to become more widely used for detection and monitoring of cancer outside of simple detection with the naked eye. These developments, some of which will be introduced in this chapter, are primarily as a result of the minimally invasive surgery revolution. This was enabled by the detection of the surgical field with a color camera system that is either mounted on the proximal end of an endoscope or in more recent times miniaturized and placed at the tip. The color responses of the red-green-blue image data are well matched to the human eye and can be presented on a color display to the surgeon in the operating theater for direct visual guidance of the intervention. Imaging the reflected white light properties of the tissue is an approach that is purely based on providing a visually recognizable picture of the tissue to the surgeon and misses the potential of light to reveal otherwise invisible tissue information. It is the potential of biophotonics techniques to guide diagnosis and interventions that will be explored in this chapter, together with descriptions of how these additional signals may be collected and understood in detail.

This content of this chapter is reproduced in a contemporaneous chapter on "Interventional Imaging: Biophotonics" submitted to another book *Handbook of Medical Image Computing and Computer Assisted Intervention* [1].

1.1 A brief introduction to light-tissue interactions and white light imaging

It is well known that light interacts strongly with tissue, with the two principal mechanisms being absorption and scattering (see Fig. 1A). The absorption will depend on the concentration and extinction coefficient of optically active molecules within the tissue, but is dominated by oxy- and deoxy-hemoglobin, which have distinct absorption spectra in the visible and near-infrared spectral regions [2]. Other significant absorbers within this

Bioengineering Innovative Solutions for Cancer
https://doi.org/10.1016/B978-0-12-813886-1.00002-4

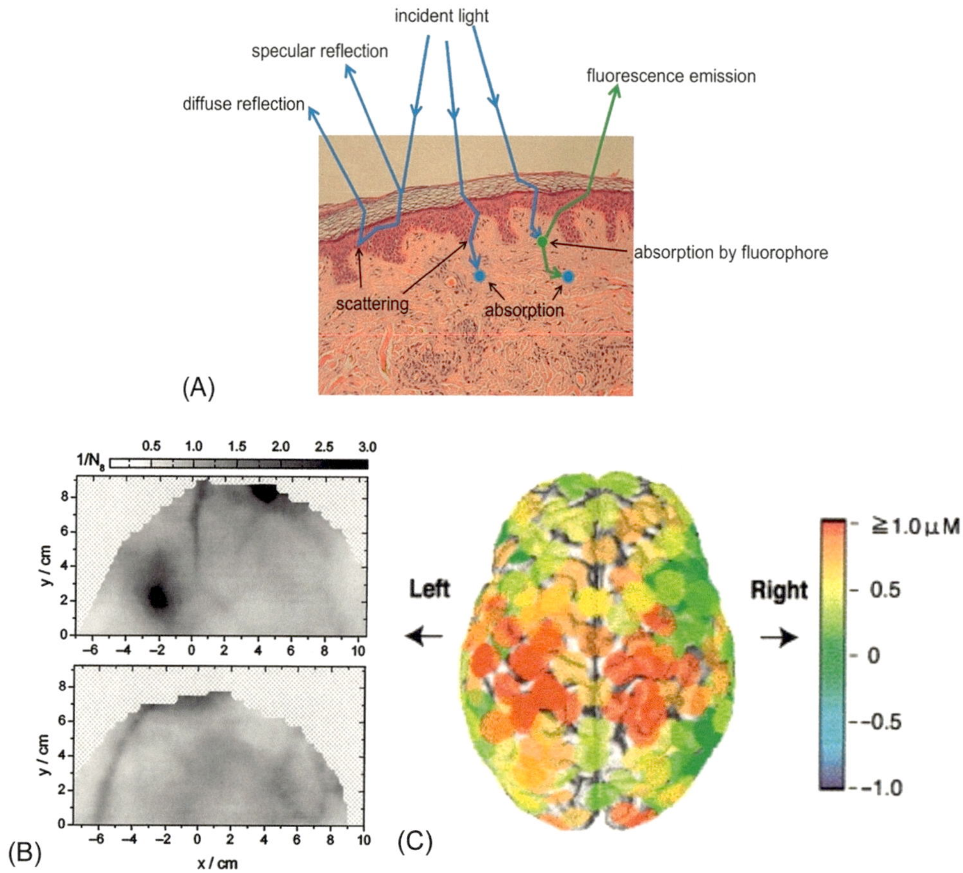

Fig. 1 (A) Summary of common light tissue interaction processes. (B) "Optical mammogram" showing normalized photon counts (related to optical absorption) in a craniocaudal projection view of the left breast (top) and right breast (bottom). Superficial blood vessels are observed in both images and a tumor at (−2 cm, 2.5 cm) in the left breast. (C) Top-down view of a sequential whole-head examination by NIRS with the activation (change in oxyhemoglobin concentration) indicated for an apple-peeling task. *(B: Adapted with permission from D. Grosenick, K.T. Moesta, H. Wabnitz, J. Mucke, C. Stroszczynski, R. Macdonald, et al., Time-domain optical mammography: initial clinical results on detection and characterization of breast tumors, Appl. Opt. 42 (2003) 3170–3186, The Optical Society; C: From M. Okamoto, H. Dan, K. Shimizu, K. Takeo, T. Amita, I. Oda, et al., Multimodal assessment of cortical activation during apple peeling by NIRS and fMRI, Neuroimage 21 (2004) 1275–1288, Elsevier.)*

spectral range include melanin and lipids, while water also absorbs strongly at longer wavelengths in the near-infrared. Since cancer often involves changes in vasculature or hemodynamics, it is to be anticipated that it could be detected by imaging the absorption properties of the tissue. For instance, optical imaging and near-infrared spectroscopy (NIRS) was proposed to detect tissue changes resulting from breast cancer (see Fig. 1B) [3], or for minimally invasive real-time monitoring of brain cortical activation (see Fig. 1C) [4]. However, although optical wavelengths in the red and near-infrared can be selected to allow a good optical transmission through centimeter path lengths, the practical application of these methods is impacted by the very strong tissue optical scattering [2].

Tissue optical scattering arises from the molecular tissue content, and at microscopic and macroscopic scales it manifests itself whenever there is a change in refractive index. Tissues comprise numerous such inhomogeneities, from macromolecules at submicron length scales, through to fibrous collagen structures, cell membranes, and layered tissue structures at microscopic to macroscopic scales. The compound result is that the scattering coefficient of tissue is high, meaning that light has a very short (~100 μm) mean free path between scattering events, although it does vary with wavelength, being reduced in the near infrared [2]. The large length scale of many scattering inclusions compared to the wavelength of light produces mainly "Mie" type scattering, which is preferentially in the forwards direction, mitigating the full effect of the high scattering coefficient. Although the tendency for forward scattering has enabled the tomographic approaches mentioned in the last paragraph, it does limit the spatial resolution that can be achieved at depths of greater than a few mean free paths, i.e., a few hundred microns. Therefore, with the exception of a couple of techniques, namely, Optical Coherence Tomography (OCT) and Photoacoustic Tomography (PAT), this optical imaging chapter will focus on imaging the surface or superficial properties of the tissue.

Returning to the standard clinical tissue imaging that uses reflected white light viewed by naked eye or digital detection, what is generally observed is a combination of the absorption and scattering information. The primary source of coloration is the hemoglobin optical absorbers, whose color is revealed by the light that is scattered superficially in the tissue before being backscattered for detection, giving the tissue its turbid appearance. It is also a common feature of white light imaging that specular highlights appear as bright white regions or spots, which are enhanced in regions of the tissue that have a smooth surface or a mucus layer. These arise from the direct surface reflections, which carry color information about the white light source only. They indicate that at these points the tissue surface normal is oriented in a direction that bisects the illumination and detection ray paths, and this information has been used to imply geometrical information about tissues with computer vision techniques.

1.2 Summary of chapter structure

This chapter will explore how more complex information besides that mentioned earlier can be obtained from tissue during diagnostic imaging. This will begin with a description of how the spectral information can be increased by modifying the light source or the camera to achieve a narrower spectral response, namely, multi/hyperspectral imaging. Fluorescence will then be introduced, initially considering autofluorescence and then the use of contrast agents in the near-infrared spectral range that can improve the ability to detect deeper tissue as well as improve targeting accuracy. Both fluorescence and reflectance imaging can be used in a microscopic format to achieve high-resolution surface information about tissue and cellular structures, providing a further image-based mechanism for tissue diagnostics. The potential of OCT for interventional imaging of subsurface tissue will then be explored and this will be compared to photoacoustic methods. Methods that evaluate motion of tissue scatterers such as laser Doppler imaging and LASCA will be briefly introduced as well as methods for scanning optical devices to acquire data from large regions of tissue and organs.

2 Fluorescence imaging

Fluorescence offers the potential to detect and image the location of endogenous molecules with a good level of specificity, or to use extrinsic contrast agents that are able to target a specific process or biomarker within the tissue. There are some parallels with interventional nuclear/radio approaches with the benefits of not requiring ionizing radiation and potentially higher spatial resolution, but the disadvantage of worse imaging depth. Fluorescence is the process whereby molecules are excited by absorbing an incident photon, followed by a radiative decay involving the emission of a longer wavelength photon (with the energy difference accounted for as heat, see Fig. 2A) [5]. The efficiency of this process depends on the (wavelength-specific) absorption cross-section of a

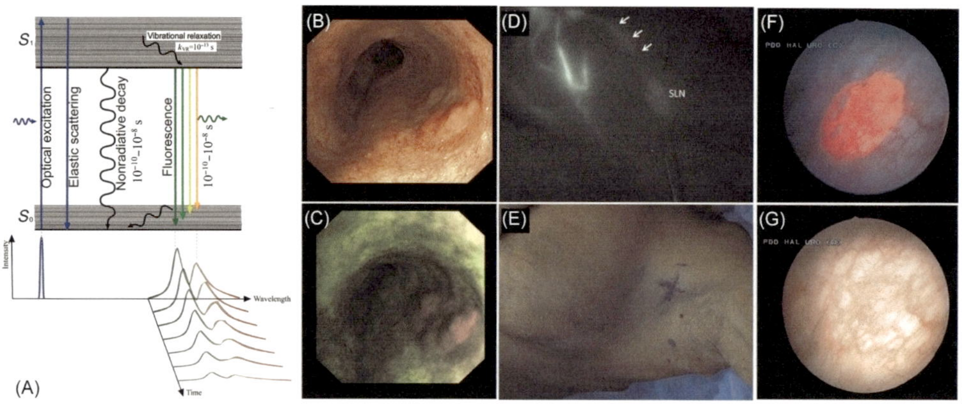

Fig. 2 Illustration of fluorescence approaches and interventional examples. (A) "Jablonski diagram" showing a set of ground states S_0 and excited states S_1 for a molecule, as well as the possible transitions that can occur after absorbing an excitation photon. (B) White light and (C) autofluorescence image of a neoplastic lesion in Barrett's esophagus showing a violet decoloration caused by a loss of fluorescence signal from the underlying tissue. (D) Near-infrared fluorescence image during breast surgery following local injection of ICG used to detect the lymphatic vessel *(arrows)* sentinel lymph node (SLN), and (E) matched white light image showing the position of the SLN marked on the tissue surface for excision. (F) Urothelial carcinoma in situ exhibiting red/pink protoporphyrins IX fluorescence, which is not visible in (G) standard white light cystoscopy. *(B, C: Elsevier copyright; D, E: Reprinted/adapted by permission from C. Hirche, D. Murawa, Z. Mohr, S. Kneif, M. Hünerbein, ICG fluorescence-guided sentinel node biopsy for axillary nodal staging in breast cancer, Breast Cancer Res. Treat. 121 (2010) 373–378, Springer Nature; F, G; Elsevier copyright.)*

molecule, which in turn is determined by the molecule's electronic energy structure, which also determines the emission spectrum (see Fig. 2A). Many biomolecules tend to be naturally fluorescent, particularly when illuminated by blue and ultraviolet light where the individual photons carry more energy. The environment within which the molecule is situated also plays a role and can affect the quantum efficiency, i.e., the probability that the molecule emits a fluorescent photon rather than decaying through a competitive nonradiative path [5].

It is known that tissue contains a number of naturally occurring fluorophores such as collagen, elastin, porphyrins, and molecules involved in metabolism, meaning that tissue exhibits the property of *autofluorescence*, particularly when illuminated in the ultraviolet or blue spectral ranges. Since some of these molecules are implicated in the progression of cancer, effort has been made to develop and translate interventional molecular detection and imaging devices that can inform clinical decisions [6, 7]. The main considerations when constructing clinical devices to image fluorescence are to (1) ensure that appropriate excitation and emission filters are selected to ensure a low non-fluorescence background (i.e., block non-frequency-shifted directly reflected photons) and maximal photon efficiency, (2) use sufficiently sensitive detectors due to the low yield of fluorescence photons, and (3) allow the light source and camera system to operate in both fluorescence and white light modes to enable standard navigation.

One commercially available example is the Olympus Lucera flexible endoscope, which has a violet excitation mode that can be selected at the flick of a switch to reveal information or contrast that is invisible to the naked eye. It is typically noted when observing epithelial tissues that the neoplastic and dysplastic thickening of this uppermost layer leads to a reduction in the green (collagen) fluorescence from the underlying stroma, and that there is potentially an increase in red fluorescence from porphyrins in lesions such as ulcerated squamous cell carcinomas [8]. Clinical studies have shown that various pathologies can be detected, including cancers of the brain [9], lung [10], esophagus [11], and colon [12] (see Fig. 2B and C).

Although autofluorescence imaging is a conceptually attractive method—requiring relatively simple device modifications and regulatory approval, provided that the wavelength is longer than potentially harmful ultraviolet range—it does not achieve the level of specificity that is required for many diagnostic applications. This may be due to the many confounding factors that can affect the image appearance, including variations in the illumination field, imaging distance, crosstalk with other fluorophores and nonspecific fluorescence background, etc. Fiber probe–based fluorescence spectroscopy devices can overcome some of these issues by recording the full fluorescence emission spectrum in a controlled measurement geometry, sometimes at multiple excitation wavelengths, providing a richer and more complete dataset with more potential for internal controls and normalization. Multiwavelength detection benefits from the hardware developments described in the section on MSI, although methods that rely on sequential images may be impractical due to the generally longer camera exposures that are required to detect the weaker fluorescence signals.

An alternative method is to record the fluorescence lifetime, where the nanosecond-scale decay of the fluorescence is imaged using ultrafast lasers or megahertz frequency-modulated sources, which can improve the specificity to certain fluorophores, or remove nonspecific fluorescence background [13] (see Fig. 2A, lower section). A further class of technologies combines reflectance and fluorescence multispectral imaging where a more detailed fluorescence spectrum is recorded, which has been used to detect cervical [14] and ovarian [15] cancers as well as guiding surgical procedures such as cholecystectomy [16]. Devices that use these technologies are mainly restricted to early clinical trials at present and have generally not been fully validated and commercialized.

The use of extrinsic fluorophore agents is also being actively pursued, where it is hoped that the high specificity and low background images obtained in microscopes for cell biology can be brought to human-scale applications. There exist many types of extrinsic labels ranging from organic fluorescent compounds through to synthetic nanoconstructs such as quantum dots, or even genetic labels such as green fluorescent protein (GFP) that render expressed proteins fluorescent. It is a natural step to ask whether these molecules could be used in a clinical context to improve the detection accuracy. However, there are safety and toxicity concerns with extrinsic markers and the targeting methods, and expensive time-consuming regulatory processes are required to pursue the commercialization of chemical probes or the associated hardware. Furthermore, when designing a fluorescent molecule for clinical application, it is desirable to use infrared fluorophores so that there is less crosstalk with the spectra of the endogenous fluorophores, and so that the achievable sensing depth of the fluorescence is increased to 5–10 mm due to reduced scattering and absorption.

For these reasons, many investigators and industry have focused on using an infrared fluorophore called indocyanine green (ICG), which received regulatory approval for systemic administration to measure perfusion in the eye and other organs [17]. ICB binds to blood proteins such as serum albumin, resulting in an excitation peak at 805 nm with emission at 830 nm [18]. Although it is frequently used without further functionalization and is rapidly excreted, it has been observed to accumulate in certain tumors thanks to the enhanced permeability and retention (EPR) effect, where the increased number and leaky angiogenic blood vessels pass the protein-bound ICG, while at the same time the compromised lymphatics result in its accumulation. The EPR effect has been used to image breast cancer [17] and others. Direct intravascular ICG injection can also allow the perfusion of downstream blood vessels to be imaged, allowing organ perfusion to be assessed including in the gastrointestinal (GI) tract [19] or in skin flaps [20]. ICG may be applied topically to tissue with the intent of imaging the time course of the fluorescence as the ICG drains into the lymphatic system. This can allow the identification of lymph nodes provided that they are within a few millimeters of the tissue surface (see Fig. 2D and E) [21]. The sentinel nodes can then be excised under image guidance for subsequent histopathological examination for the signs of the spread of cancer, an important prognostic indicator.

Various approved commercial imaging systems are available with 510(k) clearance for near-infrared fluorescence imaging, including Novadaq Technologies SPY imaging system, the Quest Medical Spectrum, and the Hamamatsu Photonics PDE Photodynamic Eye [22]. Endoscopic devices are manufactured by the major endoscope manufacturers Karl Storz, Richard Wolf, Stryker, Olympus, etc. Clinical trials and preclinical imaging studies have investigated improvements to the targeting of ICG by binding it to antibodies or peptides [23]. There are also promising results reported from other near-infrared fluorophores such as IRDye 800CW [24], which is brighter than ICG and may be targeted for use in margin assessment during breast cancer surgery [25].

Although increasingly popular, not all fluorescence studies use near-infrared contrast agents. For example, some visible fluorescence molecules are approved for clinical use, including fluorescein, which is excited in the green spectral region and is used particularly with intraocular imaging and endoconfocal microscopy (described in a later section of this chapter). It has been used for large-area surgical guidance, investigated together with folate targeting for ovarian cancer staging and debulking using a multiplexed camera multispectral imaging device [26]. Another visible fluorophore, protoporphyrin IX, arises after a precursor molecule 5-aminolevulinic acid is delivered to the tissue and is metabolized to the fluorophore as part of heme biosynthesis pathway. This process however is altered in tumors compared with healthy tissues, resulting in increased protoporphyrin IX and a visible fluorescence signal. This mechanism has been applied to the image-guided surgery of bladder cancer using this agent under the marketing name Hexvix (see Fig. 2F and G), where the papillary or sessile carcinomas can be visualized when the surgeon switches to an endoscopic "blue light" mode [27,28]. A related agent under the marketing name Gliolan has also been approved for imaging of glioma and uses a fluorescence-enabled operating microscope for image-guided intervention [29]. The applications of protoporphyrin IX continue to develop, including the use of its second excitation band in the red, which can reduce the interference from autofluorescence and increase the sensing depth.

Finally, it is worth noting that a second near-infrared region can be used to image fluorophores with wavelengths in the range 1.3–1.4 μm, where absorption remains relatively low but scattering is also reduced [30]. Imaging depths of 2 mm have been shown in vivo, including through mouse skull [30], and a range of new fluorophores are in development, including single-walled carbon nanotubes and quantum dots [31]. Further information on the fluorophores and imaging systems can be found in an article by DSouza [22].

3 Multispectral imaging

Multispectral imaging is the extension of white light imaging to incorporate better spectral resolution resulting in improved ability to distinguish different chromophores in the tissue. As with other applications of multi/hyperspectral imaging (MSI), such as remote sensing or production line inspection, the use of this technology requires a compromise to be struck between the spatial, spectral, and temporal resolutions with the ideal interventional acquisition consisting of a detailed spectral signature for each pixel in the image at video rate.

A simple example of how these resolutions may be selected is in a commercially available endoscopic technique called narrow band imaging (NBI), which achieves video rate imaging by retaining a camera but modifying the light source for limited multiwavelength illumination. During NBI, a filter wheel controls the light spectrum that is incident on the tissue, restricting it to three narrow (20–30 nm) bands of light in the red, green, and blue spectral regions (Fig. 3C) [32]. The principle is that the short wavelength bands, being more strongly absorbed by the hemoglobins, are absorbed by the superficial blood vessels, improving their visibility and contrast through the removal of other wavelengths that would result in blurring. Clinical examples of the use of NBI are numerous due to the general availability of the technique on Olympus endoscopes, including for instance in flexible endoscopy to improve the detection of adenomas or hyperplasia through better imaging of the pit pattern or microvascular tortuosity (Fig. 3B) [33], as well as in the esophagus for supporting identification of Barrett's tissue (Fig. 3A) [32]. To overcome the variable accuracy and efficacy of NBI techniques for different users, as well as the effect of the learning curve, there have recently been attempts to standardize classification systems. For instance, methods have been proposed to predict the presence or absence of dysplasia in Barrett's esophagus [34] or to stage colorectal cancer [35].

While NBI compromises potential spectral resolution in favor of high-resolution imaging at video rate, the technique of diffuse reflection spectroscopy (DRS) takes a different approach, recording high spectral resolution data at video rate, but usually from only a single point, by placing a fiber-based probe directly in contact with the tissue surface [36]. This is relevant to this chapter on interventional *imaging* because (1) this type of data can serve to guide the design of multispectral imaging instruments, particularly in the selection of the spectral bands that optimize the diagnostic sensitivity, and (2) because endoscopic scanning methods can be used to build up an image point by point. DRS has been shown to achieve high accuracy in identifying tissue pathology in a variety of different tissues [37] and the average depth of photon penetration into the tissue can be controlled by adjusting the source-detector fiber separation.

NBI and DRS are optimized for high spatial and high spectral resolution, respectively, and can acquire data at video rate or higher. Other multi- and hyperspectral instruments attempt to acquire a spectral hypercube (Fig. 3D) and may either acquire over longer time periods to maintain high spectral/spatial resolution, or may compromise spatial/spectral resolution to allow imaging in a single snapshot. One example is the use of tunable optical filters or arrays of filters placed in front of the light source or camera (Fig. 3G) [38,39]. In this case, the spectral resolution and the acquisition time simply scales with the number

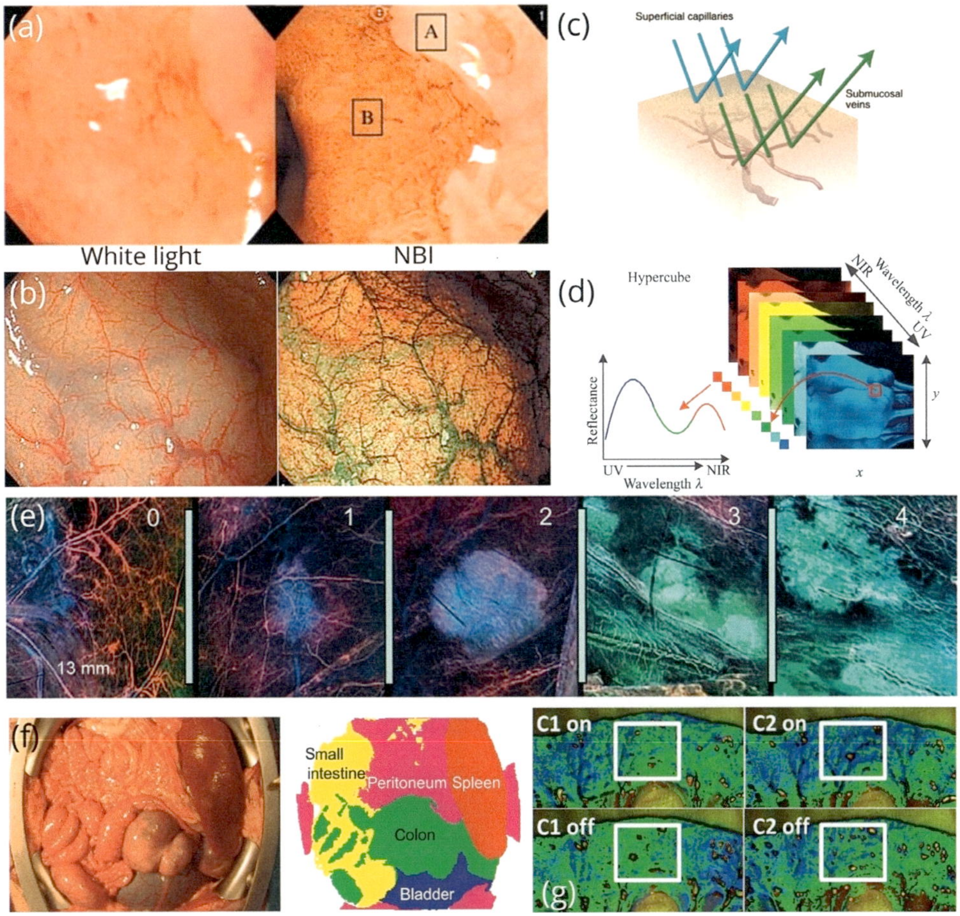

Fig. 3 Illustration of different MSI approaches and interventional examples. (A) Conventional (left) and NBI (right) images of normal ("A") and Barrett's ("B") esophagus showing enhanced pit pattern visualization for NBI. (B) Conventional (left) and NBI (right) images of sigmoid colon showing improved superficial vessel visualization. (C) Depiction of NBI blue and green light interacting with superficial vessels. Permission here. (D) The principle of MSI, showing a stack of images acquired at different reflection wavelengths (hypercube) and an illustrative spectrum extracted from one pixel. (E) Pseudocolor processed HSI examples for normal tissue (0), benign tumor (1), intraductal carcinoma (2), papillary and cribriform carcinoma (3), and carcinoma with invasion (4). (F) RGB (left) and HSI segmented (right) images of porcine abdomen classified using an artificial neural network. (G) Oxygen saturation variation showing ischemia *(blue)* during vessel occlusion in a section of porcine bowel. *(A: Adapted with permission from K. Gono, T. Obi, M. Yamaguchi, N. Oyama, H. Machida, Y. Sano, et al., Appearance of enhanced tissue features in narrow-band endoscopic imaging, J. Biomed. Opt. 9 (2004) 568–578; B: From H. Machida, Y. Sano, Y. Hamamoto, M. Muto, T. Kozu, H. Tajiri, et al., Narrow-band imaging in the diagnosis of colorectal mucosal lesions: a pilot study, Endoscopy 36 (2004) 1094–1098, © Georg Thieme Verlag KG; C: From M.J. Waldner, S. Wirtz, C. Neufert, C. Becker, M.F. Neurath, Confocal laser endomicroscopy and narrow-band imaging-aided endoscopy for in vivo imaging of colitis and colon cancer in mice, Nat. Protoc. 6 (2011) 1471–1481, originally adapted from Olympus; D: Adapted with permission from G. Lu, B. Fei, Medical hyperspectral imaging: a review, J. Biomed. Opt. 19 (2014) 010901; E: Reprinted by permission from S.V. Panasyuk, S. Yang, D.V. Faller, D. Ngo, R.A. Lew, J.E. Freeman, et al., Medical hyperspectral imaging to facilitate residual tumor identification during surgery, Cancer Biol. Therapy 6 (2007) 439–446, Taylor & Francis Ltd; F: Reprinted/adapted by permission from H. Akbari, Y. Kosugi, K. Kojima, N. Tanaka, Wavelet-based compression and segmentation of hyperspectral images in surgery, International Workshop on Medical Imaging and Virtual Reality, Springer, 2008, pp. 142–149, Springer Nature; G: Reprinted with permission from N.T. Clancy, S. Arya, D. Stoyanov, M. Singh, G.B. Hanna, D.S. Elson, Intraoperative measurement of bowel oxygen saturation using a multispectral imaging laparoscope, Biomed. Opt. Express 6 (2015) 4179–4190.)*

of filter bands that are to be used, which in surgical applications may require registration of images to account for tissue motion and deformation [40]. Another popular MSI approach that is used in production line quality control and aerial surveillance applications is to acquire spatial information about a single line, using the other dimension of the imaging detector to record spectral data, usually by dispersing the light using a prism or grating [41]. However, the slow acquisition speed limits this method for interventional applications because the line must be scanned across the tissue to build the image content in the second spatial dimension, which is then limited by the frame rate of the camera.

Another class of MSI device is snapshot imagers, which record all data simultaneously. For instance, the spatial detector may be divided by a spatially varying "Bayer" pattern of different filters to acquire more spectral bands over a larger range of wavelengths, usually at the expense of reduced spectral resolution [42]. Other methods include image mapping spectrometry that spreads the two spatial and one spectral dimensions across a single two-dimensional camera in such a way that a spectral data cube can still be reconstructed [43]. Alternatively, high spatial resolution white light imaging and low spatial resolution spectroscopy may be combined together using information processing approaches to recover higher resolution spectroscopic images [44].

When analyzing MSI data, it is possible to construct an analytical model since the majority of the tissue absorption spectra are known (Fig. 3G), allowing techniques such as spectral unmixing algorithms to recover the individual components [45]. Complexities that must be built into the model (or ignored/assumed insignificant for practical purposes) include (1) that different wavelengths of light penetrate to different depths and effectively sample a different volume, (2) that optical scattering is wavelength dependent and will affect light penetration, and (3) that the flood illumination and wide-field detection mean that the paths through which the photons travel within the tissue are less well defined compared to DRS. Therefore, as an alternative to modeling, feature extraction and classification holds promise using parallel processing hardware and machine/deep learning pathological diagnosis (Fig. 3E and F). In common with other biophotonics techniques, progress has been limited due to the requirement to test against ground truth (histology) data, which is impractical or time consuming to achieve.

There are many applications of MSI in cancer diagnosis or intervention, including for tissue monitoring prior to anastomosis following cancer resection (Fig. 3G) [39]; detecting cancerous changes in the prostate [46], stomach [46], breast [47], etc. A broader range of MSI applications is presented in a review article by Lu and Fei [45], where the different acquisition strategies are also described in more detail.

4 Microscopy techniques

Most of the techniques and examples introduced so far have focused on large area surveillance with wide-field imaging, since this is well matched to the scale of many surgical interventions. However, while many diseases will present gross changes visible at this scale, it is also the case that early-stage disease involves microscopic changes to the biochemistry and tissue architecture that may be lost for lower resolution imaging. Also, many surgical procedures require precise small-scale tissue manipulation. When imaging an exposed tissue surface, conventional image magnification optics can be used, for instance skin lesions may be evaluated by directly placing a high magnification clinical microscope system in contact with the patient. Properties such as coloration, texture, lesion borders, and size may then be used to form an indicative score that may prompt biopsy, treatment, or watch-and-wait paths [48]. In the case of open surgery, operating microscopes use a large standoff distance to allow microsurgical cancer-related procedures such as small blood vessel anastomoses or precise tissue removal or brain dissection.

In endoscopic screening and surveillance of the GI tract, neoplasia and dysplasia will result in the abnormal growth and organization of the epithelial cells as well as molecular changes, and detection or diagnosis uses biopsy and standard histology as the gold standard. Under white light imaging, these changes may appear similar to other conditions or inflammation, and it is also difficult to accurately survey a large area since multiple physical biopsies are either not possible or practical. Similar arguments apply for other epithelial tissues, as well as for the detection of microscopic metastatic deposits for instance in the peritoneal cavity, or for the detection of residual microscopic disease after a tumor has been excised in breast surgery. Therefore, various microscopic imaging devices have been proposed or are in standard clinical practice.

One way in which microscopic changes may be highlighted is through the use of chromoendoscopy, where methylene blue, toluidine blue, or other absorbing contrast agents are sprayed or applied to the tissue. Although not always applied as a microscopic technique, the microscopic cellular architectures manifest themselves as gross intensity variations in the visibility of the absorbing dye. More accurate diagnosis of ulcerative colitis has been achieved, as well as detection of intraepithelial neoplasia and cancer of the colon [49]. As with some of the autofluorescence endoscopy methods described in the section on fluorescence, there are varying results reported on the overall accuracy of the method, including for the diagnosis of Barrett's esophagus [50].

Magnified endoscopy or zoom endoscopy uses a tip attachment or modified optics to achieve a short endoscopic working distance and high magnification compared to standard imaging. Suspicious regions can then be investigated in higher resolution and using color enhancement processing techniques to show cellular architecture [51]. High magnification can also be combined with narrow band imaging so that the microvasculature can be observed in higher detail so that clinicians can look for features such as branching vessels, dark spots, or disorganized capillary patterns. This can improve performance over narrow band endoscopy alone, particularly for distinguishing sessile serrated adenomas from hyperplastic polyps (Fig. 4A and B) [52]. A low-cost endoscopic accessory has also been developed, named high-resolution microendoscopy (HRME), based on a millimeter-diameter fiber optic imaging bundle that can be placed in gentle contact with the tissue [53]. This has been evaluated in a range of gastrointestinal applications, showing high accuracy for distinguishing neoplastic polyps [54] and oral neoplasia [55], and with potential to be deployed in resource-limited settings.

The magnification methods mentioned can record high-resolution images, but are affected by out-of-focus background scattered light, which results in crosstalk between the resolution elements and contributes to blurring of the images and loss of contrast. Confocal microscopy overcomes this issue by allowing out-of-focus light to be physically rejected by a pinhole in front of the detector to produce high-contrast optically sectioned images, usually at the expense of requiring an image to be acquired point by point by beam scanning [56]. Clinical confocal microscopes have been used to record skin cancer

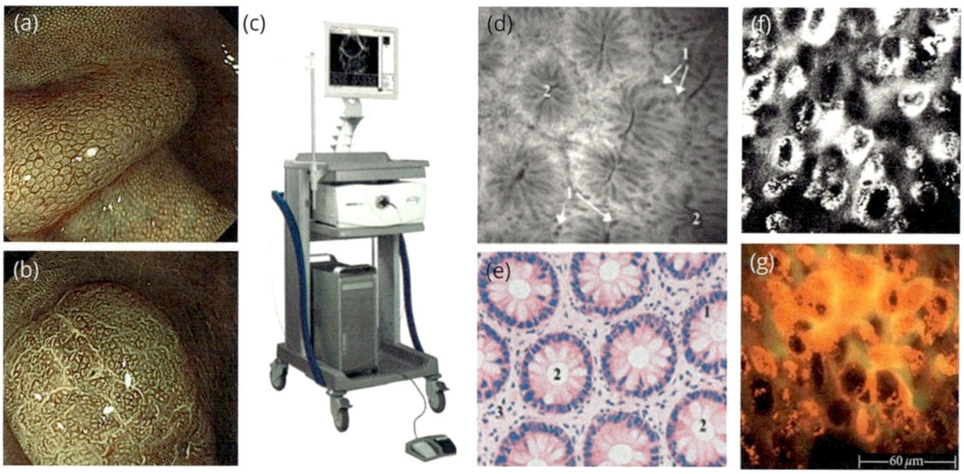

Fig. 4 Illustration of microscopic approaches and interventional examples. (A and B) Magnifying narrow-band imaging of sessile serrated adenomas/polyps (SSA/Ps) illustrating (A) regular and (B) disorganized meshed capillary patterns. (C) The Mauna Kea confocal endomicroscope system. (D) Confocal endomicroscopy images following fluorescein injection, resulted in strong staining of the cecum epithelium and (E) a closely matched histological section. 1 indicates goblet cells, 2 indicates crypt lumen. (F) A fluorescence intensity image of a malignant melanoma examined ex vivo at 40-μm depth showing an irregular distribution of pleomorphic cells. (G) A fluorescence lifetime image showing variation of lifetime between melanoma cells (short lifetime values, *yellow*). *(A, B: From M. Yamada, T. Sakamoto, Y. Otake, T. Nakajima, A. Kuchiba, H. Taniguchi, Investigating endoscopic features of sessile serrated adenomas/polyps by using narrow-band imaging with optical magnification, Gastrointest. Endosc. 82 (2015) 108–117, Elsevier; C: Courtesy: Mauna Kea Technologies; D, E: From A.L. Polglase, W.J. McLaren, S.A. Skinner, R. Kiesslich, M.F. Neurath, P.M. Delaney, A fluorescence confocal endomicroscope for in vivo microscopy of the upper-and the lower-GI tract, Gastrointest. Endosc. 62 (2005) 686–695, Elsevier; F, G: From S. Seidenari, F. Arginelli, S. Bassoli, J. Cautela, P.M. French, M. Guanti, et al., Multiphoton laser microscopy and fluorescence lifetime imaging for the evaluation of the skin, Dermatol. Res. Pract. 2012 (2012), © 2012 Stefania Seidenari.)*

architecture at high resolution across the different epithelial layers [57,58]. This can be extended to endoscopic application by miniaturizing the optics, which is usually achieved using optical fibers to simultaneously transport the light and to act as a pinhole. In the commercial Mauna Kea confocal endomicroscope, a fiber image guide is used (Fig. 4C), with distal-mounted miniature microscope optics and a proximal raster beam scanning base station [59]. The probe is inserted through the biopsy/working channel of an endoscope and is usually used to detect tissue fluorescence for blue excitation, allowing either endogenous elastin or more frequently a topically or systemically applied fluorescein contrast agent to be imaged. This endoscope has been investigated preclinically and clinically and has shown promise for detecting microscopic blood vessels and angiogenesis [60], Barrett's-associated metaplasia [61], colonic pit patterns (Fig. 4D and E) [62], and parenchymal lung disease [63]. The small imaged field means that the probe must be scanned across the tissue surface for larger area surveillance or contextualization. Image mosaicing features are a standard part of the commercial system [64], and mechanical probe scanning methods are introduced later in this chapter. Image interpretation and histological validation remains a challenge, but there is a growing database of representative images across a range of specialties available in published research papers as well as on the Mauna Kea online archive.

Since the imaging depth for confocal endoscopy reaches only around 100 μm, there is interest in extending the working depth of confocal endomicroscopy by using *nonlinear* methods such as two-photon excited fluorescence. This technique uses an illumination wavelength at approximately twice that typically used for single-photon excitation, having the benefit of reduced scattering and absorption leading to increased working depths. However, it does require more expensive and complex pulsed lasers to be used since the absorption cross section is much smaller and scales nonlinearly with the photon flux [65]. These pulsed near-infrared laser sources can also be used to record at the second harmonic, i.e., half the wavelength, and certain structures, including collagen, have a nonlinear response to intense driving electromagnetic fields. These two techniques (two-photon excited fluorescence and second harmonic generation) have been used extensively in the laboratory to image tissue and produce detailed images of unstained tissue at up to 1-mm imaging depth [66,67]. Such systems have also been adapted for the inspection of skin lesions, including using the fluorescence lifetime (Fig. 4F and G). To be able to use these more complex techniques endoscopically at depth, the short intense pulses of infrared light must be passed through a significant length of optical fiber. To overcome the issues of fiber nonlinearity and dispersion, which would destroy the short laser pulses, special fiber types such as photonic crystal or hollow core fibers are typically used [68]. Double-clad structures can also keep the excitation light tightly contained for focusing to a small spot, while the emitted fluorescence can be collected with a larger detection target [69]. Since the optical fiber is effectively a single-pixel detector, distal scanning of the beam is required for imaging, which often uses a MEMS device. Imaging of NADH and FAD together with collagen has been recorded from tissue in vivo as well as redox imaging in a perfusion animal model [70].

5 Optical coherence tomography

Optical coherence tomography (OCT) provides cross-sectional images of the optical scattering structures within a biological tissue. In early systems, the time of flight of short pulses of light was recorded as they scattered from layers or structures within the tissue, a process, which is often referred to as the optical equivalent to an ultrasound pulse-echo A-scan [71]. Due to the very fast ~10 ps ($\times 10^{-12}$ s) transit round-trip time of light over the mm length scales involved, (Michelson) interferometric detection is required that uses the detection of optical phase while a reference mirror is translated to change the interferometer arm length [72]. The maximum imaging depth is determined by the tissue scattering and absorption properties, as typically after a ~mm propagation distance the number of directly reflected photons reaching the surface becomes minimal compared to the number of diffusely scattered photons. The use of pulsed light sources has been superseded by Fourier or frequency domain methods that use broadband sources with spectrally resolved detection, offering improved sensitivity through parallel detection of an entire A-scan in a single measurement with no requirement to scan the reference mirror [73]. In more recent systems, the acquisition speed has been further improved by using frequency-swept laser sources and fast detectors to allow A-scans to be recorded at many 100's MHz frequencies [74]. Following signal extraction and processing, some form of spatial beam scanning is required to build up two- or three-dimensional images.

Due to the mm maximum depth scale, the availability of ophthalmoscopes and the layered structure of the retina, diagnostic ocular imaging has become the main application of OCT, with commercial devices found in specialist centers or even in high street opticians (see Fig. 5A–E). This has allowed retinal defects such as detachment, macular degeneration, and macular holes to be diagnosed. OCT has also been incorporated into ophthalmic image-guided therapy devices; for instance, it has allowed the guidance of microsurgical robotic devices to perform retinal surgery [75] and to allow precise manipulation of surgical forceps for membrane peeling [76]. In these studies, OCT was used for its depth-ranging capabilities to locate the precise position of the tissue surface, and this idea could be used as a surgical assistive technique beyond the retina.

A second major application area of OCT has been in detecting the layered structures within blood vessel walls, such as characterization of the thickness and vulnerability of the fibrous cap in atherosclerotic plaque or imaging of stent placement and complications [77,78]. Since OCT systems are generally implemented using optical fiber, they can be readily incorporated into a catheter device, provided that radial and axial scanning can be achieved and the blood briefly flushed from the imaging area [79]. Fast radial scanning can be achieved proximally by rotating a component on the entire length of the

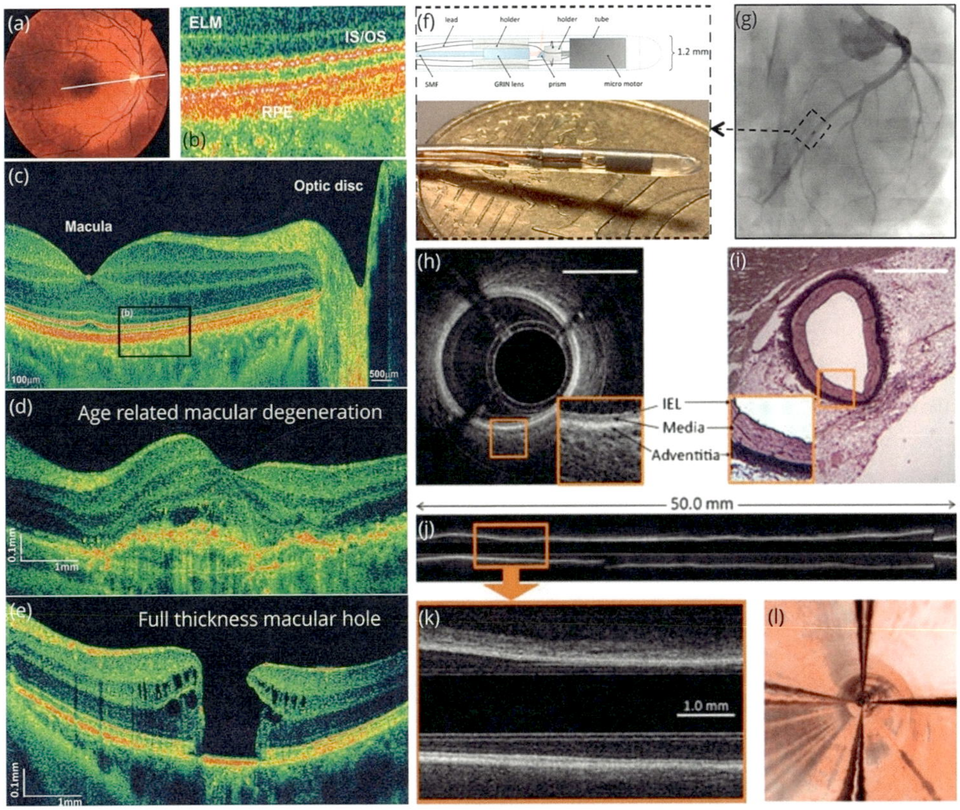

Fig. 5 Illustration of OCT images of the (A–E) retina and (F–L) blood vessel. (A) White light ophthalmoscopic image showing the location of the OCT B-scan shown in (C), with the zoomed-in region of the External Limiting Membrane (ELM), Inner Segment/Outer Segment junction (IS/OS), and Retinal Pigment Epithelium (RPE) shown in (B). (D) and (E) show two conditions of the eye that involve a disruption of the retinal structures and may be detected from OCT images. (F) An intravascular OCT catheter with distal scanning mechanism, which is visible under fluoroscopy (G). (H) and (I) Comparison of OCT and histology with arterial structures labeled, including the Internal Elastic Lamina (IEL). (J) Longitudinal rendering with a zoomed-in region (K) and reconstructed fly-through (L). *(A–E: Reprinted with permission from M. Wojtkowski, High-speed optical coherence tomography: basics and applications, Appl. Opt. 49 (2010) D30–D61, The Optical Society; F–L: Reprinted with permission from T. Wang, T. Pfeiffer, E. Regar, W. Wieser, H. Van Beusekom, T. Lancee, et al., Heartbeat OCT: in vivo intravascular megahertz-optical coherence tomography, Biomed. Opt. Express 6 (2015) 5021–5032, The Optical Society.)*

catheter, or by distally rotating a small mirror/prism using a micro motor. The latter approach has achieved more than 4000 rotational frames per second with a 100 mm-per-second pullback speed, producing complete three-dimensional datasets in a single heartbeat (see Fig. 5F–L) [80]. OCT is also being combined with autofluorescence information since the functional fluorescence signal can complement the structural OCT data, for instance distinguishing necrotic lipid pools from collagen [81]. With a standard reference guide produced by the *Journal of the American College of Cardiology* [79] and thousands of scans being performed every year, there is the potential for scaling to millions of scans per year for comprehensive longitudinal plaque characterization or OCT-based stent selection [82].

The uses of OCT for detection and diagnosis of cancer have been more modest to date, although there is much promising current research in this area. Reports suggest that morphological changes such as epithelial thickening can be detected in vivo, although the available axial and lateral resolutions are not sufficient to clearly see the cellular architecture when working at depth. There have been relatively few in vivo diagnostic studies with statistics reported and it has been suggested that new techniques that improve the spatial resolution and allow adoption into surgical formats will improve uptake [83]. Performance can also be improved by using automated image analysis approaches [84] such as texture analysis [85,86]. Other applications and findings in the GI tract are summarized in a review article [87], while OCT has also been used during robotic prostatectomy to identify the neurovascular bundles [88] and detection of bladder lesions and tumor penetration [89] among many other cancer interventions [83]. The minimally invasive use of OCT has been enhanced through the use of needle-based probes that can obtain circumferential scans along the insertion path, as shown for prostate cancer detection [90], breast cancer diagnosis [91], guidance of transbronchial needle aspiration [92], and needle insertion guidance systems [93]. The depth-ranging abilities of OCT have also been used as a method of precise surgical guidance.

6 Photoacoustic methods

The interventional imaging techniques in these sections can record data in real time for detection, diagnosis, and guidance, although they are primarily only able to sense the very superficial tissue properties, extending to around 1 mm for OCT or a few millimeters for near-infrared fluorescence. There are also optical tomographic techniques using absorption or fluorescence contrast [2,94], which have low spatial resolution and are mainly used either in a small animal investigational setting or for diagnostic/preoperative imaging (not discussed further in this chapter). As an alternative approach, the combination of optical absorption contrast with ultrasound detection has potential to bridge this gap, allowing optical contrast but with a resolution determined by ultrasound. In a typical tomographic set-up, short (nanosecond) pulses of light are applied to a tissue in flood illumination, resulting in a rapid thermoelastic expansion and the creation of an acoustic wave. The temporal evolution of this wave can be detected by multiple ultrasound transducer elements at the tissue surface and reconstruction algorithms used for identifying the three-dimensional absorption (Fig. 6A–D) [95]. As with MSI, there is intrinsic absorption contrast from blood, although many other absorbing contrast agents can be used, including clinically approved agents such as ICG and fluorescein as well as other dyes and

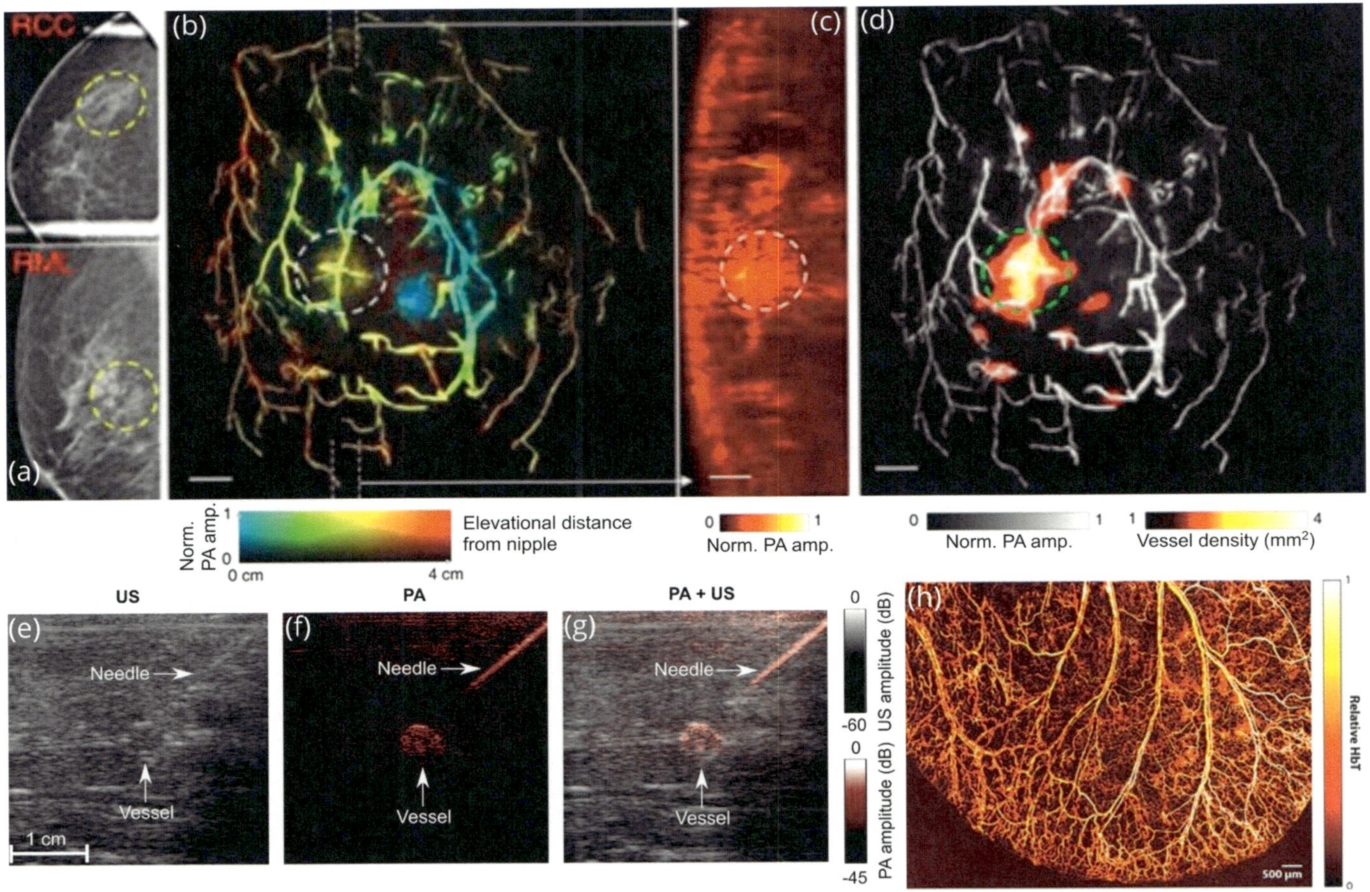

Fig. 6 Illustration of photoacoustic approaches and interventional examples. (A–D) Images of breast tissue and a tumor showing (A) X-ray mammograms of right cranial-caudal (RCC) and right mediolateral (RML) projections, (B) depth-encoded angiograms where the tumor is indicated by a *white circle*, (C) maximum amplitude sagittal projection within the volume indicated by the two *dashed lines* in (B), and (D) automatic tumor detection *(red)* displayed with the vasculature *(grayscale)*. (E–G) Ultrasound, photoacoustic, and combined images during needle insertion toward a vessel in a tissue phantom. (H) Photoacoustic microscopy image of relative total hemoglobin concentration in a mouse ear showing vascular and capillary anatomy. *(A–D: Adapted from L. Lin, P. Hu, J. Shi, C.M. Appleton, K. Maslov, L. Li, et al., Single-breath-hold photoacoustic computed tomography of the breast, Nat. Commun. 9 (2018b) 2352, under a Creative Commons Attribution 4.0 International License, http://creativecommons.org/licenses/by/4.0/; E–G: Adapted from W. Xia, M. Kuniyil Ajith Singh, E. Maneas, N. Sato, Y. Shigeta, T. Agano, et al., Handheld real-time LED-based photoacoustic and ultrasound imaging system for accurate visualization of clinical metal needles and superficial vasculature to guide minimally invasive procedures, Sensors 18 (2018) 1394, under a Creative Commons Attribution 4.0 International License, http://creativecommons.org/licenses/by/4.0/; H: Reprinted with permission from S. Hu, K. Maslov, L.V. Wang, Second-generation optical-resolution photoacoustic microscopy with improved sensitivity and speed, Opt. Lett. 36 (2011) 1134–1136, The Optical Society.)*

emerging optical contrast agents or nanoconstructs [96–98]. Multispectral absorption contrast uses multiple illumination wavelengths to distinguish multiple chromophores within the tissue [97]. As with conventional ultrasound, the image resolution scales inversely with the depth, with higher resolution achievable for more superficial sensing (see Fig. 6). At greater depths, the resolution is reduced and photon flux is also reduced by the absorption in shallower layers, and direct fiber delivery to deeper layers has been proposed using needles [99]. At more superficial depths, optical focusing can allow much higher resolutions, similar to microscopy but with acoustic detection of the signal (Fig. 6H) [95].

As a diagnostic tool, photoacoustic tomography is being applied for imaging of human breast vasculature and tissue oxygenation (Fig. 6A–D) as a nonharmful alternative to X-ray mammography for identifying breast cancer. Photoacoustic imaging has also been proposed as an interventional device guidance tool, for instance for needle guidance in sentinel lymph node biopsy enabled by ICG contrast [100] or during the detection or avoidance of blood vessels for needle insertion (Fig. 6E–G) [101,102]. Light delivered by a photoacoustic needle has also been used to locally characterize cancerous lesions and improve the guidance of biopsy needles [99]. Furthermore, brachytherapy seeds have been imaged during an animal model of prostate therapy [103]. The field is developing rapidly with new applications emerging regularly, such as multispectral discrimination between nerves and tendon [104], robotic-assisted surgical guidance [105], or for catheter tracking [106].

7 Optical perfusion imaging

It is not just the absorption from hemoglobin that can provide an optical signature for blood, but it is also possible to detect the signal scattered from moving red blood cells. This is used to sense flow within vessels or perfusion in bulk tissue, and hence may be used to record or monitor tissue conditions such as erythema, psoriasis, burn recovery, and skin patch viability. Light that is scattered by the moving blood cells is shifted in optical frequency due to the Doppler effect, which results in interference at detector between the frequency-shifted light and the light backscattered from stationary structures. When illuminated by a coherent laser, scattering from motionless structures and tissues forms a high contrast speckle pattern on an imaging detector due to randomly spatially varying constructive or destructive interference [106a]. However, in regions of the image where there is tissue motion or blood flow then the local speckle contrast is reduced. The contrast information it typically converted into a false color image representing the blood flow rate (Fig. 7A–C) and this technique is termed laser speckle contrast imaging (LSCI). Similar information may also be found by recording a stack of images with time and analyzing the temporal variation in the signal intensity on a per-pixel basis [106a]. In a closely related technique, the temporal evolution between the light scattered from stationary and

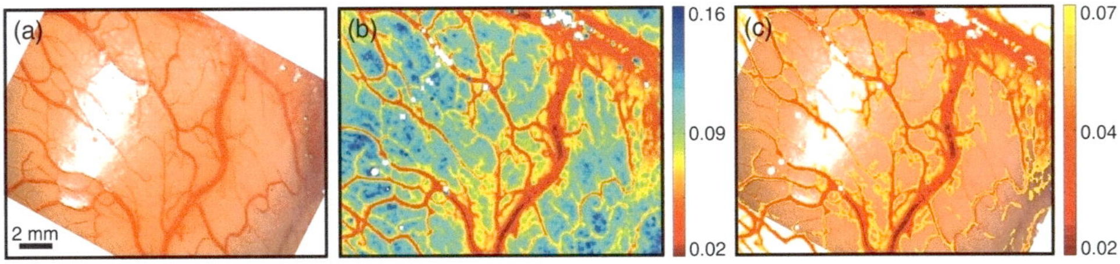

Fig. 7 (A) Standard white light image of brain tissue under operating microscope. (B) LSCI image and (C) overlay of (A) and (B) illustrating high (*red*) and low (*blue*) blood flows intraoperatively. *Reproduced under a Creative Commons Attribution 3.0 Unported License, https://doi.org/10.1117/1.NPh.1.1.015006 © The authors (L.M. Richards, E.L. Towle, D.J. Fox, A.K. Dunn, Intraoperative laser speckle contrast imaging with retrospective motion correction for quantitative assessment of cerebral blood flow, Neurophotonics 1 (1) (2014) 1–11. https://doi.org/10.1117/1.NPh.1.1.015006.).*

moving structures can be recorded at a single point with a fast detector, called laser Doppler flowmetry. If this point is scanned across the tissue surface then the technique is referred to as laser Doppler imaging.

8 Macroscopic scanning of optical systems and visualization

This section will consider the bridge between what surgeons see by naked eye or standard endoscopy, and the information that may be acquired using optical techniques. There are four scenarios to consider: (1) For many implementations of MSI or fluorescence imaging, the imaged field is identical to the white light image and may be directly registered or used to augment it; (2) For some OCT, confocal and photoacoustic devices, mechanical scanning, or reconstruction is inherent to the design of the instrument, and produces standalone visualizations that the operator must learn to interpret and use as a complement to vision and endoscopy; (3) In microscopic methods and some OCT probes, the imaged field must be tracked and understood with respect to the position on the bulk tissue surface, potentially including image mosaicking if the device is scanned across the surface; (4) Where single-point (spectroscopic) sensing is performed, tracking and scanning across the tissue surface is essential to produce image data. For the first two cases, there are a range of image tracking and registration tools that may be applied that are covered elsewhere in the literature, including the use of augmented reality headgear and immersive displays. This section will mainly consider cases 3 and 4, including how they may be practically implemented into interventional devices.

For single-point spectroscopy, where the data contain no image information, the scanning may be performed manually providing the position of the probe can be tracked in real time. As described elsewhere, this could be achieved by visual instrument tracking, magnetic tracking, or by using optical markers [1]. Visual tracking is suited to the application of probe-based techniques when using the instrument channel of a flexible endoscope, since the onboard camera can be used to locate the instrument position or even reconstruct the tissue surface shape using SLAM [107,108]. Where multiple examinations are performed at different times, the visual tracking can also help with retargeting the probe location for serial examination [109]. A similar concept can also be applied to bench-top devices, to allow a clinician to scan a spectroscopic probe across the tissue surface and have this diagnostic information displayed on a white light video that is recorded from a stationary camera. A similar concept has been used for ex vivo breast cancer specimen scanning using fluorescence lifetime spectroscopy ([110]). Besides the potential benefit of such approaches for surgical decision making, they can also act as an important link between the excised specimen and the histology, which can then allow optical spectroscopic methods to be validated more easily and accurately (see Fig. 8) [111]. Where the optical probe is acquiring microscopic images, it is possible to use the stack of image fields acquired during probe motion to create a mosaic (Fig. 8). In this case, any independently

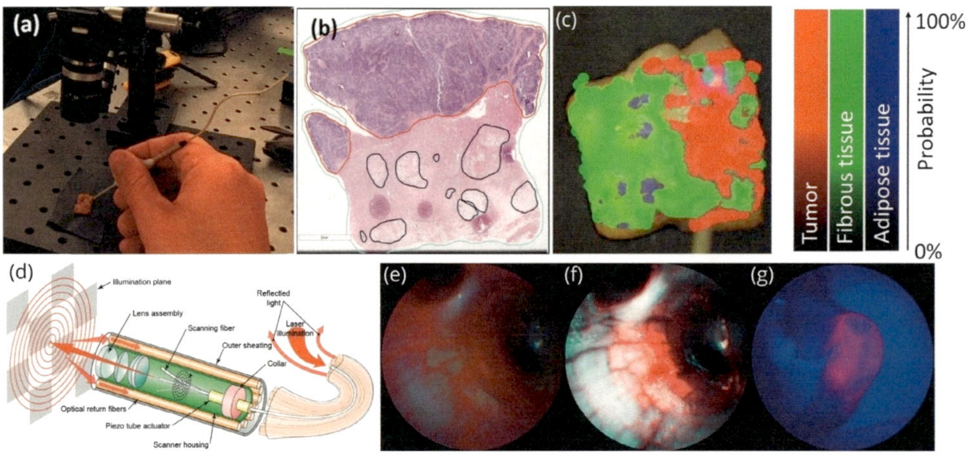

Fig. 8 Illustration of point-scanning approaches. (A–C) Acquisition and display of single-point time- and spectrally resolved fluorescence data after a probe is scanned over a tissue surface (A) Image data record the position of the probe to create a registered overlay of the classified data (C). This can then be compared with histology with a classification made by the histologist (B). (D) Schematic of a distal fiber-scanning method using a piezoelectric actuator and cantilevered fiber to trace a spiral scan pattern. (E–G) Three different imaging modes of bronchoscopic images ((E) white light, (F) a form of narrow-band imaging, and (G) fluorescence) acquired at 30 Hz. Blood vessels and inflamed tissue are better visualized in (F), and the red fluorescence in (G) shows hypericin localization within a renal cell carcinoma. *(A–C: Adapted with permission from J. Unger, C. Hebisch, J. Phipps, M. Darrow, R. Bold, L. Marcu, Real-time visualization of tumor margins in breast specimen using fluorescence lifetime imaging. Clinical and Translational Biophotonics, Optical Society of America, 2018a, CTu4B. 4, The Optical Society; E–G: From C.M. Lee, C.J. Engelbrecht, T.D. Soper, F. Helmchen, E.J. Seibel, Scanning fiber endoscopy with highly flexible, 1 mm catheterscopes for wide-field, full-color imaging, J. Biophoton. 3 (2010b) 385–407, Creative Commons.)*

acquired positional data can be used to support the mosaic process or to help with the path closure problem [112].

While manual scanning has the advantage of giving the clinician the ability to self-select the probed points on the tissue surface, mechanized or robotic scanning can play a role to either assist the targeting where manual probe control is cumbersome or to permit systematic scanning over a larger tissue surface area. For example, scanning a curved tissue surface while maintaining a specified probe-surface contact pressure can be difficult during endoscopy, and this may be supported by closed-loop force control [113,114]. More autonomous mechanisms can be used to comprehensively scan over extended tissues, with potential for surveying whole residual tumor beds [115] or for screening tubular organs such as the colon [116].

Scanning methods can also help to reduce the size of the imaging optics for ultraminimal access surgery. While there has been a drive toward smaller tip-mounted camera systems, their dimensions are ultimately limited by the desired resolution and a minimum pixel size. For very narrow endoscopes, it is still common to find coherent fiber optic bundles in use, although the resolution of these is also limited to the number of fibers in the bundle. An alternative is to use a single optical fiber and miniaturize a distal-mounted scanning mechanism for rapid pointwise reflectance data, and it is possible to use similar hardware to OCT. For instance, piezoelectric activation of the tip of an optical fiber can scan patterns on the tissue surface, using three lasers for white light imaging and with the possibility of acquiring other modalities such as confocal or fluorescence [117]. Spectral encoding of the image field has also been proposed for single fiber imaging [118] and it is possible to reconstruct images by imaging through multimode fiber although the calibration required has so far prevented the technique from being used in realistic scenarios [119].

9 Summary

A selection of optical imaging methods have been described in this chapter, many of which are applied during cancer diagnosis and intervention. The direct white light optical imaging of tissue is the primary interventional guidance method and relies on the scattering and absorption properties of the tissue to provide visible contrast to the clinician. It is the scattering that limits the imaging depth of most of the techniques that were presented, with the majority able to sense within the top 100 μm superficial layer. In this superficial regime, high resolution microscopic images can be acquired and contrast can exploit absorption from hemoglobin, or fluorescence and Raman processes to create molecular-sensitive contrasts. Optical coherence tomography pushes the imaging depth limit up to a depth of a few mm, where the number of directly backscattered photons reduces beneath detectable levels. Beyond this, diffuse optical tomography and photoacoustic tomography are able to image absorption or scattering contrast for ~5 cm tissue depths.

It remains a challenge to validate optical methods, which usually have to be compared against histology as the gold standard, but this involves a preparation method that can be difficult and time-consuming to orient and interpret with respect to the optical data. It is even harder to validate against clinical outcomes during larger trials due to the number of patients required and a lack of standardization between different studies. Other promising technologies such as near infrared fluorescence require new contrast agents to reach their full potential, although the high investment to prove safety and efficacy combined with the small established market is not driving this process. Overall, and despite these limitations, there is great promise for optical imaging and biophotonics methods to be used for interventional diagnosis and guidance, thanks to the potential for nonharmful, real-time, and highly detailed information to be acquired.

References

[1] G. Fichtinger, S. Kevin Zhou, D. Rueckert, Handbook of Medical Image Computing and Computer Assisted Intervention, Elsevier/Academic Press, 2019.

[2] D.A. Boas, D.H. Brooks, E.L. Miller, C.A. Dimarzio, M. Kilmer, R.J. Gaudette, Q. Zhang, Imaging the body with diffuse optical tomography, IEEE Signal Process. Mag. 18 (2001) 57–75.

[3] D. Grosenick, K.T. Moesta, H. Wabnitz, J. Mucke, C. Stroszczynski, R. Macdonald, P.M. Schlag, H. Rinneberg, Time-domain optical mammography: initial clinical results on detection and characterization of breast tumors, Appl. Opt. 42 (2003) 3170–3186.

[4] M. Okamoto, H. Dan, K. Shimizu, K. Takeo, T. Amita, I. Oda, I. Konishi, K. Sakamoto, S. Isobe, T. Suzuki, Multimodal assessment of cortical activation during apple peeling by NIRS and fMRI, NeuroImage 21 (2004) 1275–1288.

[5] J.R. Lakowicz, Principles of Fluorescence Spectroscopy, second ed., Kluwer Academic/Plenum Publishers, New York, 1999.

[6] E. Te Velde, T. Veerman, V. Subramaniam, T. Ruers, The use of fluorescent dyes and probes in surgical oncology, Eur. J. Surg. Oncol. 36 (2010) 6–15.

[7] P. Urayama, M.A. Mycek, Fluorescence lifetime imaging microscopy of endogenous biological fluorescence, in: M.A. Mycek, B.W. Pogue (Eds.), Handbook of Biomedical Fluorescence, Marcel Dekker, New York, 2003.

[8] F.N. Ghadially, Red fluorescence of experimentally induced and human tumours, J. Pathol. Bacteriol. 80 (1960) 345–351.

[9] S.A. Toms, W.-C. Lin, R.J. Weil, M.D. Johnson, E.D. Jansen, A. Mahadevan-Jansen, Intraoperative optical spectroscopy identifies infiltrating glioma margins with high sensitivity, Oper. Neurosurg. 57 (2005) ONS-382–ONS-391.

[10] S. Lam, C. Macaulay, J.C. Leriche, B. Palcic, Detection and localization of early lung cancer by fluorescence bronchoscopy, Cancer 89 (2000) 2468–2473.

[11] D.F. Boerwinkel, J.A. Holz, M.A. Kara, S.L. Meijer, M.B. Wallace, L.M.W.K. Song, K. Ragunath, H.C. Wolfsen, P.G. Iyer, K.K. Wang, Effects of autofluorescence imaging on detection and treatment of early neoplasia in patients with Barrett's esophagus, Clin. Gastroenterol. Hepatol. 12 (2014) 774–781.

[12] J. Haringsma, G.N. Tytgat, H. Yano, H. Iishi, M. Tatsuta, T. Ogihara, H. Watanabe, N. Sato, N. Marcon, B.C. Wilson, Autofluorescence endoscopy: feasibility of detection of GI neoplasms unapparent to white light endoscopy with an evolving technology, Gastrointest. Endosc. 53 (2001) 642–650.

[13] L. Marcu, P.M. French, D.S. Elson, Fluorescence Lifetime Spectroscopy and Imaging: Principles and Applications in Biomedical Diagnostics, CRC Press, 2014.

[14] J.M. Benavides, S. Chang, S.Y. Park, R. Richards-Kortum, N. Mackinnon, C. Macaulay, A. Milbourne, A. Malpica, M. Follen, Multispectral digital colposcopy for in vivo detection of cervical cancer, Opt. Express 11 (2003) 1223–1236.

[15] T.E. Renkoski, U. Utzinger, K.D. Hatch, Wide-field spectral imaging of human ovary autofluorescence and oncologic diagnosis via previously collected probe data, J. Biomed. Opt. 17 (2012).
[16] K. Mitra, J. Melvin, S. Chang, R. Xu, K. Park, A. Yilmaz, S. Melvin, Indocyanine-green-loaded microballoons for biliary imaging in cholecystectomy, J. Biomed. Opt. 17 (2012) 116025.
[17] B.E. Schaafsma, J.S.D. Mieog, M. Hutteman, J.R. Van Der Vorst, P.J. Kuppen, C.W. Löwik, J.V. Frangioni, C.J. Van De Velde, A.L. Vahrmeijer, The clinical use of indocyanine green as a near-infrared fluorescent contrast agent for image-guided oncologic surgery, J. Surg. Oncol. 104 (2011) 323–332.
[18] S. Mordon, J.M. Devoisselle, S. Soulie-Begu, T. Desmettre, Indocyanine green: physicochemical factors affecting its fluorescencein vivo, Microvasc. Res. 55 (1998) 146–152.
[19] J. Watanabe, M. Ota, Y. Suwa, S. Suzuki, H. Suwa, M. Momiyama, A. Ishibe, K. Watanabe, H. Masui, K. Nagahori, Evaluation of the intestinal blood flow near the rectosigmoid junction using the indocyanine green fluorescence method in a colorectal cancer surgery, Int. J. Color. Dis. 30 (2015) 329–335.
[20] B.T. Lee, M. Hutteman, S. Gioux, A. Stockdale, S.J. Lin, L.H. Ngo, J.V. Frangioni, The FLARE™ intraoperative near-infrared fluorescence imaging system: a first-in-human clinical trial in perforator flap breast reconstruction, Plast. Reconstr. Surg. 126 (2010) 1472.
[21] C. Hirche, D. Murawa, Z. Mohr, S. Kneif, M. Hünerbein, ICG fluorescence-guided sentinel node biopsy for axillary nodal staging in breast cancer, Breast Cancer Res. Treat. 121 (2010) 373–378.
[22] A.V. Dsouza, H. Lin, E.R. Henderson, K.S. Samkoe, B.W. Pogue, Review of fluorescence guided surgery systems: identification of key performance capabilities beyond indocyanine green imaging, J. Biomed. Opt. 21 (2016).
[23] J.E. Bugaj, S.I. Achilefu, R.B. Dorshow, R. Rajagopalan, Novel fluorescent contrast agents for optical imaging of in vivo tumors based on a receptor-targeted dye-peptide conjugate platform, J. Biomed. Opt. 6 (2001) 122–134.
[24] H.S. Choi, S.L. Gibbs, J.H. Lee, S.H. Kim, Y. Ashitate, F. Liu, H. Hyun, G. Park, Y. Xie, S. Bae, Targeted zwitterionic near-infrared fluorophores for improved optical imaging, Nat. Biotechnol. 31 (2013) 148.
[25] L.E. Lamberts, M. Koch, J.S. De Jong, A.L. Adams, J. Glatz, M.E. Kranendonk, A.G.T. Van Scheltinga, L. Jansen, J. De Vries, M.N. Lub-De Hooge, Tumor-specific uptake of fluorescent bevacizumab–IRDye800CW microdosing in patients with primary breast cancer: a phase i feasibility study, Clin. Cancer Res. 23 (2017) 2730–2741.
[26] G.M. Van Dam, G. Themelis, L.M. Crane, N.J. Harlaar, R.G. Pleijhuis, W. Kelder, A. Sarantopoulos, J.S. De Jong, H.J. Arts, A.G. Van Der Zee, Intraoperative tumor-specific fluorescence imaging in ovarian cancer by folate receptor-α targeting: first in-human results, Nat. Med. 17 (2011) 1315.
[27] S.P. Lerner, A. Goh, Novel endoscopic diagnosis for bladder cancer, Cancer 121 (2015) 169–178.
[28] J. Schmidbauer, F. Witjes, N. Schmeller, R. Donat, M. Susani, M. Marberger, Hexvix PCB301/01 Study Group, Improved detection of urothelial carcinoma in situ with hexaminolevulinate fluorescence cystoscopy, J. Urol. 171 (2004) 135–138.
[29] S. Zhao, J. Wu, C. Wang, H. Liu, X. Dong, C. Shi, C. Shi, Y. Liu, L. Teng, D. Han, Intraoperative fluorescence-guided resection of high-grade malignant gliomas using 5-aminolevulinic acid–induced porphyrins: a systematic review and meta-analysis of prospective studies, PLoS ONE 8 (2013).
[30] G. Hong, S. Diao, J. Chang, A.L. Antaris, C. Chen, B. Zhang, S. Zhao, D.N. Atochin, P.L. Huang, K.I. Andreasson, Through-skull fluorescence imaging of the brain in a new near-infrared window, Nat. Photonics 8 (2014) 723.
[31] G. Hong, J.T. Robinson, Y. Zhang, S. Diao, A.L. Antaris, Q. Wang, H. Dai, In vivo fluorescence imaging with Ag2S quantum dots in the second near-infrared region, Angew. Chem. Int. Ed. 51 (2012) 9818–9821.
[32] K. Gono, T. Obi, M. Yamaguchi, N. Oyama, H. Machida, Y. Sano, S. Yoshida, Y. Hamamoto, T. Endo, Appearance of enhanced tissue features in narrow-band endoscopic imaging, J. Biomed. Opt. 9 (2004) 568–578.

[33] H. Machida, Y. Sano, Y. Hamamoto, M. Muto, T. Kozu, H. Tajiri, S. Yoshida, Narrow-band imaging in the diagnosis of colorectal mucosal lesions: a pilot study, Endoscopy 36 (2004) 1094–1098.

[34] P. Sharma, J.J. Bergman, K. Goda, M. Kato, H. Messmann, B.R. Alsop, N. Gupta, P. Vennalaganti, M. Hall, V. Konda, Development and validation of a classification system to identify high-grade dysplasia and esophageal adenocarcinoma in Barrett's esophagus using narrow-band imaging, Gastroenterology 150 (2016) 591–598.

[35] Y. Sano, S. Tanaka, S.E. Kudo, S. Saito, T. Matsuda, Y. Wada, T. Fujii, H. Ikematsu, T. Uraoka, N. Kobayashi, Narrow-band imaging (NBI) magnifying endoscopic classification of colorectal tumors proposed by the Japan NBI Expert Team, Dig. Endosc. 28 (2016) 526–533.

[36] U. Utzinger, R.R. Richards-Kortum, Fiber optic probes for biomedical optical spectroscopy, J. Biomed. Opt. 8 (2003) 121–148.

[37] A.J. Radosevich, N.N. Mutyal, J. Yi, Y. Stypula-Cyrus, J.D. Rogers, M.J. Goldberg, L. Bianchi, S. Bajaj, H.K. Roy, Ultrastructural alterations in field carcinogenesis measured by enhanced backscattering spectroscopy, J. Biomed. Opt. 18 (2013).

[38] M.A. Afromowitz, J.B. Callis, D.M. Heimbach, L.A. Desoto, M.K. Norton, Multispectral imaging of burn wounds: a new clinical instrument for evaluating burn depth, IEEE Trans. Biomed. Eng. 35 (1988) 842–850.

[39] N.T. Clancy, S. Arya, D. Stoyanov, M. Singh, G.B. Hanna, D.S. Elson, Intraoperative measurement of bowel oxygen saturation using a multispectral imaging laparoscope, Biomed. Opt. Express 6 (2015) 4179–4190.

[40] N.T. Clancy, D. Stoyanov, D.R.C. James, A.D. Marco, V. Sauvage, J. Clark, G.-Z. Yang, D.S. Elson, Multispectral image alignment using a three channel endoscope in vivo during minimally invasive surgery, Biomed. Opt. Express 3 (2012) 2567–2578.

[41] H. Akbari, Y. Kosugi, K. Kojima, N. Tanaka, Wavelet-based compression and segmentation of hyperspectral images in surgery, in: International Workshop on Medical Imaging and Virtual Reality, Springer, 2008, pp. 142–149.

[42] A.S. Luthman, S. Dumitru, I. Quiros-Gonzalez, J. Joseph, S.E. Bohndiek, Fluorescence hyperspectral imaging (fHSI) using a spectrally resolved detector array, J. Biophotonics 10 (2017) 840–853.

[43] L. Gao, R.T. Kester, N. Hagen, T.S. Tkaczyk, Snapshot image mapping spectrometer (IMS) with high sampling density for hyperspectral microscopy, Opt. Express 18 (2010) 14330–14344.

[44] J. Lin, N.T. Clancy, J. Qi, Y. Hu, T. Tatla, D. Stoyanov, L.M. Hein, D.S. Elson, Dual-modality endoscopic probe for tissue surface shape reconstruction and hyperspectral imaging enabled by deep neural networks, Med. Image Anal. 48 (2018) 162–176.

[45] G. Lu, B. Fei, Medical hyperspectral imaging: a review, J. Biomed. Opt. 19 (2014).

[46] H. Akbari, K. Uto, Y. Kosugi, K. Kojima, N. Tanaka, Cancer detection using infrared hyperspectral imaging, Cancer Sci. 102 (2011) 852–857.

[47] L.L. Randeberg, I. Baarstad, T. Løke, P. Kaspersen, L.O. Svaasand, Hyperspectral imaging of bruised skin, in: Photonic Therapeutics and Diagnostics II, International Society for Optics and Photonics, 2006, p. 60780O.

[48] H. Kittler, H. Pehamberger, K. Wolff, M. Binder, Diagnostic accuracy of dermoscopy, Lancet Oncol. 3 (2002) 159–165.

[49] R. Kiesslich, J. Fritsch, M. Holtmann, H.H. Koehler, M. Stolte, S. Kanzler, B. Nafe, M. Jung, P.R. Galle, M.F. Neurath, Methylene blue-aided chromoendoscopy for the detection of intraepithelial neoplasia and colon cancer in ulcerative colitis, Gastroenterology 124 (2003) 880–888.

[50] M.B. Sturm, T.D. Wang, Emerging optical methods for surveillance of Barrett's oesophagus, Gut 64 (2015) 1816–1823.

[51] K. Yao, A. Iwashita, T. Yao, Early gastric cancer: proposal for a new diagnostic system based on microvascular architecture as visualized by magnified endoscopy, Dig. Endosc. 16 (2004) S110–S117.

[52] M. Yamada, T. Sakamoto, Y. Otake, T. Nakajima, A. Kuchiba, H. Taniguchi, S. Sekine, R. Kushima, H. Ramberan, A. Parra-Blanco, Investigating endoscopic features of sessile serrated adenomas/polyps by using narrow-band imaging with optical magnification, Gastrointest. Endosc. 82 (2015) 108–117.

[53] T.J. Muldoon, S. Anandasabapathy, D. Maru, R. Richards-Kortum, High-resolution imaging in Barrett's esophagus: a novel, low-cost endoscopic microscope, Gastrointest. Endosc. 68 (2008) 737–744.

[54] N.D. Parikh, D. Perl, M.H. Lee, B. Shah, Y. Young, S.S. Chang, R. Shukla, A.D. Polydorides, E. Moshier, J. Godbold, In vivo diagnostic accuracy of high-resolution microendoscopy in differentiating neoplastic from non-neoplastic colorectal polyps: a prospective study, Am. J. Gastroenterol. 109 (2014) 68.
[55] T.J. Muldoon, D. Roblyer, M.D. Williams, V.M. Stepanek, R. Richards-Kortum, A. M. Gillenwater, Noninvasive imaging of oral neoplasia with a high-resolution fiber-optic microendoscope, Head Neck 34 (2012) 305–312.
[56] C. Sheppard, D. Shotton, C. Sheppard, Confocal Laser Scanning Microscopy, BIOS Scientific Publishers, Oxford, 1997.
[57] P. Calzavara-Pinton, C. Longo, M. Venturini, R. Sala, G. Pellacani, Reflectance confocal microscopy for in vivo skin imaging, Photochem. Photobiol. 84 (2008) 1421–1430.
[58] M. Rajadhyaksha, A. Marghoob, A. Rossi, A.C. Halpern, K.S. Nehal, Reflectance confocal microscopy of skin in vivo: from bench to bedside, Lasers Surg. Med. 49 (2017) 7–19.
[59] G. Le Goualher, A. Perchant, M. Genet, C. Cavé, B. Viellerobe, F. Berier, B. Abrat, N. Ayache, Towards optical biopsies with an integrated fibered confocal fluorescence microscope, in: International Conference on Medical Image Computing and Computer-Assisted Intervention, Springer, 2004, pp. 761–768.
[60] A. Meining, M.B. Wallace, Endoscopic imaging of angiogenesis in vivo, Gastroenterology 134 (2008) 915–918.
[61] H. Pohl, T. Roesch, M. Vieth, M. Koch, V. Becker, M. Anders, A.C. Khalifa, A. Meining, Miniprobe confocal laser microscopy for the detection of invisible neoplasia in patients with Barrett's oesophagus, Gut 57 (2008) 1648–1653.
[62] A.L. Polglase, W.J. McLaren, S.A. Skinner, R. Kiesslich, M.F. Neurath, P.M. Delaney, A fluorescence confocal endomicroscope for in vivo microscopy of the upper-and the lower-GI tract, Gastrointest. Endosc. 62 (2005) 686–695.
[63] R.C. Newton, S.V. Kemp, G.-Z. Yang, D.S. Elson, A. Darzi, P.L. Shah, Imaging parenchymal lung diseases with confocal endomicroscopy, Respir. Med. 106 (2012) 127–137.
[64] T. Vercauteren, A. Perchant, G. Malandain, X. Pennec, N. Ayache, Robust mosaicing with correction of motion distortions and tissue deformations for in vivo fibered microscopy, Med. Image Anal. 10 (2006) 673–692.
[65] W.R. Zipfel, R.M. Williams, W.W. Webb, Nonlinear magic: multiphoton microscopy in the biosciences, Nat. Biotechnol. 21 (2003) 1369.
[66] W. Denk, J.H. Strickler, W.W. Webb, Two-photon laser scanning fluorescence microscopy, Science 248 (1990) 73–76.
[67] S. Seidenari, F. Arginelli, S. Bassoli, J. Cautela, P.M. French, M. Guanti, D. Guardoli, K. König, C. Talbot, C. Dunsby, Multiphoton laser microscopy and fluorescence lifetime imaging for the evaluation of the skin, Dermatol. Res. Pract. 2012 (2012).
[68] L. Fu, M. Gu, Fibre-optic nonlinear optical microscopy and endoscopy, J. Microsc. 226 (2007) 195–206.
[69] L. Fu, A. Jain, H. Xie, C. Cranfield, M. Gu, Nonlinear optical endoscopy based on a double-clad photonic crystal fiber and a MEMS mirror, Opt. Express 14 (2006) 1027–1032.
[70] W. Liang, G. Hall, B. Messerschmidt, M.-J. Li, X. Li, Nonlinear optical endomicroscopy for label-free functional histology in vivo, Light Sci. Appl. 6 (2017).
[71] D. Huang, E.A. Swanson, C.P. Lin, J.S. Schuman, W.G. Stinson, W. Chang, M.R. Hee, T. Flotte, K. Gregory, C.A. Puliafito, Optical coherence tomography, Science 254 (1991) 1178–1181.
[72] J.G. Fujimoto, Optical coherence tomography for ultrahigh resolution in vivo imaging, Nat. Biotechnol. 21 (2003) 1361.
[73] R. Leitgeb, C. Hitzenberger, A.F. Fercher, Performance of fourier domain vs. time domain optical coherence tomography, Opt. Express 11 (2003) 889–894.
[74] S. Chinn, E. Swanson, J. Fujimoto, Optical coherence tomography using a frequency-tunable optical source, Opt. Lett. 22 (1997) 340–342.
[75] M. Balicki, J.-H. Han, I. Iordachita, P. Gehlbach, J. Handa, R. Taylor, J. Kang, Single fiber optical coherence tomography microsurgical instruments for computer and robot-assisted retinal surgery, in: International Conference on Medical Image Computing and Computer-Assisted Intervention, Springer, 2009, pp. 108–115.

[76] H. Yu, J.-H. Shen, R.J. Shah, N. Simaan, K.M. Joos, Evaluation of microsurgical tasks with OCT-guided and/or robot-assisted ophthalmic forceps, Biomed. Opt. Express 6 (2015) 457–472.

[77] H.G. Bezerra, M.A. Costa, G. Guagliumi, A.M. Rollins, D.I. Simon, Intracoronary optical coherence tomography: a comprehensive review: clinical and research applications, J. Am. Coll. Cardiol. Intv. 2 (2009) 1035–1046.

[78] I.-K. Jang, B.E. Bouma, D.-H. Kang, S.-J. Park, S.-W. Park, K.-B. Seung, K.-B. Choi, M. Shishkov, K. Schlendorf, E. Pomerantsev, Visualization of coronary atherosclerotic plaques in patients using optical coherence tomography: comparison with intravascular ultrasound, J. Am. Coll. Cardiol. 39 (2002) 604–609.

[79] G.J. Tearney, E. Regar, T. Akasaka, T. Adriaenssens, P. Barlis, H.G. Bezerra, B. Bouma, N. Bruining, J.-M. Cho, S. Chowdhary, Consensus standards for acquisition, measurement, and reporting of intravascular optical coherence tomography studies: a report from the International Working Group for Intravascular Optical Coherence Tomography Standardization and Validation, J. Am. Coll. Cardiol. 59 (2012) 1058–1072.

[80] T. Wang, T. Pfeiffer, E. Regar, W. Wieser, H. Van Beusekom, C.T. Lancee, G. Springeling, I. Krabbendam, A.F. Van Der Steen, R. Huber, Heartbeat OCT: in vivo intravascular megahertz-optical coherence tomography, Biomed. Opt. Express 6 (2015) 5021–5032.

[81] G.J. Ughi, H. Wang, E. Gerbaud, J.A. Gardecki, A.M. Fard, E. Hamidi, P. Vacas-Jacques, M. Rosenberg, F.A. Jaffer, G.J. Tearney, Clinical characterization of coronary atherosclerosis with dual-modality OCT and near-infrared autofluorescence imaging, JACC Cardiovasc. Imaging 9 (2016) 1304–1314.

[82] B.E. Bouma, M. Villiger, K. Otsuka, W.-Y. Oh, Intravascular optical coherence tomography, Biomed. Opt. Express 8 (2017) 2660–2686.

[83] L. Van Manen, J. Dijkstra, C. Boccara, E. Benoit, A.L. Vahrmeijer, M.J. Gora, J.S.D. Mieog, The clinical usefulness of optical coherence tomography during cancer interventions, J. Cancer Res. Clin. Oncol. 144 (2018) 1–24.

[84] X. Qi, Y. Pan, M.V. Sivak, J.E. Willis, G. Isenberg, A.M. Rollins, Image analysis for classification of dysplasia in Barrett's esophagus using endoscopic optical coherence tomography, Biomed. Opt. Express 1 (2010) 825–847.

[85] P.B. Garcia-Allende, I. Amygdalos, H. Dhanapala, R.D. Goldin, G.B. Hanna, D.S. Elson, Morphological analysis of optical coherence tomography images for automated classification of gastrointestinal tissues, Biomed. Opt. Express 2 (2011) 2821–2836.

[86] K.W. Gossage, T.S. Tkaczyk, J.J. Rodriguez, J.K. Barton, Texture analysis of optical coherence tomography images: feasibility for tissue classification, J. Biomed. Opt. 8 (2003) 570–576.

[87] T.S. Kirtane, M.S. Wagh, Endoscopic optical coherence tomography (OCT): advances in gastrointestinal imaging, Gastroenterol. Res. Pract. 2014 (2014).

[88] M. Aron, J.H. Kaouk, N.J. Hegarty, J. Colombo, J. Roberto, G.-P. Haber, B.I. Chung, M. Zhou, I.S. Gill, Second prize: preliminary experience with the Niris™ optical coherence tomography system during laparoscopic and robotic prostatectomy, J. Endourol. 21 (2007) 814–818.

[89] M.J. Manyak, N.D. Gladkova, J.H. Makari, A.M. Schwartz, E.V. Zagaynova, L. Zolfaghari, J.M. Zara, R. Iksanov, F.I. Feldchtein, Evaluation of superficial bladder transitional-cell carcinoma by optical coherence tomography, J. Endourol. 19 (2005) 570–574.

[90] B.G. Muller, D.M. De Bruin, M.J. Brandt, W. Van Den Bos, S. Van Huystee, D. Faber, D. Savci, P.J. Zondervan, T.M. De Reijke, M.P. Laguna-Pes, Prostate cancer diagnosis by optical coherence tomography: first results from a needle based optical platform for tissue sampling, J. Biophotonics 9 (2016) 490–498.

[91] M. Villiger, D. Lorenser, R.A. McLaughlin, B.C. Quirk, R.W. Kirk, B.E. Bouma, D.D. Sampson, Deep tissue volume imaging of birefringence through fibre-optic needle probes for the delineation of breast tumour, Sci. Rep. 6 (2016) 28771.

[92] J. Li, B.C. Quirk, P.B. Noble, R.W. Kirk, D.D. Sampson, R.A. McLaughlin, Flexible needle with integrated optical coherence tomography probe for imaging during transbronchial tissue aspiration, J. Biomed. Opt. 22 (2017) 106002.

[93] M.-C. Kao, Y.-T. Wu, M.-Y. Tsou, W.-C. Kuo, C.-K. Ting, Intelligent epidural needle placement using fiber-probe optical coherence tomography in a piglet model, Biomed. Opt. Express 9 (2018) 3711–3724.

[94] A. Corlu, R. Choe, T. Durduran, M.A. Rosen, M. Schweiger, S.R. Arridge, M.D. Schnall, A.G. Yodh, Three-dimensional in vivo fluorescence diffuse optical tomography of breast cancer in humans, Opt. Express 15 (2007) 6696–6716.

[95] L.V. Wang, L. Gao, Photoacoustic microscopy and computed tomography: from bench to bedside, Annu. Rev. Biomed. Eng. 16 (2014) 155–185.

[96] C.J.H. Ho, G. Balasundaram, W. Driessen, R. McLaren, C.L. Wong, U. Dinish, A.B.E. Attia, V. Ntziachristos, M. Olivo, Multifunctional photosensitizer-based contrast agents for photoacoustic imaging, Sci. Rep. 4 (2014) 5342.

[97] D. Razansky, C. Vinegoni, V. Ntziachristos, Multispectral photoacoustic imaging of fluorochromes in small animals, Opt. Lett. 32 (2007) 2891–2893.

[98] Y. Wang, X. Xie, X. Wang, G. Ku, K.L. Gill, D.P. O'Neal, G. Stoica, L.V. Wang, Photoacoustic tomography of a nanoshell contrast agent in the in vivo rat brain, Nano Lett. 4 (2004) 1689–1692.

[99] D. Piras, C. Grijsen, P. Schutte, W. Steenbergen, S. Manohar, Photoacoustic needle: minimally invasive guidance to biopsy, J. Biomed. Opt. 18 (2013).

[100] C. Kim, T.N. Erpelding, K.I. Maslov, L. Jankovic, W.J. Akers, L. Song, S. Achilefu, J.A. Margenthaler, M.D. Pashley, L.V. Wang, Handheld array-based photoacoustic probe for guiding needle biopsy of sentinel lymph nodes, J. Biomed. Opt. 15 (2010).

[101] W. Xia, M. Kuniyil Ajith Singh, E. Maneas, N. Sato, Y. Shigeta, T. Agano, S. Ourselin, J.S. West, E.A. Desjardins, Handheld real-time LED-based photoacoustic and ultrasound imaging system for accurate visualization of clinical metal needles and superficial vasculature to guide minimally invasive procedures, Sensors 18 (2018) 1394.

[102] W. Xia, E. Maneas, D.I. Nikitichev, C.A. Mosse, G.S. Dos Santos, T. Vercauteren, A.L. David, J. Deprest, S. Ourselin, P.C. Beard, Interventional photoacoustic imaging of the human placenta with ultrasonic tracking for minimally invasive fetal surgeries, in: International Conference on Medical Image Computing and Computer-Assisted Intervention, Springer, 2015, pp. 371–378.

[103] M.A.L. Bell, N.P. Kuo, D.Y. Song, J.U. Kang, E.M. Boctor, In vivo visualization of prostate brachytherapy seeds with photoacoustic imaging, J. Biomed. Opt. 19 (2014) 126011.

[104] J.M. Mari, W. Xia, S.J. West, A.E. Desjardins, Interventional multispectral photoacoustic imaging with a clinical ultrasound probe for discriminating nerves and tendons: an *ex vivo* pilot study, J. Biomed. Opt. 20 (2015) 110503.

[105] N. Gandhi, M. Allard, S. Kim, P. Kazanzides, M.A.L. Bell, Photoacoustic-based approach to surgical guidance performed with and without a da Vinci robot, J. Biomed. Opt. 22 (2017) 121606.

[106] A. Cheng, Y. Itsarachaiyot, Y. Kim, H.K. Zhang, R.H. Taylor, E.M. Boctor, Catheter tracking in an interventional photoacoustic surgical system, in: Medical Imaging 2017: Image-Guided Procedures, Robotic Interventions, and Modeling, International Society for Optics and Photonics, 2017, p. 1013527.

[106a] D.A. Boas, A.K. Dunn, Laser speckle contrast imaging in biomedical optics, J. Biomed. Opt. 15 (2010) 011109.

[107] M. Giannarou, D.S. Elson, G.Z. Yang, Tracking of spectroscopic and microscopic optical probes in endoscopy using the endoscope image field, in: Optical Tissue Image Analysis in Microscopy, Histopathology and Endoscopy (OPTIMHisE), a Satellite Workshop at MICCAI, MICCAI, London, 2009.

[108] P. Mountney, S. Giannarou, D. Elson, G.Z. Yang, Optical biopsy mapping for minimally invasive cancer screening, in: G.Z. Yang, D. Hawkes, D. Rueckert, A. Nobel, C. Taylor (Eds.), Medical Image Computing and Computer-Assisted Intervention, Springer, London, 2009, pp. 483–490.

[109] M. Ye, S. Giannarou, A. Meining, G.-Z. Yang, Online tracking and retargeting with applications to optical biopsy in gastrointestinal endoscopic examinations, Med. Image Anal. 30 (2016) 144–157.

[110] J. Unger, C. Hebisch, J. Phipps, M. Darrow, R. Bold, L. Marcu, Real-time visualization of tumor margins in breast specimen using fluorescence lifetime imaging, in: Clinical and Translational Biophotonics, Optical Society of America, 2018. CTu4B. 4.

[111] J. Unger, T. Sun, Y.-L. Chen, J.E. Phipps, R.J. Bold, M.A. Darrow, K.-L. Ma, L. Marcu, Method for accurate registration of tissue autofluorescence imaging data with corresponding histology: a means for enhanced tumor margin assessment, J. Biomed. Opt. 23 (2018).

[112] B. Rosa, M.S. Erden, T. Vercauteren, B. Herman, J. Szewczyk, G. Morel, Building large mosaics of confocal endomicroscopic images using visual servoing, IEEE Trans. Biomed. Eng. 60 (2012) 1041–1049.

[113] W.T. Latt, R.C. Newton, M. Visentini-Scarzanella, C.J. Payne, D.P. Noonan, J. Shang, G.-Z. Yang, A hand-held instrument to maintain steady tissue contact during probe-based confocal laser endomicroscopy, IEEE Trans. Biomed. Eng. 58 (2011) 2694–2703.

[114] D.P. Noonan, C.J. Payne, J. Shang, V. Sauvage, R. Newton, D. Elson, A. Darzi, G.-Z. Yang, Force adaptive multi-spectral imaging with an articulated robotic endoscope, in: International Conference on Medical Image Computing and Computer-Assisted Intervention, Springer, 2010, pp. 245–252.

[115] P. Giataganas, M. Hughes, C. Payne, P. Wisanuvej, B. Temelkuran, G.-Z. Yang, Intraoperative robotic-assisted large-area high-speed microscopic imaging and intervention, IEEE Trans. Biomed. Eng. 66 (2018) , 208–216. https://10.1109/TBME.2018.2837058.

[116] F.B. Avila-Rencoret, G.P. Mylonas, D.S. Elson, Robotic Wide-Field Optical Biopsy Endoscopy. Clinical and Translational Biophotonics, Optical Society of America, 2018. CF1B. 5.

[117] C.M. Lee, C.J. Engelbrecht, T.D. Soper, F. Helmchen, E.J. Seibel, Scanning fiber endoscopy with highly flexible, 1 mm catheterscopes for wide-field, full-color imaging, J. Biophotonics 3 (2010) 385–407.

[118] G. Tearney, M. Shishkov, B. Bouma, Spectrally encoded miniature endoscopy, Opt. Lett. 27 (2002) 412–414.

[119] Y. Choi, C. Yoon, M. Kim, T.D. Yang, C. Fang-Yen, R.R. Dasari, K.J. Lee, W. Choi, Scanner-free and wide-field endoscopic imaging by using a single multimode optical fiber, Phys. Rev. Lett. 109 (2012) 203901.

Further reading

[120] S. Hu, K. Maslov, L.V. Wang, Second-generation optical-resolution photoacoustic microscopy with improved sensitivity and speed, Opt. Lett. 36 (2011) 1134–1136.

[121] L. Lin, P. Hu, J. Shi, C.M. Appleton, K. Maslov, L. Li, R. Zhang, L.V. Wang, Single-breath-hold photoacoustic computed tomography of the breast, Nat. Commun. 9 (2018) 2352.

[122] S.V. Panasyuk, S. Yang, D.V. Faller, D. Ngo, R.A. Lew, J.E. Freeman, A.E. Rogers, Medical hyperspectral imaging to facilitate residual tumor identification during surgery, Cancer Biol. Ther. 6 (2007) 439–446.

[123] M. Wojtkowski, High-speed optical coherence tomography: basics and applications, Appl. Opt. 49 (2010) D30–D61.

[124] L.M. Richards, E.L. Towle, D.J. Fox, A.K. Dunn, Intraoperative laser speckle contrast imaging with retrospective motion correction for quantitative assessment of cerebral blood flow, Neurophotonics 1 (1) (2014) 1–11. https://doi.org/10.1117/1.NPh.1.1.015006.

[125] M.J. Waldner, S. Wirtz, C. Neufert, C. Becker, M.F. Neurath, Confocal laser endomicroscopy and narrow-band imaging-aided endoscopy for in vivo imaging of colitis and colon cancer in mice, Nat. Protoc. 6 (2011) 1471–1481.

CHAPTER 2.2

Ultrasound imaging

Charles A. Sennoga
Auto entrepreneur & Visiting Researcher UMR Inserm U930 - Imagerie et cerveau Université François Rabelais de Tours France

1 Introduction

Ultrasound (US) has been used to image the human body for well over half a century. Today, US is one of the most widely used imaging modalities—offering as it does the ability to see the internal organs of the human body in real time, and as a result provide instantaneous visual guidance for many interventional procedures. US is increasingly multifunctional and versatile, portable and relatively inexpensive when compared with other imaging modalities, such as magnetic resonance and computed tomography. In addition, US waves are nonionizing, thus presenting a lower risk of tissue damage to both patient and examiner during use. Investigations into the biological effects of US on human tissue are ongoing; although, as of this writing, no harmful effect has been documented as a result of US diagnostic examinations performed within clinical guidelines (e.g., [1]). As a result, US imaging is considered to be safe. US is used widely for the imaging of adult reproductive systems and assessing fetal viability, in addition to its many routine uses, such as assessing cardiac valve malfunctions and liver perfusion during the identification of abdominal tumors, to name but a few. Its excellent safety record, notwithstanding a regular appraisal of potential harmful biological effects (see, e.g., [2–5]), is maintained by national and international US regulatory authorities and prudent expert clinical centers.

Today, medical US falls into two distinct categories, namely, diagnostic and therapeutic US. Diagnostic US, also known as sonography or ultrasonography, is a noninvasive imaging technique that uses sound above the threshold of human hearing to create images of internal structures and organs of the human body. Diagnostic US can be further subdivided into anatomical and functional US. Anatomical US produces tomographic cross-sectional images of internal structures and organs. Functional US combines information such as the movement and velocity of tissue or blood, the softness or hardness of tissue, and other physical characteristics, with anatomical images to create information maps. These maps are used by medical interventionists to evaluate and act upon apparent changes or differences in function within those structures or organs.

Bioengineering Innovative Solutions for Cancer
https://doi.org/10.1016/B978-0-12-813886-1.00007-3

Therapeutic US employs similar technologies but does not produce images. Its purpose is to interact with tissues in the body leading to their modification or destruction. US-induced thermal destruction of malignant tumors, fragmentation of blood clots and kidney stones, and site-specific mediation of drug delivery to locations in the body are typical examples. These destructive and delivery functions are most commonly achieved using high-intensity sound beams that can destroy diseased or abnormal tissues such as malignant tumors. The main advantage of US therapies is that, in the majority of cases, surgical incisions or cuts are generally not required, thus eliminating surgical wounds as well as reducing the duration of post-treatment aftercare.

Therapeutic US techniques are explored elsewhere in this book and will be discussed only in passing. Instead this chapter will focus on the basic understanding of the physical principles involved in US image generation, which is essential for the accurate interpretation of US images, especially when applied to cancer imaging. These principles can be somewhat complicated and sometimes counterintuitive.

2 Ultrasound generation and physical characteristics

Two basic principles are important for US generation, transmission and ultimately US image production. The first is the piezoelectric effect, which explains how US waves are generated from excitable crystals by the conversion of mechanical stimulus to electrical activity, i.e., mechanoelectrical transduction. The second is the pulse-echo principle, which explains how the image is generated. These and other physical characteristic of US are discussed. US-tissue interactions are also discussed, separately.

2.1 Definition and units

Sound is a type of mechanical wave. Described as oscillations or vibrations, waves are one of the ways in which energy is transferred. Unlike electromagnetic radiation, which can propagate in vacuum, mechanical waves need a medium (solid, liquid, or gas) in which to propagate. For example, when sound transmitted, say, from a loudspeaker, is detected by the human ear, it does so by inducing air particles to vibrate back and forth about a rest position. In the absence of a medium, sound cannot be transmitted, adding credence to the cinematic tagline, "In space no one can hear you scream."

Whereas vibrations with a pitch (or rate) from 20 to about 20,000 cycles/second are detected by the human ear as sound, vibrations above the audible limit,[a] i.e., > 20,000 cycles/second is classified as US. In US medical applications, vibrations in the range 20,000 to 50,000,000 cycles/second are used. The term frequency instead of

[a] The audible limit varies from person to person. However, research shows that the range of vibrations that can be heard by healthy young individuals varies from approximately 20 to 20,000 Hz. Ultrasound is characterized by vibrations above this range. The human ear cannot detect all vibrations.

rate of vibration is employed and the SI unit hertz [Hz] is preferred to cycles/s (1 cycle/second = 1 hertz = 1 Hz). Therefore, frequencies from 20 kilohertz (20 kHz) to 50 megahertz (50 MHz) are employed for US medical applications.

2.2 The piezoelectric effect

In medical applications, US waves are generated and detected by a piezoelectric transducer, which is a device capable of converting electrical energy into vibrational energy and vice versa. US transducers comprise either a single, or multiple piezoelectric ceramic crystals, which are electronically interconnected. When an electric current is applied to the crystals, i.e., voltage is delivered through a cable to the transducer, the piezoelectric crystals deform[b] rapidly and vibrate in response to the applied electrical current. These rapid deformations change the pressure at the surface of the transducer, thus yielding an US beam. The mechanical waves produced propagate longitudinally (compression and expansion), but not transversely (side-to-side motion) through adjacent medium where they duly create alternating areas of compression and rarefaction in that medium.

The US waves generated can be modeled in terms of a 3-dimensional matrix of intrinsic particles separated by perfect elastic springs representing interparticle forces (Fig. 1A). When a particle is pushed a distance from its neutral position, a disturbance or force is transmitted to the adjacent particle by the spring, thus creating a pressure fluctuation in the medium. If the driving force is oscillating, the particles will respond by oscillating in the same way. It is important to remember that while the particles of the medium vibrate during wave propagation, mass is not transferred. This is because an internal elastic restorative force arises to resist displacement of particles from their normal configurations.

The intrinsic nature of US waves that are produced is dependent on the characteristics of applied electrical current. Applying a sudden voltage to the transducer, e.g., by intermittently discharging a loaded capacitor at very short time intervals, yields a pulse wave whereas applying an oscillating voltage to the transducer generates a continuous wave. Fig. 1B and C exhibits schematic representations of compression wave generation in a medium both as a matrix of intrinsic particles and as a pressure waveform for a pulse and continuous waves.

US waves have several physical characteristics that should be considered in order to optimize the imaging procedure for specific diagnostic applications, including tumor identifications. The most significant are frequency, wavelength, wave propagation speed, and wave amplitude.

[b] The piezoelectric effect was first described in 1880 by Pierre and Jacques Curie who subjected a cut piece of quartz to mechanical stress, generating an electric charge on the surface. Later, they demonstrated the reverse piezoelectric effect, i.e., application of electrical current to the quartz resulting in its deformation.

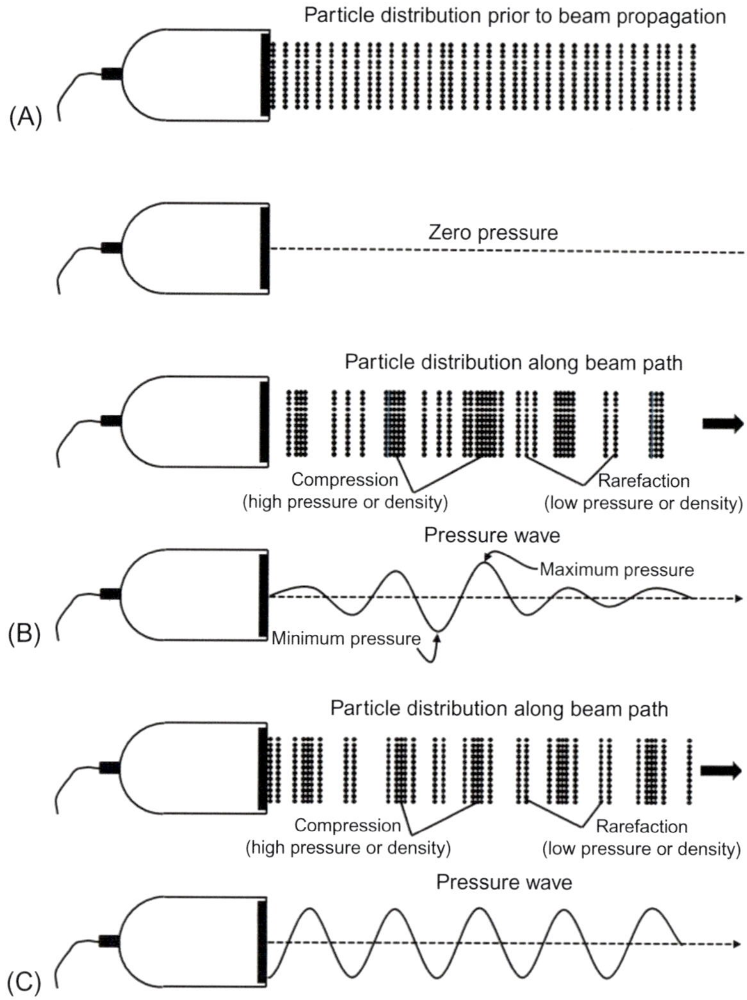

Fig. 1 Sound waves propagate by longitudinal (compression/expansion) motion, but not transverse (side-to-side) motion and can be modeled as particles connected by springs. (A) Schematic representation of a medium as a matrix of intrinsic particles in their neutral positions prior to pressure fluctuations triggered by ultrasound wave propagation. (B–C) The same medium following pressure fluctuation by a pulse (B) and continuous (C) waveform.

2.3 Ultrasound frequency and wavelength

The frequency (f) of a given wave is determined both by the properties and geometry of the source alone, and not by the medium through which the waves propagates. In medical US systems, the frequency is predetermined by the size and shape of crystals within the chosen transducer. US frequency is defined therefore by the number of vibration cycles or pressure changes that occur in 1 s. Medical US diagnostic transducers generate frequencies in the range of 1–20 MHz.

Wavelength (λ) is the distance a wave propagates over the time course (the period) of one vibration. It is the length over which a single vibration cycle occurs. It is defined by the velocity (v) of the wave divided by its frequency (f). Wavelength is determined both by the source and the medium through which the wave propagates. Because wavelength determines the size (length) of the wave, it has an effect on the quality of US image.

The frequency and wavelength of US waves are inversely related, that is, high frequency is characterized by shorter wavelength and vice versa. This relationship affects the choice of frequency used in patients undergoing US examination. Thus, the choice of transducer and therefore frequency to use is an important consideration when optimizing US image resolution. High-frequency US waves (short wavelength) generate images of high axial resolution. Increasing the number of compression and rarefaction (i.e., increasing frequency) for a given distance would more accurately discriminate between two separate structures along the axial plane of wave propagation. However, high-frequency US waves are more attenuated than lower-frequency waves for a given distance. This is because their shorter wavelength makes them unable to penetrate deeper into tissues as they are rapidly attenuated. Conversely, low-frequency US waves (long wavelength) have better penetrating power due to a lower degree of attenuation (see Table 1), but because of their longer wavelengths, offer images of lower resolution. It is for these reasons that low US frequencies (2–5 MHz) are used to image deeply embedded structures, especially in adults, while relatively higher US frequencies (10–15 MHz) are used to image superficial structures.

It is instructive to consider some numerical evaluations for some of the frequencies encountered in medical US. For example, at a frequency of 1 MHz, the penetration depth is in the range of 0.1–0.2 m and image resolution would be about 0.003 m. Increasing the frequency to 30 MHz[c] improves the resolution to about 0.0001 m.

Table 1 Relationship between attenuation and frequency of US waves.

Frequency, f [MHz]	Penetration depth [m]
1	0.40
2	0.20
3	0.13
5	0.08
10	0.04
20	0.02

[c] The typical diagnostic range is usually between 1 and 20 MHz. Here, a catheter-based 30-MHz intravascular transducer placed inside the blood vessel is used. Compared to conventional US, this is a highly invasive procedure.

However, the penetrating ability of such a wave decreases to 0.005 mm. Thus, the need for higher resolution versus more penetrating power is an important tradeoff when selecting the frequency and therefore transducer to use during US diagnostic investigations.

2.4 Ultrasound velocity and wave period

The US wave energy used in medical diagnostic equipment travels through the body in the form of a longitudinal or compression wave, that is, waves in which the particle motion is in the same direction as the wave propagation. Transverse or shear[d] waves, in which the particle motion is perpendicular to the direction of wave propagation, have not been utilized for medical diagnosis because they are rapidly attenuated in biological media at the MHz frequencies used in medical imaging. The speed (c) with which longitudinal waves propagate through a given medium is determined by the characteristics of that medium, specifically the density (ρ) and resistance to compression of the medium in question, through equation:

$$c = \sqrt{K/\rho}, \tag{1}$$

where K, the bulk modulus, is a measure of how resistant to compression that medium is. In other words, US propagation speed is determined by the density and stiffness of the medium through which the wave is propagating. Density is the concentration of matter (mass per unit volume [kg m^{-3}]). Stiffness is the resistance of a material to compression (inverse of compressibility). Therefore, the harder or incompressible a material or medium is, the higher is its sound speed. A list of sound propagation speeds in selected biological media is presented in Table 2. These data are compiled from previous reports [6–11].

The speed with which US waves propagate through most soft biological tissues is close to 1540 m/s. This is extremely fast, and suggests as shown later that US wave packets (pulses) can be transmitted and the corresponding echoes rapidly collected, thus allowing US images to be built up in as little as a fraction of a second. However, this medium-dependent variation in the speed of sound is one of the main causes of artifacts and erroneous attribution of structural observations in US images. Elsewhere, the variation in sound speed allows for the velocity of blood to be determined through the Doppler equation. These and related issues are addressed in later sections.

Whereas it is theoretically possible for the speed of sound to be frequency dependent, a property known as dispersion, extensive investigations have shown that the speed of sound is independent of frequency, at least over the frequency range used in US

[d] Recent investigations now suggest that transverse waves at low frequencies may be utilized in elastographic applications.

Table 2 Characteristic ultrasound properties of selected media.

Material	Speed, c [m/s]	Compressibility, K [10^{-12} m^2 N^{-1}]	Density, ρ [kg m^{-3}]	Attenuation coefficient at 1 MHz, α/f [dB/cm MHz^{-1}]	Characteristic impedance, Z [kg m^{-2} s^{-1}]
					($\times 10^6$)
Air at STP	330	7.65×10^6	1.2	12.000	0.0004
Water (at 20°C)	1480	456	1000	0.0025	1.48
Fat	1450	494	920	0.56	1.37
Amniotic fluid	1510			0.007	1.50
Vitreous of eye	1520			0.1	1.52
Soft tissue	1540			0.81	1.62
Liver	1550	388	1060	0.95	1.66
Spleen	1550		1060	0.52	1.65
Kidney	1560	400	1040	1.10	1.63
Blood	1570		1060	1.8	1.61
Muscle (cardiac)	1590	370	1070	1.8	1.71
Bone	4080	31	1380–1810	20.0	3.75–7.38

diagnostic imaging. Finally, it should be noted that the speed of sound, c, [m/s] is also related to frequency, f [Hz], and wavelength, λ [m], through equation

$$c = f\lambda \tag{2}$$

Thus far, US waves have been described in terms of distance alone, but waves must involve a time component too. Fig. 2 shows a waveform description of pressure fluctuation for a pulsed wave both in spatial and temporal terms. Recognizing its temporal component allows us to explore its time-dependent features, such as the period of a pulse wave.

The period (T) of a pulse wave is the duration of time required for one vibration cycle in a repeating event to occur. Therefore, the period is the reciprocal of frequency and vice versa. Increasing wave frequency decreases the period. For example, if a wave cycles at a frequency of 5,000,000 times a second (5 MHz), its period—the time interval between cycles—is 0.0000002 s. That is, 1 s divided by 5,000,000 cycles (see Eq. 3). A wave with a higher cycling frequency, e.g., 15 MHz leads to a substantial reduction of T down to 0.00000007 s.

$$T = 1/f \tag{3}$$

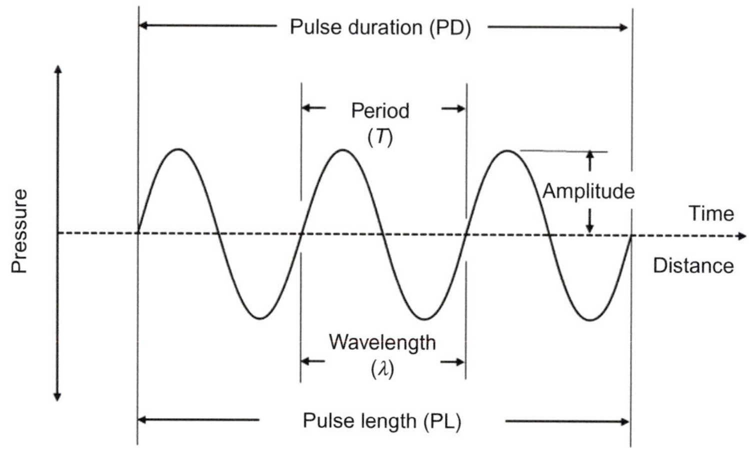

Fig. 2 Waveform representation of a pulse wave in spatial and temporal modes.

The distance between equivalent points on the waveform is the wavelength (λ) of the US wave and the maximum pressure fluctuation is the wave amplitude. The rate at which the tissue structures vibrate is the frequency and the rate at which the vibrations move through tissue is the sound speed. The space through which the wave propagates is the ultrasound field or beam. The terms "field" and "beam" are used interchangeably, although it would probably be better to restrict "field" to US emitted by the transducer and "beam" to a combination of the emitted and the returning echoes in front of the transducer.

In summary, wavelength is the ratio of velocity to frequency or the product of velocity and the period. Thus, the wavelength of an US wave is determined by the characteristics of both the transducer (through frequency) and the material through which the sound is propagating (through velocity).

2.5 The pulse-echo principle

Although continuous waves are generated and used most commonly in basic US fetal heart monitors, the majority of US diagnostic systems employ pulse waves. Here, waves are generated as short bursts or pulses (intermittent trains of pressure) consisting of two or three cycles of the same vibration frequency that are separated in time with gaps of no signal (Fig. 3). The time intervals between pulses are used to detect echoes produced within the body.

Essentially, as the pulse waves penetrate body tissues of different acoustic impedances (compressibility) along the path of beam propagation, some are reflected back to the transducer as echo signals and some continue to penetrate deeper. As the echo signals returning from sequential pulses strike the transducer, they cause the piezoelectric crystals

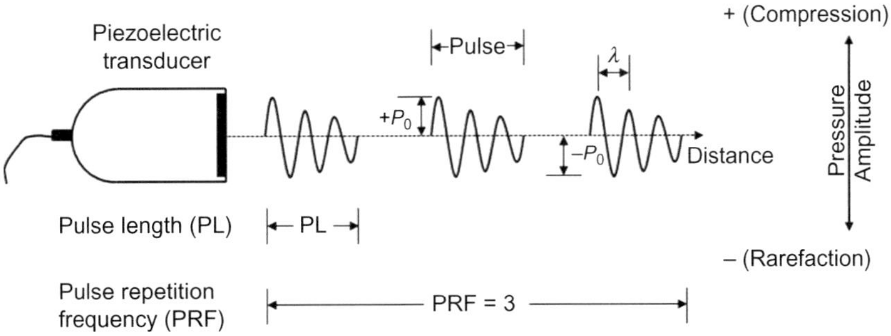

Fig. 3 Representation of ultrasound pulses in terms of their spatial propagation.

in the transducer to deform again and produce electrical signals that are then processed and combined to generate an image which is displayed most commonly as a brightness-mode (B-mode) image on the US imaging system's monitor. The transducer therefore functions both as a speaker or transmitter by generating pulse waves, and as a microphone or receiver by detecting pulse echoes. The US pulse wave in itself is quite short, but since it propagates in a straight path, it is often referred to as an US beam. Such is the efficiency of the system that the transducer generally emits US for as little as 1% of the time; the rest of the time is spent receiving the returning echoes and reconstructing it into US images. This technique exploits the pulse-echo principle.

In addition to achieving the depth of resolution required for diagnostic images, the pulse-echo principle is employed because the transducer operating both as a transmitter and receiver cannot fulfill both functions at exactly the same time. Essentially, by carefully measuring the time between the transmission of the transmit pulse and the reception of a given echo, the imaging system can calculate the distance between the transducer and the structure that caused that echo. For example, if we consider a pulse reflector at distance d from the surface, then the total distance traveled by the US pulse (from the transducer to the reflector and then from the reflector back to the transducer) is simply ($2d$). The time taken to complete this distance (and therefore the time delay between the transmit pulse and its echo) is given by $2d/c$, where c is the speed of sound for the media in question. Rearranging this expression allows us to calculate the depth of the reflector from the time delay as $d = ct/2$. In fact, this points to a proportional relationship between the arrival time of an echo (t) and the depth of the structure causing that echo (d). Note that for the imaging system to use this relationship, it must assume a value for the propagation speed c. In most medical imaging systems, an average propagation speed in soft tissue of 1540 m/s is assumed, although some imaging systems now offer the option of changing this value to more accurately reflect the speed of sound in the tissues under investigation.

The pulse repetition frequency (PRF) is the number of pulses emitted by the transducer per unit time. The pulses are typically a millisecond long and several thousands are emitted per second. Most importantly, pulses are emitted with sufficient time interval to allow the emitted signal to reach the target of interest and be reflected back to the transducer as echoes, before the next pulse is generated. In US diagnostic systems, pulses are emitted at a rate of approximately 1000 pulses per second. The pulse rate (i.e., pulses per second) should not be confused with frequency, which is the rate of vibration of the tissue within the pulse and is of the order 1–20 MHz for US imaging applications as opposed to higher frequencies in therapeutic applications.

Pulsed waves are generally employed in diagnostic imaging applications, while continuous waves are commonly employed in therapeutic applications.

2.6 Ultrasound wave amplitude

The amplitude of an ultrasound pulse is a parameter concerned with the strength of the US beam. It is the range of pressure excursions. US wave amplitude is defined as the difference between the peak pressure and average pressure values of the waveform (Fig. 3). Pressure is related to the degree of tissue displacement caused by vibration. Amplitude is related to the energy content, i.e., the loudness, if you like, of the US pulse. The amplitude of a given pulse as it emerges from the transducer is generally determined by the strength of the electrical pulses sent to the piezoelectric elements. Expressed in decibels (dB), which is a logarithmic scale, amplitude can be altered by the operator. In fact, most US diagnostic systems have a control on the pulse generator that changes the size of the electrical pulse and the US pulse amplitude, often designated as the intensity control, although different names are used by various equipment manufacturers. Amplitude decreases as the pulse wave propagates through tissue, through a process called attenuation.

For most diagnostic applications, it is usually necessary to know only the relative amplitude of US pulses. For example, it is necessary to know how much the pressure amplitude (A) of a pulse decreases as it propagates through a given thickness of tissue. Accordingly, the relative amplitude of two US pulses, or of one pulse after it has undergone an amplitude change, can be expressed as a ratio:

$$\text{Relative amplitude (ratio)} = A_2/A_1 \tag{4}$$

However, there are advantages in expressing relative pulse amplitude in terms of the logarithm of the amplitude ratio. When this is done, the relative amplitude is specified in units of decibels (dB). The relative pulse amplitude, in dB, is then related to the actual amplitude ratio through:

$$\text{Relative amplitude (dB)} = 20\log A_2/A_1 \tag{5}$$

When the amplitude ratio is >1 (e.g., when comparing a large pulse to a smaller one), the relative pulse amplitude has a positive dB value; when the ratio is <1, the dB value is negative. In other words, if the amplitude of a pulse is increased by some means, it will gain dB, and if it is reduced, it will lose dB. Note that amplitudes that differ by a factor of 2 differ by 6 dB. A reduction in amplitude of—6 dB divides the amplitude by a factor of 2, or 50%; and the doubling of a pulse's amplitude increases it by +6 dB. During its lifetime, an ultrasound pulse undergoes many reductions in amplitude as it propagates through tissue because of absorption. If the amount of each reduction is known in decibels, the total reduction can be determined by simply adding all of the dB losses. This is much easier than multiplying the various amplitude ratios.

3 Intensity and power

Both US wave intensity and power are specific measures of energy transmission. When a piezoelectric element is excited by an electrical voltage, energy is transferred from the transducer to tissue. Power is the rate of energy transfer and is expressed in units of Watts. It is the rate of energy transmission through the entire cross-sectional area of the beam. Hence, power is a quantity that provides the total energy output.

Intensity (I), on the other hand, is the rate at which power is delivered to or transmitted through a specified unit area. Thus, intensity is a ratio of the amount of power delivered per unit area. Intensity is expressed in units of Watts per square meter [W/m^2]. Hence, intensity is the rate at which US wave energy is applied to a specific location within a patient's body. When US beam power increases, intensity increases. This is why power and intensity are considered important parameters in diagnostic and therapeutic applications. When beam area decreases due to, e.g., beam focusing, intensity increases. This means more energy is concentrated in a small section of tissue. Intensity is related to the pressure amplitude of the individual pulses and the pulse rate. Since pulse rate is fixed in most ultrasound diagnostic systems, intensity is determined by pulse amplitude. However, absolute values of intensity are difficult to measure. Relative sound intensities in dB are given in Table 3.

Like amplitude, the relative intensity of two pulses, I_1 and I_2, can be expressed in the units of decibels (dB)

$$\text{Relative intensity}\,[\text{dB}] = 10 \log I_1/I_2 \tag{6}$$

Note that when intensities are being considered, a factor of 10 appears in the equation rather than a factor of 20, which is used for relative amplitudes. This is because intensity is proportional to the square of the pressure amplitude, which introduces a factor of 2 in the logarithmic relationship. Moreover, the intensity of a propagating US beam does not remain constant with respect to time nor is it uniform with respect to spatial area. For example, the intensity of a transducer is usually highest at the center of the beam and decreases progressively toward the beam periphery. Intensity also varies along the direction of beam propagation due to, e.g., focusing and attenuation. Accordingly, intensity is commonly measured at the focus of the field and this must be taken into consideration

Table 3 Relative sound intensities in decibels.

Relative intensity I/I_0 [W/m^2]	Change of level [dB]
100	20
10	10
4	6
2	3
1	0
0.5	−3
0.25	−6
0.10	−10
0.01	−20
0.001	−30
0.0001	−40

when describing intensities. Thus, it must be clearly stated whether it is the peak intensity or the average intensity that is being considered.

Power and intensity are the two main quantities that must be considered with respect to potential bioeffects and the overall safety of US imaging. From a clinical perspective, the two main bioeffects commonly encountered during US examinations of living tissue are thermal (localized heating) and mechanical, in nature. Two indices: thermal index (TI) and mechanical index (MI), are derivative measures of these bioeffects. Both indices are implemented on US imaging systems, where they are used by sonographers as a risk-benefit analysis tool aimed at ensuring that potential bioeffects are minimized.

3.1 Duty factor

Fig. 4 illustrates three sequential pulses as well as two time intervals: the pulse duration (PD) and the pulse repetition period (PRP). The ratio of PD to PRP is the duty factor.

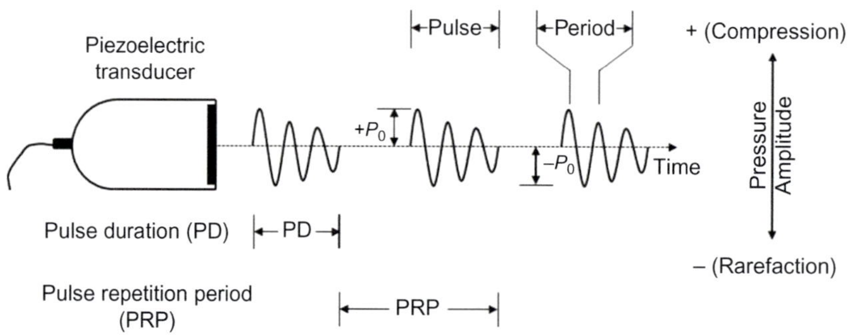

Fig. 4 Schematic representation of ultrasound pulses in terms of their temporal propagation, highlighting the key characteristics of US pulses that affect intensity values.

The duty factor is the fraction of time over which an US pulse is generated. Note that if US is generated as a continuous wave, then duty factor will have a value of 1. Intensity and power are proportional to the duty factor. For most pulsed imaging applications, duty factors are small, i.e., <0.01. With respect to time, there are three possible power (intensity) values. One is the peak power, which is associated with the time of maximum pressure. Another is the average power within a pulse. The lowest value is the average power over the PRP for an extended time.

3.2 Spatial characteristics

The energy or intensity of a pulse is not distributed uniformly over the area of an US pulse. Accordingly, it can be expressed either as the peak intensity, which is often in the center of the pulse, for the average intensity over a designated area.

4 Ultrasound—Tissue interactions and therapeutic applications

US waves interact with different tissues in a variety of ways. Some of the interactions are desirable and necessary for the formation of US diagnostic images or therapeutic applications, whereas others are undesirable and compromise diagnostic examinations by absorbing much of the US energy or produce artifacts. Specifically, as US pulses propagate through tissues, they are partly transmitted to deeper structures, reflected back to the transducer as echoes, scattered or transformed into heat. The ability to conduct and interpret the results of an US examination depends on a thorough understanding of these interactions. For therapeutic purposes, this transformation into heat energy now forms the basis of thermal ablative applications. Thermal ablation is a bioeffect of focused ultrasound, capable of inducing cell death in a targeted area (e.g., malignant tumor) with minimal damage to the surrounding tissue. For imaging purposes, we are mostly interested in the echoes reflected back to the transducer. Accordingly, attenuation and refraction are the two major types of interactions that suppress echoes.

4.1 Absorption and attenuation

As US pulses propagate through tissue, their intensity and pressure wave amplitude are progressively weakened with every additional distance propagated. This process is known as attenuation. Attenuation can be caused by pulse wave reflection, scattering, as well as frictional losses. These losses result from the induced oscillatory tissue motion produced by the pulse, which in turn causes conversion of energy from its original mechanical form into heat. This energy loss to localized heating is referred to as absorption and is the most important contributor to US attenuation.

The rate at which a pulse is attenuated depends on the US frequency used and the density of the medium through which it propagates. Therefore, higher-frequency waves and longer path length result in greater attenuation. Attenuation also varies among body tissues; i.e., for any given frequency, attenuation is greater in bone (due to its high density), moderate in muscle and solid organs (medium densities), and relatively low in blood (lower densities). US diagnostic imaging systems intrinsically compensate for the expected attenuation by automatically increasing the overall brightness or signal intensity in deeper areas of the image. Since attenuation is the term used to account for loss of wave amplitude or "signal amplitude" due to all mechanism(s), including absorption, scattering, reflection, and beam divergence, attenuation is experimentally determined by measuring the overall decrease in pressure amplitude as a wave passes through a medium of known thickness. By taking the ratio of the input and output pressure amplitudes on either side of the experimental medium, the decrease in pressure amplitude can be quoted as the attenuation coefficient in units of dB/cm. Because attenuation increases with frequency, it is necessary to specify the frequency when attenuation rates are given. Accordingly, attenuation through a thickness of material x, can be determined via Eq. (7).

$$\text{Attenuation [dB]} = \alpha f x, \tag{7}$$

where α is the attenuation coefficient [dB/cm/MHz] and f is the US frequency [MHz]. Approximate values of attenuation coefficients for various biological materials are given in Table 2 for a 1-MHz frequency. However, a careful examination of Table 2 shows that there is considerable variation in the attenuation rates from one medium to the next. For example, of all the media listed, water produces by far the least attenuation. This means that water is a very good transmitting medium for US. Indeed, water within various tissues, such as in cysts and the bladder, often forms the basis of so-called imaging windows through which underlying structures that are otherwise difficult to image, can be ultimately imaged. With the exception of fat and muscle, most of the soft tissues of the body have attenuation coefficient values of about 1 dB/cm/MHz, whereas the value for bones is much larger and ranges from around 10 to 20 dB/cm/MHz. This is one of the main obstacles limiting US imaging of brain in the intact skull and other tissues located behind bone structures such as ribs. Muscle has a range of values that depends on the direction of the US with respect to the muscle fibers. Lungs have a much higher attenuation rate than either air or soft tissue. This is because the small pockets of air in the alveoli are extremely effective in scattering US energy. Because of these air pockets, normal lung structure is extremely difficult to penetrate with US.[e] Compared to the soft tissues of the body, bone has a relatively high attenuation rate. Bone, in effect, shields some parts of the body against easy access by US.

Attenuation is frequency dependent, a fact which has significant impact on equipment design and performance. Thus the higher the US frequency, the more rapidly is the

[e] Diseased lungs such as those presented by pneumonia patients have been imaged with US due to mucosal flooding of these air pockets.

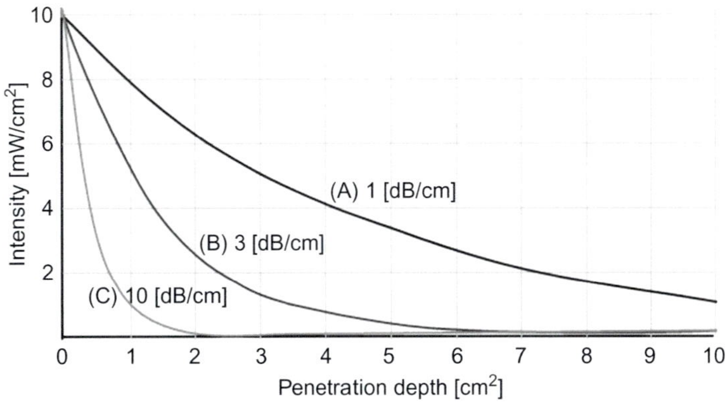

Fig. 5 Attenuation of ultrasound through a medium increases with depth and proportionally with attenuation coefficient. (A) The intensity decreases 20% for each centimeter propagated, corresponding to an attenuation coefficient of 1 dB/cm. (B) Intensity decreases by 50% corresponding to an attenuation coefficient of 3 dB/cm. (C) The attenuation coefficient is 10 dB/cm.

amplitude reduced. Attenuation increases linearly with increasing frequency hence attenuation in tissue is highly dependent on frequency.

Most importantly, attenuation of sound as it travels through a given medium increases with depth (Fig. 5).

4.2 Reflection and image speckle

Reflection of US pulses by structures within the body is the primary interaction process leading to part of this energy returning to the transducer for image generation. The portion of the US beam that is reflected and returned as echoes after encountering a tissue interface is determined by a tissue property called acoustic impedance (Z).[f] This is an intrinsic physical property of a medium that is related to its density (ρ) and elastic properties. Since the velocity (c) of US wave propagation through the medium is related to the same material characteristics, a relationship exists between tissue impedance and velocity, through the equation.

$$Z = \rho c \tag{8}$$

Impedance is a measure of how readily medium particles move under the influence of the wave pressure (compressibility); therefore, the higher the density of the medium, the greater the acoustic impedance. US pulses are reflected only when they encounter a boundary between two media that involves a change in density or compressibility or both (Fig. 6), i.e., the two materials must differ in terms of acoustic impedance. Organs that

[f] Although the traditional symbol for impedance Z is similar to that used to denote the atomic number of elements, the two quantities are not related.

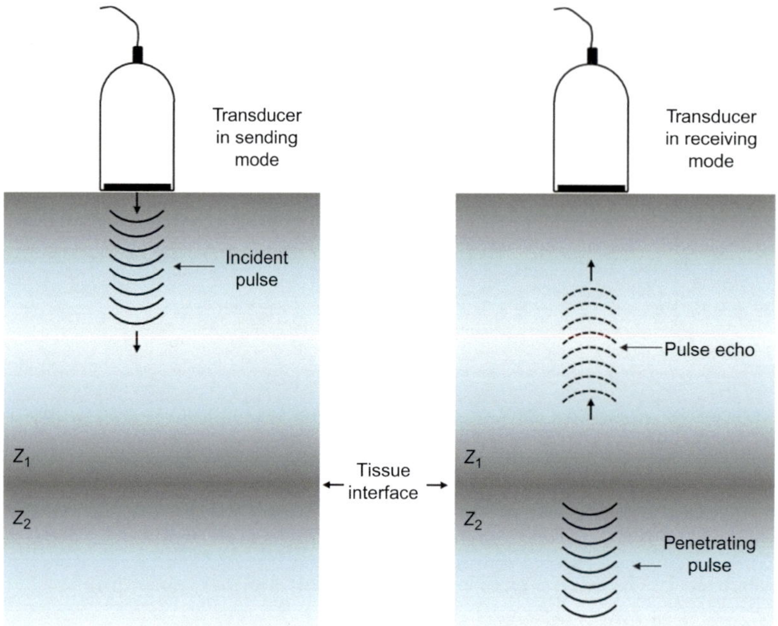

Fig. 6 Ultrasound pulses are reflected only when they encounter a boundary between two media that involves a change in density or compressibility or both.

contain air (such as lungs) have the lowest acoustic impedance, while organs with high particle density such as bone have very high-acoustic impedances (Table 2). The large difference in density and sound velocity between air, bone, and soft tissue creates a correspondingly large difference in acoustic impedance, causing almost all of the sound waves to be reflected at soft tissue-bone and soft tissue-air interfaces. On the other hand, because there is little difference in acoustic impedance between soft-tissue structures, relatively few echoes are reflected back to the transducer from these interfaces. This suggests that the intensity of a reflected echo is proportional to the difference in acoustic impedances between the two media. Media interfaces between soft tissues of similar acoustic impedances usually generate low-intensity echoes. Conversely, interfaces between soft tissue and bone or the lung generate very strong echoes due to a large mismatch in acoustic impedances.

When an incident US pulse encounters a large, smooth interface of two body tissues with different acoustic impedances, the sound energy is reflected back to the transducer. This type of reflection is called specular reflection, and the echo intensity generated is proportional to the acoustic impedance gradient between the two media. If the incident US beam intersects the linear interface at 90°, almost all of the generated echoes will be reflected back to the transducer. However, if the angle of incidence with the specular

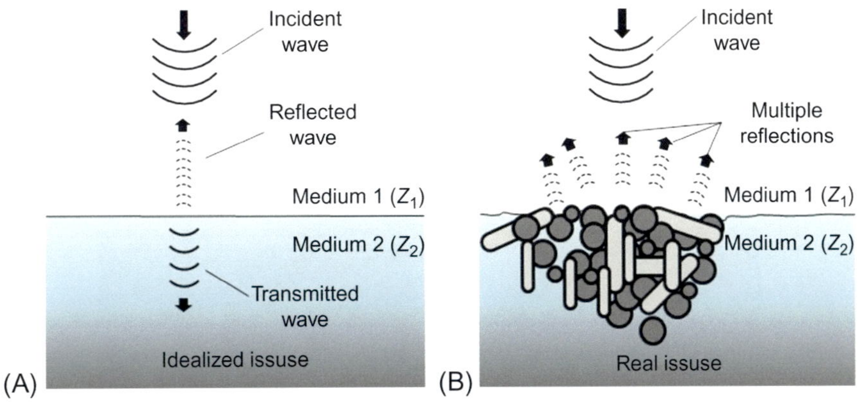

Fig. 7 (A) A flat surface perpendicular to an incident US beam will reflect sound and thus create an echo. Reflection from living tissue is much more complicated as shown in (B).

boundary is <90°, the echo will not return to the transducer, instead they are reflected at an angle equal to the angle of incidence, just like visible light reflecting off a mirror plane. These returning echoes will potentially miss the transducer and not be detected.

For an idealized incident US beam from a medium with impedance Z_1 into a medium with impedance Z_2, the propagating pulse will split into two pulses with reduced energy at planar interface. One pulse, the echo, is reflected back toward the transducer while the other propagates deeper into the rest of the material beyond the interface.

The situation in real tissues, however, is a little more complicated. This is partly due to numerous variations in the acoustic impedance per unit volume and a nonplanar interface (see schematic depiction in Fig. 7). This gives rise to many small reflections and interferences of the pulse waves. As a result, US images are characterized with a typical speckle pattern, i.e., US images comprise numerous small bright dots on a black background.

A representative abdominal US image of a normal liver is shown in Fig. 8. Although the anatomy is uniform, the speckle texture clearly dominates the image. The brightness of a structure in an US image depends on the strength of the reflection, or echo producing it. Only a portion of the US pulse is reflected at most interfaces. This in turn is dependent upon how much the two materials differ in terms of acoustic impedance. The amplitude ratio of the reflected to the incident pulse is related to the tissue impedance values by:

$$\text{Reflection loss}\,[\text{dB}] = 20\,\log(Z_2 - Z_1)/(Z_2 + Z_1), \tag{9}$$

where Z_1 is the impedance of the first medium and Z_2 that of the second medium. Significantly, only a small fraction of the pulse is reflected at interfaces of most soft tissues. As a result, the reflection process produces relatively weak echoes. At interfaces between soft tissue and materials such as bone, gallbladder stones, and gases, strong reflections are produced. The reduction in pulse amplitude during reflection for several different

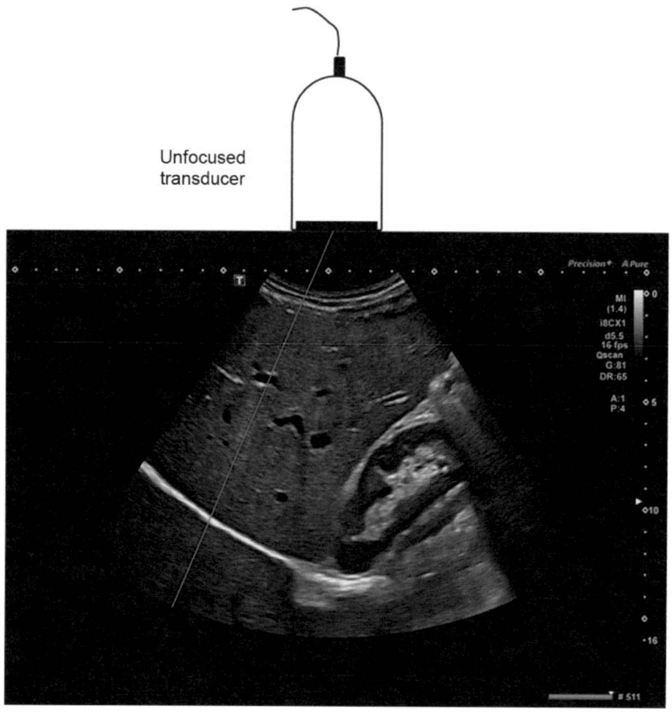

Fig. 8 A typical abdominal ultrasound image of a human liver. Note the characteristic speckle texture. To form a B-mode image, the line along the depth of sound propagation is swept across an angular range, usually by electromagnetic beam forming. The amplitude values are converted to brightness along the line and displayed on a screen. *(Main figure courtesy of Prof. Frederic Patat and M. Paul-Armand Dujardin at CHRU Bretonneau, Tours France).*

Table 4 Pulse amplitude losses due to reflection at selected media interfaces.

Interface	Amplitude loss [dB]
Ideal reflector	0.0
Tissue/air	−0.01
Bone/soft tissue	−3.8
Bone/muscle	−4.0
Fat/muscle	−20.0
Tissue/water	−26.0
Muscle/blood	−30.0
Kidney/spleen	−45.0

interfaces is given in Table 4. Strikingly, in soft tissues, the reflection coefficient varies from −20 dB (between fat and muscle) to −45 dB (between kidney and spleen). These are low-level reflections (<0.5%) so that most of the acoustic energy is transmitted through the interface and is therefore available for imaging deeper structures. In some cases, however, high levels of reflection may occur, such as bone and soft tissue interfaces, which have a reflection coefficient of −3.8 dB. In such a case, considerably less energy is transmitted, thus not very much pulse energy is available for imaging deeper structures. As a result, the amplitude of the pulse is attenuated both by absorption and reflection losses. Because of this, an echo returning to the transducer is much smaller than the original pulse produced by the transducer.

4.3 Refraction

Refraction refers to a deviation in the direction of US pulse propagation after it encounters an interface of two media with different propagation velocities. Because the US frequency is constant, in that instance, the wavelength has to change to accommodate the difference in the velocity of US propagation in the two media (Fig. 9). This results in a redirection, i.e., bending or refraction of the US pulse as it intersects the interface. If the pulse propagation angle of incidence (i.e., between the beam director and interfacial surface) is 90°, there is no beam deviation, but at all other angles a change in direction occurs. Whereas deviations of a few degrees such as those encountered at soft tissue interfaces do not represent a problem, large angle pulse deviations due to the large differences in speed of US between, e.g., soft tissue and bone, represent a severe problem and can produce undesirable artifacts.

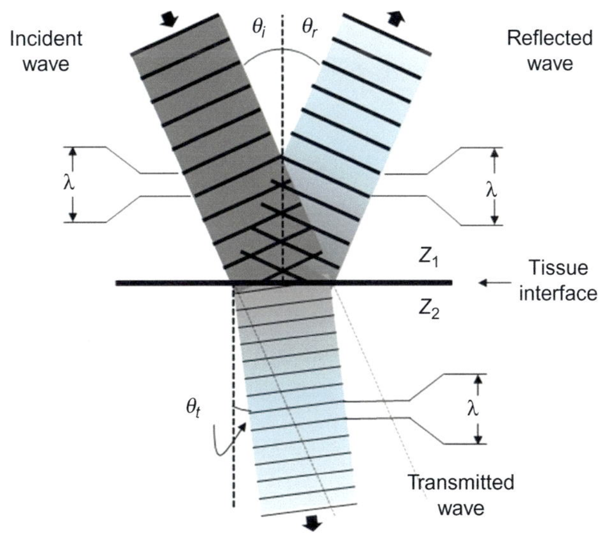

Fig. 9 Schematic representation of ultrasound pulse reflection and transmission. Notice the change in wavelength for the refracted wave.

Refraction is one of the main causes of erroneous localization and attribution of anatomical structure in US images. This type of interaction causes artifacts that need to be recognized by the operator. For example, because the speed of sound in fat (1450 m/s) and in soft tissues (1540 m/s) differ, refraction artifacts typically occur at muscle-fat interfaces. This arises because almost all US diagnostic systems measure the time of echo return after the instant of pulsed US transmission and then use the speed of sound in tissue to convert this time into the tissue depth of the reflecting structure. For a mixture of soft tissues along the pulse path, therefore a measure of depth is obtained by assuming a global average speed of 1540 m/s for conversion of time into depth.

4.4 Scattering

Scattering refers to the redirection of US as it interacts with small, rough, or uneven structures. Scattering occurs when an US pulse encounters reflectors whose dimensions are smaller than the interrogating US wavelength, or when the pulse encounters a rough, irregular tissue interface. In this case, echoes reflected through a wide range of angles result into a reduction of echo signal intensity. However, the positive result of scattering is the return of some echo to the transducer regardless of the angle of the incident pulse. Most biological tissues appear in US images as though they are filled with tiny scattering structures. The speckle signal that provides the visible texture in organs like the liver or muscle is a result of interface between multiple scattered echoes produced within the volume of the incident US pulse. This tissue interaction occurs in organ parenchyma, where there is little difference in acoustic impedance, and is responsible for producing the texture of the organ seen on the monitor. Scattering increases as frequency increases. Thus, higher-frequency transducers provide better resolution or image detail.

5 Transducers and image optimization

The diameter, which is also the width of the beam, is an important characteristic of an US pulse. The diameter of a pulse not only changes as it propagates along the beam path but also has an effect on resolution and image detail. The diameter of the pulse is determined by the characteristics of the transducer. At the transducer surface, the diameter of the pulse is the same as the diameter of the vibrating piezoelectric elements. As the pulse moves through the body, the diameter generally changes. This is determined by the focusing characteristics of the transducer.

5.1 Transducer focusing

Transducers can be designed to produce either a focused or unfocused beam (Fig. 10). A focused beam is desirable for most imaging applications because it produces pulses with a small diameter, which in turn gives better visibility of image details. Specifically, the best detail and therefore resolution is obtained for structures within the focal zone.

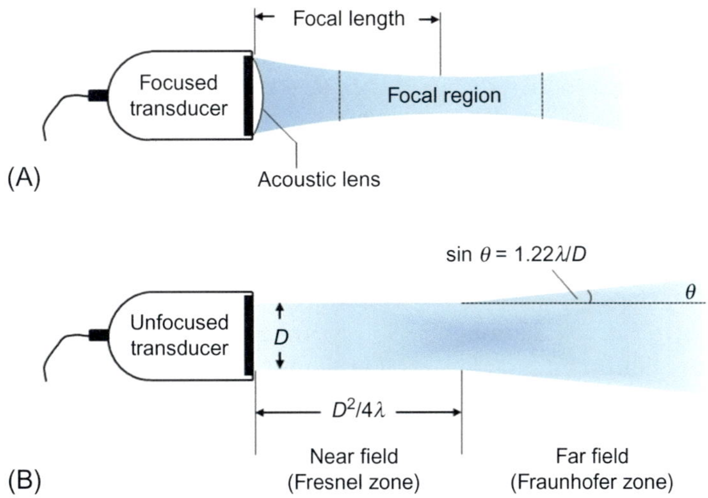

Fig. 10 Transducer focusing. (A) The beam can be focused by adding an acoustic lens to the transducer face, resulting in a narrower beam in the focal region. (B) An unfocussed disc shaped transducer leads to a diffraction beam pattern.

The distance between the transducer and the focal zone is the focal length. A focused beam is equally desirable for therapeutic ablative applications because a focused beam will direct and deliver more energy in a confined spot such as a tumor.

5.2 Unfocused transducers

An unfocused transducer produces a beam with two distinct regions (Fig. 10B). One is the so-called near field or Fresnel zone and the other is the far field or Fraunhofer zone. In the near field, a relatively constant beam diameter is maintained that can be used for imaging. Notice that the diameter of the beam is determined by the diameter of the transducer while the near-field length is related to the diameter (D) of the transducer and the wavelength (λ) of the US beam through the following expression

$$\text{Near field length} = D^2/4\lambda \tag{10}$$

Since wavelength is inversely related to frequency, for a given transducer size, the length of the near field is proportional to frequency. Another characteristic of the near field is that the intensity along the beam axis is not a constant. Rather intensity oscillates between maximum and zero several times between the transducer surface and the boundary between the near and far field. This arises because of the interference patterns created by the sound waves from the transducer surface. An intensity of zero at a point along the axis simply means that the sound vibrations are concentrated around the periphery of the beam. Indeed, a picture of the ultrasound pulse in that region would look more like concentric rings or "a torus" than a disk.

A major characteristic of the far field is that the beam diverges. This causes the US pulses to be larger in diameter but also to have less intensity along the central axis. The approximate angle of divergence is therefore related to the diameter of the transducer, (D) and the wavelength, (λ) through:

$$\text{Divergence angle [degrees]} = 70\lambda/D \tag{11}$$

Because of the inverse relationship between wavelength and frequency, divergence decreases with increasing frequency. A major advantage of using higher US frequencies (i.e., shorter wavelengths) is that the US beams are less divergent and generally produce less blur and better detail. Although all US beams are often presented as idealized US beams, in reality unfocused transducers often produce beams with side lobes. These secondary beams have the capacity to fan out around the primary beam. The principal concern is that under certain conditions echoes will be produced by the side lobes and these have the potential of inserting artifacts into an otherwise better image.

5.3 Fixed focus

Transducers can be designed to produce a focused US beam by using concaved piezoelectric elements or an acoustic lens in front of the element(s). Transducers are designed with differing degrees of focusing. Relatively weak focusing produces a longer focal zone and greater focal depth. A strongly focused transducer will have a shorter focal zone and a shorter focal depth. Unfortunately, fixed focus transducers have the obvious disadvantages of not being able to produce the same image detail at all depths within the tissues probed.

5.4 Adjustable transmit focus

The focusing of some transducers can be adjusted to a specific depth for each transmitted pulse. This is illustrated in Fig. 11. The transducer is made up of an array of several

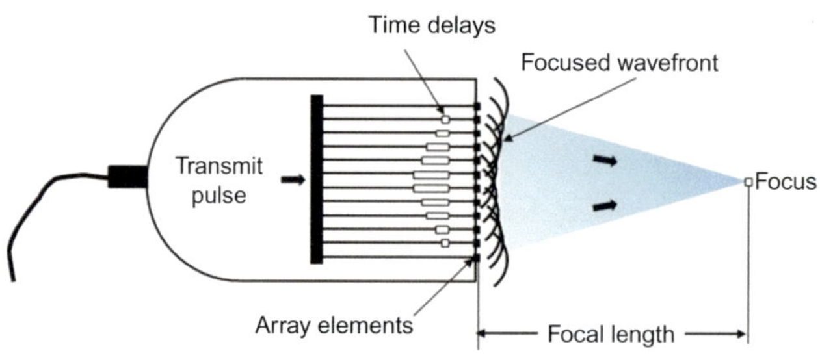

Fig. 11 Electronic focusing (transmit focus). If the outer elements are fired first with increasing delays toward the central elements, a converging wave front is formed, which produces a focused transmit beam. Conversely, in reception, delays are applied to the signals received at each element before they are added together. The delays are calculated for each target depth so that the receive focus tracks the current receive depth, to produce a much narrower receive beam.

piezoelectric elements rather than a single element as in the fixed-focus transducer. Two basic array configurations, linear and annular, are commonly used. In the linear array, the elements are arranged in either a straight or curved line. The annular array transducer consists of concentric transducer elements. Although these two designs have different clinical applications, the focusing principles are similar.

Focusing is essentially achieved by not applying the electrical pulses to all of the transducer's piezoelectric elements simultaneously. Instead, the pulse to each element is passed through an electronic delay. The sequence with which the transducer elements are pulsed is therefore regimented. Specifically, the outermost elements for either annular or linear array are pulsed first. This produces ultrasound that begins to propagate away from the transducer. The other elements are then pulsed sequentially, working toward the center of the array. The central element will receive the last pulse. In this way, the pulses from the individual elements combine in a constructive manner to create a curved composite pulse, which will converge on a focal point at some specific depth from the transducer. Here, the focal depth is determined by the time delay between the electrical pulses. This can be changed electronically to focus pulses to give good image detail at various depths within the body rather than just one depth as in the case of a fixed focus transducer. One approach is to create an image by using a sequence of pulses, each one focused to a different depth or zone within the body. A notable distinction between the two transducer designs described here is that the annular array focuses the pulse in two dimensions, whereas the linear array can only focus in the one dimension, i.e., in the plane of the transducer.

5.5 Dynamic receive focus

The focusing of an array transducer can also be changed electronically when it is in the echo-receiving mode. This is achieved by processing the electrical pulses from the individual piezoelectric elements through different time delays before they are combined to form a composite electrical pulse. The effect of this is to give the transducer a high sensitivity for echoes coming from a specific depth along the central axis of the beam. This produces a focusing effect for the returning echoes.

An important factor is that the receiving focal depth can be rapidly changed. Since echoes at different depths do not arrive at the transducer at the same time, the focusing can be swept down through a range of depths to pick up the echoes as they occur. This is the major distinction between dynamic or sweep focusing during the receive mode and adjustable transmit focus. Any one transmitted pulse can only be focused to one specific depth. However, during the receive mode, the focus can be swept through a range of depths to pick up the multiple echoes produced by one transmit pulse.

6 Image display modes

Information generated from returning echoes can be displayed in a variety of ways, called US imaging modes. The mode used to display images generated from returning echoes depends on the type of US scanner used, the information sought, and the organ examined.

6.1 A (amplitude)-mode

In the Amplitude Modulation or A-mode, the returning echoes are displayed on the monitor as spikes originating from a single vertical or horizontal baseline. The depth of the echo is determined by the position of the spike on the axis, with the top or left side of the monitor being the most superficial and the bottom or right side being farther away. The height of the spikes (y-axis) correlates to the amplitude of the echo, while the x-axis represents tissue depth. Fig. 12 shows an example of A-mode image display.

Except in ophthalmologic applications where it is used to identify defects in the cornea and optic nerve, today the A-mode is not widely used.

6.2 B (brightness)-mode

In B-mode (Fig. 13), echoes are represented by dots on a line that form the basis of a two-dimensional image. The brightness of each dot indicates the amplitude of the returning echo. Its location relative to the transducer is displayed along the vertical axis of the monitor, with the top of the monitor representing the transducer. The returning echo's location along the axis is based on the amount of time it takes for the US wave to be

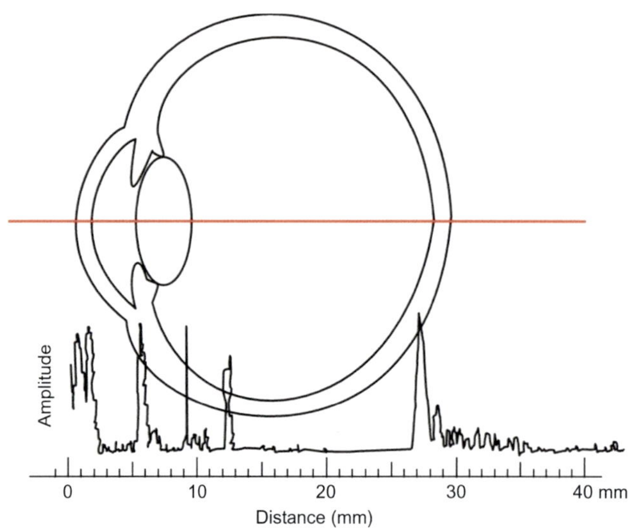

Fig. 12 Schematic transconeal A-mode ultrasound image of the eye. Note that the spikes are amplitudes generated by tissue structures more echogenic (sound reflective) in the A-mode. Where needed, the distances between differing reflective structures can be determined.

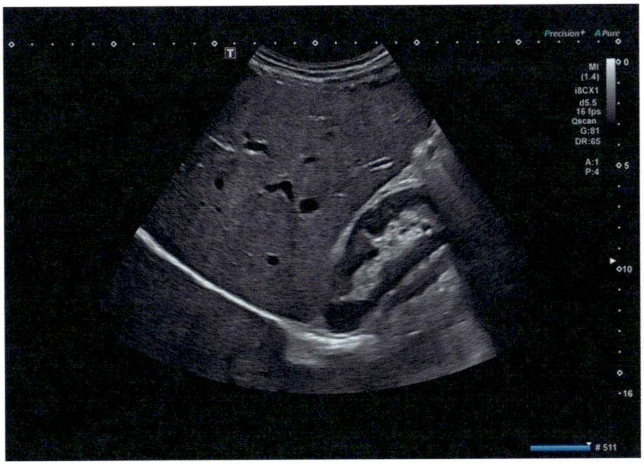

Fig. 13 Abdominal B-mode image of the human liver and kidney. *(Figure courtesy of Prof Frederic Patat and M. Paul-Armand Dujardin (CHRU Bretonneau, Tours, France).)*

transmitted from the transducer and reflected back. Echoes arising from structures in the near field (close to the transducer) take less time than those coming from the far field (farther away from the transducer) because they travel a shorter distance.

Real-time B-mode ultrasonography allows a complete, two-dimensional, cross-sectional image to be generated by using multiple B-mode lines. In real time B-mode imaging, the transducer sweeps the US beam through the patient many times a second. With each pass of the US beam, multiple lines of dots are generated on the monitor, producing a complete image. These B-mode lines remain on the monitor until the next sweep of the US beam. Because several beam sweeps are performed each second, a moving, changing, "real-time" image is generated. This is the mode most commonly used in clinical practice.

6.3 M (motion)-mode

M-mode (Fig. 14) is used in echocardiography and allows the sonographer to measure the heart to assess cardiac function and chamber size. M-mode uses a single B-mode line, with the amplitude of the echoes indicated by the brightness of the displayed dots. The difference is that the information obtained from that single line is constantly swept across the monitor so that the motion of the body part being investigated is displayed along the horizontal axis.

7 Specialized and emerging imaging techniques

In this section, the underpinning technologies of ultrasound diagnostic imaging are reviewed. Once again, the choice of which technologies to include is selective.

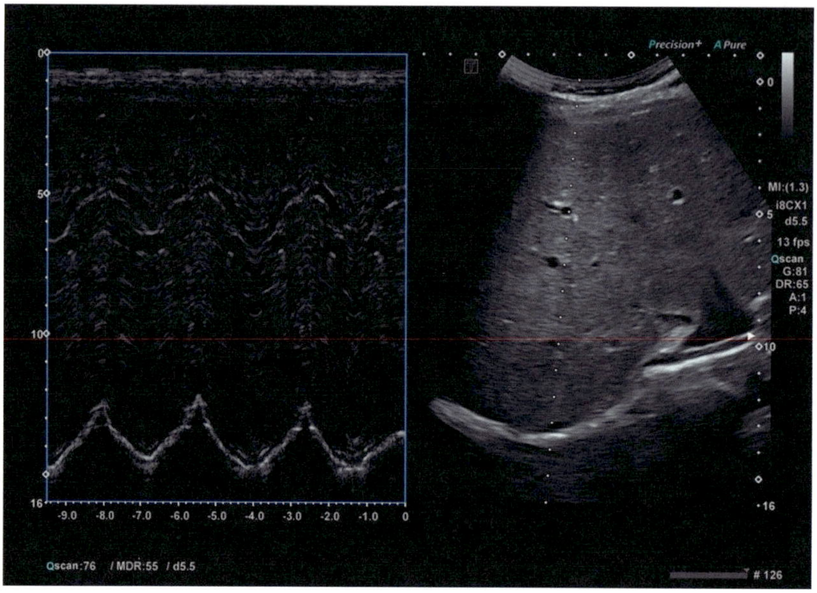

Fig. 14 Abdominal M-mode image depicting human diaphragm movement. *(Figure courtesy of Professor Frederic Patat and M. Paul-Armand Dujardin (CHRU Bretonneau, Tours, France).)*

7.1 Contrast-enhanced ultrasound imaging

Despite its diagnostic efficiency, conventional US imaging is sometimes unable to definitively differentiate, e.g., malignant from benign tissues. To augment the sensitivity of US imaging and map blood perfusion, the use of US contrast agents—small gas-filled microbubbles that are usually injected intravenously, were introduced. Contrast agents based on microbubbles typically contain air or high-molecular-weight, low-solubility gas encapsulated in a lipid or albumin shell. With a size range of 1–7 micrometers in diameter (Fig. 15), these bubbles are small enough to pass through the lungs and circulate in the blood stream for several minutes, when administered into the bloodstream. Although the diameter of these microbubbles is 1000 times smaller than the wavelength of the interrogating US, they can be made to resonate in an US beam, rapidly contracting and expanding in response to the pressure changes of the sound wave.

The resonant frequency of the bubble increases as its diameter decreases if the shell properties are unaltered. When introduced even at high dilutions, echoes from the bloodstream in which microbubbles are suspended are greatly enhanced, particularly when their oscillation resonates with the interrogating US frequency. This is due to their intrinsic compressibility, which is approximately 20,000 times greater than that of water. This in itself provides the potential to obtain additional clinical information by contrast-enhanced US imaging. Fig. 16 provides a good diagnostic example of how improvements

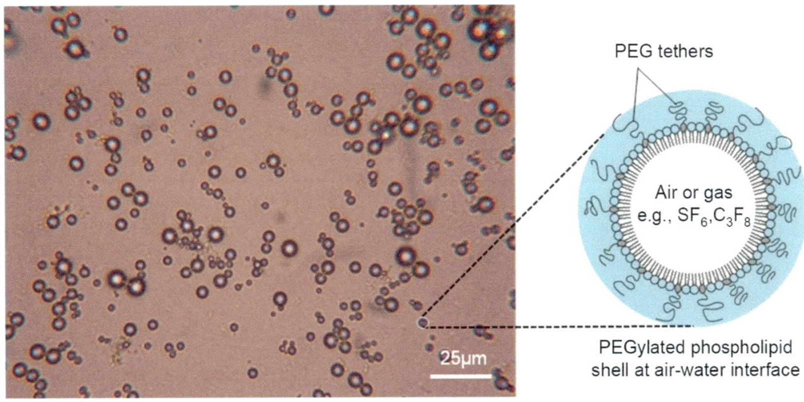

Fig. 15 Micrographic image of microbubbles as ultrasound contrast agent. (Scale bar = 25 μm). The inset shows a schematic representation of typical microbubble compositions.

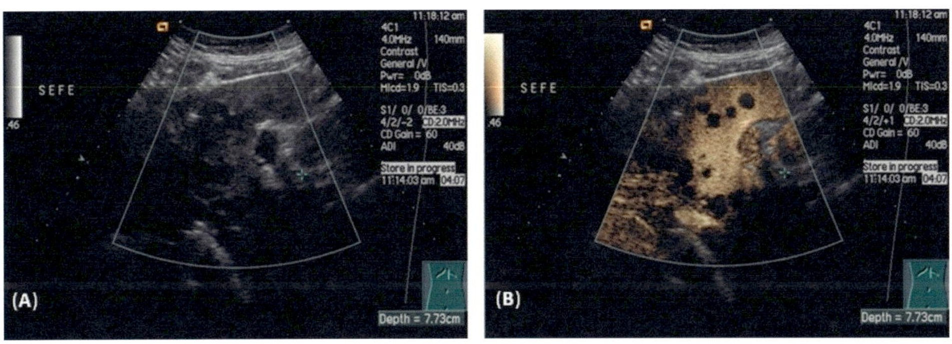

Fig. 16 Contrast-only imaging of microbubble in the bloodstream allows clinicians to distinguish malignant tumors from benign and normal tissue. For example, panel (A) shows an abdominal B-mode image of a patient with hidden liver metastases. Panel (B) shows a contrast-enhanced image of the same patient with the previously hidden metastases now inconspicuous. *(Images courtesy of the late Prof Martin J.K. Blomley, Hammersmith Hospital and Imperial College, London.)*

in US imaging sensitivity through the use of engineered microbubbles today allows clinicians to reveal tumors that would otherwise remain hidden. Moreover, it was noted soon after clinical experiments with contrast agents began that quite strong echo signals were also received from microbubbles at the second harmonic of the interrogating US frequency. Because the tissues and blood itself behave in a much more linear fashion, the echoes from all scatterers except microbubbles can be suppressed by arranging to receive echo signals in the spectrum centered at twice the transmitted frequency. This way, it becomes possible to image small blood vessels, including the microvasculature, otherwise indistinguishable from the surrounding tissue with traditional pulse-echo imaging.

An important characteristic of most microbubble contrast agents is that they can be destroyed if exposed to US of sufficient intensity. Fragmentation occurs when bubbles experience a large expansion and subsequent contraction. This is an extremely fast process, occurring on a time scale of microseconds. Thus, a population of microbubbles in a selected volume of tissue can be destroyed by the application of a brief pulse of US with a peak negative pressure typically of 2–3 MPa. This allows the time course of the subsequent refill dynamics to be observed by interventionists as this is related to the perfusion of the tissue by blood. Using the refill pattern as a guide, interventionists can differentiate benign from malignant tumors.

Two other significant mechanisms of microbubble destruction are also exploited for imaging purposes. US-enhanced diffusion can destroy microbubbles in as short a time as 50 ms, depending on the properties of the shell and the gas. In the absence of US, static diffusion typically results in bubble destruction in time scales of minutes or even hours. Second harmonic imaging with contrast agents involves a compromise between image contrast and spatial resolutions. This is because unavoidable overlap between the frequency spectrum of the transmitted beam and the bandpass of the receiver means that some of the transmitted power contaminates the received second harmonic signals. The problem can be solved, however, by transmitting sequentially inverted pulses and summing the resultant consecutive echo wave trains. In this way, the linear echoes from tissues and blood add up to zero, whereas the nonlinear echoes reinforce each other. This process called pulse inversion is now used primarily to enhance the sensitivity of Doppler color flow imaging with microbubble contrast agents.

Although the US contrast agents in current clinical use are those based on microbubbles, there is interest and potential in nongaseous agents. Perfluorocarbon colloidal nanoparticles, microdroplets, and microtubules all appear to be promising, particularly as they may be immunologically targeted to specific anatomical or pathological sites. Here, immunoglobulin antibodies with binding specificity to certain tissues or biomolecular receptors present in the blood vessels of diseased tissues or organs can be conjugated on the exterior of the contrast agents to provide strong echo enhancement, which approaches that of microbubbles, wherever they arrive.

The clinical potential of targeted US contrast agents is considerable, both for cancer diagnostic [12–14] and therapeutic drug and gene delivery [15–17] applications. Ultimately, microbubbles might one day allow US to replace much of contemporary radionuclide imaging for functional studies, thus avoiding the problems associated with radioactivity as well as deliver genes for cancer treatment.

7.2 Ultrasound molecular imaging

US molecular imaging has emerged and advanced rapidly in the last 15 years. The main components in the context of US molecular imaging are the use of targeted microbubbles and relatively high-frequency imaging systems. As already mentioned in Section 7.1, there is widespread use of US contrast agents consisting of encapsulated gas microbubbles. Cancer has the ability to spread to adjacent or distant organs, which makes it life threatening. Tumor cells can penetrate blood or lymphatic vessels, circulate through the intravascular stream, and then proliferate at another site: metastasize. Thus, metastatic spread of cancer cells depends on an adequate supply of oxygen and nutrients and the removal of waste products. Here, new blood and lymphatic vessels form through processes called angiogenesis and lymph angiogenesis, respectively. Because the vascular network of malignant tumors differs from that of benign tumors, the circulation of microbubbles offers a simpler way of identifying malignancies.

Commercially available high-frequency imaging systems operating at 40 MHz are commonly used to interrogate voxel sizes as small as $40 \times 80 \times 80\ \mu m^3$ (at the focal zone). This high spatial resolution can be useful for small animal and preclinical studies, although lower frequencies will still be required upon translation of these applications to human use. The use of targeted microbubbles provides the needed marking component. As explained earlier, microbubbles increase significantly the acoustic reflection of the volume in which they reside. Consequently, high concentration of microbubbles can be detected and be indicative of the physiological state of the tissue. The combination of high-frequency high-resolution US with the targeted microbubbles therefore offers a powerful research tool, whose potential is currently under investigation.

Targeting is commonly achieved by attaching antibodies to the shell surface of microbubbles. These antibodies serve as the homing ligand for contrast material causing the agent to accumulate in or around the target tissue such as tumors. The localized enhanced concentration of the contrast material augments the acoustic reflectivity from the diseased target making it distinguishable from nontarget tissue. A schematic description of this concept is depicted in Fig. 17. In this case, a gas-filled microbubble serves as contrast agent. Since all foreign agents such as liposomes are usually cleared from the bloodstream within a short time if left unprotected, a poly-ethylene-glycol (PEG) coating is commonly employed to make these targeted or drug-loaded contrast agents undetectable by the macrophages that would otherwise rapidly clear them from the bloodstream. Here, antibodies are attached to the PEGylated shell in order to provide targeting capability; and the gas within the bubbles due to its compressibility serves as the US reflecting material as explained in Section 3.1.

7.3 Tissue characterization and elastography

When US echoes are collected, there is often interest in learning more about the characteristics of the tissues that produced them. The goal of elasticity imaging therefore is to map tissue properties such as Young's elastic modulus, Poisson's ratio, and viscosity. These properties are now known to be modified by various disease processes. For

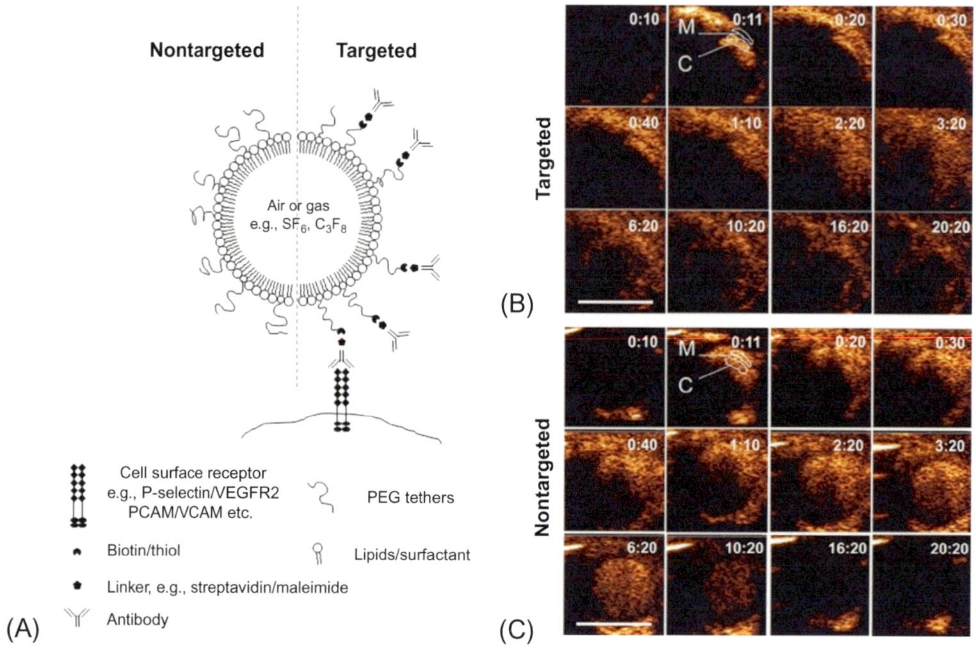

Fig. 17 Ultrasound molecular imaging using targeted microbubbles. (A) Schematic representation of a nontargeted (left) and targeted microbubble (right) adherent to a cell surface receptor. (B) Panels show the heart chamber of a diseased heart with targeted microbubbles arriving in the chamber after 10 s and ultimately accumulation of the targeted microbubbles in the heart muscle (denoted M). (C) Similar microbubbles in the control heart are cleared from the heart muscle by 20 min suggesting that targeted microbubbles do not adhere to healthy heart muscle tissues. Scale bar = 5 mm. *(Panels (B) and (C) courtesy of Drs James Yeh (London North West University Healthcare NHS Trust) and Charles Sennoga (Tours France).)*

example, malignant lesions are usually stiff and unyielding compared to surrounding benign tissues. As a result, the most obvious characterization would necessarily call for a way of distinguishing, e.g., benign from malignant tumors, fatty from fibrotic liver or indeed the grading of arterial plaque calcification.

Recent research into tissue characterization has developed elastography also known as sonoelastography as a clinical tool for such tissue characterizations. Parameters related to US phenomena described in previous sections, such as the speed of sound and attenuation coefficient, have been adopted in order to determine tissue characteristics. The characterization of tissue elasticity using conventional diagnostic frequencies is often made difficult by the degradation of the US beam by fat and muscle between the transducer and the site of interest. Invasive imaging can reduce some of these problems, for example by using catheter-based transducers to characterize blood vessel walls, which are examined with higher imaging frequencies and shorter beam paths through blood. The problems are also reduced if there is a well-specified path between the transducer and the tissue of interest, e.g., through the eye to the retina or through a water bath to superficial skin lesions.

There is substantial development at the moment in the measurement of tissue elasticity using US. When tissue is compressed by an external force, changes in its shape and size can be measured from the changes in position of the echo speckle pattern. From the size of the force and corresponding tissue distortions, quantities relating to the elasticity of the tissue can be calculated. This is analogous to manual palpation although the aim here is to develop techniques that are quantitative and applicable to tissues deep within the body. Based on the knowledge that the hardness of a biological mass can often be related to its pathology, several different methods may be employed to apply a force to tissue. These typically range from simply pushing the transducer, to using the beam radiation force to generate shear waves, which spread out around the focus of a normal US beam. It was noted earlier that when the momentum of an US beam is interrupted by reflection or absorption at a target, the target experiences a pushing force. In the arrangement shown in Fig. 18, a target tissue at the focus of the beam will experience a force

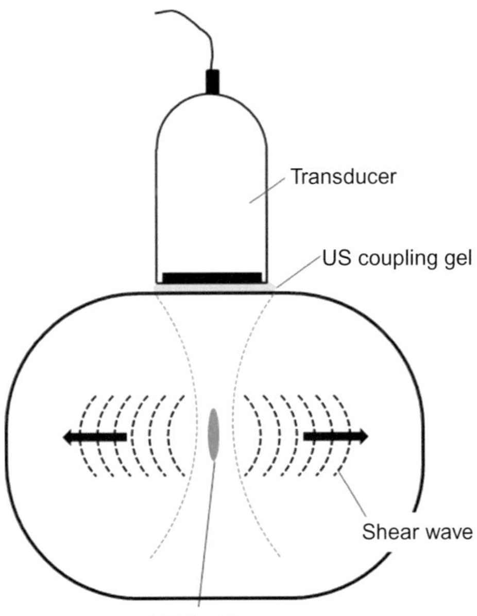

Fig. 18 In elastography, a reference image is first obtained in an unloaded condition. Additional images are then acquired during the application of pressure to the tissue of interest. Shear waves are generated by the force at the focus of an ultrasound beam. The corresponding deformations are used to display and interpret the resultant tissue elasticity.

impulse when subjected to US pulse. The resulting motion of the tissue produces a transverse shear wave, which propagates sideways from the direction of the US beam, in a manner not unlike ripples in a pond when a pebble is dropped in. If several focal points are targeted at successive depths by successively transmitted pulses, the shear waves produced interfere with each other as they travel sideways and produce two plane shear waves traveling in opposite directions at an angle to each other.

Whereas shear waves at MHz frequencies are strongly attenuated in biological tissues and as a result had not found clinical applications (Section 2.4), shear waves at frequencies of a few hundred Hz propagate effectively through tissue and today form the basis of elastography imaging. The frequencies of the shear wave are determined by the pulsing rate of the US transducer. Moreover, both the speed of shear waves (c) and the wavelength (λ) in tissue also differ from typical pulse waves employed in conventional US imaging, i.e., $c=1$ m/s and $\lambda=1$ mm for transverse US shear waves as compared to $c=1540$ m/s and $\lambda=0.5$ mm for longitudinal US waves. As the shear waves travel out from the focus (the shear wave source), tissue moves under its influence, e.g., back and forth over typically 50 μm. US imaging systems operating at very high frame rates are able to image these small motions by using signal processing methods similar to those used in Color Doppler imaging. From such images, the speed of the shear wave at each point can be calculated. Since this speed depends on the elasticity (i.e., the stiffness) of the tissue, the resultant output can be used not only to understand something about the tissues that produced them but also to diagnostically differentiate diseased (e.g., tumors in breast) from normal tissue. The outcomes may be quoted in terms of the shear modulus, i.e., a measure of tissue stiffness, in Pascal's [Pa] units. It should however be noted that considerable computation is required to provide the ultrafast imaging and elasticity images in real time. In spite of the computational demands, shear wave techniques are now well under development, and commercial equipments are now available for clinical evaluation. Efficacy studies are underway and the intensities of US used in such systems fall within regulatory guidelines. It is anticipated that in addition to elasticity, measurements of tissue viscosity will be provided in the near future.

7.4 Tissue harmonic imaging

It was noted during early experiments with US contrast agents (Section 7.1) that the echoes from the deeper parts of tissues imaged by pulse-echo techniques actually contain significant components at the second harmonic of the transmitted frequency. These harmonic components arise because a finite-amplitude transmitted pulse is distorted as it propagates through inherently nonlinear tissue. By selecting the second harmonic frequency signals for imaging in the absence of contrast agents, improvements in spatial resolution and other aspects of the image become apparent, particularly when examining what are considered technically difficult patients, with obese or dense muscle structures.

The reasons for this is that, because these harmonics develop slowly, the low-frequency transmitted beam suffers relatively little distortion when passing through the superficial tissues, most of the second harmonics develop in the main beam rather than in the lower amplitude side lobes; the side lobes of the transmitting and the receiving beams have different angular distributions, whereas the main beams are coincident.

Although tissue harmonic imaging at the second harmonic frequency is thought to be more efficient than imaging at the fundamental frequency, there is some controversy over what aspects of the imaging are better and the process certainly has some difficulties. Since the bandwidth of the transducer has to necessarily accommodate the frequency spectra of both the transmitting and receiving beams, it is inevitable that some of the high-frequency components of the spectrum of the short transmitted pulse leak into the spectrum of the received signals. In addition, the dynamic range of the received signals is reduced, because their absolute amplitude is closer to that of the system noise.

Improvement in tissue harmonic imaging is currently obtained at the expense of image frame rate, by transmitting pulses with sequentially reversed polarities and adding consecutively received echo wave trains. This pulse inversion technique effectively removes signals with odd harmonic frequencies (including echoes at the fundamental transmitted frequency) and doubles the amplitudes of those at the second harmonic and higher even numbered frequencies. In practice, harmonics higher than the second one are also produced as a result of the nonlinear propagation of the transmitted pulse. Although these are relatively weak, a usable signal can be obtained by summing the third, fourth, and even the fifth harmonics to constitute what has been called the superharmonic component of the received echoes. For this, a composite broadband transducer is required. Superharmonic imaging is characterized by high spatial and temporal resolutions and the relative absence of clutter.

7.5 Three-dimensional imaging

Most conventional US imaging systems employ handheld transducers, which produce two-dimensional images in real time. A skilled operator moves the transducer over the surface of the patient to explore the internal anatomy and thus to build up, in their mind's eye, a three-dimensional image.

There are two main approaches to display the scanned anatomy in three dimensions. Firstly, the two-dimensional imaging transducer may be moved mechanically in a predetermined trajectory or the beam may be scanned in three dimensions by means of an electrically controlled two-dimensional array so that position and orientation are directly measured and these data are used to assemble a three-dimensional image volume from the set of two-dimensional images. Secondly, the position and orientation of the two-dimensional imaging transducer, while being freely moved by the operator over the surface of the patient, may be measured by sensors with respect to a frame of reference, which is fixed in relation to the patient. By appropriate calibration of the spatial relationships in the system, a three-dimensional image can be assembled.

Because a three-dimensional image is assembled from a set of two-dimensional scans, acquisition is proportionately slower than it is for two-dimensional imaging, unless a scheme to accelerate the process is adopted. As with other medical imaging modalities, the display of three-dimensional data is challenging. Since it is usually difficult automatically to segment US images, three-dimensional rendering is problematic. Today, the easiest way to explore a three-dimensional US image volume is usually to enable the observer to rapidly display two-dimensional images in any desired orientation within the scanned volume. Of course, this is not ideal, especially if four-dimensional imaging (i.e., three time-varying spatial dimensions) is actually what is needed.

7.6 Transducers

In many respects, the transducer is the most critical component in any US imaging system. The ideal transducer for US imaging should have a characteristic acoustic impedance that is perfectly matched to that of human tissue, and achieve good efficiency in the transduction process, i.e., have high efficiency as a transmitter and high sensitivity as a receiver, a wide dynamic range, and a wide frequency response for pulse operation.

The great majority of medical US imaging systems employ transducers based on piezoelectric materials, of which the artificial ferroelectric ceramic, lead zirconate titanate (PZT) is the most common. PZT exhibits higher characteristic impedance than that of water, but it can be made to perform well by the cautious use of matching layers consisting of materials with intermediate characteristic impedances. Better performance can be obtained by embedding small particles or shaped structures of PZT in a plastic to form a composite material: this exhibits lower characteristic impedance than that of PZT alone, although it has similar ferroelectric properties.

Polyvinylidene difluoride (PVDF) is a plastic, which can be polarized so that it has piezoelectric properties. The piezoelectric effect can be enhanced by the addition of small quantities of appropriate chemicals. The advantages of this material are that it has a relatively low characteristic impedance and broad frequency bandwidth; it is fairly sensitive as a receiver but rather inefficient as a transmitter.

Piezoelectric transducers are normally operated over a band of frequencies centered at their resonant frequency. The resonant frequency of a transducer occurs when it is half a wavelength in thickness. Typically, a PZT transducer resonant at a frequency of, for example, 3 MHz is about 650 μm thick; meaning that it is mechanically robust for simple, even manual, fabrication techniques. Higher-frequency transducers are proportionally thinner and, consequently, more fragile.

The potential of capacitive micromachined ultrasound transducers (cMUTs) for at least partially replacing PZT and PVDF devices is a subject of ongoing research.

A cMUT transducer typically consists of a micromachined capacitor that is typically mounted on a silicon substrate and with a thin electrode membrane as the other plate of the capacitor (the active surface of the transducer). When a direct current voltage is applied between the plates of the device, an alternating current voltage causes the membrane to transmit a corresponding oscillatory force, on one hand, while on the other hand a received wave causes a corresponding change in the spacing between the plates, thus generating an electrical signal. cMUTs are adequately sensitive as receivers, but need high voltages to be effective transmitters. Some of the potential advantages of these devices are that they can be fabricated into arrays with integrated electronics and, if manufactured in large quantities, could be relatively inexpensive.

Although some simple probes contain single-element transducers (e.g., one element for transmitting and one for receiving, in a continuous-wave Doppler system), most modern imaging systems use arrays of transducer elements for beam forming as previously discussed (Section 5.4).

8 Ultrasound therapeutic applications

Ultrasound therapeutic applications are discussed here only in passing since this chapter focuses primarily on imaging applications.

8.1 High-intensity focused ultrasound

High-intensity focused ultrasound (HIFU) is an emerging technology for noninvasive surgery. HIFU has become popular only in the past few years, even though it was first proposed in the 1940s. This re-emergence is directly attributed to the integration of HIFU with other more advanced imaging modalities. Recently, image-guided HIFU has been offered as a noninvasive alternative to conventional lumpectomy [18] in the breast and other organs [19]. For the HIFU technique, a US transducer (or an array of transducers that are focally aligned with the region of interest), is used as a virtual "knife" to noninvasively destroy a small volume of tissue within the body. The focused high US energy induces a rapid and significant temperature rise at the focal point leading to localized tissue ablation.

The basic principles of HIFU-based intervention are schematically depicted in Fig. 19. The HIFU transducer, placed externally to the body, is acoustically coupled to the patient's skin by water or acoustic gel. The focal zone, which has a typical cigar shape, is aimed toward the target tissue, such as a tumor, by using another imaging modality (MRI or US) as guidance. The wave intensity, i.e., energy per unit area, increases as the wave converges toward the focal zone. Most importantly, the intensity at the focal zone can reach several hundred Watts/cm^2. As a result, the temperature at the focal zone increases rapidly. Instant cell death is assumed when the induced temperature exceeds 57°C. The HIFU transducer is then steered mechanically or electronically to ablate

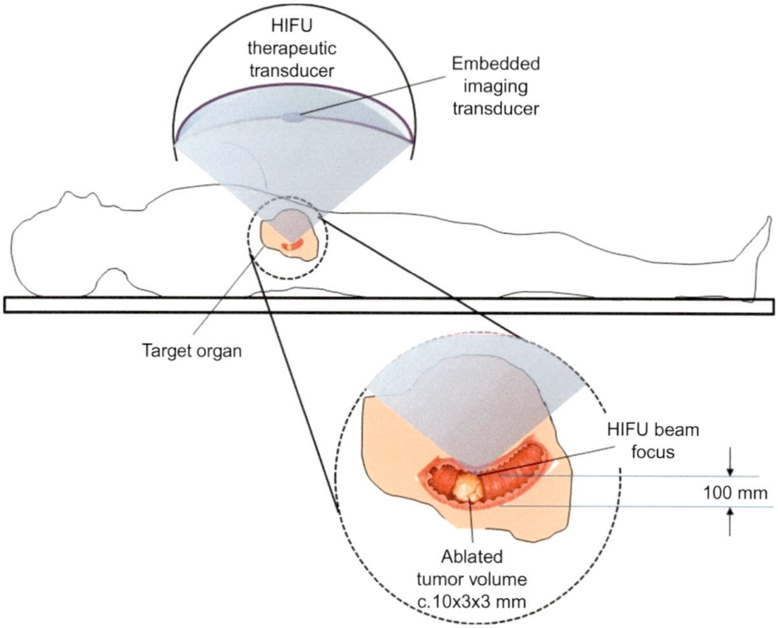

Fig. 19 Principles of high-intensity focused ultrasound (HIFU) therapy. HIFU from an external source is conveyed deep into tissues with a large convergence angle. The focal zone of high intensity US beam is aimed and moved within the target tissue in order to achieve complete thermal ablation of the malignant tissue. The focus is oval in shape with a short axis of approximately 3 mm and a long axis of 10 mm. Typical therapeutic transducers have a multielement array and concave focusing. The angle of convergence is typically 80°. The size of the focal point is commonly 3 mm × 3 mm × 10 mm.

additional zones within the target tissue, until complete destruction of the target has been achieved. Although a temperature of 57°C is considered sufficient to kill the tissue in the focal zone, current systems aim to reach temperatures ranging from between 70°C and 80°C. Unfortunately, this results in a larger treatment zone primarily due to heat dissipation.

HIFU is currently the only clinically practiced radiation free method for noninvasive surgery. Despite its success, there remain many technical issues that need to be resolved with this technology.

8.2 Drug delivery and gene therapy

Drug delivery and gene therapeutic applications exploit the ability of US to temporarily open gateways in tissue membranes. The exposure of living biological cells to intensive US waves for a sufficiently long duration alters the cells membrane properties by creating transient nonlethal perforations and aids the ingress of large molecules and particles into the cells in a process called sonoporation. In this way, cell membranes subjected to US fields with a prescribed energy intensity become highly permeable for a certain amount of time before reversibly closing the pore structures again. While this requires US power that is substantially higher than that permitted for medical imaging, the power needed is greatly reduced when microbubbles are present. This is because microbubbles, acting as cavitation sites [20], lower the amount of energy necessary for cavitation, the process in which extreme oscillations induced by US pulses lead to microbubble collapse. The resultant membrane permeability can be utilized for drug delivery and gene transfection. In an exemplary application, a transdermal drug-delivery device [21], based on an US transducer, was attached to the abdominal exterior and allowed to transmit waves with a frequency of 55 kHz. This induced the formation of cavitation bubbles within the skin layer. The collapse of these bubbles led to transdermal micropathways. These micropathways ultimately allowed the transmission of fluids and chemical substances into the body. This device, which remains under development, is aimed at helping diabetic patients but also serves as a predicate for cancer treatment.

Extensive research is also underway with the specific aim of delivering drugs through the blood–brain-barrier (BBB) to brain tissue. Intensive US pulses transmitted to the brain under MRI guidance and aimed at reversibly disrupting the BBB, may allow large drug molecules, that would not under normal conditions cross the BBB, to reach specific targets in brain tissue comprising tumors for example. Another approach utilizes thermosensitive liposomes to encapsulate drugs [22]. As explained in Section 7.2, these liposomes can be targeted by conjugating immunoglobulin antibodies with binding specificity to specific tissues or sites within the vasculature. In this way, the circulation and capture of these targeted drug-loaded liposomes can be followed with US imaging, and can be subsequently thermally disrupted by an acoustic beam, thereby releasing their therapeutic payload after accumulating in or at the target tissue.

The potential of this approach in gene therapy has already been demonstrated [16] by using viral vectors. However, the use of viral vectors presents an increased risk of undesirable side effects. Moreover, the therapeutic genes commonly provided in a plasmid form need to be delivered into the nucleus of the target cells without damaging or compromising the cells themselves. US used in combination with microbubbles on the other hand may provide a safer option and as was previously shown [23] can achieve gene transfection by using physiotherapy-level intensities.

8.3 Ultrasound physiotherapy

The basic mechanism, which governs US physiotherapy, is the conversion of mechanical energy into heat. When a sound wave propagates through tissue, its amplitude is progressively reduced. While some of the energy is simply scattered, most of it is converted into heat. As a result, the gap between the US energy entering a given tissue element and the

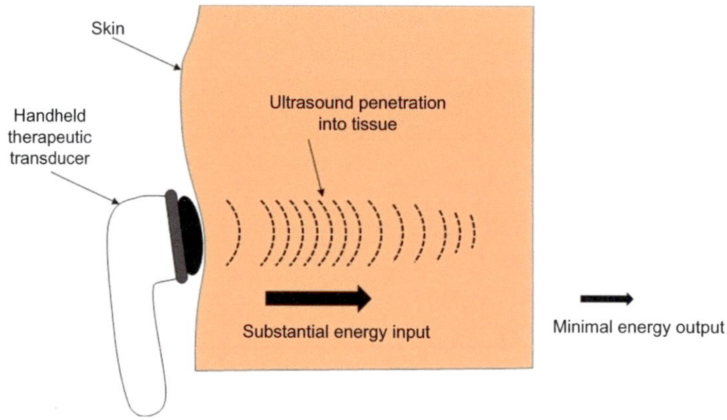

Fig. 20 Attenuation of ultrasound waves generates heat within the treated tissue. This resultant heat is then utilized for therapeutic applications.

energy leaving it serves as the energy source for the therapeutic process. As mentioned in Section 4, this resultant heat energy is used to destroy cells in pathologies such as malignant liver tumors. This is shown schematically in Fig. 20. For safety reasons, the intensities used in physiotherapy are usually kept as low as 3 W/cm^2. Indeed, the temperature rise within the tissue depends on several factors, including the transmitted intensity, the local attenuation coefficient, which as previously described depends on the frequency of the transmitted US beam, the specific heat and heat conductance coefficients of the tissue, and the local heat dispersion rate through blood perfusion. Accordingly, temperature rise in US physiotherapy is typically of the order 1–3°C.

References

[1] British Medical Ultrasound Society, Guidelines for the safe use of diagnostic ultrasound equipment, https://www.bmus.org/static/uploads/resources/BMUS-Safety-Guidelines-2009-revision-FINAL-Nov-2009.pdf, 2009. (Accessed 22 August 2018).
[2] J.S. Abramowicz, S.B. Barnett, F.A. Duck, P.D. Edmonds, K.H. Hynynen, M.C. Ziskin, Fetal thermal effects of diagnostic ultrasound, J. Ultrasound Med. 27 (2008) 541–559.
[3] D.L. Miller, Safety assurance in obstetrical ultrasound, Semin. Ultrasound CT MR 29 (2008) 156–164.
[4] NCRP, Exposure Criteria for Medical Diagnostic Ultrasound: II. Criteria Based on All Known Mechanisms (Report No. 140), National Council on Radiation Protection and Measurements, Bethesda, MD, 2002.
[5] M.E. Stratmeyer, J.F. Greenleaf, D. Dalecki, K.A. Salvesen, Fetal ultrasound: mechanical effects, J. Ultrasound Med. 27 (2008) 597–605.
[6] J.C. Bamber, C.R. Hill, Ultrasonic attenuation and propagation speed in mammalian tissues as a function of temperature, Ultrasound Med. Biol. 5 (1979) 149–157.
[7] F.A. Duck, Physical Properties of Tissue, Academic Press, London, 1990.
[8] F.A. Duck, Propagation of sound through tissue, in: G. Ter Haar, F.A. Duck (Eds.), The Safe Use of Ultrasound in Medical Diagnosis, The British Institute of Radiology, London, 2000, pp. 4–15.
[9] G.W.C. Kaye, T.H. Laby, Tables of Physical & Chemical Constants, National Physics Laboratory, UK, 2018. https://web.archive.org/web/20190202035258/http://www.kayelaby.npl.co.uk/general_physics/2_4/2_4_1.html. (Accessed 15 September 2018).

[10] B. Park, A.D. Whittaker, R.K. Miller, D.S. Hale, Predicting intramuscular fat in beef longissimus muscle from speed of sound, J. Anim. Sci. 72 (1994) 109–116.
[11] P.N.T. Wells, Biomedical Ultrasonics, Academic Press, New York, 1977.
[12] J.R. Lindner, Microbubbles in medical imaging: current applications and future directions, Nat. Rev. Drug Discov. 3 (2004) 527–533.
[13] A.L. Klibanov, Targeted delivery of gas-filled microspheres, contrast agents for ultrasound imaging, Adv. Drug Deliv. Rev. 37 (1999) 139–157.
[14] J.S. Yeh, C.A. Sennoga, E. McConnell, R.J. Eckersley, M.X. Tang, S. Nourshargh, J.M. Seddon, D.O. Haskard, P. Nihoyannopoulos, A targeting microbubble for ultrasound molecular imaging, PLoS One 10 (2015) e0129681.
[15] M.W. Miller, Gene transfection and drug delivery, Ultrasound Med. Biol. 26 (Suppl 1) (2000) S59–S62.
[16] R.V. Shohet, S. Chen, Y.-.T. Zhou, Z. Wang, R.S. Meidell, R.H. Unger, et al., Echocardiographic destruction of albumin microbubbles directs gene delivery to the myocardium, Circulation 101 (2000) 2554–2556.
[17] H.-L. Liu, C.-H. Fan, C.-Y. Ting, C.-K. Yeh, Combining microbubbles and ultrasound for drug delivery to brain tumors: current progress and overview, Theranostics 4 (2014) 432–444.
[18] J. Randal, High intensity focused ultrasound makes its debut, J. Natl. Cancer Inst. 94 (13) (2002) 962–964.
[19] K. Hynynen, O. Pomeroy, D. Smith, P. Huber, N. McDannold, J. Kettenbach, J. Baum, S. Singer, F. Jolesz, MR imaging-guided focused ultrasound surgery of fibroadenomas in the breast: a feasibility study, Radiology 219 (2001) 176–185.
[20] M.J. Blomley, J.C. Cooke, E.C. Unger, M.J. Monaghan, D.O. Cosgrove, Microbubble contrast agents: a new era in ultrasound, BMJ 322 (2001) 1222–1225.
[21] I. Lavon, J. Kost, Ultrasound and transdermal drug delivery, Drug Discov. Today 9 (2004) 670–676.
[22] S. Dromi, V. Frenkel, A. Luk, B. Traughber, M. Angstadt, M. Bur, J. Poff, J. Xie, S.K. Libutti, K. C. Li, B.J. Wood, Pulsed-high intensity focused ultrasound and low temperature-sensitive liposomes for enhanced targeted drug delivery and antitumor effect, Clin. Cancer Res. 13 (2007) 2722–2727.
[23] J. Alter, C.A. Sennoga, D.M. Lopes, R.J. Eckersley, D.J. Wells, Microbubble stability is a major determinant of the efficiency of ultrasound and microbubble mediated in vivo gene transfer, Ultrasound Med. Biol. 35 (2009) 976–984.

Further reading

[24] K.W. Beach, 1975–2000: a quarter century of ultrasound technology, Ultrasound Med. Biol. 18 (1992) 377–388.
[25] H. Becher, P.N. Burns, Handbook of Contrast Echocardiography, Springer, Berlin, Germany, 2000.
[26] C.B. Merritt, Ultrasound safety: what are the issues? Radiology 173 (1989) 304–306.
[27] P.N.T. Wells, Physics and engineering: milestones in medicine, Med. Eng. Phys. 23 (2001) 147–153.

CHAPTER 2.3

Magnetic resonance imaging

Neal K. Bangerter[a], **Glen Morrell**[b], **Matthew Grech-Sollars**[c]
[a]Department of Bioengineering, Imperial College London, London, United Kingdom
[b]Department of Radiology and Imaging Sciences, University of Utah, Salt Lake City, UT, United States
[c]Department of Surgery & Cancer, Imperial College London, London, United Kingdom

1 Introduction

The 20th and early 21st centuries have seen advances in the ability to form images of the living human body. The discovery in 1946 of the phenomenon of nuclear magnetic resonance (NMR) in condensed matter by Felix Bloch and Edward Purcell laid the theoretical framework for a new medical imaging technique, commonly referred to as magnetic resonance imaging (MRI). In 1973, the first anatomical images were obtained by Paul Lauterbur. Over the next decade, the technique was expanded and improved. By the late 1980s, MRI was in widespread clinical usage, and in 2002, more than 60 million MRI examinations were performed worldwide. Felix Bloch and Edward Purcell were awarded the Nobel Prize in Physics in 1952 for their discovery. Paul Lauterbur and Peter Mansfield, another early pioneer in the field, shared the 2003 Nobel Prize in Physiology or Medicine for their contributions to magnetic resonance imaging.

The use of MRI has continued to expand. Approximately 121 MRI scans were performed for every 1000 Americans in 2016, 56 for every 1000 Canadians, and 43 per 1000 persons in Australia. One significant issue with the use of MRI is its cost compared with other imaging modalities, and extreme disparities exist in MRI costs across countries and health systems. The headline of a Washington Post story from 2012 was "Why an MRI Costs $1,080 in America and $280 in France." The situation has changed very little in the interim.

The development of magnetic resonance imaging has been exciting for a number of reasons. First, MRI does not make use of ionizing radiation and is generally noninvasive (except with the introduction of contrast agents, which will be discussed later in this chapter). Imaging relies on a strong uniform polarizing magnetic field, smaller "gradient" magnetic fields, and radio-frequency magnetic fields, none of which damages the tissues of the body. Furthermore, many different imaging parameters and methods are available to change image contrast. Excellent soft-tissue anatomical contrast can be obtained, as well as images of metabolic activity and even blood flow velocities.

One of the drawbacks of many magnetic resonance imaging techniques is the length of time needed to acquire an image and the costs associated with siting and maintaining the strong superconducting magnets and other hardware required. Patient comfort and

Bioengineering Innovative Solutions for Cancer
https://doi.org/10.1016/B978-0-12-813886-1.00008-5

the ability to maintain a stationary position can be concerns during long scans. Likewise, magnet utilization costs can be significant and disparate, as mentioned. In addition to research on new contrast generation mechanisms, significant effort has been expended in speeding up image acquisition.

This chapter is divided into two broad sections. The aim of the first section is to provide background on the fundamental concepts and physics that make magnetic resonance imaging possible. While the treatment may seem irrelevant to those with a clinical focus, we believe that a basic understanding of the physics underpinning MRI is essential for those who wish to effectively exploit magnetic resonance imaging techniques for cancer research. Emphasis has been placed on presenting an intuitive overview of the basic concepts, rather than on developing a rigorous mathematical framework for MRI, which would be beyond the scope of this book. There are several excellent books on the subject that provide mathematical treatments of magnetic resonance imaging [1, 2], and the interested reader is referred to these.

The second section of this chapter seeks to build on this basic intuitive framework to provide a description of the most common uses of MRI for cancer in both the clinic and the laboratory. Where further mathematical development is required, it will be treated in context of the specific imaging technique under discussion.

2 Fundamentals of magnetic resonance imaging

2.1 Spin physics

2.1.1 Polarization

Protons and neutrons possess an inherent quality called spin angular momentum. In an atomic nucleus with an even combined number of protons and neutrons, the net magnetic moment of the nucleus is zero. However, nuclei with an odd combined number of protons and neutrons have an unpaired component of spin angular momentum. This gives rise to a small net magnetic moment: the nucleus acts as a tiny magnetic dipole.

The hydrogen nucleus in its most abundant form contains a single proton and therefore exhibits a small dipole moment. Hydrogen is very abundant in organic tissue, in the form of water and other organic compounds. It is this happy abundance of hydrogen nuclei that makes magnetic resonance imaging as we know it possible.

Consider a body containing a large number of hydrogen nuclei. In the absence of an external magnetic field, the orientation of the magnetic moment of each nucleus will be more or less random. Macroscopically, this leads to an internal cancelation of the small moments of each nucleus, and no net polarization is seen across the body. This is illustrated in Fig. 1A.

However, if a magnetic field B_0 is applied to the body, the spins align either parallel or antiparallel to the field, as shown in Fig. 1B. The parallel orientation is slightly lower in energy than the antiparallel state. Although spins tend to occupy the lowest energy state

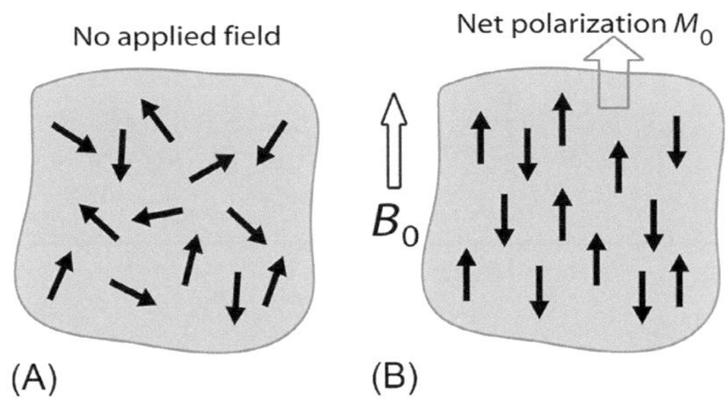

Fig. 1 In the absence of an external magnetic field, the magnetic moments of hydrogen nuclei are randomly oriented, as shown in (A), and no net polarization is seen across the body. In the presence of a magnetic field B_0, the spins align either parallel or antiparallel to the direction of the applied field, as illustrated in (B). Slightly more spins align in the parallel direction, resulting in a small net magnetization M_0.

available, thermal energy is sufficient to keep the two states almost evenly populated. However, slightly more spins still tend to align parallel, or longitudinal, to the applied field B_0, and a small net magnetic moment M_0 remains in the direction of B_0.

Going forward, this macroscopic net magnetic moment M_0 across a localized collection of nuclei with similar properties will be termed a "spin." References will often be made to a "spin" of a particular tissue at a particular location. This is a useful simplification: what is actually meant is a collection of nuclei exhibiting a net magnetic moment or polarization. As seen in the figures, this magnetic moment is graphically depicted by an arrow in the direction of polarization.

2.1.2 Precession

Assume now that a spin, in the presence of the polarizing field B_0, is made to point in a direction other than longitudinal to B_0. Classically, a torque on the dipole moment (or spin) is expected. The spin begins to precess, or nutate, about the direction of B_0 (the longitudinal direction, or z-axis in the diagrams). This is illustrated graphically in Fig. 2 and is analogous to a spinning top precessing in a gravitational field.

The frequency at which the spin precesses is linearly proportional to the applied magnetic field, in this case B_0. The proportionality constant relating the applied magnetic field to the frequency of precession is called the gyromagnetic ratio and is denoted by γ. The gyromagnetic ratio depends on the type of nucleus. The actual frequency at which a spin precesses is termed the Larmor frequency. For a hydrogen nucleus subjected to a polarizing field strength of 3.0 T, the Larmor frequency is 127.73 MHz. Most MRI machines in current clinical usage operate at main polarizing field strengths of 1.5 or 3 T, although 7-T whole-body MRI systems are becoming common in research environments. The advantages of higher field strengths are typically increased image quality

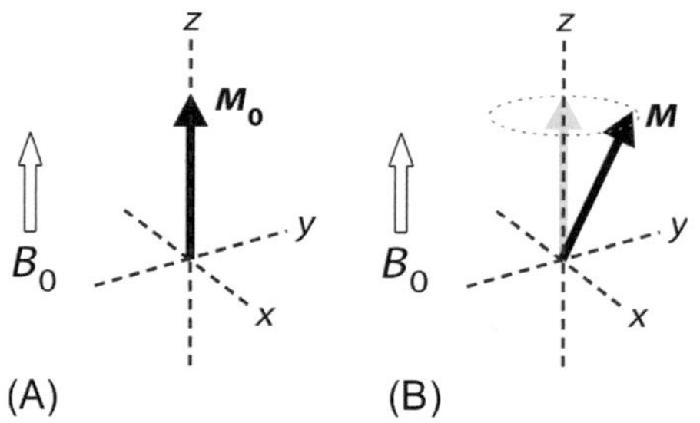

Fig. 2 (A) In equilibrium, the net magnetization M_0 (or spin) is oriented in the direction of the polarizing field B_0. (B) If the spin is caused to move out of its equilibrium position, it begins to precess about the direction of B_0 at the Larmor frequency.

and resolution, although there are accompanying technical challenges that make imaging at progressively higher field strengths more difficult.

2.1.3 Excitation

Spins can be tipped off axis, or excited, by an incident radio-frequency field tuned to the Larmor frequency and polarized in the plane transverse to the direction of B_0. This is illustrated in Fig. 3A. In the figure, the incident RF wave is polarized in the xy plane. The spin M exhibits nuclear magnetic resonance at the Larmor frequency. The applied radio-frequency field causes the spin to spiral downward, yielding a component of the magnetization in the transverse plane.

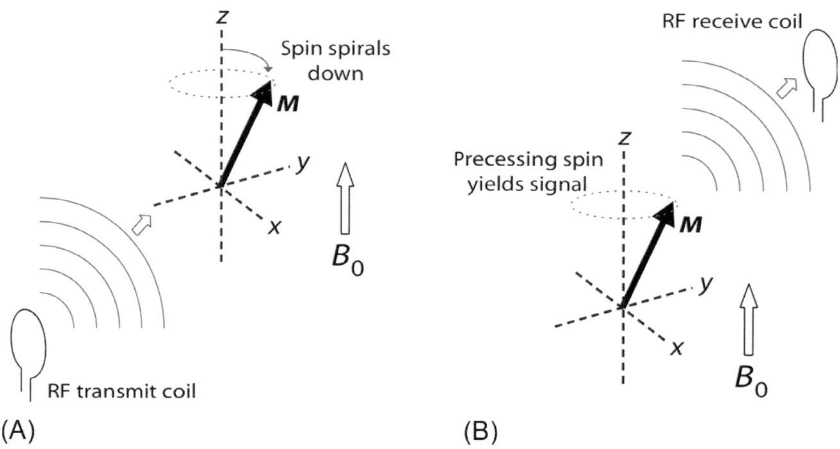

Fig. 3 (A) A radio-frequency field tuned to the Larmor frequency and polarized in the plane transverse to the direction of B_0 causes the spin to spiral out of its equilibrium position, effectively tipping it out of its equilibrium direction. (B) After the spin is tipped off axis, spin precession leads to a detectable RF signal at the Larmor frequency.

Once tipped, the spin will continue to precess at the Larmor frequency. This precessing dipole moment generates a fluctuating electromagnetic field, which leads to a detectable radio-frequency signal (Fig. 3B). This signal can be detected with a receive coil and ultimately used to form an image. The strength of the generated radio-frequency signal is proportional to the component of M in the transverse plane, M_{xy} in the diagram. The exciting radio-frequency field is often termed B_1.

In the absence of any continued radio-frequency signal, the spin will gradually return to its equilibrium state longitudinal to B_0, and the MR signal will decay away. This phenomenon, known as relaxation, will be described in more detail in the succeeding text. First, however, the configuration of a typical MRI scanner is described and an overview given of how the MR signal is used to form an image.

2.2 MRI hardware configuration

An actual MR machine typically creates the polarizing field B_0 with a large superconducting solenoidal magnet (Fig. 4). The person or body to be imaged is placed in the bore of the magnet. An RF transmit and receive coil is generally fitted concentric to the main solenoid inside the bore. Several other conventional electromagnets, termed gradient magnets, are also fitted just inside the magnet bore. The main superconducting magnet is designed to be extremely uniform (generally to within a few parts per million) across an imaging region in the center. The gradient magnets produce slight variations to the polarizing field B_0, as will be discussed in more detail later.

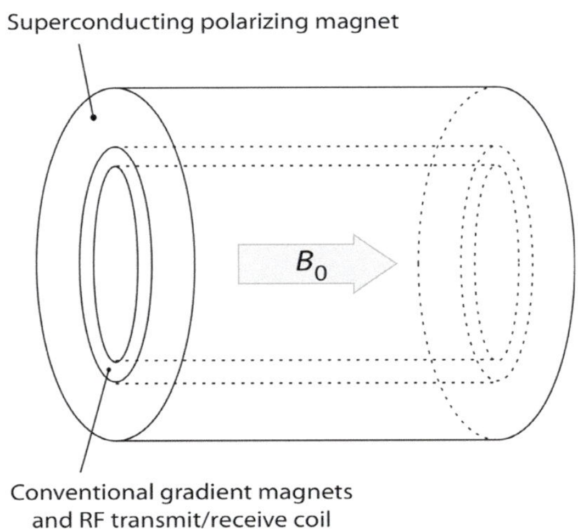

Fig. 4 The polarizing field B_0 is typically created with a large superconducting solenoidal magnet. Conventional magnets to create field gradients are fitted inside the bore of the superconducting magnet, along with an RF transmit and receive coil. The body to be imaged is placed in the bore of the magnet.

2.3 Spatial encoding

When a body to be imaged is placed in the bore of the scanner and an RF field B_1 applied to tip the spins, the spins begin to precess and emit a signal at the Larmor frequency, as discussed. However, all of the spins are now emitting signal at the same frequency. One would like to be able to distinguish the signal from spins at different locations within the body to form an image; in other words, one would like to perform some type of spatial encoding. Spatial encoding in MRI is accomplished by some combination of three mechanisms: slice-selective excitation, frequency encoding, and phase encoding.

2.3.1 Slice-selective excitation

Slice-selective excitation is the ability to selectively excite, or tip, spins in a certain slice or slab of the body to be imaged. When the RF field B_1 is applied in the presence of the polarizing field, all of the spins in the imaging volume that are subjected to the RF pulse tip off axis and begin to precess, yielding signal. If, however, the gradient magnets are used to produce a gradient in the magnetic field along a certain direction, spins at different spatial locations will resonate at different frequencies, depending on the local magnetic field at their locations.

This can be exploited to excite only spins in a certain slice or slab. For example, suppose that the gradient magnets are used to produce a small linear gradient in the magnetic field along the magnet bore, which we will call the z-axis. While this gradient is applied, the applied RF pulse is tuned to the Larmor frequency of the main field B_0. Due to the varying magnetic field along the z-axis, only spins in the vicinity of the z location where the local magnetic field is approximately B_0 will be excited and begin to emit signal. Spins at other locations, where the gradient magnetic field alters their resonant frequency, will not resonate and will consequently remain unexcited. This is illustrated in Fig. 5.

The thickness of the slab may be controlled either by varying the bandwidth of the RF pulse or by varying the magnitude of the slice-selective gradient field. A larger gradient field will produce a thinner excited slab if the RF pulse bandwidth is held constant. Decreasing the bandwidth of the RF pulse will likewise lead to a thinner slab excitation if the gradient field is held constant.

In this way, a three-dimensional spatial localization problem can be reduced to two dimensions (e.g., imaging a thin slice of the volume), or a smaller three-dimensional volume (a slab) excited within the body. Two-dimensional magnetic resonance imaging makes extensive use of slice-selective excitations. Many 3D MRI techniques make use of thicker slab-selective excitation to limit the signal-yielding volume.

2.3.2 Frequency encoding

Consider a two-dimensional array of excited spins in the polarizing field B_0 (as is the case when a thin slice-selective excitation has been performed). Now, suppose a spatially varying gradient field is applied, while the MR signal is sampled, as shown in Fig. 6A.

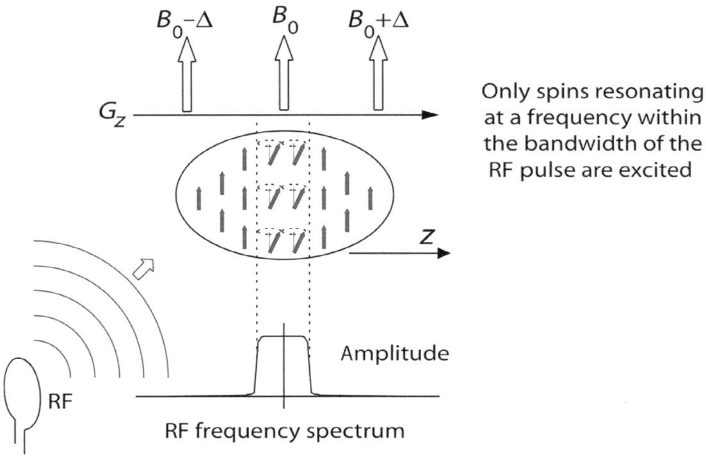

Fig. 5 It is often useful to excite spins in a slice or slab, rather than exciting the whole volume. An applied magnetic field gradient G_z varies the resonant frequency of spins along the z-axis. If a radio-frequency (RF) signal centered at the Larmor frequency with limited bandwidth is played in conjunction with such a gradient field, only spins resonating at a frequency within the bandwidth of the RF pulse will be excited.

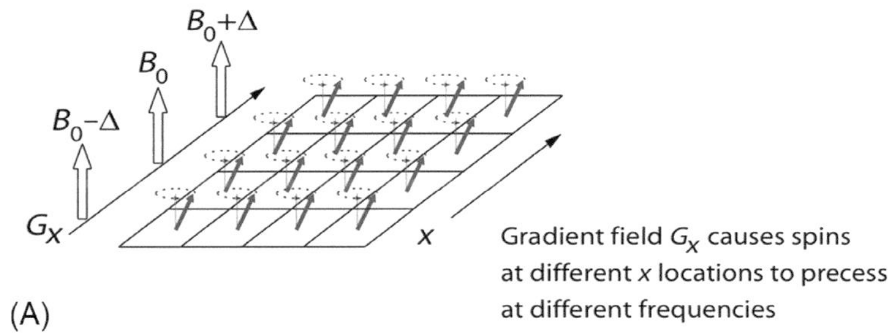

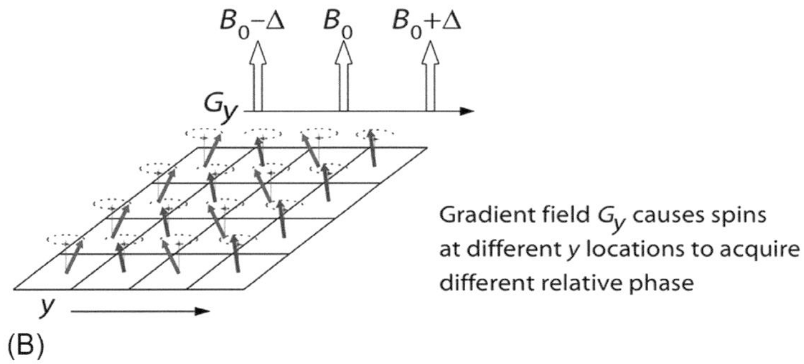

Fig. 6 (A) A gradient field G_x played during signal sampling (or readout) causes spins at different x locations to resonate at different frequencies, encoding x spatial location in the frequency of the sampled signal. (B) A gradient field G_y played for a fixed period of time prior to readout causes spins at different y locations to accrue different relative phase. This encodes y spatial location in the phase of the sampled signal.

In this example, the excited slice is an xy plane, and the gradient field (denoted G_x) induces a linear variation in magnetic field in the x direction.

Spins at different x locations now see different local magnetic fields and consequently precess at different frequencies. By analyzing the frequency components of the sampled MR signal, the amount of signal arising from spins at different x locations may be ascertained. In other words, x spin location has been encoded in the signal frequency. This is referred to as *frequency encoding*. One spatial dimension can be encoded in frequency and subsequently resolved from the detected MR signal.

2.3.3 Phase encoding

If the gradient field is applied for a fixed period of time *prior* to signal sampling, a different effect is observed. Suppose the gradient field G_y induces a linear variation in magnetic field in the y direction (Fig. 6B). Spins at different y locations precess at different frequencies during the duration of the gradient field. When the gradient field G_y is turned off, spins at different y locations will have acquired different relative phases. Spin location along y has now been encoded in phase.

2.4 Image formation

Repeated application of excitation (possibly slice selective), phase encoding, and signal sampling concurrent with frequency encoding eventually yield all of the data necessary to reconstruct an image. How many such repetitions are required depends on the dimensionality of the image (i.e., 2D or 3D) and the image matrix size. While the mathematical complexities of image formation will not be discussed here, a set of basic rules governing scan repetitions are given.

In general, one dimension is fully resolved by frequency encoding during signal sampling after each excitation. Formation of a 1D image therefore requires only a single excitation, followed by a sampling period. A frequency-encoding, or readout, gradient is applied during the signal sampling period. If N voxels are to be resolved along the single dimension, the signal must be sampled at least N times during the sampling period.

To form a 2D image, repeated excitation/phase-encode/readout steps must be performed, each with a different phase-encoding gradient played prior to signal readout. If an $M \times N$ image is required, M excitations must be performed, each followed by a readout with at least N sample points. Three-dimensional images can be formed by adding even more repetitions. An $L \times M \times N$ image would require $L \times M$ excitation/phase-encode/readout steps.

For a detailed discussion of MR signal sampling criteria, phase-encoding gradient requirements, and a more mathematical description of MR image formation, please see Ref. [3].

2.5 Relaxation and the Bloch equation

If all spins gave identical MR signal and the signal did not decay significantly before or during sampling, image formation with phase and frequency encoding would yield contrast based on the density of spins at different locations. This is possible and is referred to as *proton density (PD) contrast*. However, spins gradually return to their equilibrium (nonprecessing) orientation following excitation, leading to a decay in the MR signal. Furthermore, spins in different tissues return to equilibrium at different rates.

The magnitude of the MR signal from a spin corresponds to its transverse component M_{xy}. The transverse component of M decays exponentially with a time constant known as T_2. For example, after a 90-degree excitation (where the entire spin is tipped into the transverse plane), M_{xy} is given by

$$M_{xy} = M_0 e^{-t/T_2}.$$

This signal decay as a function of time is illustrated in Fig. 7A. T_2 is highly tissue dependent. Some values for various body tissues are shown in Table 1.

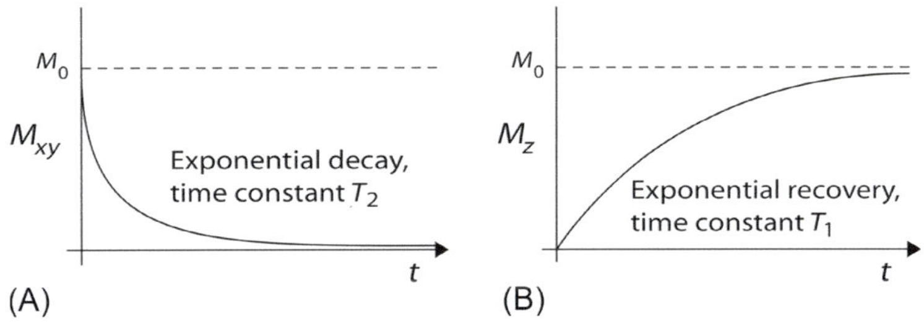

Fig. 7 (A) The magnitude of the RF signal from a spin is proportional to its transverse component M_{xy}. After excitation, the spin relaxes back to its equilibrium position M_0. The transverse component of the spin decays exponentially with time constant T_2. (B) The longitudinal component recovers exponentially with time constant T_1.

Table 1 Selected values at 1.5 T of T_1 and T_2 for relevant tissues.

Tissue	T_1 (ms)	T_2 (ms)
Gray matter	920	100
White matter	790	92
MS lesion	1300	200
CSF/synovial fluid	4000	2000
Arterial blood	1000	200
Venous blood	1000	100
Muscle	870	47
Fat	270	85

Note that values are approximate. There can be significant variations in T_1 and T_2 based on location in the body, physiological function, pathology, temperature, and a host of other factors.

M_z, the longitudinal component of M, recovers exponentially to its equilibrium value of M_0. Following a 90-degree excitation, M_z is given by

$$M_z = M_0\left(1 - e^{-t/T_1}\right).$$

This is depicted in Fig. 7B. T_1, like T_2, is tissue dependent. T_1 is also dependent on field strength. Values for a selection of tissues at a field strength of 1.5 T are shown in Table 1.

The dynamics of spin precession and relaxation can be captured in a single phenomenological differential equation known as the Bloch equation. Let γ represent the gyromagnetic ratio; **B** the magnetic field vector seen by a spin; T_1 and T_2 the longitudinal and transverse relaxation time constants, respectively; and M_0 the equilibrium magnetization due to the main polarizing field. Then,

$$\frac{d\mathbf{M}}{dt} = \mathbf{M} \times \gamma\mathbf{B} - \frac{M_x\mathbf{i} + M_y\mathbf{j}}{T_2} - \frac{(M_z - M_0)\mathbf{k}}{T_1},$$

where **i**, **j**, and **k** are unit vectors in the x, y, and z directions, respectively.

Precession of the spin at the Larmor frequency is described by the cross product term in the Bloch equation, while transverse and longitudinal relaxation are described by the terms containing T_2 and T_1, respectively.

2.6 Basic contrast generation mechanisms

MRI is one of the most flexible imaging modalities for generating different types of image contrast between soft tissues in the body. MRI images are almost always displayed as simple grayscale images, with black corresponding to the lowest signal value one wishes to display and white corresponding to the brightest. We often use the term *hyperintense* (brighter) to indicate areas or voxels of stronger signal (more white), while the term *hypointense* (darker) indicates areas of lower signal. We also often speak of *regions of interest* (ROIs) in an image. Differences in the density of 1H in the tissue (or *proton density*), T_1, T_2, and a variety of other tissues properties can be exploited to generate MR image contrast.

2.6.1 Proton density contrast

For example, suppose that a long repetition time TR is used between excitations, sufficient for near-complete longitudinal (T_1) recovery between excitations. Suppose further that we sample the signal as rapidly as possible following excitation, leaving very little time for the signal to decay irrespective of the tissue T_2. We then achieve what is known as *proton density* contrast (*PD* contrast). The rapidity of signal decay for each tissue (given by the tissue's T_2) is effectively removed from the contrast due to the rapid signal sampling after acquisition, and the long repetition time TR between excitation effectively removes any differences in signal that might be introduced from different longitudinal recovery times T_1 across tissues. Signal differences then depend almost entirely on the density of or protons in each voxel. Stronger signal is typically depicted as brighter (whiter) on grayscale images.

2.6.2 T_2 contrast

Now, let's repeat the aforementioned experiment, preserving the very long TR so each tissue achieves near-complete longitudinal (T_1) recovery between excitations. However, we now sample the signal during a window centered around a delay of time TE after each excitation, as illustrated in Fig. 8. TE is referred to as the echo time. By varying the length of TE, different kinds of contrast based on tissue T_2 can be generated.

If the signal is sampled after a relatively long TE, only spins with relatively long T_2 will still be producing signal and will have a much stronger weighting in the image than short T_2 tissues. By varying TE, different degrees of so-called T_2-weighted contrast can be generated. Note that in these T_2-weighted images, tissues with longer T_2 are typically displayed as brighter (whiter in grayscale imaging, or hyperintense) than tissues with shorter T_2. However, it is important to note that, while these images contain this T_2-weighting, the presence of proton density weighting is also always present. Furthermore, the intensity of a given voxel is a complicated nonlinear function of the T_2 and proton density of tissues in that voxel, so simple T_2-weighting is not a truly quantitative imaging method: that is, a quantitative value of T_2 cannot be directly extracted from a simple T_2-weighted image, although broad differences in T_2 between tissues can be ascertained.

2.6.3 T_1 contrast

Consider now the case where the repetition time TR is shortened to allow only partial longitudinal recovery between excitations. After a series of excitations, one would expect to detect signal only from tissues with relatively rapid longitudinal recovery, or T_1. Signal from longer-T_1 tissues would be attenuated due to the incomplete longitudinal recovery between excitations. This is known as T_1-weighted contrast. In this case, TR is generally

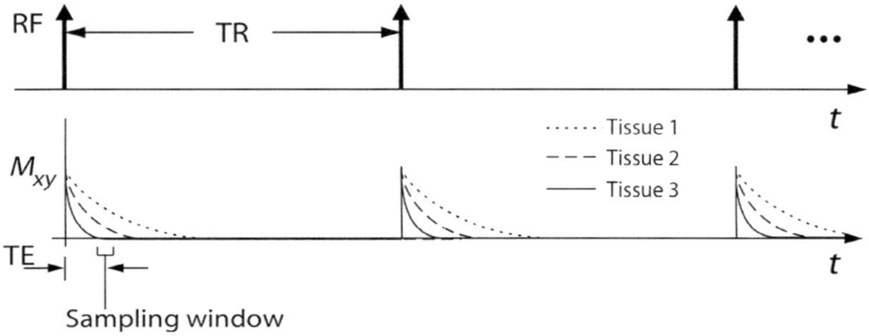

Fig. 8 A long repetition time TR between excitations allows near-complete spin relaxation. The delay after each excitation prior to signal sampling (or echo time TE) can then be adjusted to yield varying degrees of T_2-weighted contrast.

chosen to be on the order of T_1 of the tissues one wishes to see, and the echo time TE is typically kept short to avoid the introduction of T_2 weighting. Again, however, the signal will always be scaled by the proton density in a given voxel. Like T_2-weighted imaging, T_1-weighted imaging does not give a quantitative measure of T_1 in any given voxel, and the intensity of the signal for a given voxel is again a nonlinear function of proton density and tissue T_1. It is important to note, however, that, in T_1-weighted imaging, tissues with longer T_1 tend to appear darker (closer to black, or hypointense) on grayscale images. T_1-weighted imaging might more aptly be described as *inverse* T_1-weighted imaging.

Proton density (PD) contrast, T_1, and T_2 contrast describe just the most basic contrast mechanisms possible in MRI. As will be seen in future sections, the fundamental physics of MRI allows the generation of contrast from a large variety of tissue characteristics, most of which have been exploited in various ways for the imaging of cancer. These include the imaging of magnetic susceptibility differences between tissues (often induced, by, e.g., the presence of iron), the use of contrast agents that alter T_1 and/or T_2, the imaging of both flow and diffusion characteristics of tissues and molecules, and a host of other mechanisms. Some of these relevant to cancer imaging are describe in the following section.

2.7 Advanced contrast generation mechanisms

2.7.1 MRI contrast agents

The introduction of chemical *contrast agents* into tissues being imaged can be used to improve the visibility of certain structures in MRI; this is known as *contrast-enhanced imaging*.

Clinically, the most commonly used compounds for contrast enhancement are gadolinium based. Gadolinium-based contrast agents are paramagnetic and typically shorten the T_1 relaxation time of hydrogen nuclei within body tissues. They thus are most often used in conjunction with T_1-weighted imaging sequences. In cancer, degradation of the blood-brain barrier can cause gadolinium to leak into the tumor and cause enhancement on a T_1-weighted image. Gadolinium is also widely used for depiction of vessels (angiography). Contrast agents can be introduced orally, intravenously, or through direct injection into tissues of interest (e.g., into the joint capsule for some musculoskeletal imaging applications).

While gadolinium-based contrast agents are by far the most widely used in current clinical practice, the development and application of contrast agents remain an active area of research. There are many different compounds that affect the NMR signal in various ways, and some lend themselves to selective targeting of certain pathologies and/or tissues. In addition to gadolinium, there are other paramagnetic contrast agents under development (e.g., manganese-based contrast agents). Superparamagnetic iron oxide (SPIO) consists of nanoparticles of iron oxide in a suspended colloid. These SPIOs have a significant shortening effect on the T_2^* of tissues, and images with T_2^* contrast can typically be very rapidly acquired. SPIO-based contrast agents have successfully been used for liver tumor enhancement, among other applications.

When time series of images are acquired before and after the introduction of a contrast agent, it is called *dynamic contrast-enhanced* (DCE) MRI. A distinction is then often made between the *pre- and postcontrast* images (i.e., those images acquired before the introduction of the contrast agent and those acquired following the introduction of the contrast agent). Clinically useful information can often be extracted from the distribution of an injected contrast agent (based on its effect on the MRI images) and the temporal dynamics of the changes (i.e., how rapidly the contrast changes occur and then wash out in certain tissues). This is known as the study of *contrast kinetics*.

2.7.2 Quantitative relaxometry

As described earlier, T_1 and T_2 weighting, although based on the relaxation time constants T_1 and T_2, are by no means providing a quantitative map of the actual T_1 and T_2 values of the tissues being imaged. *Quantitative relaxometry* or *relaxometry mapping* refers to the process of creating MR images that are actual maps of relaxation (or, more precisely, average observed relaxation properties) at each voxel. While there are a host of techniques that have been proposed to acquire quantitative relaxometry maps of T_1 and T_2, almost all require the repeated acquisition of multiple T_1- or T_2-weighted images, followed by mathematical fitting on a voxel-by-voxel basis to signal models that enable an estimate of the T_1 or T_2 of that voxel.

One can also perform quantitative mapping of other relaxation properties. These include $T_2{}^*$ (a measure of signal decay that is typically more rapid than T_2 given inhomogeneity in magnetic field across single voxels and other nonidealities) and a variety of more complicated relaxation properties (e.g., T_1-rho). In addition, for many tissues and for voxels in which multiple tissues with different relaxation properties are present, biexponential or multiexponential signal decay models are often a more accurate description of the signal relaxation properties T_2 and $T_2{}^*$.

2.7.3 Fat suppression and fluid attenuation (FLAIR)

Critical to many imaging applications is the suppression of signal from tissues that might obscure the pathology of interest. The most common tissues suppressed in MRI are fat (lipids) and fluids [such as cerebrospinal fluid (CSF) in the brain and spinal column or synovial fluid in the joints].

There are multiple techniques in common usage for the suppression of fat. The most common of these are *fat saturation* (*fat sat* in common parlance), *water excitation* (a technique in which only water in nonlipid tissue is excited), so-called *Dixon techniques* [e.g., 2-point Dixon, 3-point Dixon, and *iterative decomposition of water and fat with echo asymmetry and least-squares estimation* (*IDEAL*)], and *short TI inversion recovery (STIR)*. A variety of hybrid techniques for fat suppression also exist. The suppression of fluids, or fluid attenuation, is typically accomplished in T_2-weighted images (where the fluid would normally be hyperintense due to its very long T_2) through the use of a technique known as *fluid-attenuated inversion recovery (FLAIR)*. It is important to note that T_1-weighted images are naturally fluid attenuated, as the long T_1 of fluid appears hypointense on these images. T_2-weighted FLAIR images are commonly used in neurological imaging, where the presence of bright fluid signal from CSF could obscure pathology of interest.

2.7.4 Diffusion MRI

The freedom with which water molecules diffuse through tissue is a characteristic which is often altered in disease. A method of MRI known as diffusion weighted imaging (DWI) can reveal changes in the diffusion of water molecules. In DWI, after spin excitation a strong magnetic gradient field (called the positive diffusion gradient) is applied for a period of time. This field causes spins to accumulate an amount of phase which varies according to their spatial position within the gradient field. A second gradient field is then applied which is the same as the positive diffusion gradient but opposite in sign (called the negative diffusion gradient) which has the effect of undoing the phase accumulated during the positive diffusion gradient. For spins which have not moved between the positive and negative diffusion gradients, the phase accumulated during the positive diffusion gradient is exactly cancelled or undone by the phase accumulation during the negative diffusion gradient. However, spins which have diffused through the tissue in the short time interval between the positive and negative diffusion gradients are in a different spatial location during the negative diffusion gradient than during the positive diffusion gradient. For these spins, the phase accumulation due to the negative diffusion gradient does not exactly cancel the phase accumulation from the positive diffusion gradient. These spins accumulate net phase which is proportional to their displacement in the tissue between the two diffusion gradients. Phase accumulation causes signal loss. Therefore, diffusing water molecules which are moving through the tissue give less signal and appear less intense on images than stationary water molecules. Images showing this variation of signal intensity caused by motion of water molecules are referred to as "diffusion weighted." Diffusion can be quantified if multiple diffusion weighted images are obtained with different diffusion gradient strengths or durations. The pattern of signal loss with varying diffusion gradient strength can be modeled to give a quantity known as the apparent diffusion coefficient (ADC) for each image pixel, which quantifies the freedom of diffusion of water molecules within the tissue. For water in free solution, diffusion occurs randomly in all directions without any preferential direction. In some tissues, water may diffuse more freely in one direction than another due to the organization of the tissue. For instance, in the brain water may diffuse preferentially along the long axis of a fiber tract. An extension of DWI known as diffusion tensor imaging (DTI) reveals the preferential direction of diffusion, giving insight into the underlying tissue organization. In DTI, diffusion imaging is performed repeatedly with different diffusion gradient directions. The signal level observed depends not only on the freedom of diffusion of water molecules but also on the direction of diffusion relative to the diffusion gradient direction. The degree to which diffusion shows a preference for direction is quantified as fractional anisotropy (FA). FA measured by DTI has a range of 0 to 1 and represents the degree of spatial organization of tissue.

Diffusion-based MRI techniques are becoming increasingly common in the imaging of cancer and are discussed in more detail in the second section of this chapter.

2.7.5 MR spectroscopy (MRS)

Increasingly, the ability to form not only an image at each voxel but also an actual nuclear magnetic resonance (NMR) spectrum of the tissues for each imaged voxel is being exploited for cancer imaging. This is known as *MR spectroscopy*, or *MRS*, and allows some of the chemical structure and possible presence of known markers of cancer at each voxel to be ascertained. MRS techniques are very useful but are also very time consuming and are an active area of research.

2.8 Variations from perfection

This chapter has thus far ignored any imperfections in the imaging process that could degrade image quality. There are, in reality, numerous factors that pull magnetic resonance imaging systems away from the ideal. These include effects from nonstationary spins (which can, incidentally, also be used to advantage for flow imaging and diffusion imaging contrasts mentioned earlier), signal decay during the sampling interval, limitations to transmit and receive coil sensitivities, noise in the electrical systems and body, and a host of confounding factors relating to spins that do not precess at exactly the appropriate frequency (broadly termed off-resonance). Off-resonance sources include variations in the main polarizing field, field variations induced by susceptibility differences between tissues and between tissue and air, and something known as chemical shift (the basis for NMR spectroscopy and the MRS methods mentioned earlier).

3 MRI of cancer in the clinic

3.1 Current clinical role of MRI

Three tasks of cancer imaging are *detection* of tumors, *characterization* of tumors, and *evaluation of response to therapy*. Along with computed tomography (CT) and positron emission tomography (PET), MRI is widely used for each of these tasks. The unique ability of MRI to depict intrinsic tissue characteristics such as T_1, T_2, and vascularity makes MRI a sensitive and accurate modality for imaging of many cancers.

3.1.1 Detection

Detection and characterization of cancer with MRI are based on changes in vascularity and changes of fundamental tissue characteristics in tumors that are manifested as changes in T_1 and T_2 time constants. Although there is great potential for quantitative evaluation of tissue characteristics with MRI, most current cancer imaging still relies primarily on subjective evaluation of images by a radiologist. Evaluation of the stage of cancer and progression or regression in response to therapy is mostly based on tumor size. The aim of a clinical cancer MRI examination is therefore to maximize the conspicuity of tumors.

Although specific MRI protocols vary between institutions according to the preference of individual radiologists, evaluation of cancer always includes T_1- and T_2-weighted scans without intravenous contrast agent administration and postinjection (contrast-enhanced) T_1-weighted scans. The T_1- and T_2-weighted images provide contrast between the tumor and adjacent normal tissue based on intrinsic tissue properties. The actual factors contributing to the T_1 and T_2 of tissue are multiple and interrelated, so there is no simple rule for detection of cancer based on image intensity. However, differences in T_1 and T_2 in tumors compared with surrounding normal tissue make tumors visible. For instance, in many organs, cancerous tissue appears brighter on T_2-weighted images than normal surrounding tissue due to increased tissue water content in tumors.

T_1 signal intensity is more variable but similarly useful in depiction of many tumors. Pre- and postinjection (contrast-enhanced) T_1-weighted images depict the blood flow to tissues. Vascularity of tumors is often different than surrounding tissues. Most cancers have abnormally increased metabolic activity that results in increased oxygen demand. As cancerous tissue grows without normal restraints, it recruits increased blood supply from surrounding tissue, often secreting angiogenic factors that cause growth of abnormal or deranged blood vessels. This effect can be seen on contrast-enhanced T_1-weighted images as increased signal intensity in the tumor compared with surrounding tissue. As growth of a tumor progresses, the interior of the tumor may become oxygen starved as the tissue in the periphery avidly consumes the available oxygen. This may lead to tissue necrosis, manifest on MRI as altered T_1 and T_2 and decreased enhancement. The combination of T_1- and T_2-weighted noncontrast sequences and postinjection contrast-enhanced T_1-weighted sequences provides contrast between normal and cancerous tissue that allows diagnosis and measurement of tumors by the radiologist.

Within this broad framework, examinations are tailored to individual body parts. For instance, because abdominal organs move with each breath, abdominal imaging protocols rely on rapid T_1-weighted scans performed in a single breath hold and T_2-weighted scans employing respiratory gating (synchronization of data acquisition with the patient's breathing to reduce motion artifact). In contrast, when imaging the brain, longer high-resolution scans are possible because the brain is easily held immobile in the MRI scanner.

Other body parts have their own particular constraints on imaging protocols. Additional techniques such as fat suppression may be useful depending on the body part being imaged and the imaging sequence in use. Diffusion-weighted MRI, described earlier, has also become widely used in clinical imaging protocols. Many tumors show decreased diffusion that makes them conspicuous on diffusion-weighted images.

3.1.2 Characterization

In addition to detection of lesions, MRI can often characterize a lesion as benign or malignant and can often discriminate between different types of malignant tumors. Subjective evaluation of T_1- and T_2-weighted and contrast-enhanced images often provides an accurate tumor diagnosis. For example, the lower row of Fig. 9 shows a lesion in the liver that has increased T_2-weighted signal intensity compared with adjacent normal liver and decreased T_1-weighted intensity. These findings are suggestive of a tumor not originating from the liver itself. Postinjection contrast enhancement is also different in the tumor than surrounding normal liver. These characteristics suggest metastatic disease from another organ, in this case esophageal cancer.

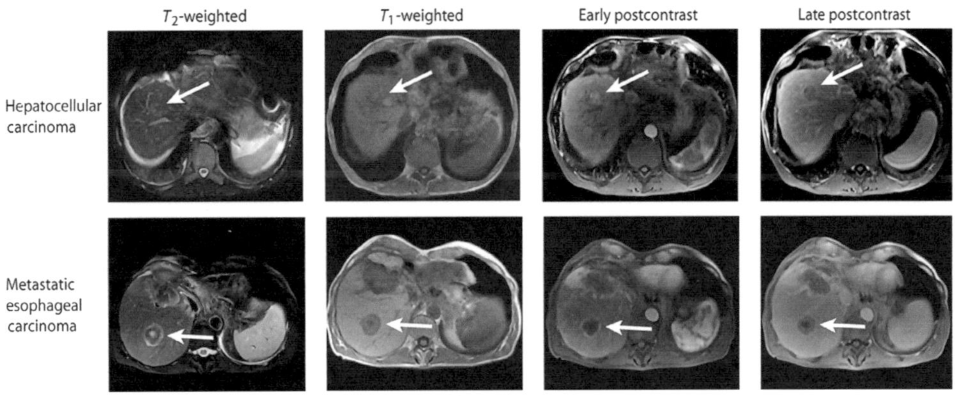

Fig. 9 MRI signal aids the characterization of cancerous lesions. The top row shows single axial images at the same level of the liver dome with four different MRI pulse sequences in a case of hepatocellular carcinoma (HCC), a cancer that arises from liver tissue. The tumor is barely visible on T_1- and T_2-weighted images (a bright lesion on the T_1-weighted image adjacent to the tumor is a benign regenerative nodule). On postcontrast images, the HCC enhances in an early postinjection phase and then becomes less intense than surrounding liver on 3- to 5-min delayed images. This is the characteristic appearance of HCC. In contrast, the lower row of images shows the same pulse sequences in a case of metastatic esophageal carcinoma in the liver. This tumor is conspicuous on the T_1- and T_2-weighted images and shows prominent peripheral enhancement on early postcontrast images with gradual filling in toward the center on more delayed images. This appearance is characteristic of metastatic disease. The primary site of cancer (the esophagus in this case) could not be determined from the MRI appearance, as metastases from other organs such as the colon would have a similar appearance.

In contrast, primary tumors of the liver such as the hepatocellular carcinoma shown in the top row of Fig. 9 often show noncontrast T_1- and T_2-weighted signal intensity similar to surrounding liver, reflecting their origin as liver tissue, but can be detected based on altered vascularity (hyperenhancing in early postinjection phases, with subsequent contrast washout greater than normal tissue causing hypoenhancement on delayed phases).

This is just one specific example of how MRI signal attributes are useful in the characterization of tumors. Each organ has its own characteristic pattern of normal signal intensities, and various benign and malignant lesions that occur in each organ have their own patterns as well. The wide range of appearances of both benign and malignant tissues makes cancer imaging a complex and challenging aspect of radiology. Some benign lesions have an MRI appearance so characteristic that they can be confidently diagnosed based solely on MRI. For the majority of cancerous lesions, diagnosis is confirmed with needle biopsy before treatment begins. However, some tumors such as diffuse midline glioma may be clinically challenging to biopsy due to their location. Although more recently biopsies are becoming part of standard of care in these tumors, some institutions may decide to rely solely on radiological diagnosis.

3.1.3 Response to therapy

MRI is frequently used to evaluate the effectiveness of cancer therapies such as chemotherapy and radiation. Serial scans over several months demonstrate changes in the size and appearance of tumors that are indicative of response to treatment. Although there is great potential for advanced MRI methods such as quantitative diffusion imaging to give information about tumor biology, currently response to therapy is predominantly assessed solely by tumor size [4]. That said, 2018 guidelines from the National Institute for Health and Care Excellence (NICE) on the clinical management of brain tumors recommend consideration of advanced MR techniques discussed in subsequent sections: diffusion, perfusion, and spectroscopy.

3.2 Promising future directions for cancer imaging with MRI

Current MRI protocols for cancer imaging are focused primarily on maximizing tumor conspicuity to enable detection of new cancers and to facilitate accurate evaluation of tumor size. Several MRI techniques have been developed that have the potential to go beyond simple depiction of anatomy to give information about tumor biology, including assessment of tissue cellularity and perfusion. These methods may have significant future impact on cancer imaging.

3.2.1 Tumor biology

A number of biological processes are associated with cancer [4a]. Tumor growth and development are dependent on both the property of the cancer cell itself (e.g., the cell's ability to resist cell death) and the properties of its microenvironment (e.g., blood supply). The tumor microenvironment thus contributes to the growth of cancer cells.

In imaging cancer with MRI, each voxel consists of a comparatively large region (e.g., 1 mm^3), which will constitute both cancer cells and the tumor microenvironment. This implies that, within each voxel of the MRI, we are analyzing a heterogeneous region of a heterogeneous tumor. While different imaging techniques may be used to assess and quantify different biological parameters, it is necessary to study these in light of the biological complexity that exists within tumors.

In addition to the cancer cells and the microenvironment biology, certain tumors will have biology specific to the site of the disease or cells from which they are derived. In the brain, for example, there exists a barrier between the blood and the brain parenchyma known as the blood-brain barrier. This stops chemicals, such as MRI gadolinium-based contrast agents, from reaching the brain. As previously noted, these contrast agents shorten the T_1 values of the tissue, giving increased signal to T_1-weighted images. While this would have no effect on an image of a brain with an intact blood-brain barrier, some brain tumors will disrupt this barrier and can therefore be easily visualized using contrast-enhanced T_1-weighted images. Such images are therefore part of the standard clinical imaging protocols in brain tumors.

While the effect of tumors on the blood-brain barrier may be significant in assessing these tumors, the effect on other imaging techniques may be undesired. Hence, some imaging techniques' use, such as DCE imaging discussed earlier, may be more widespread in other organs than in the brain.

3.2.2 Imaging tumor microstructure: Diffusion imaging revisited

Diffusion imaging was described in broad intuitive terms in Section 1 of this chapter. However, given its emerging importance in the imaging of cancer, a more detailed treatment is warranted for the interested student and more information can be found in [32].

The microstructure of the tumor and its surrounding environment can be visualized and quantified using diffusion-weighted imaging. As previously described, the MRI acquisition is modified to measure the diffusion (small random motion) of water molecules within voxels. These random motions are a result of collisions with other molecules, and in the absence of any other barriers, the mean-squared displacement $\langle r^2 \rangle$ of the molecules is described by Einstein's equation:

$$\langle r^2 \rangle = 6D\tau$$

where D is the diffusion coefficient and τ is the diffusion time.

The diffusion coefficient D can be measured in MRI using the basic pulse sequence developed by Stejskal and Tanner in 1965 [5]. Fig. 10 graphically depicts this basic diffusion-weighted pulse sequence and provides an illustration of the dephasing of

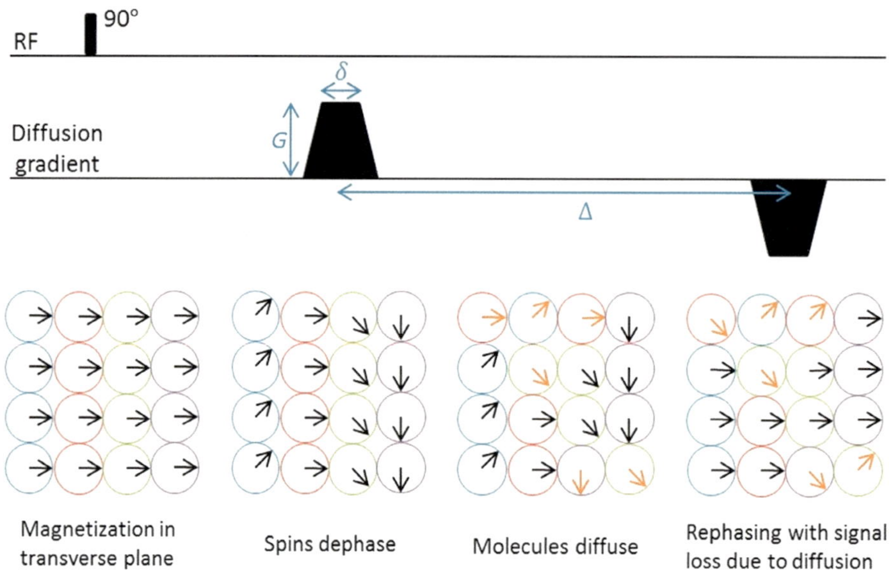

Fig. 10 Basic acquisition for diffusion-weighted imaging. Consider each box of 4 × 4 molecules shown as one voxel. When the RF pulse is applied, all spins are aligned in the transverse plane. Following the first gradient, the spins will dephase depending on location. If no diffusion occurs during time Δ, then the second gradient will rephase the signal such that all spins will once again be in phase with each other. However, due to diffusion, some molecules experiencing dephasing due to the first gradient lobe may have moved to a new location (*yellow spins*) within the voxel and thus will not rephase completely when the second gradient is applied. Therefore, diffusion is measured in terms of a signal loss. *(Reproduced with permission from M. Grech-Sollars, Diffusion MRI for characterising childhood brain tumours, 2014, UCL (University College London).)*

moving spins as previously described. Diffusion-weighted imaging is measured in terms of a signal loss, and the extent of that loss is dependent on the diffusion coefficient D and the diffusion weighting, which in MRI is given by a quantity known as a b-value:

$$b = \gamma^2 G^2 \delta^2 \left(\Delta - \frac{\delta}{3} \right)$$

where γ is the gyromagnetic ratio of protons, G is the gradient amplitude, δ is the gradient duration, and Δ is the gradient separation (Fig. 10). The diffusion time is given by the quantity $\left(\Delta - \frac{\delta}{3}\right)$.

The signal resulting from a given b-value, S_b, is given by

$$S_b = S_0 e^{-bD}$$

where S_0 is the signal with no diffusion weighting.

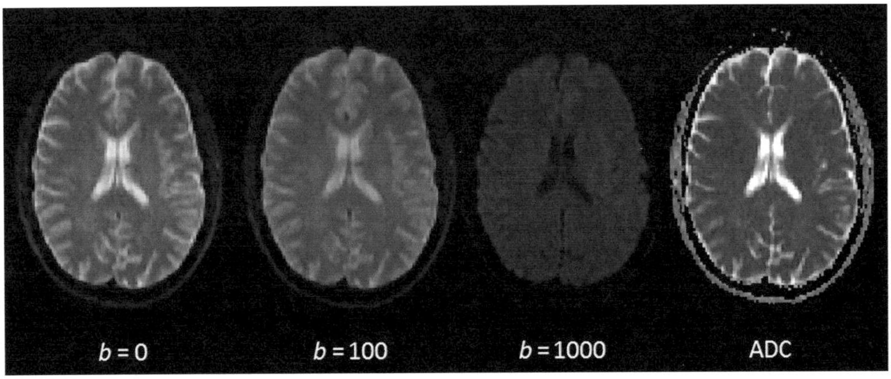

Fig. 11 Diffusion-weighted images of the brain. A note around the structure of the brain: the central part of the image shows the cerebrospinal fluid (CSF), which is a clear liquid. This is surrounded by white and gray matter. The first image on the left ($b = 0$) shows the image with no diffusion weighting, and this image is a T_2-weighted image. The next two images show two diffusion weightings with both of these images having both a T_2 and a diffusion weighting. The higher the b-value, the larger the diffusion weighting, and hence the greater the signal loss. In addition to the diffusion weighting, the higher the diffusion coefficient, the greater the signal loss, and in these images, the regions of the CSF will appear much darker than the surrounding gray and white matter as the b-value increases. The ADC is measured from a linear fit across all three images and eliminated the T_2 effect on the images. The ADC shows a high diffusion coefficient in areas of CSF. *(Reproduced with permission from M. Grech-Sollars, Diffusion MRI for characterising childhood brain tumours, 2014, UCL (University College London).)*

To measure the diffusion coefficient, we therefore need to acquire a minimum of two images, one diffusion-weighted image and one with no diffusion weighting. In MRI, we refer to measuring the apparent diffusion coefficient (ADC). An example of images with different diffusion weightings and the derived ADC is shown in Fig. 11.

In describing the measurement of ADC, we have so far focused on one gradient direction. However, we can apply our gradients in the x, y, or z directions or indeed in any arbitrary direction by simultaneously applying gradient fields along the primary axes with different strengths. In an isotropic medium, the direction in which we apply our measurement is not important. However, tissue is anisotropic. We often therefore measure the diffusion coefficient along the three primary axes and average these to obtain the ADC.

ADC and tumor biology

One of the first studied relationships between ADC and tumor biology was the correlation with cellularity, which was first presented in brain tumors by Sugahara in 1999 [6]. The principle behind this correlation is that, in a more cellular tumor, the diffusion path is more tortuous, which will result in a lower ADC as shown in Fig. 12. While multiple studies have investigated this relationship and observed a negative correlation between ADC and cellularity [7, 8], others did not [9, 10]. In reality, the relationship

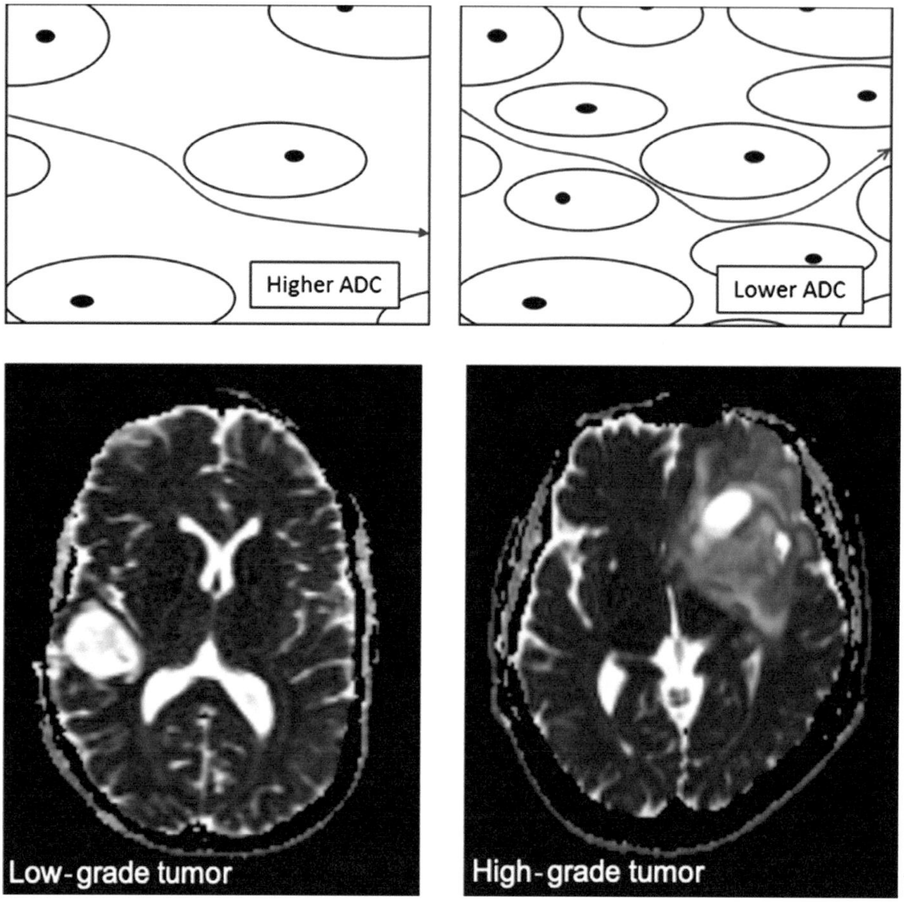

Fig. 12 ADC and tumor biology. In a more cellular tumor *(right)*, the diffusion path is more tortuous, and hence, the ADC is expected to be lower. The relationship between ADC and tumor biology is however more complex than just cellularity and the extracellular space. While cellularity has an effect on the value of ADC, it is not the sole component of tumor biology to do so. Other factors to have an effect on ADC include necrosis and microcysts. *(Reproduced with permission from M. Grech-Sollars, Diffusion MRI for characterising childhood brain tumours, 2014, UCL (University College London).)*

between ADC and tumor biology is more complex than simply cellularity, and the intracellular component can also contribute to the measurement of ADC [11]. While ADC has been identified as a biomarker that can be used in determining treatment response in tumors [12], confounders have been identified that suggest a deeper understanding of tumor biology, and how the physics is affected by this biology need to be taken into account [13]. To tackle the complexity between ADC and tumor biology, more advanced models have been developed [14].

Diffusion tensor imaging

We have so far discussed the ADC, which is a measured of the diffusion coefficient, averaged over three directions. With diffusion tensor imaging, the diffusion in each of the directions and the anisotropy can be measured to determine the tissue microstructure. The measured parameters can be defined by the "diffusion ellipsoid" shown in Fig. 13.

In anisotropic diffusion, six parameters are required to define a unique ellipsoid, and therefore a minimum of six measurements, in addition to the $b=0$ measurement, are required to construct what is known as the diffusion tensor $\overline{\overline{D}}$ from the diffusion coefficients along the orthogonal axes x, y, and z:

$$\overline{\overline{D}} = \begin{bmatrix} D_{xx} & D_{xy} & D_{xz} \\ D_{yx} & D_{yy} & D_{yz} \\ D_{zx} & D_{zy} & D_{zz} \end{bmatrix}$$

Due to the symmetric property of the diffusion tensor, the eigenvalues λ_1, λ_2, λ_3 correspond to the diagonal of the matrix. The diffusion coefficient is now known as the mean diffusivity (MD):

$$\text{MD} = \frac{\lambda_1 + \lambda_2 + \lambda_3}{3} = \frac{D_{xx} + D_{yy} + D_{zz}}{3}$$

The second measurement that is commonly calculated and output by scanners when running a DTI protocol is the fractional anisotropy (FA) that is a measure of the anisotropy

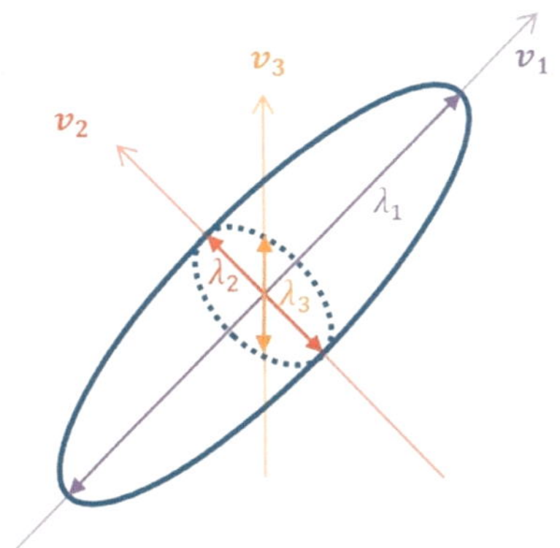

Fig. 13 The diffusion ellipsoid. To measure the anisotropy in a voxel, we need to define the characteristics that define an ellipsoid. The ellipsoid can be defined using six parameters: three eigenvalues describing the strength of the diffusion in each direction and three eigenvectors describing the main directions of diffusion. *(From M. Grech-Sollars, Diffusion MRI for characterising childhood brain tumours, 2014, UCL (University College London).)*

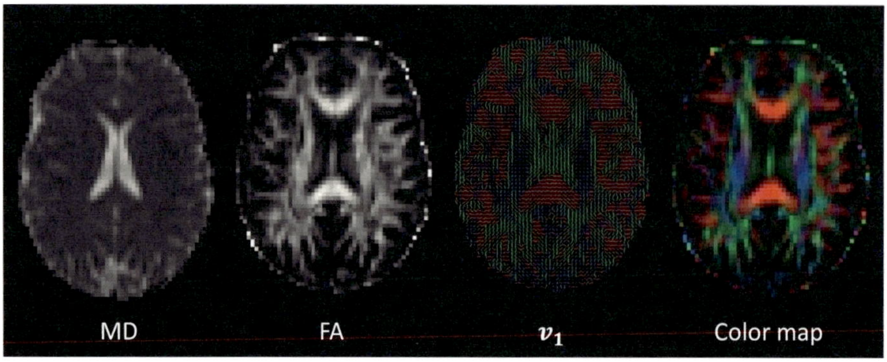

Fig. 14 Standard DTI parameters. The standard DTI parameters output by the scanner are the mean diffusivity (MD), which is similar to the ADC, and the fractional anisotropy (FA). The FA is a measure of the anisotropy with 0 referring to an isotropic medium and 1 referring to maximum anisotropy. By combining the primary eigenvector v_1 and the FA, a color map can be built, which represents the directionality of the white matter tracts. In the color map image, the corpus callosum, which connects the left and right side of the brain, can be seen in *red*. Tracts going between the anterior and posterior parts of the brain are shown in *green*, and those between the inferior and superior part of the brain are shown in *blue*. *(From M. Grech-Sollars, Diffusion MRI for characterising childhood brain tumours, 2014, UCL (University College London).)*

that exists in a given voxel, with 0 representing isotropy and 1 representing maximum anisotropy. The FA is given by

$$FA = \sqrt{\frac{3}{2}\left(\frac{(\lambda_1 - MD)^2 + (\lambda_2 - MD)^2 + (\lambda_3 - MD)^2}{\lambda_1^2 + \lambda_2^2 + \lambda_3^2}\right)}$$

The images commonly output by the scanner from a DTI scan are shown in Fig. 14. Importantly, a color map can be constructed from the eigenvectors that indicates the direction of the main eigenvalue and the strength of the anisotropy. The eigenvectors are used in the process of constructing a 3D image of white matter pathways, known as tractography.

The main use of tractography for cancer patients in the clinic is in planning neurosurgery. White matter tracts can be pushed to a side as brain tumors grow, they can be infiltrative within the tumor, or they can also be destroyed. In the case of infiltrative tracts within the tumor, it is likely that surgical excision of the tumor would result in loss of brain function, which could, for example, paralyze the patient if in the motor cortex or damage their ability to speak if in the language pathway. In the case where the white matter tracts are pushed to a side by the tumor, the surgeon would benefit from knowing the location of such tracts to ensure these are not damaged.

3.3 Tumor vascularity

Tumors rely on the blood supply for essential nutrients to grow. Quantitative measurement of vascularity is therefore an essential part of cancer imaging. Various techniques can be employed to measure vascularity. The most widely accepted techniques are dynamic contrast-enhanced MRI (DCE-MRI) and dynamic susceptibility contrast MRI (DSC-MRI). However, both of these techniques use gadolinium-based contrast agents. Safety concerns around the use of gadolinium are spearheading the development of noncontrast-based methods such as intravoxel incoherent motion (IVIM) and arterial spin labeling (ASL). Each of these techniques is discussed in the succeeding text.

3.3.1 *Intravoxel incoherent motion*

Intravoxel incoherent motion (IVIM) is a diffusion-weighted imaging-based technique. In healthy tissue, capillaries are oriented in an ordered and structured fashion. However, in tumor, capillaries are randomly oriented, and the blood flowing through them can be measured in terms of pseudodiffusion. This random pseudodiffusion within capillaries is faster than the water diffusion measured by standard clinical imaging protocols where a b-value of around 1000 is normally applied in the brain. To measure IVIM, we need to measure the fast diffusion component at b-values <200. When measuring these low b-values, particularly those <50, a biexponential fit to the diffusion data is required. In doing so, the slow diffusion coefficient D, the fast diffusion coefficient D^*, and the perfusion fraction f can be measured as shown in Fig. 15. IVIM is modeled from the equation

$$\frac{S_b}{S_0} = (1-f)e^{-bD} + fe^{-b(D+D^*)}$$

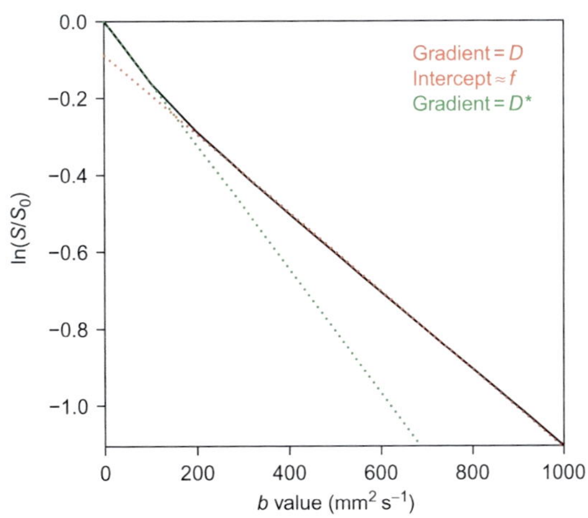

Fig. 15 IVIM parameters. *(From M. Grech-Sollars, Diffusion MRI for characterising childhood brain tumours, 2014, UCL (University College London).)*

3.3.2 Dynamic contrast-enhanced MRI

Dynamic contrast-enhanced (DCE) MRI is one of the perfusion techniques that uses gadolinium-based contrast agents to measure perfusion-related parameters. In DCE-MRI, multiple T_1-weighted images with a short temporal resolution ($\approx$2–3 s) are acquired consecutively. After the first few acquisitions, the contrast agent is injected while continuing to acquire repeated images for a duration of time depending on the acquisition protocol. As gadolinium reduces the T_1 value, the concentration of the contrast agent, C, flowing through a given voxel can be measured from the T_1-weighted images. Detailed descriptions of the method can be found in Ref. [15]. Once the data are acquired, models are used to fit the data. Model selection is an important process in DCE-MRI (Fig. 16), and, while specific models may be more appropriate for specific acquisitions or tumor types, multiple models may be able to best describe one heterogeneous tumor [16,16a]. Outputs from a DCE-MRI exam often include a measure of the permeability of the contrast agent from the plasma to the interstitial space, called k^{trans}. DCE-MRI has been shown to discriminate between low- and high-grade brain tumors [17]. It has also been considered as a biomarker of common genetic mutations that relate to angiogenesis, such as the epidermal growth factor receptor [18]. It is a commonly used perfusion technique in assessing body tumors such as breast cancer [19].

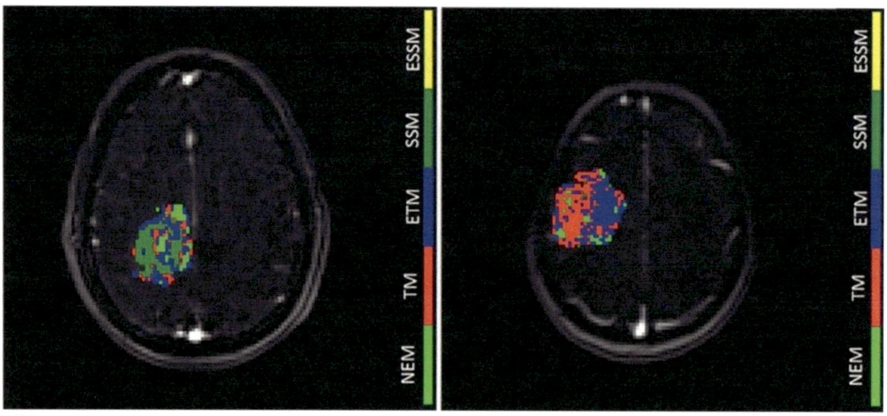

(A) (B)

Fig. 16 DCE-MRI model selection in a contrast-enhanced (A) and nonenhancing (B) glioma. A number of models may be required to fit the data within a single tumor. In the diagram each color represents a different model: *NEM*, no-exchange model; *TM*, Tofts' model; *ETM*, extended Tofts' Model; *SSM*, shutter speed model; *ESSM*, extended shutter speed model. *(Reproduced with permission from M. Inglese, K.L. Ordidge, L. Honeyfield, T.D. Barwick, E.O. Aboagye, A.D. Waldman, M. Grech-Sollars, Reliability of dynamic contrast-enhanced magnetic resonance imaging data in primary brain tumours: a comparison of Tofts and shutter speed models, Neuroradiology (2019) 1–12.)*

3.3.3 Dynamic susceptibility contrast MRI

Dynamic susceptibility contrast (DSC) MRI is another perfusion technique that uses contrast agents, typically gadolinium based, to measure perfusion-related parameters. In this case, T_2* images are acquired. DSC-MRI is commonly used in assessing brain tumors. Parameters output from DSC imaging include measures of cerebral blood flow (CBF) and cerebral blood volume (CBV). One of the reasons for its increased use in the brain, as compared with DCE and other perfusion imaging techniques, is that most MRI vendors provide software tools that will analyze the data acquired and output CBF and CBV maps without too much involvement from radiologists or clinical scientists. The use of DSC-MRI is in fact a good example of how vendors can influence the use of advanced MRI tools in the clinic. Its use in the clinic is primarily for assessing differences between low- and high-grade tumors [20] as was done in the case of the patient shown in Fig. 17.

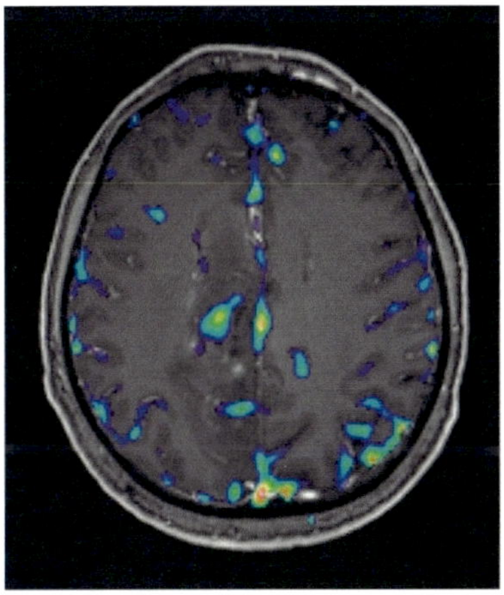

Fig. 17 DSC dataset showing the rCBV in a patient with a brain tumor. One section of the tumor showed a relatively high rCBV. Following biopsy, this patient was found to have anaplastic transformation from WHO Grade II to WHO Grade III glioma.

3.3.4 Arterial spin labeling

Arterial spin labeling (ASL) is one of the newer perfusion MRI technique and does not require the use of contrast agents. It produces a value for the cerebral blood flow (CBF). Details of the technique can be found in the consensus paper [21]. As it is one of the newest techniques, further work needs to be carried out to determine its clinical applicability to cancer. However, studies have so far shown promise in differentiating between tumor grades [22]. The main benefit of ASL lies in the fact that it is able to quantify blood flow without requiring the use of a contrast agent and hence is a safer technique, particularly in patients with impaired renal function that might be at risk from the administration of gadolinium-based contrast agents.

3.4 Tumor proliferation

One of the hallmarks of cancer is the ability to sustain proliferation. To replicate, cells require choline as part of cell membrane synthesis. Choline (Cho)-containing compounds can be measured using MR spectroscopy (MRS) in relation to creatine (Cr). In tumor diagnosis and follow-up, Cho/Cr ratio is one of the biomarkers used to determine whether a tumor is likely to be aggressive or not. MRS has been studied in terms of grading gliomas (a primary brain tumor), assessing transformation from a low- to a high-grade brain tumor, and also in differentiating between pseudoprogression and true progression [23–25]. In the clinic, single voxel spectroscopy is most commonly used. However, 2D CSI (Fig. 18) can be used to determine the choline levels across a heterogeneous tumor and help identify more aggressive tumor regions [33].

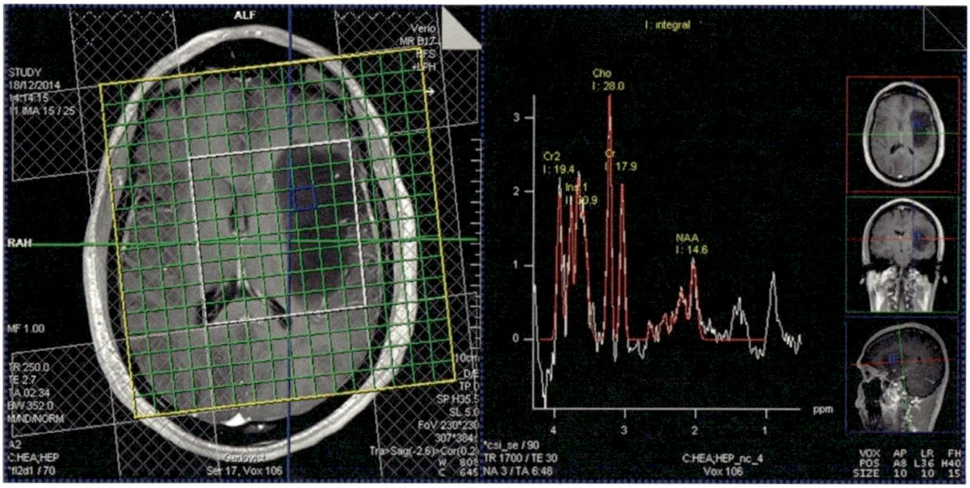

Fig. 18 Two-dimensional MR spectroscopy. In 2D MRS, data from large voxel regions (around 1 cm^3) are acquired. A spectrum for the concentration of metabolites (*right*) can be processed for each voxel region (such as that shown by the *blue box* on the *left*). *(Reproduced with permission from M. Grech-Sollars, B. Vaqas, G. Thompson, T. Barwick, L. Honeyfield, and K. O'Neill, and A.D. Waldman, An MRS-and PET-guided biopsy tool for intraoperative neuronavigational systems, J. Neurosurg. 127 (4) (2017) 812–818.)*

3.5 Standard clinical cancer imaging

3.5.1 Whole-body diffusion (detection of metastatic disease)

In patients with known or suspected metastatic disease, whole-body diffusion-weighted imaging (WB DWI) is gaining popularity for screening and detection of metastases. PET/CT has conventionally been considered the reference standard for imaging of metastatic disease but exposes the patient to radiation, is not readily available in many hospitals, and is often not the most cost-effective modality. Other conventional CT and MRI techniques are also widely used for screening for metastatic disease, and, while achieving good sensitivity and specificity, CT also exposes the patient to ionizing radiation, and conventional whole-body MRI techniques have extremely long scan times and often involve the administration of contrast agents. Whole-body MRI exams have been developed using diffusion-weighted imaging that are rapid, don't require the administration of contrast agents, and don't expose the patient to ionizing radiation. While some studies have shown that WB DWI is inferior to conventional techniques, it can still be an attractive and reliable alternative for whole-body screening for a variety of metastases, such as hepatic and skeletal metastases [26].

3.5.2 MRI in the operating theater

In addition to its use in diagnosing tumors, MRI is used in the clinic to guide some surgeries and is in routine use in biopsy and surgical resection of brain tumors (Fig. 19). Basic image-guided operating theater equipment would include a mechanism to indicate to neurosurgeons the location of their surgical tools in relation to MRI images that would have been acquired presurgery. These images are registered to patient space generally using markers known as fiducials that are attached to the patient's head before their MRI scan. Using a system of triangulation, the neuronavigational system is able to show to the neurosurgeons the location they are in within the brain on the MRI structural images.

While the extent of surgical resection is a prognostic factor in brain tumors, this can only be achieved if the location of the tumor is not in an eloquent part of the brain. In resecting tumors, surgeons must take care to avoid brain regions that could cause surgery-related comorbidities such as loss of vision, speech, or movement. To address this, neurosurgery neuronavigational systems may incorporate the use of advanced MR

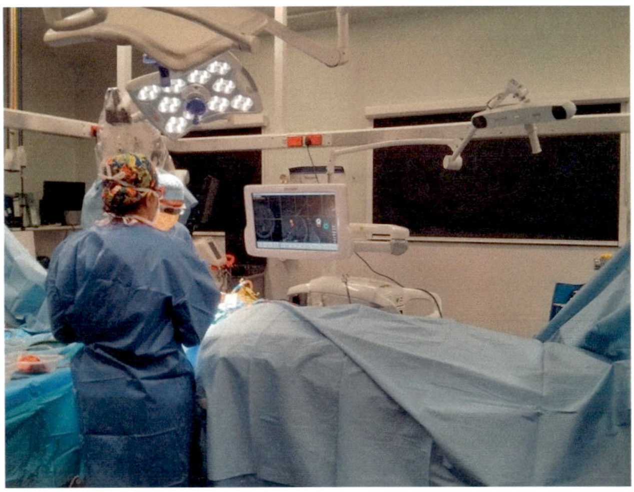

Fig. 19 MRI in the operating theater.

techniques such as MR tractography. In this case, the system can show the important white matter tracts that the surgeon needs to avoid in resecting the tumor.

One of the issues with these tools is that, in opening the skull, the brain may shift by 1 cm or more [27, 28] making the MRI inaccurate. Alternative ways exist that try to deal with this. One method is to use ultrasound images obtained during the surgery overlaid on top of the MRI image to indicate to surgeons the real location of the tumor. Another option is to use intraoperative MRI. The latter requires more complex equipment that is currently only available in specialist centers. It provides updated MRI images during the surgery itself and requires that all surgical tools are MR safe. Using intraoperative MRI has been shown to increase the extent of gross tumor resection [29] and to reduce the risks of surgery-related comorbidities such as visual deficits that could lead to the loss of a driving license [30].

3.6 Engineering solutions: Development of tools for assessing MRI in cancer

The development of advanced MRI techniques and their implementation in the clinic needs to take into account how both patients and clinicians experience these tools.

While MRI is considered to be a safe imaging technology, patients may experience anxiety due to claustrophobia or other concerns during an MRI scan. This could mean that patients fail to stay still in the scanner long enough to get images of good diagnostic quality, or they may only be able to tolerate the first few scans. In addition to claustrophobia, scan time is a major concern for patients and their families. A long scan may not be tolerated well, particularly by cancer patients who are severely ill. Therefore, in designing MRI scan protocols, these limitations need to be taken into account by, for example, ensuring that the most important sequences are captured first. In addition, issues around claustrophobia and patient movement may be mitigated by improving the patient experience using background music or allowing patients to watch a movie or TV program while being scanned. More recent concepts have also been explored to improve the patient experience, such as the virtual reality application that allows children to experience what an MRI scan is like before they undergo the procedure.

4 Clinical perspective: Summary

MRI tools need to be able to give clinicians the information required with the most ease. A tool that requires heavy offline processing of data is unlikely to succeed as a clinical tool. As such, engineering solutions for integrating advanced MRI tools with the systems in use within the hospitals are required. One good example is the use of DSC imaging over DCE imaging within the brain. DSC sequences are generally readily available on clinical scanners and are integrated with scanner software for analysis of data. DCE sequences are harder to set up and tend to require offline processing using tools that may not be easily integrated with hospital computing systems. This can therefore result in clinicians favoring one tool over the other based on availability rather than clinical applicability.

In addition, a lack of quantitative and reproducible parameters can also result in a reduction of the use of MRI techniques in the clinic. Attempts to use quantified imaging measures for tumor diagnosis have been limited by the variability in the measurements and the heterogeneity that exists within tumors. While some parameters such as ADC, MD, and FA in diffusion imaging have been found to be reproducible in a clinical setting, provided certain parameters (such as the highest *b*-value) are kept the same, others such as the perfusion fraction in IVIM have been found to vary much more [31]. In general, the reproducibility of parameters suggested for use in diagnosing cancer needs to be studied and the robustness of the techniques improved. Standardization of MRI sequence parameters and analysis methods are critical to integrate the engineering tools developed into the clinic.

References

[1] Z.P. Liang, P.C. Lauterbur, Principles of Magnetic Resonance Imaging: A Signal Processing Perspective, Wiley-IEEE Computer Society PR, 1999.
[2] E.M. Haacke, R.W. Brown, M.R. Thompson, R. Venkatesan, Magnetic Resonance Imaging: Physical Principles and Sequence Design, John Wiley and Sons, 1999.
[3] D.G. Nishimura, Principles of Magnetic Resonance Imaging, Stanford University, 2010.
[4] E.A. Eisenhauer, P. Therasse, J. Bogaerts, L.H. Schwartz, D. Sargent, R. Ford, et al., New response evaluation criteria in solid tumours: revised RECIST guideline (version 1.1), Eur. J. Cancer 45 (2) (2009) 228–247.
[4a] D. Hanahan, R.A. Weinberg, Hallmarks of cancer: the next generation, Cell 144 (5) (2011) 646–674.
[5] E.O. Stejskal, J.E. Tanner, Spin diffusion measurements: spin echoes in the presence of a time-dependent field gradient, J. Chem. Phys. 42 (1965) 288–292.
[6] T. Sugahara, Y. Korogi, M. Kochi, I. Ikushima, Y. Shigematu, T. Hirai, et al., Usefulness of diffusion-weighted MRI with echo-planar technique in the evaluation of cellularity in gliomas, J. Magn. Reson. Imaging 9 (1) (1999) 53–60.
[7] B.M. Ellingson, M.G. Malkin, S.D. Rand, J.M. Connelly, C. Quinsey, P.S. LaViolette, et al., Validation of functional diffusion maps (fDMs) as a biomarker for human glioma cellularity, J. Magn. Reson. Imaging 31 (3) (2010) 538–548.
[8] P.D. Humphries, N.J. Sebire, M.J. Siegel, O.E. Olsen, Tumors in pediatric patients at diffusion-weighted MR imaging: apparent diffusion coefficient and tumor cellularity, Radiology 245 (3) (2007) 848–854.
[9] M.D. Jenkinson, D.G. du Plessis, T.S. Smith, A.R. Brodbelt, K.A. Joyce, C. Walker, Cellularity and apparent diffusion coefficient in oligodendroglial tumours characterized by genotype, J. Neurooncol. 96 (3) (2010) 385–392.
[10] M.I. Yoshikawa, S. Ohsumi, S. Sugata, M. Kataoka, S. Takashima, T. Mochizuki, et al., Relation between cancer cellularity and apparent diffusion coefficient values using diffusion-weighted magnetic resonance imaging in breast cancer, Radiat. Med. 26 (4) (2008) 222–226.
[11] Y. Matsumoto, M. Kuroda, R. Matsuya, H. Kato, K. Shibuya, M. Oita, et al., In vitro experimental study of the relationship between the apparent diffusion coefficient and changes in cellularity and cell morphology, Oncol. Rep. 22 (3) (2009) 641–648.
[12] B.A. Moffat, T.L. Chenevert, T.S. Lawrence, C.R. Meyer, T.D. Johnson, Q. Dong, et al., Functional diffusion map: a noninvasive MRI biomarker for early stratification of clinical brain tumor response, PNAS 102 (15) (2005) 5524–5529.
[13] M. Grech-Sollars, D.E. Saunders, K.P. Phipps, R. Kaur, S.M. Paine, T.S. Jacques, et al., Challenges for the functional diffusion map in pediatric brain tumors, J. Neurooncol. 16 (3) (2014) 449–456.
[14] E. Panagiotaki, S. Walker-Samuel, B. Siow, S.P. Johnson, V. Rajkumar, R.B. Pedley, et al., Noninvasive quantification of solid tumor microstructure using VERDICT MRI, Cancer Res. 74 (7) (2014) 1902–1912.

[15] P.B. Barker, X. Golay, G. Zaharchuk, Clinical Perfusion MRI: Techniques and Applications, Cambridge UP, Cambridge, 2013.
[16] H. Bagher-Ebadian, R. Jain, S.P. Nejad-Davarani, T. Mikkelsen, M. Lu, Q. Jiang, et al., Model selection for DCE-T1 studies in glioblastoma, Magn. Reson. Med. 68 (1) (2012) 241–251.
[16a] M. Inglese, K.L. Ordidge, L. Honeyfield, T.D. Barwick, E.O. Aboagye, A.D. Waldman, M. Grech-Sollars, Reliability of dynamic contrast-enhanced magnetic resonance imaging data in primary brain tumours: a comparison of Tofts and shutter speed models, Neuroradiology (2019) 1–12.
[17] T.F. Patankar, H.A. Haroon, S.J. Mills, D. Baleriaux, D.L. Buckley, G.J.M. Parker, et al., Is volume transfer coefficient (K_{trans}) related to histologic grade in human gliomas? Am. J. Neuroradiol. 26 (10) (2005) 2455–2465.
[18] J. Arevalo-Perez, A.A. Thomas, T. Kaley, J. Lyo, K.K. Peck, A. Holodny, et al., T1-weighted dynamic contrast-enhanced MRI as a noninvasive biomarker of epidermal growth factor receptor vIII status, Am. J. Neuroradiol. 36 (12) (2015) 2256–2261.
[19] K. Pinker, T.H. Helbich, E.A. Morris, The potential of multiparametric MRI of the breast, Br. J. Radiol. 90 (1069) (2017).
[20] A.F. Delgado, A.F. Delgado, Discrimination between glioma grades II and III using dynamic susceptibility perfusion MRI: a meta-analysis, Am. J. Neuroradiol. 38 (7) (2017) 1348–1355.
[21] D.C. Alsop, J.A. Detre, X. Golay, M. Gunther, J. Hendrikse, L. Hernandez-Garcia, et al., Recommended implementation of arterial spin labeled perfusion MRI for clinical applications: a consensus fo the ISMRM perfusion study group and the European consortium for ASL in dementia, Magn. Reson. Med. 73 (1) (2015) 102–116.
[22] L. Kong, H. Chen, Y. Yang, L. Chen, A meta-analysis of arterial spin labelling perfusion values for the prediction of glioma grade, Clin. Radiol. 72 (3) (2017) 255–261.
[23] H. Poptani, R.K. Gupta, R. Roy, R. Pandey, V.K. Jain, D.K. Chhabra, Characterization of intracranial mass lesions with in vivo proton MR spectroscopy, Am. J. Neuroradiol. 16 (8) (1995) 1593–1603.
[24] G. Tedeschi, N. Lundbom, R. Raman, S. Bonavita, J.H. Duyn, J.R. Alger, et al., Increased choline signal coinciding with malignant degeneration of cerebral gliomas: a serial proton magnetic resonance spectroscopy imaging study, J. Neurosurg. 87 (4) (1997) 516–524.
[25] T. Tamiya, K. Kinoshita, Y. Ono, K. Matsumoto, T. Furuta, et al., Proton magnetic resonance spectroscopy reflects cellular proliferative activity in astrocytomas, Neuroradiology 42 (5) (2000) 333–338.
[26] S.M. Paruthikunnan, R. Kadavigere, L.H. Karegowda, Accuracy of whole-body DWI for metastases screening in a diverse group of malignancies: comparison with conventional cross-sectional imaging and nuclear scintigraphy, Am. J. Roentgenol. 209 (2017) 477–490.
[27] D.W. Roberts, A. Hartov, F.E. Kennedy, M.I. Miga, K.D. Paulsen, Intraoperative brain shift and deformation: a quantitative analysis of cortical displacement in 28 cases, Neurosurgery 43 (4) (1998) 749–758.
[28] P.M. Black, E. Alexander 3rd, C. Martin, T. Moriary, A. Nabavi, T.Z. Wong, et al., Craniotomy for tumor treatmnet in an intraoperative magnetic resonance imaging unit, Neurosurgery 45 (3) (1999) 423–431.
[29] D. Kuhnt, M.H.A. Bauer, C. Nimsky, Brain shift compensation and neurosurgical image fusion using intraoperative MRI: current status and future challenges, Crit. Rev. Biomed. Eng. 40 (3) (2012) 175–185.
[30] G.P. Winston, P. Daga, M.J. White, C. Micallef, A. Miserocchi, L. Mancini, et al., Preventing visual field deficits from neurosurgery, Neurology 83 (7) (2014) 604–611.
[31] M. Grech-Sollars, P.W. Hales, K. Miyazaki, F. Raschke, D. Rodriguez, M. Wilson, et al., Multi-centre reproducibility of diffusion MRI parameters for clinical sequences in the brain, NMR Biomed. 28 (4) (2015) 468–485.
[32] M. Grech-Sollars, Diffusion MRI for characterising childhood brain tumours, UCL (University College London), 2014.
[33] M. Grech-Sollars, B. Vaqas, G. Thompson, T. Barwick, L. Honeyfield, K. O'Neill, A. D. Waldman, An MRS-and PET-guided biopsy tool for intraoperative neuronavigational systems, J. Neurosurg. 127 (4) (2017) 812–818.

PART 3

Cancer therapy

CHAPTER 3.1

Nanotechnology and drug delivery

Derfogail Delcassian[a,b], Asha K. Patel[c]
[a]Koch Institute for Integrative Cancer Research, Massachusetts Institute of Technology, Cambridge, MA, United States
[b]Division of Regenerative Medicine and Cellular Therapies, University of Nottingham, Nottingham, United Kingdom
[c]National Heart & Lung Institute, Imperial College London, London, United Kingdom

1 Introduction

There are two main therapeutic approaches in treating cancer: (1) eliminating preexisting tumors and (2) controlling further cancer growth or metastasis. For people with late-stage cancer, it may not be possible to eliminate cancerous tissue, and instead the focus is on controlling further spread of the disease to alleviate symptoms and extend life. Clinicians typically use a combination of approaches to target cancer tumors and cancerous cells. These treatments can be roughly grouped into drug delivery (discussed in this chapter), radiotherapy to damage cancer cells and shrink tumors, immunotherapy to activate the immune system to combat cancer (Chapter 3.2), and surgery to excise cancerous tissue (Chapter 3.3). The type of cancer therapy a patient receives depends on a number of factors, including tumor type, location, severity of disease, and any preexisting conditions.

Drug delivery plays an important role in treating many patients with cancer [1–4]. For example, patients with invasive disease that is difficult to surgically remove due to volume or location (such as deep in the brain) require other strategies to treat cancerous tissue. Chemotherapeutic drugs are routinely used to treat widespread, systemic disease and can be used in combination with other therapies to increase treatment efficacy [5–7]. Administering chemotherapy drugs before other treatments is termed neoadjuvant chemotherapy, and this approach can help to sensitize cancer cells before radiotherapy. Postoperative chemotherapy can also help to eliminate small tumor deposits and individual cancer cells that may have been missed during surgical removal of cancerous tissue.

Unfortunately, many cancer drugs are unable to discriminate between healthy and cancerous tissue, causing acute toxicity to healthy cells. To combat this, bioengineers have developed ways to improve delivery of these drugs to cancerous tissue. New drugs that target specific organs, cells, or biochemical pathways can help to increase specificity. Additionally, packaging drugs in various materials enables controlled therapeutic delivery to specific biological targets [1, 2, 8–10]. By carefully designing systems that help to deliver drugs to specific targets, engineers and clinicians can work together to maximize therapeutic benefits and minimize side effects.

Bioengineering Innovative Solutions for Cancer
https://doi.org/10.1016/B978-0-12-813886-1.00009-7

In this chapter, we will briefly describe the types of pharmaceutical drugs that can be used for cancer treatment and will then focus on exploring some of the engineering challenges that must be overcome to deliver these drugs to the right target at the right time. Finally, we will highlight successful examples of recent therapies that have moved from the research bench to clinical use in patients.

2 Pharmaceutical therapies for cancer

Tumors are abnormal growths of cancerous cells, surrounded by a tumor microenvironment. This microenvironment typically consists of cancerous cells, healthy cells, immune cells, and vasculature to supply oxygen and nutrients to the cells (Fig. 1) [11]. Clinicians can use drugs to target and interrupt biological processes in the tumor microenvironment to treat cancer. These drug-based therapies aim to create a hostile environment to limit cancer cell survival and tumor growth [12–16]. For example, chemotherapeutic drugs can be used to prevent the rapid division of cells [17], causing cancer cell death. In an alternative strategy, antiangiogenic drugs help to prevent the formation of new blood vessels, leading to oxygen starvation as the tumor grows. A brief summary of the types of pharmaceutical therapies and some examples are provided in the succeeding text.

2.1 Drugs which act as angiogenic inhibitors

The tumor microenvironment typically contains blood vessels and a lymphatic drainage system, which cancer cells can use to spread to secondary tumor sites. As tumors grow larger in size, their oxygen and nutrient demands also increase [18]. Cancer cells can secrete vascular growth factors (i.e., VEGF and PDGF) that stimulate vessel growth in the tumor microenvironment. Blood vessels newly formed in the proximity of the tumor increase blood flow to the tumor. This helps to meet the metabolic demands of growing cancer cells and also provides the cancer with local vasculature that helps it migrate and metastasize. Blocking this neovascularization and preventing stabilization of leaky vessels are therefore a promising strategy to limit cancer growth. Angiogenic inhibitors such as Bevacizumab are drugs typically given in combination with chemotherapeutics to kill tumor cells and limit vessel formation.

2.2 Hormone therapies

Cancers from gynecological or urological tumors are often hormone sensitive. These cancers frequently express hormone receptors on their cell surface, and cell growth is affected by the level of hormone present in the tumor microenvironment [19, 20]. For example, many breast cancer cells express either estrogen or progesterone receptors, which can bind the hormones estrogen or progesterone, respectively. If the levels of these hormones rise, they can bind to the surface of the cancer cells and help them to divide

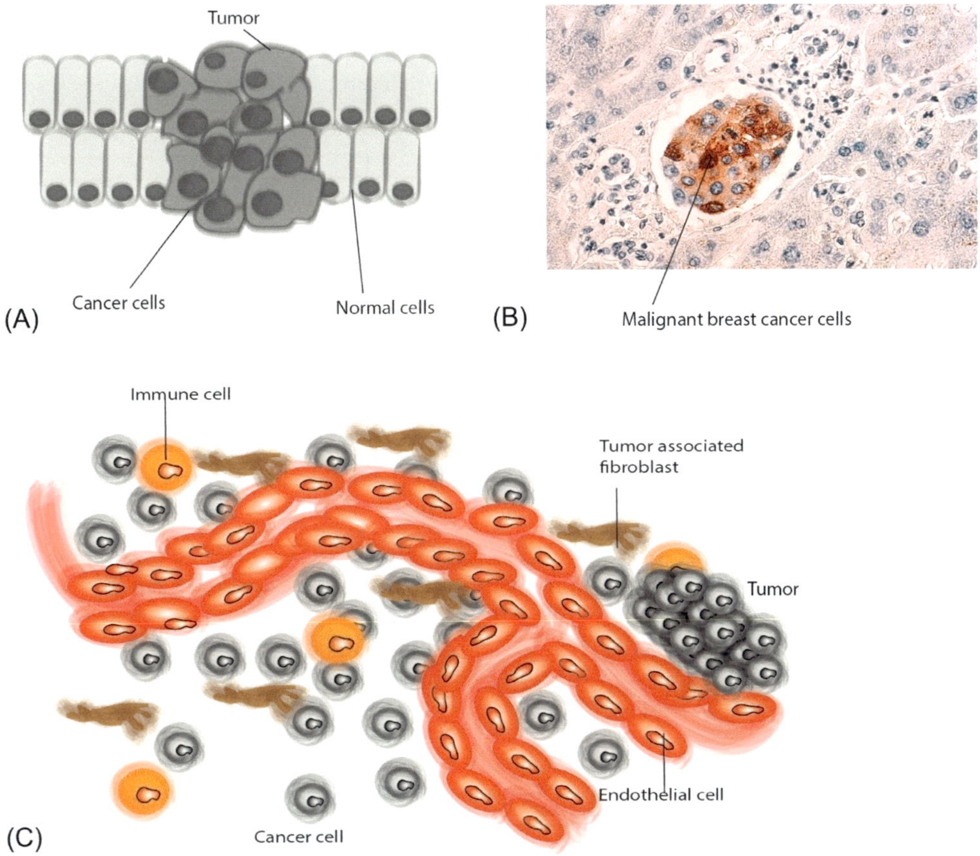

Fig. 1 Targets in the tumor microenvironment. (A) Schematic of a tumor formation. (B) Histological section of breast cancer metastasis in the liver. (C) Schematic of local blood vessels and immune cells in the tumor microenvironment. *(Panel B: Adapted from the National Cancer Institute archives image AV-8100-3062.)*

more quickly. To treat these cancers, clinicians aim to keep estrogen levels low throughout the whole body or prevent estrogen from binding to the estrogen receptor. Fulverstan is an example of an antiestrogen molecule that binds to the estrogen receptor present in some cancer cells. This prevents estrogen from binding to the receptor and can slow cancer growth. A similar strategy is used for many hormone receptor-positive cancers, and bioengineers have a key role to play in designing new antibodies that can block specific receptor targets.

2.3 Immunotherapy

Many of the latest cancer therapies focus on redirecting the patient's own immune system to better recognize and target cancerous tissue. Strategies include delivering small-molecule agents that can activate local immune cells or antibodies that can block the cell surface receptors that cancer cells typically use to evade the immune system [12, 14, 15, 21, 22]. Cancer vaccines can also be delivered to train the immune system to recognize and respond to specific cancer epitopes [23]. In the latest therapies a patient's own immune cells can be extracted from tissue biopsy, sorted for the most potent anticancer cells, then expanded, and reinfused. Alternatively a patient's T cells can be extracted and genetically engineered to express a chimeric antigen receptor, which can help cells to identify and target cancerous cells. These strategies are collectively termed immunotherapy and are discussed in more detail in Chapter 3.2.

2.4 Chemotherapy and the cell life cycle

Chemotherapeutic agents are the best-known class of drugs used to treat cancer. Normally, cells divide to produce two, identical daughter cells in a process called mitosis. Although all cells progress through this cycle, cancer cells often do this at a much faster rate than normal cells. By carefully selecting appropriate pharmaceutical agents, oncologists aim to block or interrupt specific phases of cell division and prevent cell growth. There are several distinct phases in the cell life cycle, shown in Fig. 2. In the growth (G1 and G2) phase, cells increase in size and generate additional organelles (such as mitochondria) to prepare for cell division. In the S phase, DNA replication occurs; DNA unwinds, and an identical copy is made using an enzyme called topoisomerase. Finally, during mitosis (M phase), the duplicated DNA and cellular organelles organize along fibrils in the cell called microtubules, before the cell divides in two to produce two identical cells.

Cytotoxic chemotherapeutic drugs interrupt the cell cycle by targeting distinct cellular processes occurring in each of the phases of cell division. There are five main types of chemotherapy drug: alkylating agents, antimetabolites, enzyme inhibitors, antitumor antibiotics, and antimicrotubule agents, summarized in Table 1. Many of these

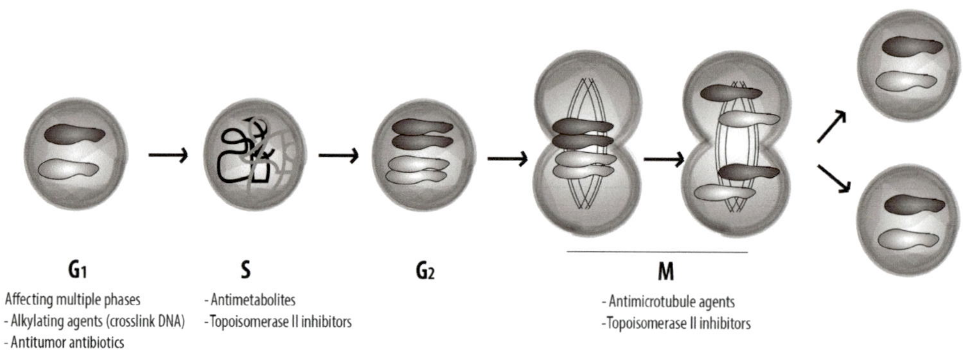

Fig. 2 A schematic of the cellular life cycle and cell cycle-specific cytotoxic chemotherapeutic agents.

Table 1 Summary of chemotherapeutic cancer drugs that target the cell cycle.

Drug classes, mechanism of action, and example	Examples
Alkylating agents: 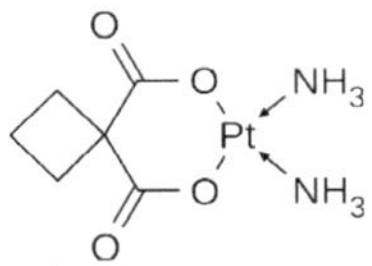 Chemically react with DNA to prevent replication	• Mustard gas derivatives: ifosfamide, cyclophosphamide, chlorambucil • Nitrosoureas: carmustine, lomustine, streptozocin • Hydrazine and triazine: altremaine, dacarbazine, temozolomide • Metal salts; carboplatin, cisplatin, oxaliplatin • Ethylenimines: thiotepa and hexamethylmelamine • Alkyl sulfonates: busulfan
Antimetabolites: 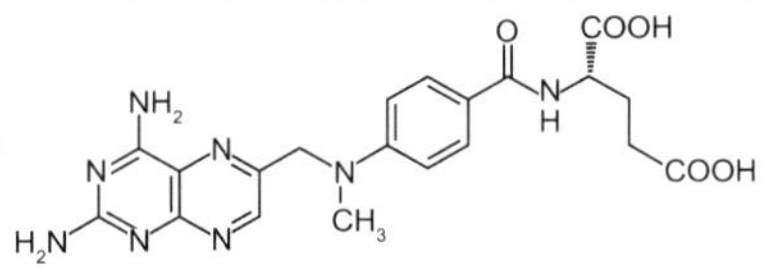 Cell cycle-specific inhibitors: these are similar to endogenous substances that, however, prevent cellular metabolism when incorporated into the cell	• Folic acid antagonist: methotrexate • Pyrimidine antagonist: 5-fluorouracil, floxuridine, cytarabine, capecitabine, and gemcitabine • Purine antagonist: 6-mercaptopurine and 6-thioguanine • Adenosine deaminase inhibitor: cladribine, fludarabine, nelarabine, and pentostatin
Enzyme inhibitors: 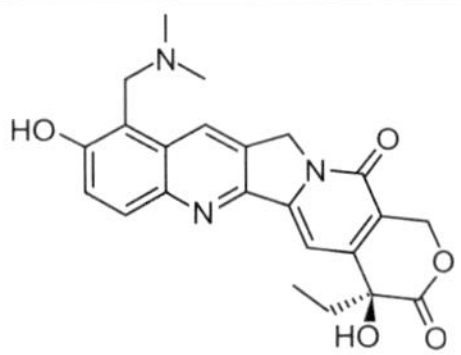 Typically, topoisomerase inhibitors that inhibit topoisomerase I and II, enzymes crucial for DNA replication	• Topoisomerase I inhibitors: irinotecan, topotecan • Topoisomerase II inhibitors: amsacrine, etoposide, etoposide phosphate, teniposide
Antitumor antibiotics: 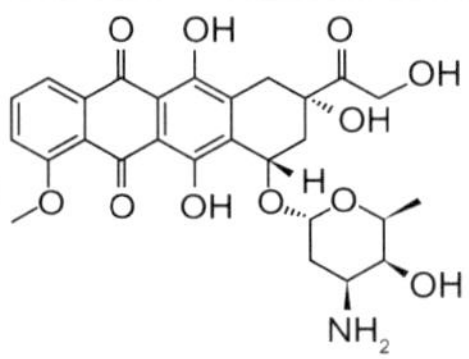 Chemotherapeutic treatments typically produced by *Streptomyces* species, which inhibit several cell cycle phases	• Anthracyclines: doxorubicin, daunorubicin, epirubicin, mitoxantrone, and idarubicin • Chromomycins:dactinomycin and plicamycin • Miscellaneous: mitomycin and bleomycin
Antimicrotubule agents: 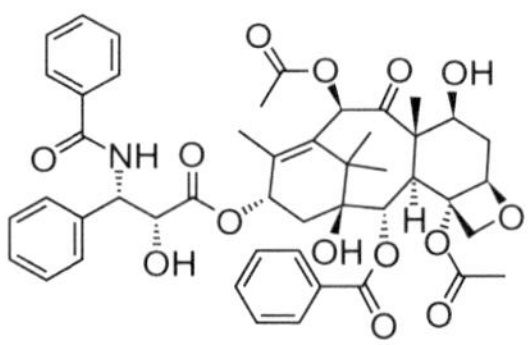 Typically plant alkaloids that inhibit microtubule formation during the S phase of cell division	• Vinca alkaloids: vincristine, vinblastine, and vinorelbine • Taxanes: paclitaxel and docetaxel

chemotherapy drugs interrupt DNA replication. For example, alkylating agents chemically react with DNA to prevent DNA unwinding, while enzyme inhibitors block the activity of the replication enzyme (topoisomerase inhibitors) to prevent DNA from being copied in the cell. These classes of drugs are widely used and include platinum-based drugs, which intercalate with DNA to prevent it from unwinding. Carboplatin is one example of this class and is often the first-line chemotherapeutic used to treat many types of cancer, demonstrating the widespread use of this strategy.

Other chemotherapeutic approaches focus on inhibiting the synthesis of new DNA using antimetabolites. Antimetabolites are molecules that are very similar to substances used to build DNA, normally found in the cell. However, they are slightly different to endogenous substances and prevent the DNA from being used properly, causing cell death. Finally, alkaloids extracted from plants can be effective at preventing cell division by disrupting the formation of the microtubules in the cell. Cells require these tubules to align the newly synthesized DNA before division; preventing microtubule formation helps to slow cancer growth.

Although almost all cells progress through this cycle, cancer cells often do this at a much faster rate than normal cells. By spacing chemotherapeutic treatments appropriately, oncologists try to ensure that chemotherapeutic drugs interrupt cell division only in rapidly dividing cells, leaving the majority of normal cells unharmed. Unfortunately, this often means that healthy cells that divide rapidly (i.e., hair, nails, and white blood cells) are also affected, leading to unwanted side effects such as low white blood cell counts and hair loss. To reduce these side effects, bioengineers can design better delivery systems to ensure the right drug reaches the right target.

3 Biological barriers in drug delivery

Once an appropriate drug has been identified, the next challenge lies in delivering a therapeutic dose to the right location. Drugs must first cross the body's natural physical barriers to reach the systemic circulation for effective delivery [24]. Drugs are typically administered via ingestion, inoculation, inhalation, or topical adsorption through mucus membranes (Fig. 3). Orally administered drugs travel via the stomach to the intestines, where they are absorbed through the highly adapted microvilli in the gastrointestinal tract (GI tract) [25]. There, they are metabolized and cleared through the liver and enter systemic circulation. Drugs that are administered via infusion/inoculation typically bypass the liver metabolism and can provide increased levels of circulating, bioavailable drugs compared with orally delivered pharmacologics of the same dose [26–29]. Once in the circulation, drugs must then cross location-specific barriers (such as the blood-brain barrier) to reach their target cancer cell.

Drug delivery is therefore a crucial step for cancer therapeutics, and many engineering strategies have been explored to improve drug delivery. These strategies typically focus on

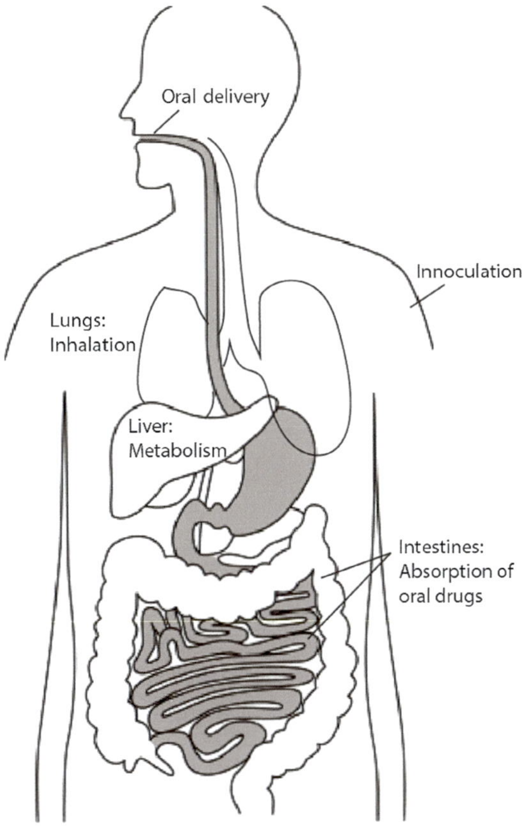

Fig. 3 Key locations for drug delivery and adsorption across biological barriers.

(1) increasing the therapeutic yield and circulation time of pharmaceutical agents delivered across biological barriers into the systemic circulation [30, 31]; (2) developing continuous, controlled release systems to regulate drug delivery [32–37]; and (3) engineering targeted delivery strategies for site-specific therapeutic action [33, 38–40]. Over the next few sections, we will discuss the engineering challenges in each of these steps and highlight successful examples.

4 Increasing systemic circulation of pharmaceutical agents

The design of delivery vehicles that can carry drugs across biological barriers more effectively is an active field of bioengineering research [24, 41]. Delivery vehicles are packages for (or modifications to) pharmaceuticals, which enable improved delivery across biological barriers or enhance delivery yield. Here, we'll highlight three approaches to increase delivery to the systemic circulation: (A) targeted delivery to the intestinal region of the GI tract, (B) strategies to enhance the solubility and cell penetration of chemotherapeutic drugs, and (C) modifications that can extend circulation times of absorbed drugs.

4.1 Enhancing oral delivery by targeting the intestines

Orally delivered drugs must first pass through and be absorbed through the GI tract to become bioavailable. The GI tract consists of the stomach, small intestine, and large intestine, with drug absorption primarily in the small intestine [25]. The chemical environment of the stomach often poses a challenge for orally administered drugs due to low pH. Many protein and peptide drugs (such as targeted immunotherapy drugs or antibodies used to treat cancer) are denatured at low pH, and so, oral delivery is not an appropriate administration route. Many other drugs are also pH sensitive. To protect these drugs from the chemical environment of the stomach, devices that can control delivery to distinct chemical environments within the GI tract have been developed [42–45].

Ingestible delivery vehicles typically package drugs inside a polymer casing or shell to protect drugs from the low-pH environment in the stomach. Selective delivery of drugs to the intestine can also be increased by encapsulating chemotherapeutics within pH-sensitive materials. For example, drugs can be coated in materials that only dissolve in the alkaline environment of the intestine [42–45]. Methotrexate is a chemotherapy drug that has similar bioavailability whether administered IV or orally and is frequently given to patients in tablet form. Methotrexate tablets are designed with an enteric coating, which only dissolves in the alkaline environment of the lower GI tract. This prevents drugs from being exposed to the low pH in the stomach and facilitates dissolution of the tablet and absorption of the drug in the GI tract.

4.2 Enhancing solubility and cell penetration using nanotechnology

Small-molecule chemotherapeutics are often poorly absorbed across the GI tract, and these drugs do not typically penetrate biological barriers such as cell membranes. To increase the adsorption of these drugs across the GI tract, drugs can be optimized for delivery by tuning hydrophilicity and size. Drugs can also be reformulated with agents that increase solubility or can be encapsulated within carriers that are able to penetrate these barriers more easily. Several of the delivery challenges associated with chemotherapeutic drugs can be addressed by encapsulating drugs within nanocarriers or by using nanotechnology [46, 47]. This can aid delivery across several biological barriers, including mucus membranes and the blood-brain barrier, to facilitate cell penetration. Nanomedicine has enabled improvements in drug stability, drug distribution, tissue targeting, and toxicity, with many nanocarrier-based therapeutics currently approved for clinical use (Table 2). In medicine, nanotechnology typically refers to systems or therapies that are on the order of 1 to several hundred nanometers in at least one dimension.

Liposomes are one of the most established nanocarrier systems used for chemotherapy [53]. Inspired by the phospholipid bilayer structure of cell membranes, liposomes generally consist of lipids that have a polar head group and hydrophobic tail that self-assemble into spheres. These spheres enable encapsulation of both polar or nonpolar cargo (Fig. 4).

Table 2 Approved nanotherapeutics for the treatment of cancers.

Brand name	Active drug	Carrier	Indication	Approval
Lipid				
Doxil/Caelyx	Doxorubicin	HSPC, cholesterol, PEG2000-DSPE	Breast, ovarian cancer, Kaposi's sarcoma	USA 1995 EU 1996
DaunoXome	Daunorubicin	DSPC, Cholesterol	Kaposi's sarcoma	11 EU countries USA 1996[a]
Marqibo [48]	Vincristine	Sphingomyelin and Cholesterol	Leukemia	USA 2012
Onivyde	Irinotecan	methoxyPEG-2000-DSPE, DSPC, Cholesterol	Pancreatic cancer	USA 2015 EU 2016
Polymer				
Genexol-PM	Paclitaxel	Polylactide-methoxy PEG block copolymer	Breast, ovarian, lung cancer	South Korea 2007
Eligard	Leuprolide	Poly(lactide-co-glycolide)	Prostate cancer	USA 2004
Conjugate				
Abraxane	Paclitaxel	Albumin	Breast cancer	USA 2005
Oncaspar [49]	L-Asparaginase	PEG	Leukemia	US 1994 EU 2016
Zinostatin (SMANCS) [50]	Neocarzinostatin	Poly(styrene-co-maleic acid)	Hepatocellular carcinoma	Japan 1994
Viral				
Gendicine [51]	Human p53 gene	Adenovirus	Head, neck cancer	China 2003
Rexin G [52]	Human cyclin G1 gene	Retrovirus	Solid tumors	Philippines 2007

[a] Discontinued in the United States.

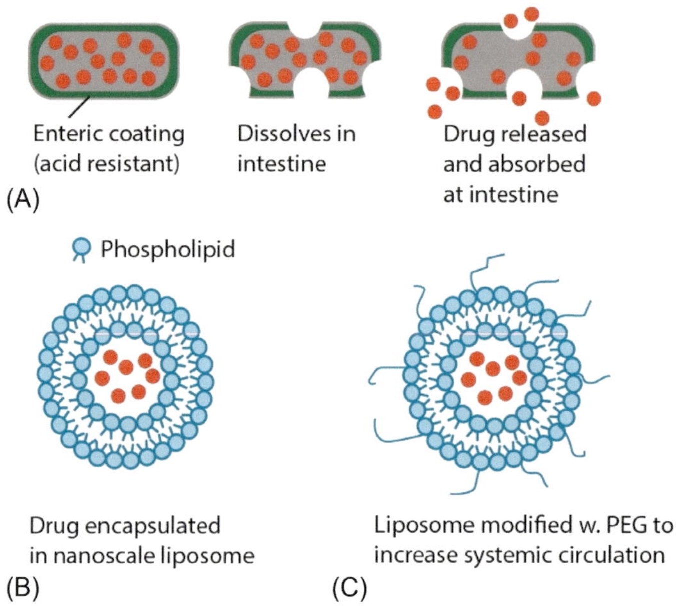

Fig. 4 Strategies to increase systemic drug delivery. (A) Enteric coatings for pH sensitive drug delivery. (B) Encapsulation of drugs within phospholipid liposomes to improve solubility of drugs across biological barriers. (C) Modified liposomes which resist phagocytic clearance and enhance systemic circulation times.

Drugs that have poor aqueous solubility can be encapsulated within liposomes to increase delivery. Other strategies for nanoformulation of chemotherapeutics include polymeric micelles or drug-polymer conjugates (Table 2). For example, paclitaxel is poorly water soluble unless it is solubilized in certain solvents. Many of these solvents are toxic to humans in high doses, and so the amount of paclitaxel that can normally be delivered is limited. By formulating the drug with a biodegradable polymer, nanoscale micelles of the drug have been developed (Genexol-PM). This formulation can be delivered systemically to patients with reduced toxicity and facilitates delivery of the normally poorly soluble paclitaxel across cell membranes.

4.3 Enhancing systemic drug circulation

Although nanocarriers such as liposomes enhance delivery across biological barriers, they are subject to rapid clearance depending on their size and route of administration. For example, intravenous delivery of carriers smaller than 5 nm are typically filtered by the kidneys for renal excretion, whereas larger nanocarriers 200–500 nm may be recognized as foreign bodies and undergo clearance by the immune system via the mononuclear phagocytic system primarily carried out by the phagocytic cells of the liver and spleen [41]. Engineering strategies to enhance the therapeutic circulation time of drugs and prevent uptake by the wrong cells is an important consideration in chemotherapeutic drug delivery. Due to the small size of chemotherapeutic nanoformulations, immune cells such as phagocytes can engulf drug nanoparticles and prevent them from reaching their target. To reduce phagocytosis and enhance therapeutic efficacy, nanosystems can be conjugated with chemical ligands that prevent immune cells from uptaking the nanoparticles [31, 54, 55]. This, in turn, enhances the drug circulation time and increases the chances that drugs can reach the tumor target from the systemic circulation.

In one example, liposomes have been decorated with "self" peptides such as CD47 [56], which act as a "don't eat me" marker. This prevents immune cells from recognizing the drug vehicle and from engulfing it. In an alternative approach, so-called "stealth" liposomes were developed that utilize polyethylene glycol (PEG) coatings. These coatings reduce protein adsorption onto the liposome, reduce phagocytic uptake by the immune system, and therefore increase the circulation time of drug-containing liposomes in the blood (Fig. 4) [54]. For example, after delivery, free doxorubicin has an area under the curve (AUC) of only 4 μg h/mL. This is increased to 45 μg h/mL using an uncoated liposome formulation (Myocet) and to 900 μg h/mL using a PEG-liposomal formulation of doxorubicin, Caelyx (Doxil) [57]. These studies demonstrate that packaging the free drug doxorubicin in different liposomal formulations can affect the systemic circulation time of anticancer drugs and that the inclusion of PEG can dramatically increase this circulation time. To design more effective delivery systems for chemotherapeutic drugs, bioengineers need to identify new techniques to enhance delivery across biological barriers and enable selective uptake by desired cell types.

5 Continuous and controlled systems for drug delivery

Drug-based cancer treatment plans can vary dramatically between patients. Type, location, and stage of disease impact the selection of an appropriate chemotherapeutic drug alongside the number of doses, interval between these, and duration of drug treatment for each patient. In some cases, continuous delivery of pharmaceutical agents is required for periods of a few weeks to a few months, and daily infusion or oral delivery may not be an option for every patient. Bioengineers can design systems that facilitate the continuous and controlled delivery of chemotherapeutic agents to patients. Ideally, these systems should be administered in a single application to reduce patient burden. By implanting these systems close to tumor sites, local delivery of pharmaceuticals can be also achieved to reduce off-target side effects. The development of new drug delivery systems offers the potential for controlled and continuous drug delivery directly in tumor sites.

5.1 Systemic drug delivery systems

Long-term and continuous drug delivery systems have been used to deliver contraceptives, antiinflammatory drugs, and antibiotics since the 1970s [1, 58–63]. Many of these devices are based on implants made of polymer-drug composites coated in porous membranes. By tuning the membrane porosity and drug loading, the rate of drug diffusion can be controlled. These systems typically enable fixed-rate drug diffusion on the order of months to years and when implanted enable long-term drug delivery into the systemic circulation. Due to the toxicity associated with most chemotherapeutic agents, long-term and systemic continuous chemotherapy delivery via implants is often undesirable. However, these systems are more suited to nontoxic pharmaceuticals, such as long-term hormone delivery to treat hormone-sensitive cancers.

Similarly, transdermal patches are another well-accepted system used for the controlled and continuous delivery of small molecules, most commonly associated with nicotine delivery. More recently, transdermal patches have been used to deliver hormones such as estradiol and estrone [27], which can also help to treat hormone-sensitive cancers. The use of a dermal patch to deliver hormones can maintain systemic circulation of hormones and keep hormone-sensitive cancers under control, and these systems are currently used in clinic. Drug delivery profiles range from a few days to a few weeks due to limitations in drug loading capacity within the patches, and patches must remain attached to the skin for the duration of delivery.

The discovery of biodegradable polymers has enabled the development of alternative, implantable controlled release drug delivery systems with tunable kinetic release profiles. These systems can be loaded with a drug and injected or implanted within the body where they elute a continuous, low dose of drugs into their local environment. A variety of polymers have been used to fabricate these systems, and the choice of polymer directly impacts the specific release kinetics of the device. For example, poly(lactide-co-glycolic-acid) (PLGA) is a biodegradable polymer that is FDA approved for clinical use in a range of applications, including chemotherapeutic drug delivery and hormone release. This polymer is widely used for the delivery of drugs as the release profile can be tuned to continuously deliver therapeutic load for shorter durations (of 2 days) or for up to 8 weeks, depending on the lactide-to-glycolide ratio [19, 64, 65] within the polymer. Zoladex, for example, is a subcutaneous implant consisting of goserelin acetate dispersed in PLGA, which is implanted into patients and releases goserelin. This drug helps to maintain testosterone suppression in prostate cancer patients, [66] and the device enables drug delivery for between 1 and 3 months.

5.2 Localized delivery to the tumor microenvironment

Many of the systemic controlled release implants described here can also be used for more localized delivery. By surgically implanting these next to the tumor site, therapeutic

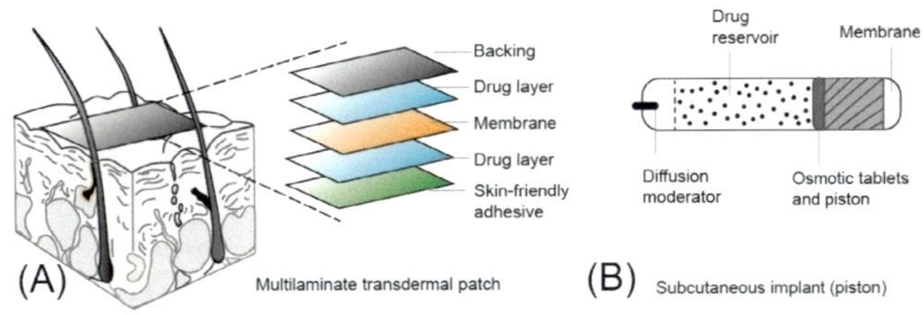

Fig. 5 Controlled drug delivery systems. (A) A schematic of a typical multilaminate transdermal patch for the continuous delivery of small molecules and hormones in cancer treatment. (B) A schematic of a piston-driven subcutaneous implant for the delivery of continuous therapeutic chemotherapy doses.

delivery in the vicinity of the tumor can be enhanced. For example, a current treatment for glioblastoma involves implanting biodegradable wafers that elute the chemotherapeutic carmustine (Gliadel) into the brain [7]. After surgical resection of brain tumors, surgeons can pack the cavity with these wafers and close the incision. The wafers will continue to elute the chemotherapeutic into the surgical site for 2–3 weeks. As the blood-brain barrier is typically difficult to penetrate, this ensures drugs are at a high concentration directly where needed.

These strategies require a surgical procedure to place the implant within the body. In some cases, tumors may be surgically inaccessible and surgical implantation of proximal drug delivery systems may not be possible. Additionally, many of these systems are only suitable for the delivery of nontoxic pharmaceuticals which have well-controlled off-target effects. For patients with inaccessible tumors, or those who are prescribed toxic pharmaceutical agents to control cancer cell growth, drugs that can be administered systemically but inherently target cells in the tumor microenvironment are desired (Fig. 5).

6 Delivering drugs to the right target

Oncologists typically want to deliver high concentrations of anticancer drugs to the tumor microenvironment while avoiding nontarget cells and environments. Localized delivery strategies, such as local injection, inhalation for lung targeting, implantation of proximal devices, or topical administration, are simple ways to ensure an increased concentration of therapeutic in the target area [67]. For hematological cancers, delivery of therapeutic agents directly into the blood stream helps ensure drugs reach malignant immune cells circulating in the blood. For solid tumors, it is more challenging to target a specific organ within the body or multiple organs if cancer has metastasized. These tumors are often not readily accessible, and so, alternative strategies to increase the local concentration of pharmaceuticals in the tumor environment are required (Fig. 6).

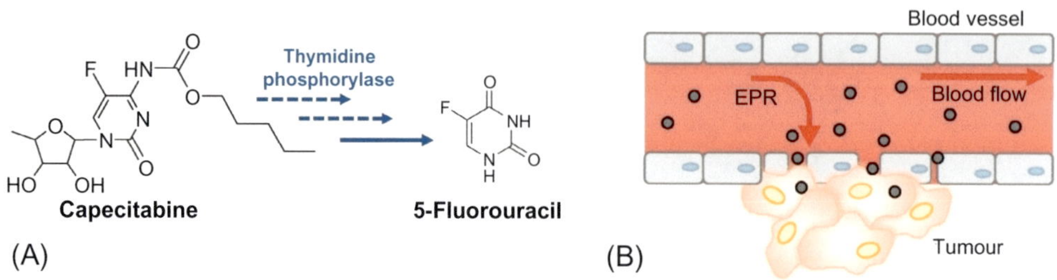

(A) (B)

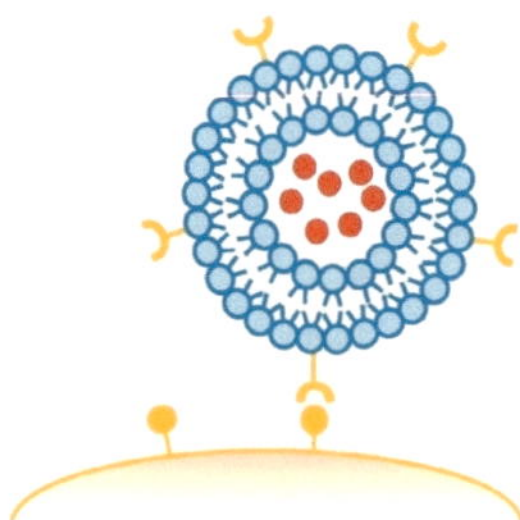

Target	Drug	Targeting moiety	Carrier
LHRH receptor	Paclitaxel	LHRH peptide	Poly(amidoamine)
HER2 receptor	Emtansine	HER2 antibodies eg Trastuzumab	Polylactide
Transferrin receptor	Oxaliplatin	Transferrin	Liposome

(C)

Fig. 6 Targeted drug delivery approaches. (A) Prodrug conversion to active drug by enzyme activity in the tumor environment. (B) Passive targeting to tumor environment through enhanced permeation and retention (EPR). (C) Nanocarriers containing chemotherapeutics can be decorated with targeting ligands that can actively bind to receptors on tumor cell surfaces and encourage cell uptake and payload delivery into the cell. Examples of such targeting ligands used to actively deliver the chemotherapeutics paclitaxel [68], emtansine [69], and oxaliplatin [70] are shown. *LHRH*, luteinizing hormone-releasing hormone; *HER2*, human epidermal growth factor receptor 2.

6.1 Systemic administration and passive accumulation in the tumor microenvironment

Although systemic administration of pharmaceuticals does not specifically target cancer cells, or even the cancer microenvironment, some drug formulations naturally accumulate there. Passive targeting is thought to occur when nanoparticle drug formulations accumulate in the tumor environment due to the enhanced permeability and retention (EPR) effect [17]. The EPR effect describes the increased permeability of endothelium surrounding tumors. Nanoparticles circulating in the blood are small enough to cross the leaky blood-tumor barrier resulting in accumulation at the tumor site. The consequence is an increased drug concentration at the tumor site and relatively reduced drug concentrations at other sites resulting in reduced systemic toxicity. While this passive accumulation can lead to improved local activity of some drugs (therapeutic index), the EPR effect can vary greatly between different tumors and patients. Additionally, the EPR effect may be overestimated in animal models compared with humans where its contribution in enhancing therapeutic activity of nanoparticle drugs remains under debate with some studies suggesting that less than 1% of administered drug passively accumulates in the tumor [71]. It is therefore necessary to design therapeutics that can be selectively delivered to and accumulate within the only cells or tissues of interest.

6.2 Targeted delivery to the tumor microenvironment

To deliver drugs directly to the target site, advanced strategies focus on modifying pharmaceuticals directly to make them act only in the tumor sites or packaging pharmaceuticals in delivery vehicles that are better able to cross biological barriers and target tumors. Identifying specific receptor ligand pairs that are overexpressed in cancer cells, such as the transferrin receptor [72], can aid tumor targeting and enhance delivery. For example, doxorubicin is a potent chemotherapeutic; however, it is challenging to deliver to specific biological sites or across the blood-brain barrier. To address this, nanoparticles that preferentially bind apolipoproteins allowing transport across the blood-brain barrier (polysorbate 80-modified polybutylcyanoacrylate nanoparticles) were used to deliver doxorubicin to the brain [47]. In another example, HER2 antibodies have been conjugated to liposomal doxorubicin to mediate targeted binding in breast cancer [72a]. The HER2-coated liposome binds to tumor cells, where it is endocytosed and then acts more effectively to target tumors.

By designing drug nanocarriers, which present specific ligands, bioengineers can redirect cellular uptake. For example, Abraxane is an albumin-bound nanoparticle formulation of paclitaxel. This formulation shows enhanced intratumoral concentrations and antitumor activity, likely due to an albumin receptor (gp60)-mediated endothelial transcytosis [73–75] that enables the drug to pass through endothelial cell walls in tumor microvessels and into cells. This is also an attractive strategy for intracellular targeting of newer treatment modalities such as gene therapy. Recently, viral vectors that deliver the human p53 gene and cyclin G1 gene have been approved as cancer therapies [51]. These therapeutics require intracellular gene expression to be effective, and so, nanoparticle delivery systems that facilitate intracellular delivery have been chosen as packaging agents. As the mechanism for cellular entry and endosomal escape becomes better understood, new ligands for targeted therapeutic delivery will be identified to enhance intracellular targeting.

7 Next-generation and on-demand drug delivery

To enable a truly patient-specific treatment regime, oncologists need to move toward on-demand and targeted delivery of personalized cancer pharmaceuticals. Typically, histological and biomolecular analysis of tumor tissue allows oncologists to select a combination of pharmaceuticals most likely to work for a patient. Recently, new polymer implants have been developed that enable in situ testing of pharmaceutical agents and identification of personalized chemotherapy options. Small, implantable devices have been developed to identify patient-specific responses to drug combinations [76, 77]. These devices contain over 100 individual wells of anticancer agents or combinations of anticancer agents and can be inserted into tumor tissue. The device can be removed using a biopsy needle, and the tissue can be examined to determine tumor regression in response to each of the therapeutic drugs and drug combinations tested. In this way, surgeons and oncologists can develop a personalized drug treatment plan for cancer patients based on positive responses to specific drug combination therapies [76, 77], tested using the microchip.

Separately, stimulus-responsive systems have been developed to facilitate chemotherapeutic delivery to local environments [33, 78–82]. These materials typically encapsulate a systemically administered chemotherapeutic drug in an inactive form. Application of a specific stimulus, such as heat, light, pH, magnetic fields, and ultrasound, changes the properties of the materials. The change in material properties activates the release of the drug. This enables drugs to be delivered systemically without toxic side effects, as drugs are only activated when the specific stimuli are encountered.

This can allow spatiotemporal control of drug delivery to maximize targeted delivery of selected pharmaceuticals. For example, ThermoDox liposomes include a thermosensitive lipid with a transition temperature of between 40°C and 45°C to create a temperature-sensitive doxorubicin-PEGylated liposome. These new-generation liposomes are being developed to release their encapsulated drug where local tissue temperatures are elevated to 39–42°C by application of radiofrequency, a technique referred to as radiofrequency ablation. For ThermoDox, this technology allows concentration of the drug up to 25 times more in the treatment area than IV doxorubicin and several fold the concentration of other liposomally encapsulated doxorubicins [42, 43, 83]. Recent innovation in ultrasound and microbubble technology enables spatiotemporal control of therapeutics by injection of microbubbles or nanoparticles containing drug followed by local and targeted ultrasound to increase penetration through vasculature in the region of interest or to encourage microbubble disruption (cavitation) to trigger drug release [84].

In an alternative approach, prodrugs have also been developed, which aim to reserve therapeutic activity until a specific cleavage or conversion event occurs in the tumor microenvironment to avoid widespread toxicity. A number of enzyme-cleavable drug delivery systems have been developed for the controlled delivery of dyes or anticancer therapeutics [9, 85]. Capecitabine is a prodrug of 5-fluorouracil (5-FU) developed to reduce intestinal side effects compared with orally delivered 5-FU. Capecitabine is not cytotoxic until it is converted to 5-FU in tumors. However, this conversion also occurs in the liver, so while this strategy has helped to reduce intestinal toxicity during oral administration, systemic toxicity including cardiotoxicity remains due to the

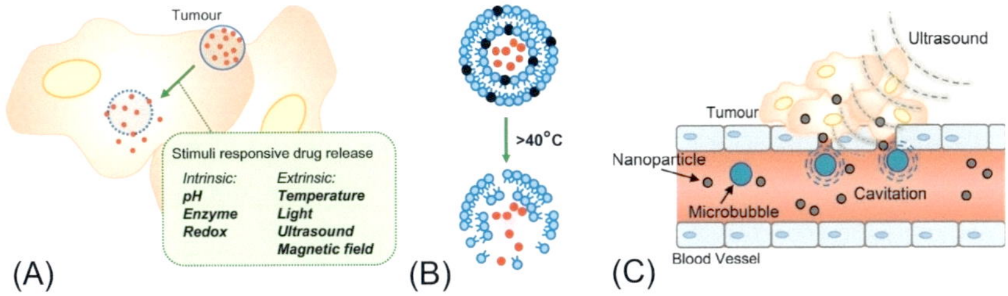

Fig. 7 Stimulus-responsive systems for cancer therapeutic delivery. (A) An overview of strategies for increasing drug delivery to the tumor site by using intrinsic and extrinsic stimuli that can trigger drug release. (B) An example of temperature-triggered drug delivery. Chemotherapeutics such as doxorubicin can be encapsulated in liposomes that have been formulated with temperature sensitive lipids. ThermoDox is one such formulation where the lipid can undergo a gel-to-liquid phase transition at elevated temperatures applied locally using an external heat source and trigger release of drug. (C) Ultrasound-mediated drug delivery can enable drug-loaded nanoparticles (NPs) to be locally delivered using microbubbles that are disrupted by application of focused ultrasound.

widespread conversion to 5-FU by the liver that is then systemically distributed. To combat systemic toxicity, localized, stimulus-responsive drug delivery systems can be used. For example, stimulus-responsive microneedle patches for topical application have recently been developed [86]. These systems interact with cancerous and noncancerous cells in the tumor margins and however only release their payload following enzymatic cleavage in cancer cells. This facilitates the selective delivery of therapeutic agents to the cancer cells (Fig. 7).

8 Translation: The final hurdle

Modern cancer drug therapies include small molecules, proteins, and genetic engineering-based strategies, and these therapies often have precise biological therapeutic targets. As therapeutic targets become increasingly specific and therapies become increasingly potent, delivering site-specific therapeutic action with minimal side effects becomes a more critical design feature. In the future, selection of a personalized chemotherapeutics will be combined with stimulus-sensitive drug delivery to ensure the right drug is delivered to the right target. Control over both dosing and targeting of anticancer pharmaceuticals can improve patient safety, quality of life, and compliance by lessening the stringent requirements of strict medication schedules and reducing off-target effects. However, translating bench-based research to patients poses a significant challenge.

For drug delivery systems to move from the research bench toward patients, drugs must be appropriately regulated and licensed by the correct agency. Region-specific regulatory bodies typically define healthcare products, which primarily differentiate between drug, medical device, advanced biologics, or combination products [38]. In the United States, biologics are regulated by the Center for Biologics Evaluation and Research within the US FDA. Within the European Union the European Medicine Agency is responsible for regulating therapeutics. A brief overview of the development pipeline and regulatory framework for ATMP therapeutic drug delivery systems is shown schematically in Fig. 8.

Developing a commercial product from laboratory research is a significant challenge [87]. Once a product has been licensed, it must be continually evaluated to determine whether it still provides benefits to patients. Earlier, we described how the addition of PEG into liposomes can increase circulation times, enhancing therapeutic efficacy of encapsulated drugs. However, recently, it has been suggested that repeated administration of PEGylated formulations can actually lead to an increased clearance rate of subsequent doses of PEGylated liposomes. This is thought to occur due to IgM antibodies raised after first exposure [88] or preexisting IgG antibodies to PEG [89] facilitating elimination of the PEGylated liposome. In addition, some PEGylated drugs can accumulate in the epidermis causing hand-foot syndrome, characterized by tingling,

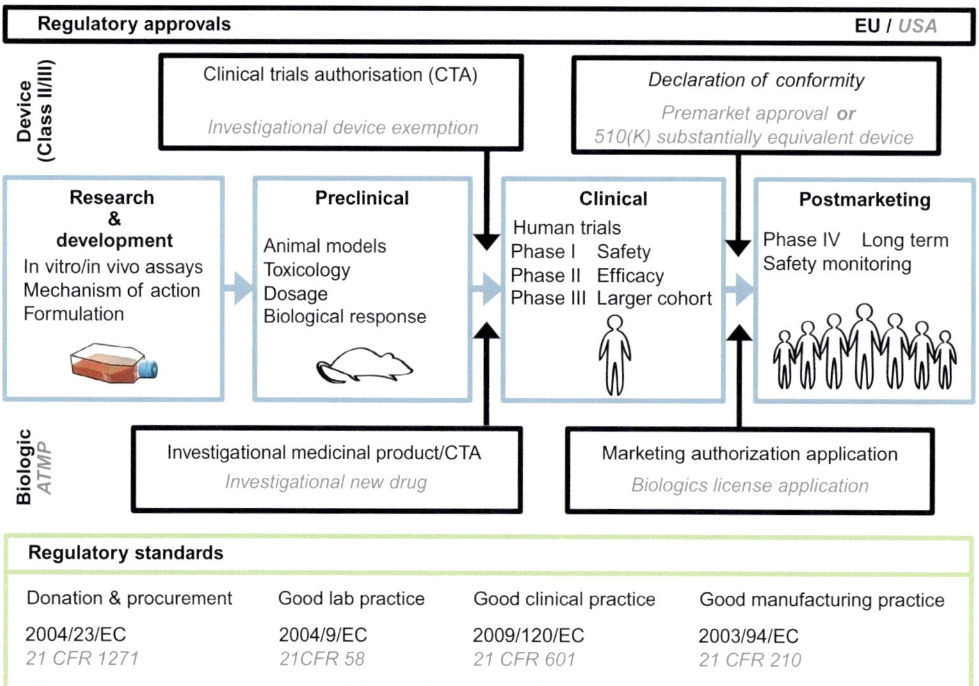

Fig. 8 Regulatory pathways in the UK and US pathway for translation of medicinal products. *(Adapted from Q.A. Rafiq, et al., The early career researcher's toolkit: translating tissue engineering, regenerative medicine and cell therapy products, Regen. Med. 10 (8) (2015) 989–1003.)*

numbness, swelling, and redness [90]. Although these side effects may be tolerated in patients with life-threatening diseases, as new therapies reach the market, they must be continuously monitored to ensure drugs are beneficial and do not compromise patient safety.

9 Conclusion

As drug delivery systems become more sophisticated, enabling stimulus-responsive and personalized chemotherapy options, strategies that can aid targeted and selective delivery of anticancer drugs to cancerous cells are increasingly required. This is a dynamic process; drugs and delivery systems are continuously evolving, and it sometimes takes many years for side effects to be identified and solutions to be explored. These side effects and scientific questions about how best to deliver drugs pose additional design and delivery challenges for bioengineers to address in the next generation of anticancer therapeutics. Bioengineers can use expertise across the whole pipeline of cancer therapeutics—from biomolecular design and computational identification of new drugs and drug targets, to formulation design for improved delivery, to stimulus-responsive and controlled release delivery systems. Bioengineers have a crucial role to play in developing these next-generation drug delivery systems across macro- to nanolengthscales, to enable targeted delivery of anticancer therapies.

References

[1] S.A. Chew, S. Danti, Biomaterial-based implantable devices for cancer therapy, Adv. Healthcare Mater. 6 (2) (2017) 22.
[2] H. Derakhshankhah, et al., Colon cancer and specific ways to deliver drugs to the large intestine, Anti Cancer Agents Med. Chem. 17 (10) (2017) 1317–1327.
[3] H.F. Liang, et al., Paclitaxel-loaded poly(gamma-glutamic acid)-poly(lactide) nanoparticles as a targeted drug delivery system for the treatment of liver cancer, Biomaterials 27 (9) (2006) 2051–2059.
[4] D.W. Zhu, et al., Docetaxel (DTX)-loaded polydopamine-modified TPGS-PLA nanoparticles as a targeted drug delivery system fore the treatment of liver cancer, Acta Biomater. 30 (2016) 144–154.
[5] G.M. Keating, Rituximab: a review of its use in chronic lymphocytic leukaemia, low-grade or follicular lymphoma and diffuse large B-cell lymphoma, Drugs 70 (11) (2010) 1445–1476.
[6] N. Kemeny, et al., A randomized trial of intrahepatic infusion of fluorodeoxyuridine with dexamethasone versus fluorodeoxyuridine alone in the treatment of metastatic colorectal-cancer, Cancer 69 (2) (1992) 327–334.
[7] M. Westphal, et al., Gliadel (R) wafer in initial surgery for malignant glioma: long-term follow-up of a multicenter controlled trial, Acta Neurochir. 148 (3) (2006) 269–275.
[8] A. Arora, M.R. Prausnitz, S. Mitragotri, Micro-scale devices for transdermal drug delivery, Int. J. Pharm. 364 (2) (2008) 227–236.
[9] B.W. Barry, Novel mechanisms and devices to enable successful transdermal drug delivery, Eur. J. Pharm. Sci. 14 (2) (2001) 101–114.

[10] M. Kumar, Nano and microparticles as controlled drug delivery devices, J. Pharm. Pharm. Sci. 3 (2) (2000) 234–258.
[11] S. Pradhan, et al., A three-dimensional spheroidal cancer model based on PEG-fibrinogen hydrogel microspheres, Biomaterials 115 (2017) 141–154.
[12] J. Maher, E.T. Davies, Targeting cytotoxic T lymphocytes for cancer immunotherapy, Br. J. Cancer 91 (5) (2004) 817–821.
[13] M.C. Milone, C.H. June, Adoptive immunotherapy: new ways to skin the cat? Clin. Immunol. 117 (2005) 101–103.
[14] C.M. Capitini, C.L. Mackall, A.S. Wayne, Immune-based therapeutics for pediatric cancer, Exp. Opin. Biol. Ther. 10 (2) (2010) 163–178.
[15] L.E. Huye, G. Dotti, Designing T cells for cancer immunotherapy, Discov. Med. 9 (47) (2010) 297–303.
[16] M. Cartellieri, et al., Chimeric antigen receptor-engineered T cells for immunotherapy of cancer, J. Biomed. Biotechnol. (2010).
[17] Y. Matsumura, H. Maeda, A new concept for macromolecular therapeutics in cancer-chemotherapy—mechanism of tumoritropic accumulation of proteins and the antitumor agent SMANCS, Cancer Res. 46 (12 Part 1) (1986) 6387–6392.
[18] T.U. Luu, W.F. Liu, Regulation of macrophages by extracellular matrix composition and adhesion geometry, Regener. Eng. Transl. Med. 4 (4) (2018) 238–246.
[19] H. Parmar, et al., Response to D-TRP-6-Luteinising hormone-releasing hormone (decapeptyl) microcapsules in advanced ovarian-cancer, Br. Med. J. 296 (6631) (1988) 1229.
[20] E.J. Small, et al., Immunotherapy of hormone-refractory prostate cancer with antigen-loaded dendritic cells, J. Clin. Oncol. 18 (23) (2000) 3894–3903.
[21] A.A. Chekmasova, R.J. Brentjens, Adoptive T cell immunotherapy strategies for the treatment of patients with ovarian cancer, Discov. Med. 9 (44) (2010) 62–70.
[22] B. Jena, G. Dotti, L.J. Cooper, Redirecting T-cell specificity by introducing a tumor-specific chimeric antigen receptor, Blood 116 (7) (2010) 1035–1044.
[23] W.A. Li, D.J. Mooney, Materials based tumor immunotherapy vaccines, Curr. Opin. Immunol. 25 (2) (2013) 238–245.
[24] M. Schneider, et al., Crossing biological barriers for advanced drug delivery, Eur. J. Pharm. Biopharm. 84 (2) (2013) 239–241.
[25] D. Dahlgren, et al., Regional intestinal permeability of three model drugs in human, Mol. Pharm. 13 (9) (2016) 3013–3021.
[26] M.R. Prausnitz, R. Langer, Transdermal drug delivery, Nat. Biotechnol. 26 (11) (2008) 1261–1268.
[27] R.J. Chetkowski, et al., Biologic effects of transdermal estradiol, N. Engl. J. Med. 314 (25) (1986) 1615–1620.
[28] S.R. Davis, et al., Postmenopausal hormone therapy: from monkey glands to transdermal patches, J. Endocrinol. 185 (2) (2005) 207–222.
[29] M.R. Prausnitz, S. Mitragotri, R. Langer, Current status and future potential of transdermal drug delivery, Nat. Rev. Drug Discov. 3 (2) (2004) 115–124.
[30] X. Guo, et al., Investigation of a new injectable thermosensitive hydrogel loading solid lipid nanoparticles, Pharmazie 66 (12) (2011) 948–952.
[31] T. Simon-Yarza, et al., PEGylated-PLGA microparticles containing VEGF for long term drug delivery, Int. J. Pharm. 440 (1) (2013) 13–18.
[32] K.B. Sutradhar, C.D. Sumi, Implantable microchip: the futuristic controlled drug delivery system, Drug Deliv. 23 (1) (2016) 1–11.
[33] C.B. Fox, et al., Micro/nanofabricated platforms for oral drug delivery, J. Control. Release 219 (2015) 431–444.
[34] G.T. Franzesi, et al., A controlled-release strategy for the generation of cross-linked hydrogel microstructures, J. Am. Chem. Soc. 128 (47) (2006) 15064–15065.
[35] F. Ganji, E. Vasheghani-Farahani, Hydrogels in controlled drug delivery systems, Iran. Polym. J. 18 (1) (2009) 63–88.
[36] A.R.D. Reeves, et al., Controlled release of cytokines using silk-biomaterials for macrophage polarization, Biomaterials 73 (2015) 272–283.

[37] J. Rodriguez Villanueva, et al., Optimising the controlled release of dexamethasone from a new generation of PLGA-based microspheres intended for intravitreal administration, Eur. J. Pharm. Sci. 92 (2016) 287–297.
[38] R. Goffredo, D. Accoto, E. Guglielmelli, Swallowable smart pills for local drug delivery: present status and future perspectives, Expert Rev. Med. Devices 12 (5) (2015) 585–599.
[39] A.C.R. Grayson, et al., Multi-pulse drug delivery from a resorbable polymeric microchip device, Nat. Mater. 2 (11) (2003) 767–772.
[40] C.J. Kearney, D.J. Mooney, Macroscale delivery systems for molecular and cellular payloads, Nat. Mater. 12 (11) (2013) 1004–1017.
[41] E. Blanco, H. Shen, M. Ferrari, Principles of nanoparticle design for overcoming biological barriers to drug delivery, Nat. Biotechnol. 33 (9) (2015) 941–951.
[42] L.C. Dong, A.S. Hoffman, A novel-approach for preparation of pH-sensitive hydrogels for enteric drug delivery, J. Control. Release 15 (2) (1991) 141–152.
[43] A. Goyanes, et al., Development of modified release 3D printed tablets (printlets) with pharmaceutical excipients using additive manufacturing, Int. J. Pharm. 527 (1-2) (2017) 21–30.
[44] S. Rujivipat, R. Bodmeier, Improved drug delivery to the lower intestinal tract with tablets compression-coated with enteric/nonenteric polymer powder blends, Eur. J. Pharm. Biopharm. 76 (3) (2010) 486–492.
[45] H.J. Park, et al., Colon-targeted delivery of solubilized bisacodyl by doubly enteric-coated multiple-unit tablet, Eur. J. Pharm. Sci. 102 (2017) 172–179.
[46] J. Huwyler, D.F. Wu, W.M. Pardridge, Brain drug delivery of small molecules using immunoliposomes, Proc. Natl. Acad. Sci. U. S. A. 93 (24) (1996) 14164–14169.
[47] J. Kreuter, et al., Apolipoprotein-mediated transport of nanoparticle-bound drugs across the blood-brain barrier, J. Drug Target. 10 (4) (2002) 317–325.
[48] G.J. Schiller, et al., Marqibo®, Vincristine sulfate liposome injection, for the treatment of advanced, relapsed or refractory philadelphia chromosome-negative (Ph-) acute lymphoblastic leukemia (ALL) in an adolescent young adult (AYA) population, Blood 126 (23) (2015) 1291.
[49] L.B. Silverman, et al., Intravenous PEG-asparaginase during remission induction in children and adolescents with newly diagnosed acute lymphoblastic leukemia, Blood 115 (7) (2010) 1351.
[50] H. Maeda, SMANCS and polymer-conjugated macromolecular drugs: advantages in cancer chemotherapy, Adv. Drug Deliv. Rev. 46 (1) (2001) 169–185.
[51] W.-W. Zhang, et al., The first approved gene therapy product for cancer Ad-p53 (Gendicine): 12 years in the clinic, Hum. Gene Ther. 29 (2) (2018) 160–179.
[52] E.M. Gordon, F.L. Hall, Rexin-G, a targeted genetic medicine for cancer, Expert. Opin. Biol. Ther. 10 (5) (2010) 819–832.
[53] U. Bulbake, et al., Liposomal formulations in clinical use: an updated review, Pharmaceutics 9 (2) (2017).
[54] A.L. Klibanov, et al., Amphipathic polyethyleneglycols effectively prolong the circulation time of liposomes, FEBS Lett. 268 (1) (1990) 235–237.
[55] B. Naeye, et al., PEGylation of biodegradable dextran nanogels for siRNA delivery, Eur. J. Pharm. Sci. 40 (4) (2010) 342–351.
[56] P.L. Rodriguez, et al., Minimal "Self" peptides that inhibit phagocytic clearance and enhance delivery of nanoparticles, Science 339 (6122) (2013) 971–975.
[57] P.K. Working, et al., Pharmacokinetics, biodistribution and therapeutic efficacy of doxorubicin encapsulated in Stealth® Liposomes (Doxil®), J. Liposome Res. 4 (1) (1994) 667–687.
[58] M. Polaneczky, et al., The use of Levonorgestrel implants (Norplant) for contraception in adolescent mothers, N. Engl. J. Med. 331 (18) (1994) 1201–1206.
[59] D. Shoupe, D.R. Mishell, Norplant—subdermal implant system for long-term contraception, Am. J. Obstet. Gynecol. 160 (5) (1989) 1286–1292.
[60] I. Sivin, International experience with Norplant and Norplant-2 contraceptives, Stud. Fam. Plan. 19 (2) (1988) 81–94.
[61] H. Brem, et al., Interstitial chemotherapy with drug polymer implants for the treatment of recurrent gliomas, J. Neurosurg. 74 (3) (1991) 441–446.
[62] H. Gollwitzer, et al., Antibacterial poly(D,L-lactic acid) coating of medical implants using a biodegradable drug delivery technology, J. Antimicrob. Chemother. 51 (3) (2003) 585–591.

[63] S.S. Lee, et al., Biodegradable implants for sustained drug release in the eye, Pharm. Res. 27 (10) (2010) 2043–2053.
[64] D. Goldray, et al., Decreased bone-density in elderly men treated with the gonadotropin-releasing-hormone agonist decapeptyl (D-TRP6-GNRH), J. Clin. Endocrinol. Metab. 76 (2) (1993) 288–290.
[65] L.G.J. De Leede, et al., Novel controlled-release Lemna-derived IFN-alpha 2b (Locteron): pharmacokinetics, pharmacodynamics, and tolerability in a phase I clinical trial, J. Interferon Cytokine Res. 28 (2) (2008) 113–122.
[66] L. Illum, et al., Novel chitosan-based delivery systems for the nasal administration of a LHRH-analogue, Stp Pharma Sci. 10 (1) (2000) 89–94.
[67] Y. Brudno, D.J. Mooney, On-demand drug delivery from local depots, J. Control. Release 219 (2015) 8–17.
[68] T. Minko, et al., LHRH-targeted nanoparticles for cancer therapeutics, in: S.R. Grobmyer, B.M. Moudgil (Eds.), Cancer Nanotechnology: Methods and Protocols, 2010, 281–294.
[69] L. Rong, et al., Trastuzumab-modified DM1-loaded nanoparticles for HER2+ breast cancer treatment: an in vitro and in vivo study, Artif. Cell Nanomed. Biotechnol. 46 (8) (2018) 1708–1718.
[70] K.K. Sankhala, et al., A phase I pharmacokinetic (PK) study of MBP-426, a novel liposome encapsulated oxaliplatin, J. Clin. Oncol. 27 (15_suppl) (2009) 2535.
[71] S. Wilhelm, et al., Analysis of nanoparticle delivery to tumours, Nat. Rev. Mater. 1 (2016) 16014.
[72] T.R. Daniels-Wells, M.L. Penichet, Transferrin receptor 1: a target for antibody-mediated cancer therapy, Immunotherapy 8 (9) (2016) 991–994.
[72a] D. Nellis, et al., Preclinical manufacture of anti-HER2 liposome-inserting, scFv-PEG-lipid conjugate, Biotechnol. Prog. 21 (2005) 221–232.
[73] N. Desai, V. Trieu, Z. Yao, Increased antitumor activity, intratumor paclitaxel concentrations, and endothelial cell transport of cremophor-free, albumin-bound paclitaxel, ABI-007, compared with cremophor-based paclitaxel (vol 12, pg 1317, 2006), Clin. Cancer Res. 12 (12) (2006) 3869.
[74] N.P. Desai, et al., Improved effectiveness of nanoparticle albumin-bound (nab) paclitaxel versus polysorbate-based docetaxel in multiple xenografts as a function of HER2 and SPARC status, Anti-Cancer Drugs 19 (9) (2008) 899–909.
[75] S.S.W. Ng, et al., Influence of formulation vehicle on metronomic taxane chemotherapy: albumin-bound versus cremophor EL-based paclitaxel, Clin. Cancer Res. 12 (14) (2006) 4331–4338.
[76] O. Jonas, et al., First in vivo testing of compounds targeting group 3 medulloblastomas using an implantable microdevice as a new paradigm for drug development, J. Biomed. Nanotechnol. 12 (6) (2016) 1297–1302.
[77] O. Jonas, et al., An implantable microdevice to perform high-throughput in vivo drug sensitivity testing in tumors, Sci. Transl. Med. 7 (284) (2015).
[78] Y.H. Deng, et al., Preparation, characterization, and application of multistimuli-responsive microspheres with fluorescence-labeled magnetic cores and thermoresponsive shells, Chem. Eur. J. 11 (20) (2005) 6006–6013.
[79] H. Li, et al., Magnetic actuated pH-responsive hydrogel-based soft micro-robot for targeted drug delivery, Smart Mater. Struct. 25 (2) (2016) 9.
[80] P. Wanakule, et al., Nano-inside-micro: disease-responsive microgels with encapsulated nanoparticles for intracellular drug delivery to the deep lung, J. Control. Release 162 (2) (2012) 429–437.
[81] E.M. White, et al., Advances in smart materials: Stimuli-responsive hydrogel thin films, J. Polym. Sci. B Polym. Phys. 51 (14) (2013) 1084–1099.
[82] L. Zorzetto, et al., From micro- to nanostructured implantable device for local anesthetic delivery, Int. J. Nanomed. 11 (2016) 2695–2709.
[83] P.S. Yarmolenko, et al., Comparative effects of thermosensitive doxorubicin-containing liposomes and hyperthermia in human and murine tumours, Int. J. Hyperth. 26 (5) (2010) 485–498.
[84] S. Mullick Chowdhury, T. Lee, J.K. Willmann, Ultrasound-guided drug delivery in cancer, Ultrasonography (Seoul, Korea) 36 (3) (2017) 171–184.
[85] J.C. Yu, et al., Microneedle-array patches loaded with hypoxia-sensitive vesicles provide fast glucose-responsive insulin delivery, Proc. Natl. Acad. Sci. U. S. A. 112 (27) (2015) 8260–8265.

[86] C. Chiappini, et al., Biodegradable silicon nanoneedles delivering nucleic acids intracellularly induce localized in vivo neovascularization, Nat. Mater. 14 (5) (2015) 532–539.
[87] Q.A. Rafiq, et al., The early career researcher's toolkit: translating tissue engineering, regenerative medicine and cell therapy products, Regen. Med. 10 (8) (2015) 989–1003.
[88] T. Ishida, et al., Accelerated blood clearance of PEGylated liposomes upon repeated injections: effect of doxorubicin-encapsulation and high-dose first injection, J. Control. Release 115 (3) (2006) 251–258.
[89] N.J. Ganson, et al., Pre-existing anti-polyethylene glycol antibody linked to first-exposure allergic reactions to pegnivacogin, a PEGylated RNA aptamer, J. Allergy Clin. Immunol. 137 (5) (2016) 1610–1613.
[90] N. Yokomichi, et al., Pathogenesis of hand-foot syndrome induced by PEG-modified liposomal doxorubicin, Hum. Cell 26 (1) (2013) 8–18.

CHAPTER 3.2

Immunotherapy

Chensu Wang, Murillo Silva, Leyuan Ma
Koch Institute for Integrative Cancer Research, Massachusetts Institute of Technology, Cambridge, MA, United States

1 Introduction

Despite the great amount of progress made in cancer treatment with conventional methods, including surgical removal, chemotherapy and radiation therapy, cancer still remains the leading cause of human death across the world [1,2]. The incredible heterogeneity of cancer is likely responsible for the partial responsiveness to traditional nontargeted therapies. As genomic sequencing techniques advance, personalized medicine with the aid of targeted therapy and combinational approaches has revolutionized the way we combat cancer [3]. Within this shifting paradigm, immunotherapy that stimulates patients' own immune system has enjoyed its golden era of discoveries and validations over the past decade.

Fundamental work by Robert Schreiber has shown that the immune system functions as an effective tumor-suppressor system [4]. Under normal conditions, the immune system is constantly surveying our body to destroy infections and abnormal cellular growth (Fig. 1). Therefore, tumors need to find a way to escape or overcome this natural defense system to survive and proliferate. The progression of cancer can be viewed as a failure of immune surveillance [5]. The goal of immunotherapy is to restore the balance in favor of the immune system, and to achieve the level of specificity that surpass conventional chemotherapy and radiation therapy by leveraging the natural immune recognition mechanisms that have evolved over millions of years. In general, antigen-presenting cells (APCs), mainly dendritic cells (DCs) and macrophages (MΦ), are responsible for sampling tumor-associated antigens (TAAs) and presenting them through major histocompatibility complex (MHC) molecules to prime T cell response in secondary lymphoid organs such as lymph nodes and the spleen.

Once activated, antigen-specific cytotoxic T cells migrate to the tumor site where they recognize and kill tumor cells that present mutated antigens on their surfaces. While CD4+ effector T cells, cytotoxic CD8+ T cells and other activated innate immune cells release proinflammatory cytokines and chemokines to facilitate the killing process, regulatory T cells (Tregs) and other suppressive cells of the immune system constrain overt immune responses by inhibiting the activation of effector and cytotoxic T cells through the release of antiinflammatory cytokines. Other key inhibitory pathways

Bioengineering Innovative Solutions for Cancer
https://doi.org/10.1016/B978-0-12-813886-1.00011-5

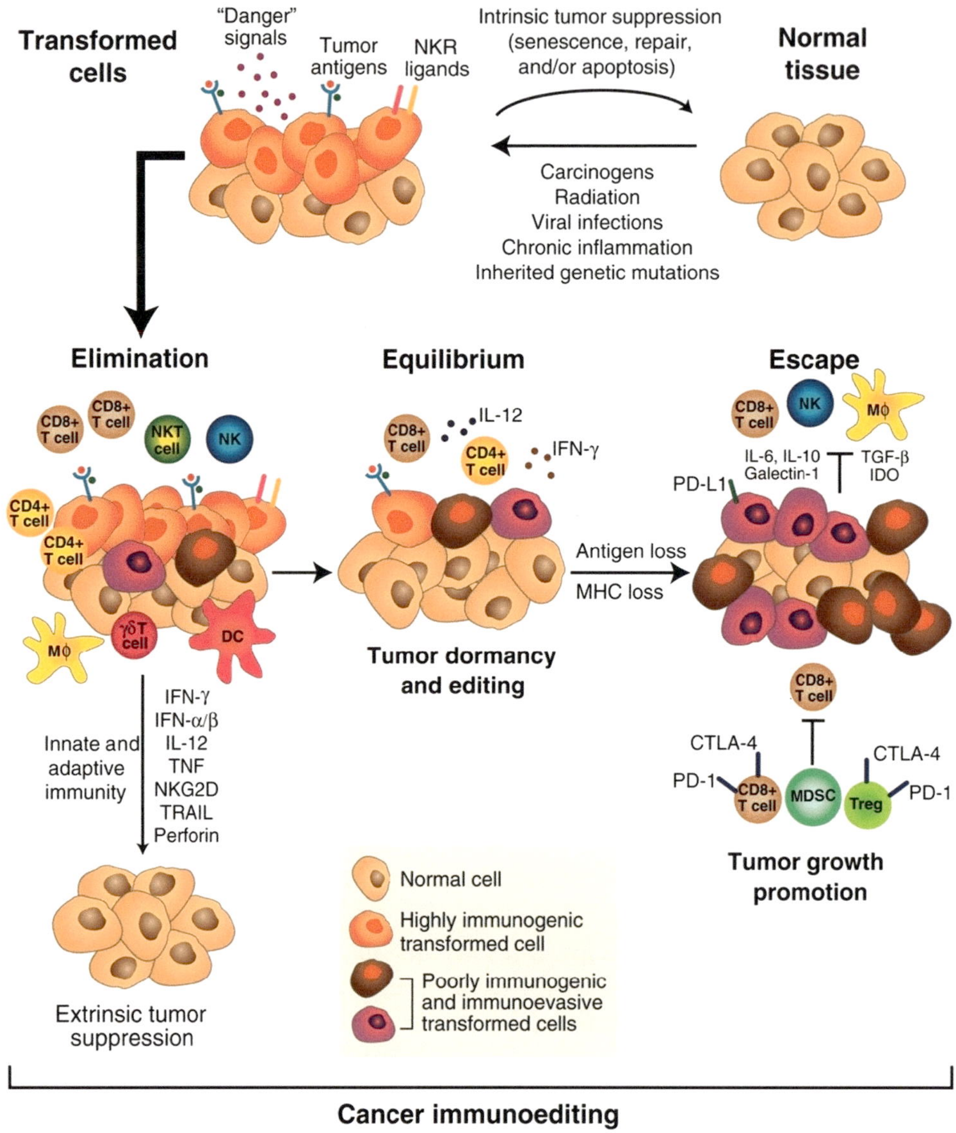

Fig. 1 Theory of cancer immunoediting. Cancer immunoediting consists of three general phases: elimination, equilibrium, and escape. During the elimination phase, innate and adaptive cells of the immune system cooperate to identify mutations in cancerous cells and trigger their destruction. In rare cases, tumor cells can avoid elimination and enter the equilibrium phase where full tumor outgrowth is prevented by immunologic mechanisms. T cells, and pro-inflammatory cytokines such as IL-12, and IFN-γ ensure the immune system keeps cancerous cells in check. During this equilibrium phase however, genetically unstable tumor cells may mutate and develop the ability to evade immune detection. These tumor cells may then enter the escape phase, in which their outgrowth is no longer held in check by the immune system [4a].

commonly exploited by tumors to escape from immune surveillance are immune checkpoints, such as cytotoxic T-lymphocyte antigen 4 (CTLA-4) [6] and programmed cell death protein 1 (PD-1) [7] and their ligands that are expressed on the surface of innate immune cells or tumor cells. Tumor-associated macrophages (TAMs) and myeloid-derived suppressive cells (MDSCs) that have been reprogrammed by the tumor microenvironment (TME) can also suppress immune cell infiltration and functions through multiple pathways [8,9].

A variety of immunotherapeutic strategies have been developed, ranging from anticancer vaccines that target DCs [10–12], to immune checkpoint blockades [6,7] that restore and improve T cell functionality, to chimeric antigen receptor (CAR) T cell therapies [13, 13a] that enhance patients' tumor-specific cytotoxic T lymphocyte (CTL) response through genetic modification ex vivo. Among these, the use of anticancer vaccines containing antigenic materials, such as TAA and/or immune-boosting adjuvants, has been explored for several decades. Unlike successful traditional vaccines targeting foreign pathogens, hurdles in choosing the correct antigenic material and the poor accumulation of the vaccines in lymphatic systems largely confound the therapeutic benefits that anticancer vaccines can provide in clinical settings. So far, checkpoint blockades, such as antibodies against CTLA-4 and PD-1, are the most broadly efficacious immunotherapies and have shown unprecedented clinical responses in melanoma, nonsmall-cell lung cancer, head and neck cancer, renal cancer, Hodgkin's lymphoma, among many others [14]. Unfortunately, issues such as severe side effects and limited response rates still exist [15–18]. In general, the nonspecific activation of the immune system is responsible for the adverse events observed in patients. As a result, a key challenge is to achieve targeted activation of the right immune cells to improve tumor-specific immune responses and to minimize systemic exposure to the potent therapeutics.

2 Overcoming the immunosuppressive tumor microenvironment

A number of recent studies have revealed that favorable prognosis is directly correlated to the extent of tumor-infiltrating cytotoxic T cells [1, 2]. The potential of immunotherapy for cancer is however limited by various immunosuppressive escape mechanisms present in the tumor microenvironment (TME) [3]. The clinical efficacy of checkpoint blockade therapies such as Yervoy (ipilimumab; CTLA-4 blocking antibody) and Keytruda (Pembrolizumab; PD1 blocking antibody) showcases the importance of immune regulation and suppression in tumor progression (Fig. 2). Under homeostatic conditions, immune checkpoints are critical for the maintenance of self-tolerance and for the prevention of collateral damage during responses to a foreign pathogen. These negative regulatory cues however can be hijacked by tumors to essentially cloak these cells from immune destruction [4].

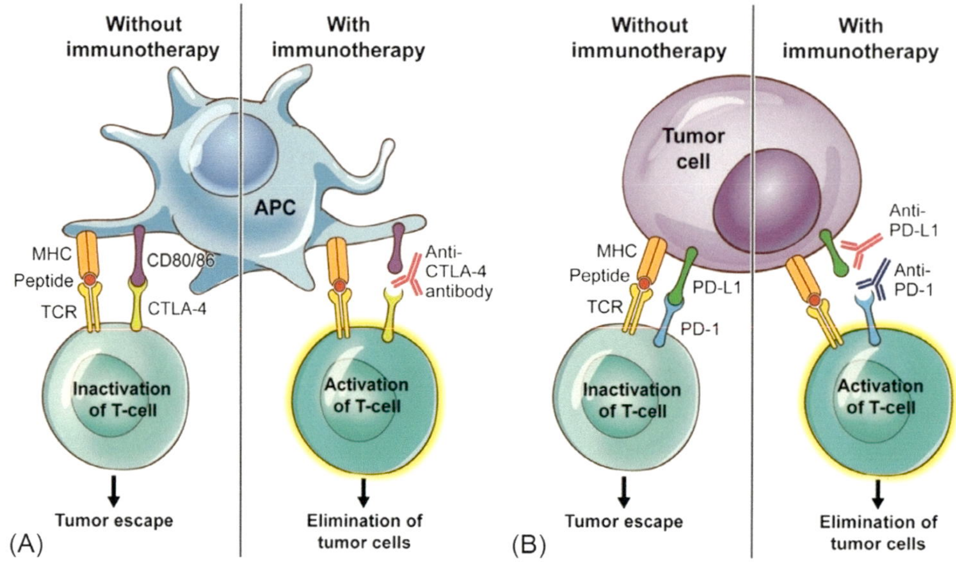

Fig. 2 Effect of checkpoint blockade on immune response: (A) CTLA-4 signaling represents an important control mechanism to prevent T-cell overactivation. This pathway is activated upon T-cell interactions with APCs (B) PD1 functions mainly to limit T cell activity at the site of inflammation. PD1 ligands are commonly overexpressed by tumor cells as an active process to evade T-cell-mediated killing. Antibody blockade of both these immune checkpoints has been shown to clinically improve survival in cancer patients [4a].

T cells have been a major focus of tumor immunotherapies given their ability to selectively recognize small mutational differences in tumors down to the single amino acid level, and to directly eradicate cells it recognizes. T-cell-mediated immunity however involves a complex series of events that require tumor-specific antigen acquisition by APCs, antigen presentation to and activation of cognate CD8 and CD4 T cells, clonal selection, and expansion of antigen-specific cells which ultimately culminates in the destruction of tumor or infected cells [19]. This activation process typically occurs in secondary lymphoid organs such as lymph nodes and the spleen. These specialized environments directly orchestrate the interaction between APC and T cells and are highly efficient in initiating an immune response. In the case of solid tumors however activated T cells must migrate to the tumor microenvironment to carry out their effector function. At this location, a myriad of regulatory ligands or metabolic products, expressed by infiltrating innate immune cells or tumor cells themselves, inhibit tumor cell lysis. This limitation was first identified more than 50 years ago when Hellstrom and colleagues noticed that extracted lymphocytes from neuroblastoma patients could eradicate their respective tumor cells in an in vitro setting but failed to mount an appreciable response in vivo [20].

CTLA-4 was the first immunosuppressive checkpoint to be clinically targeted and validated. Seminal work by Jim Allison nearly 20 years before Yervoy's (ipilimumab)

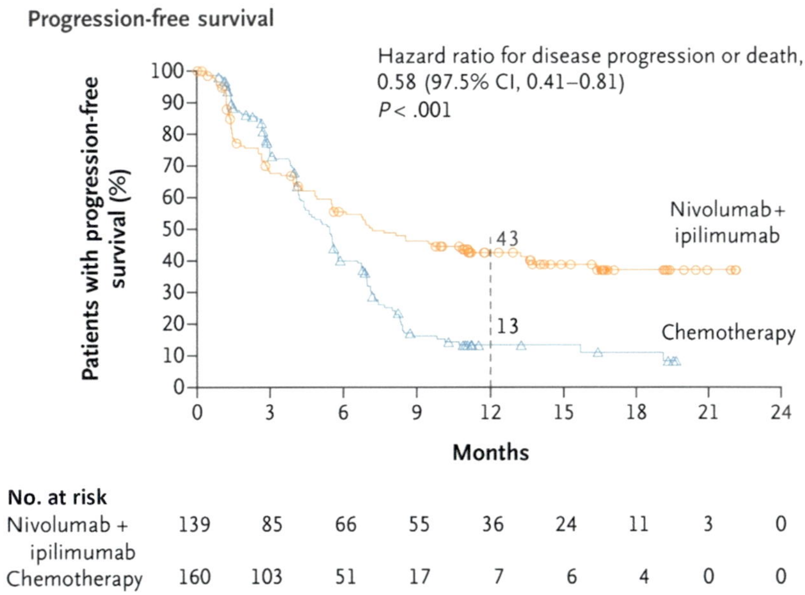

Fig. 3 Efficacy of Nivolumab (PD1 blockade) plus Ipilimumab (CTLA-4 blockade) versus chemotherapy in patients with a high tumor mutational burden [20a].

eventual approval identified that this immunological checkpoint was a critical aspect of immune evasion [6]. CTLA-4 primarily regulates early stages of T cell activation and provides homeostatic cues to prevent T cell overactivation. Thus, as a general control mechanism of the immune system, targeting this receptor as a potential therapeutic strategy was initially dismissed. In fact, mice deficient in this receptor experience severe T cell proliferative disorders and autoimmune disease. Clinical trials however showed that there is a therapeutic window where CTLA-4 can be efficacious, despite the prevalent occurrence of adverse events [15] (Fig. 3).

Shortly after the success of CTLA-4 checkpoint blockade, PD1 blocking antibodies also gained approval as an antitumor therapy. PD1 functions mainly to limit T-cell activity at the site of inflammation presumably to prevent collateral tissue damage [19]. Unlike CTLA-4, ligands for PD1 have been found to be overexpressed by tumor cells and likely illustrate active measure of immune evasion. Blockade of this signaling molecule with a monoclonal antibody permits T cells to remain functional in the TME and has led to significant clinical efficacy [21]. In fact, PD1 blockade is now used as a first line of treatment for nonsmall-cell lung cancer.

The efficacy of checkpoint therapy is directly correlated to the mutational burden of tumors [22]. Importantly, these studies have shown that T cells identify unique mutational differences, neoantigens, found in tumor cells highlighting the exquisite ability of the immune system to discriminate self from nonself [23]. Although it is a daunting problem, identification of these neoantigens will enable the development of personalized cancer vaccines where the mutational differences in a patient's specific tumor can be exploited for therapy [24]. Advancements in genetic screening and predictive algorithms designed to assess whether or not a patient's immune system is capable of responding to the identified neoantigen will facilitate further development.

Despite the promising efficacy of checkpoint blockades, many patients still do not respond to therapy [21]. This is perhaps due to the inhibitory effects of other mediators in the TME. Targeting inhibitory innate immune cells (TAMs and MDSCs) and their immunosuppressive mediators such as TGF-b, IL-10, IDO, and extracellular adenosine may be required to further improve efficacy. The hypoxic microenvironment of tumors has also been shown to be highly immunosuppressive [25,26].

3 Engineering T cells for enhanced adoptive cell therapy

Adoptive cell therapy (ACT) enables ex vivo manipulation of patients' own T cells by either selecting for strong tumor-reactive T cells or redirecting nontumor-reactive T cells to specifically target tumors with additional genetic modifications [27]. Adoptive cell therapy using T lymphocytes dates back to 1960s, when researchers first observed mild tumor suppression after transferring vaccine boosted syngeneic T cells to tumor-bearing animals [28,29]. Since then, three different types of adoptive cell therapy using T lymphocytes have been developed, namely, tumor-infiltrating lymphocytes (TIL) therapy, TCR-transgenic T-cell therapy, and chimeric antigen receptor (CAR) T-cell therapy [27].

Successful ACT using tumor-infiltrating lymphocytes (TIL) was built on two major findings that systematic depletion of patients' lymphocytes (lymphodepletion) prior to ACT enhances the antitumor efficacy of transferred T cells and that cytokine Interleukin-2 (IL-2) could support ex vivo growth of T cells [30,31]. In 1988, Rosenberg and colleagues demonstrated for the first time that ex vivo expanded autologous TILs can substantially suppress the progression of metastatic melanoma in patients [32]. However, favorable clinical response to TIL therapy has mostly been reported in metastatic melanoma and certain epithelial cancers [27,33]. Cancer genome mutation analysis across different cancer types reveals huge variations in mutational burden, with melanoma and lung cancer ranking at the top [34]. These observations indicate a strong correlation between mutation burden and TIL therapy outcome. Tumor-specific mutations often lead to the production of nonself proteins, i.e., tumor neoantigens, and elicit host T-cell responses, leading to the infiltration of activated T lymphocytes with the capability of recognizing a range of these tumor neoantigens [35]. Systemic evaluation of mutation-possessing peptides has been carried out in vitro by stimulating T cells to identify mutations recognized by TILs [22,36]. Surprisingly, every mutation was recognized by distinct TILs. This suggests the contribution of each individual's genetic background and lifestyle in shaping tumor mutations and the accompanying TIL repertoire. In spite of the robust clinical efficacy of TIL therapy, it is limited to certain cancer types and requires intensive care support as well as long manufacturing lead time to generate sufficient cells for therapy (~5–6 weeks), which cannot meet the need of most cancer patients [27]. Thus, more accessible and affordable approaches are needed.

Certain mutations have been shared among patients and different cancer types, thus making them attractive targets for ACT. One strategy is targeting shared tumor-specific mutations using T cells genetically modified to express mutation-specific T cell receptors (TCRs) [37]. T cells have the ability to recognize and attack target cells with super low antigen density on the cell surface via their TCRs [38]. TCR is a multimeric structure composed of an *alpha* and a *beta* TCR chain associated with CD3 complexes [39] (Fig. 4). Optimal CD8 T-cell activation and expansion in vivo are achieved through TCR recognizing a short peptide of ~9 amino acids presented by major histocompatibility complex I (MHCI) on professional antigen-presenting cells (APCs) [40]. Activated T cells will enter circulation and attack encountered tumor cells expressing the same

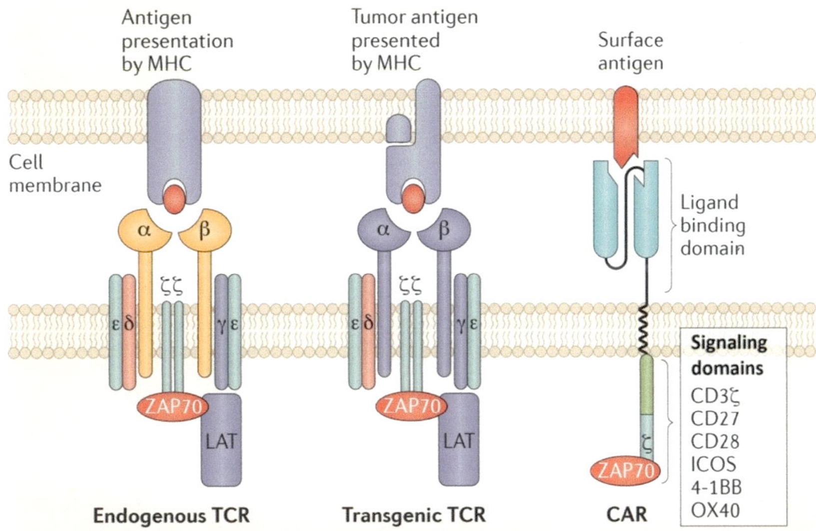

Fig. 4 Schematic structures of T-cell receptor (TCR) and chimeric antigen receptor (CAR). Both endogenous and transgenic TCRs require recruitment of downstream signaling molecules such as linker for activation of T-cell family member 1 (LAT) and ζ-associated protein of 70 kDa (ZAP70) to transmit the activation signals. They recognize intracellularly processed antigens that must be presented in the context of the major histocompatibility complex (MHC) and require costimulatory signals for optimal T-cell activation. In contrast, a CAR recognizes tumor surface antigen independent of MHC through a single-chain variable fragment derived from immunoglobulin heavy-chain variable (V_H) and Ig light-chain variable (V_L) domains. This scFV is fused through a transmembrane domain to an intracellular costimulatory domain and an intracellular CD3ζ chain domain. *(From A.D. Fesnak, C.H. June, B.L. Levine, Engineered T cells: the promise and challenges of cancer immunotherapy, Nat. Rev. Cancer 16 (9) (2016) 566–581.)*

antigen with the ability to detect single peptide-MHC complex on tumor cell surface [41,42] (Fig. 4). The potency and sensitivity of TCRs make them the natural arsenal for precision cancer therapy.

The first TCR-transgenic T-cell therapy trial utilizes a low-avidity TCR recognizing an epitope from the melanoma/melanocyte differentiation antigen, MART-1 (melanoma-associated antigen recognized by T cells 1), that is restricted by human lymphocyte antigen (HLA)-A2, a human MHC class I molecule. An objective mild response was observed in 2 out of 17 patients [43]. A subsequent trial utilized a high-avidity version of MART-1 TCR, intending to target tumor cells with lower MART-1 expression [44]. Although objective tumor regression occurred in an increased number of patients, severe toxicity was seen in the skin, eyes, and ears where TCR-transgenic T cells also attacked low-MART-1-expressing normal melanocytes. Such on-target, off-tumor toxicity highlights the potency of TCR-transgenic T cell therapy but also warrants extreme caution in targeting tumor-associated antigens that are shared with normal cells. Toxicity was also observed in trials targeting cancer-testis antigen, MAGE-A3, resulting from unknown crossreactivity with antigen MAGE-A12, which is expressed by normal tissues [45]. In contrast, targeting another cancer testis antigen, NY-ESO-1, with T cells expressing high-avidity TCR produced favorable clinical responses with minimal toxicity [46]. These results provide an early demonstration of the feasibility of TCR-transgenic T cells for cancer therapy. Therapeutic efficacy and toxicity profile largely depend on the antigen being targeted. Therefore, ideal targets for TCR-transgenic T cell would be tumor neoantigens as well as tumor-associated antigens but expressed in nonessential tissues.

One downside of TCR-transgenic T-cell-based therapy is that the artificially introduced exogenous TCR may mispair with the endogenous TCR, compromising its surface expression and function as well as leading to the creation of de novo heterodimeric TCRs with unknown off-target reactivity [47]. To improve the pairing of cognate TCR chains, a portion of constant regions in human TCRs has been replaced with their murine counterparts, which improves the stability of paired TCRs [48]. In another approach, additional cysteine residues were introduced into both *alpha*-chains and *beta*-chains to form a covalent disulfide bond and prevent exogenous TCR chains from dissociating and pairing with endogenous TCR chains [49]. By harnessing the advanced genome editing technology, e.g., CRISPR-Cas9, researchers have created more potent TCR transgenic T cells by knocking out endogenous TCRs and simultaneously replacing them with cancer-reactive TCRs [50]. This leads to markedly increased surface expression as well as up to 1000-fold enhancement in sensitivity toward cancer antigens.

While the transgenic TCR-based ACT approach could efficiently redirect host T cells against tumor cells, its application is largely limited to patients with an HLA haplotype that is specific for the transgenic TCR, i.e., HLA restriction. Among the efforts to broaden the application of ACT, the one that spurs most excitement would be the aforementioned chimeric antigen receptor (CAR) T cells [51]. A CAR typically comprises a signal peptide followed by an extracellular antigen-binding domain, which often is a single-chain variable fragment (scFv) derived from a tumor antigen-specific antibody, a transmembrane domain, and intracellular signaling domains from TCRζ chain and costimulatory receptors, such as CD28, CD137, and OX40 [51] (Fig. 4).

The concept of CAR was pioneered by Kuwana and Eshhar in the 1980s, showing for the first time that this type of synthetic receptor could relieve T cells from HLA restriction, endowing them with the ability to recognize any desired tumor surface antigen [52,53]. This property is extremely advantageous that it offers a way to target tumor cells that have lost their surface MHC expression during immunoevasion [54].

Clinical trial results are largely disappointing with the first-generation CARs, which only possess a CD3ζ signaling domain [55]. Studies in preclinical animal models revealed insufficient signaling for initiating effective immune responses [56]. Natural T-cell activation by APCs requires TCR-MHC interaction, costimulatory signals as well as cytokine support. Incorporation of a second signaling domain from costimulatory molecules into CAR, dubbed second-generation CAR, leads to both potent activation and expansion of CAR T cells upon engagement with target cells [57]. Preclinical work using the second-generation CAR demonstrated that incorporation of different costimulatory domains leads to different responses of T cells, including cytokine secretion profile, T-cell proliferation as well as memory formation [58]. For example, signaling domain from costimulatory molecule CD28 promotes T-cell expansion, stronger cytotoxicity but is compromised with early exhaustion. In contrast, CAR T with signaling domain from 41BB has a less prominent effect, but the increased memory formation ensures longer persistence in circulation and stronger therapeutic efficacy. The third-generation CAR T has thus been constructed by incorporating both CD28 and 41BB signaling domains, aiming to encompass the advantages from both costimulations [59].

Clinical trials with this second-generation CAR T-targeting CD19, a B-cell-specific surface antigen, showed pronounced response in patients with B-cell malignancies, especially refractory B cell-acute lymphoblastic leukemia (B-ALL) [60]. A large portion of patients still possesses circulating CAR T cells after years of complete remission. Notably, successful CD19 CAR T therapy is often accompanied by severe B-cell aplasia [61]. This is because both malignant B cells and normal B cells express a high level of CD19 antigen, thereby both being eliminated by CD19 CAR T cells. Another caveat is that massive CAR T expansion and activation often results in a life-threatening cytokine storm, a surge of multiple inflammatory cytokines in circulation. But this threat can be mitigated by intravenous injection of a neutralizing antibody for cytokine interleukine-6 (IL-6), a major culprit of the CAR T-induced cytokine storm [62].

The immense success with CD19 CAR T sets a lead paradigm for current CAR T therapy. A successful CAR T therapy not only requires a potent CAR design but also entails the proper selection of CAR antigen. Similar to that for TCR-transgenic T cells, an ideal CAR antigen would be exclusively expressed on the tumor cell surface; however, it is often very rare except for those tumor neoantigens created by mutation, such as

EGFRvIII in glioblastoma patients [63]. In the case of CD19 CAR, B cell is an important but nonessential part of the immune system. Its ablation could be rescued with replacement therapies using intravenous immunoglobulin (IVIG), thus also making it a viable target for CAR T therapy.

Although the initial successful CAR design was solely empirical, further optimization of CAR entails delicate design. Synthetic biology has played an important role in applying engineering concepts to enhance CAR specificity as well as mitigating toxicity [64]. Simultaneous recognition of two tumor-associated surface antigens offers an opportunity to avoid on-target off-tumor toxicity. One approach is splitting a single CAR into two CARs (Fig. 5A). One CAR will recognize a tumor-associated antigen and possesses the

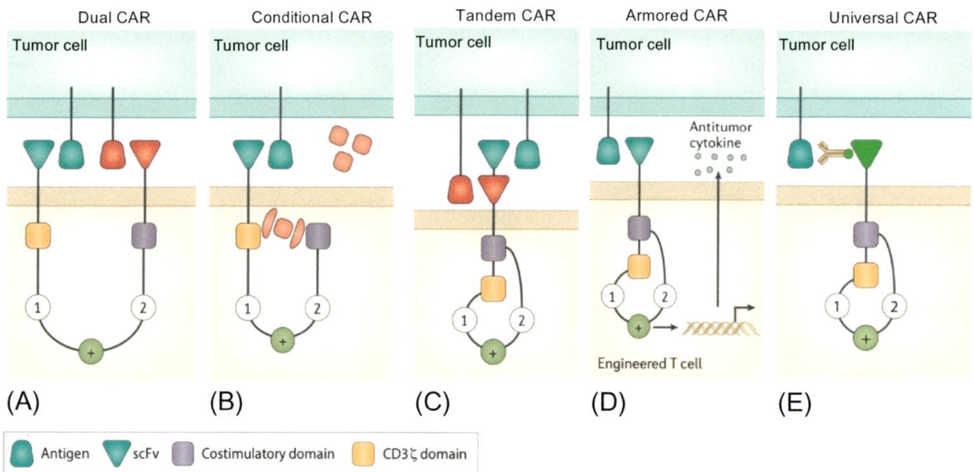

Fig. 5 Next-generation chimeric antigen receptor design concepts. (A) Dual CAR T cells express two separate CARs with different ligand binding targets; one CAR includes only the CD3ζ domain and the other CAR includes only the costimulatory domain(s). Dual CAR T-cell activation requires engagement of both targets on the tumor cells. (B) Conditional CAR cells are by default unresponsive, or switched "off," until the addition of a small molecule to complete the circuit, enabling full transduction of both signals 1 and 2, thereby activating the CAR T cells. (C) Tandem CAR T cells express a single CAR consisting of two linked single-chain variable fragments (scFvs) that have different target specificities fused to intracellular costimulatory domain(s) and a CD3ζ domain. Tandem CAR T-cell activation is achieved when tumor cells express either target. (D) Armored CAR T cells co-express a chimeric antigen receptor (CAR) and an antitumor cytokine. Cytokine expression may be constitutive or induced by T-cell activation (for example, interleukin-12 (IL-12)). Targeted by CAR specificity, localized production of proinflammatory cytokines recruits endogenous immune cells to tumor sites to potentiate an antitumor response. (E) Universal CAR T cells are engineered to express an adaptor-specific receptor with affinity for subsequently administered secondary antibodies directed at target antigen. For example, the receptor could be a scFv recognizing a small hapten molecule (e.g., FITC) or a short peptide (e.g., peptide neoepitope (PNE)) conjugated to an antitumor antibody. *(From A.D. Fesnak, C.H. June, B.L. Levine, Engineered T cells: the promise and challenges of cancer immunotherapy, Nat. Rev. Cancer 16 (9) (2016) 566–581.)*

CD3-signaling domain, the other CAR will recognize a second tumor-associated antigen but only contains the costimulatory domain [65]. Only the simultaneous engagement of both antigens will trigger the optimal CAR T activation. This split design is further refined by Lim and colleagues to be remotely controllable via a small-molecule-mediated dimerization [66] (Fig. 5B). The same group also developed a synthetic Notch (SynNotch) receptor platform, which allows separating CAR T engagement into a two-step process [67]. These SynNotch-CAR T cells will engage one tumor antigen via a chimeric scFV-notch receptor, the intracellular domain with an orthogonal transcription factor will subsequently get cleaved, trafficked to the nucleus to drive expression of a CAR, which then gets transported onto cell surface to engage the second tumor antigen. By tuning the affinity of the scFV on SynNotch receptor, slow-on/quick-off rate of first engagement and postponed second-step CAR engagement may preferentially select for target cells with higher antigen expression, thus offering additional benefit to reduce on-target off-tumor toxicity. However, the enhanced selectivity comes at a cost of tumor escape, and loss of either antigen will result in therapy failure.

Even in clinical trials where a single antigen is targeted, resistance to CAR T therapy has been reported, and this often inevitably involves loss of CAR antigen [68]. In the case of B-cell leukemia, to avoid the loss of CD19 antigen, a bispecific CAR has been constructed with a tandem scFV targeting both CD19 and CD20 [69] (Fig. 5C). It is shown that this bispecific CAR could effectively eliminate leukemia cells with the loss of either CD19 or CD20 single antigen expression.

In spite of the dramatic clinical responses in hematologic malignancies, CAR T therapy in solid tumors has always been disappointing. Many barriers to treatment of solid tumors have been identified and these include poor persistence of the CAR cells, dysregulated metabolism, and T-cell exhaustion [70,71]. It has been shown that anti-PD1/PD-L1 checkpoint blockade immunotherapy in solid tumor revives exhausted tumor-infiltrating T cells. Abrogation of these endogenous inhibitory signals using genome editing or RNA interference will facilitate the development of enhanced CAR T cells [72,73]. Additional genetic modifications such as inducing CAR T cells to also express immune potentiating cytokines in an autocrine manner may also enhance the efficacy of CARs in a solid tumor setting (Fig. 5D). Effector cells expressing IL-15 and IL-12 are currently being evaluated in clinical trials (NCT03294954 and NCT02498912, respectively). These modified CARs are likely to have improved clinical efficacy but may also induce more systemic toxicities due to their constitutive secretion of proinflammatory cytokines.

A universal CAR T platform has also been reported to further increase tunability and flexibility of CAR T therapy [74,75]. The universal CAR is composed of an adaptor scFV recognizing a surrogate ligand, such as a FITC molecule or an orthogonal peptide tag. Subsequent infusion of FITC or peptide-tagged Fab or intact antibody will bridge the interaction of CAR T with target cancer cells (Fig. 5E). As current CAR T therapy costs can reach upwards of half a million dollars, the potential cost saving of a universal CAR T platform is immense and could make treatment more accessible to the general public.

Most recently, Darrell Irvine's group developed a "backpack" strategy by conjugating nanoparticles containing proinflammatory mediators directly onto the cytotoxic T cell or CAR T surface [76]. Cytokines such as IL-15 are chemically crosslinked with a redox-sensitive linker to form a nanogel structure that can then be conjugated onto the surface of CAR T cells (Fig. 6). This nanomaterial-based approach offers a unique way of transiently manipulating CAR T cells without additional genetic modifications. An important benefit of this strategy is that these immune potentiating agents are selectively released only upon antigen recognition. T cells contain high surface redox activity upon activation; therefore, once a nanoparticle-coated CAR T cell engages its cognate target, the redox-sensitive crosslinker breaks down releasing its cargo at the site of interest. The local release of therapeutic cytokines minimizes the adverse events associated with systemic injections.

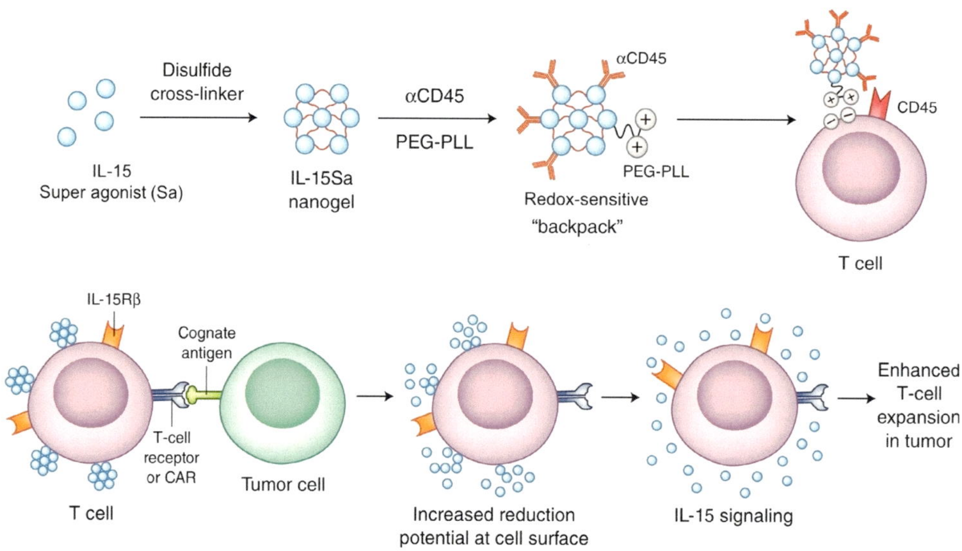

Fig. 6 IL-15 "backpacks" enhance adoptive T-cell therapy. IL-15 super agonist molecules are aggregated into a nanogel with a synthetic crosslinker. PEG-PLL and anti-CD45 antibodies are adsorbed or conjugated onto the nanogel surface to facilitate electrostatic binding to T-cell membranes and to prevent nanogel internalization, respectively. The nanogel backpacks are then loaded onto T cells. The backpacks remain inert at the T cell surface until the T cell encounters a tumor cell. Activation of the T cell induces a wave of thiol emission at the cell surface, creating a reduction potential that breaks the disulfide bonds in the synthetic crosslinkers and releases the IL-15 constituent molecules. IL-15 can then bind to cytokine receptors on the T cell surface to enhance T cell activity at the tumor site. *(From T. Shum, H.E. Heslop, A backpack revs up T-cell activity, Nat. Biotechnol. 36(8) (2018) 702–703.)*

4 Using biomaterials for cancer immunotherapy

Engineered biomaterials can overcome many of the barriers associated with tumor immunotherapy. For example, biomaterials can enhance targeted drainage of antigens, immune stimulants, or immunosuppressive blockers into lymph nodes (LNs), and into the tumor itself. The physicochemical properties of biomaterials, including shape, size, surface charge or hydrophobicity can all be finetuned to enhance loading capability and improve immune response. Additionally, implantable scaffolds, microneedles, and hydrogels can provide a local proinflammatory environment that recruits effector cells, especially APCs, and activate them with antigens and adjuvants. The key benefit that biomaterials offer is the control over release kinetics depending on various parameters, such as pore size and degradability, which helps achieve optimal immune responses.

4.1 Biomaterial-enhanced anticancer vaccine

Anticancer vaccines consist of tumor associated antigens (TAAs) and adjuvants targeting the antigen presentation process in DCs (a typical type of APCs) in order to stimulate antigen-specific T cell responses. TAAs loaded on MHC molecules are usually short peptides that can be in the form of peptide fragments or even the whole protein in an anticancer vaccine. However, the presence of TAAs is not enough to fully engage cytotoxic T lymphocytes (CTLs). It was discovered in late 19th century that a mixture of killed bacteria injected directly into patients' tumors could result in boosted antitumor immune responses [77]. Now it is widely accepted that adjuvants, usually mimic parts of pathogens' structures or functions, need to be included in anticancer vaccines to serve as a danger signal to the body. In general, adjuvants like DNA or RNA fragments bind to Toll-like receptors (TLRs), which facilitate the maturation of DCs, including upregulation of costimulatory molecules and production of proinflammatory cytokines. In this way, T-cell anergy can be avoided upon binding to the antigen-MHC complexes.

The first and only FDA-approved anticancer vaccine is a DC-based strategy, called Sipuleucel-T (Provenge, Dendreon Pharmaceuticals LLC.), which uses autologous APCs from patients' peripheral blood pulsed with a prostatic acid phosphatase (PAP, as a TAA) and granulocyte-macrophage colony-stimulating factor (GM-CSF, as an adjuvant) fusion protein ex vivo [78,79] (Fig. 7A). The cells are then adoptively transferred back to the patient. Despite the fact that Sipuleucel-T has been approved for treating asymptomatic or minimally symptomatic metastatic castration-resistant prostate cancer, its widespread adoption has been hampered by the expensive and time-consuming manufacturing process, limited availability of patients' cells as well as modest clinical benefit (Fig. 7B). While soluble TAAs and adjuvants can be administered directly as anticancer vaccines, their poor accumulation in lymphatic tissues and nontargeted/synchronized delivery into APCs remain the major challenges in their development and translation. To overcome the difficulties that current strategies are facing, more

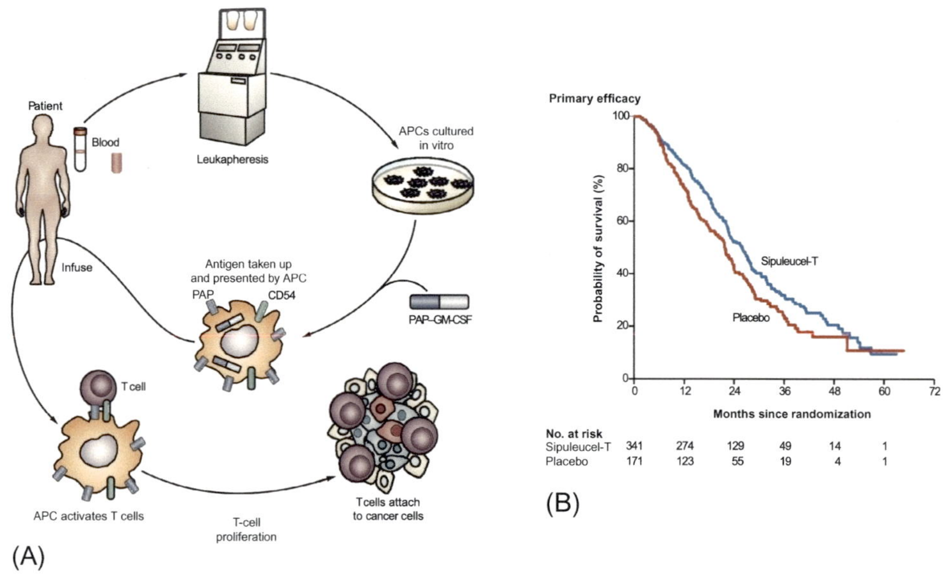

Fig. 7 (A) Schematic of sipuleucel-T's mechanism of action *(bottom)* and the process of cell therapy *(top)*. (B) Results of the primary efficacy analysis of treatment with sipuleucel-T as compared with placebo among men with metastatic castration-resistant prostate cancer (hazard ratio for death in the sipuleucel-T group, 0.78; 95% confidence interval [CI], 0.61–0.98; $P=0.03$). *((A) From G. Di Lorenzo, C. Buonerba, P.W. Kantoff, Immunotherapy for the treatment of prostate cancer, Nat. Rev. Clin. Oncol. 8(9) (2011) 551; (B) From J. Barar, Y. Omidi, Translational approaches towards cancer gene therapy: hurdles and hopes, Bioimpacts 2 (2012) 127 and P.W. Kantoff, et al., Sipuleucel-T immunotherapy for castration-resistant prostate cancer, N. Engl. J. Med. 363 (2010) 411–422.)*

and more research has been geared toward engineered biomaterials that enable prolonged and concurrent delivery of antigens and adjuvants into targeted cells, such as DCs.

4.2 Nanomaterial anticancer vaccines

Biomaterials, such as liposomes, self-assembled proteins, polymeric or inorganic nanoparticles, have key properties that make them well suited for simultaneous delivery of various payloads to lymphatic tissues with improved retention time. It has been found that molecules with small sizes (<5 nm in diameter) and low molecular weight (<10 kDa) tend to quickly enter systematic circulation and be cleared by the kidney. In contrast, properly sized (20–100 nm in diameter) and weighted (>40 kDa) biomaterial vehicles can efficiently drain to and remain in the lymph nodes when injected subcutaneously [80]. This ensures that DCs get sufficient amount of antigens and adjuvants and it also extends the antigen presentation by DCs to mount subsequent T-cell response. For example, synthetic high-density lipoprotein-mimicking nanodiscs loaded with antigen peptides and adjuvants remarkably enhanced their codelivery into lymphoid organs [81] (Fig. 8A). This resulted in 47-fold stronger antigen-specific T-cell response over soluble vaccines and even 31-fold higher response than a standard adjuvant used in clinical trials.

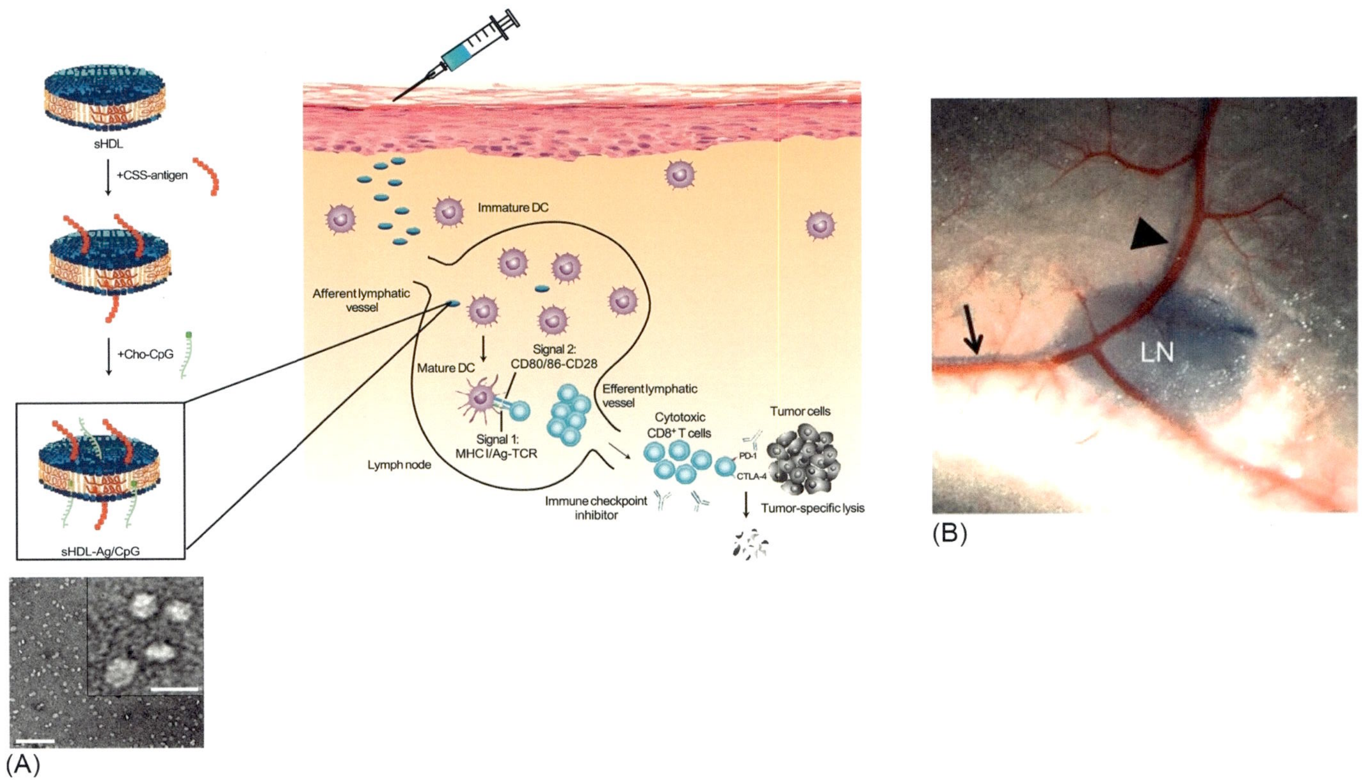

Fig. 8 (A) Schematic showing the synthetic high-density lipoprotein (sHDL) nanodisc as a personalized cancer vaccine platform. The lower left electronic microscopic image of the sHDL nanodiscs shows their disc-like morphology with an average size of 10.5 nm in diameter. Scale bar = 100 nm. (B) Anesthetized 6-week-old mouse was injected with 5% Evans Blue dye in the tailbase and in the rear dorsal toe. The *arrows* indicate the afferent LV from the gluteal LN to the inguinal LN. The *arrowheads* indicate the efferent LV that travels along the milk line from the inguinal LN to the axillary LNs. *(From R. Kuai, L.J. Ochyl, K.S. Bahjat, A. Schwendeman, J.J. Moon, Designer vaccine nanodiscs for personalized cancer immunotherapy. Nat. Mater. 16 (2017) 489–496, https://doi.org/10.1038/nmat4822 and L.M. Habenicht, S.B. Kirschbaum, M. Furuya, M.I. Harrell, A. Ruddell, Tumor regulation of lymph node lymphatic sinus growth and lymph flow in mice and in humans, Yale J. Biol. Med. 90 (2017) 403.)*

Biomaterials can passively drain through interstitial fluid to lymph nodes in a size-dependent manner. LN is the place where T cells get primed by APCs and then proliferate and eventually generate long-term immunological memory. Therefore, an LN-draining vaccine generally shows better immune responses than a distal vaccination to peripheral tissues. In general, materials that are smaller than 5 nm in diameter are more than likely to enter blood vessels as the blood clears 10-fold more fluid from peripheral tissues than lymph. In contrast, large particles (>100–200 nm in diameter) tend to be trapped by extracellular matrix at the injection site and are not able to passively drain to the blood or lymph. However, they can be phagocytosed by monocytes or DCs in the surrounding tissue and potentially be processed and presented in LNs. The optimal size of particles for LN drainage has been thoroughly studied. In one example, a series of PEGylated poly (propylene sulfide) nanoparticles with diameters ranging from 20 to 100 nm were compared. Nanoparticles with a diameter of 20–45 nm were found to drain to LN efficiently and also to induce the strongest immune response. Others have also found that although large particles are less efficient in LN drainage than smaller particles, once inside the LN, they tend to exhibit a longer retention time while smaller particles are not necessarily trapped in the LN. The aspect ratio of nanoparticles also affects passive drainage and cell-mediated transport toward LNs. For example, cylindrical nanoparticles were reported to induce higher levels of LN drainage and phagocytosis by macrophages than spherical nanoparticles [82]. Another LN-targeting strategy is to hijack endogenous LN-draining proteins, such as albumins. Albumin has a hydrodynamic size of ~5 nm, and is the most prevalent protein in blood and interstitial fluid. It can efficiently drain to LNs, so small-molecule dye that has a high affinity for serum albumin, such as Evans blue, has been used in clinics for LN mapping (Fig. 8B). Inspired by this, antigen peptides and adjuvants were conjugated to albumin-binding lipids and showed much higher accumulation in LNs and elicited stronger CTL response than soluble vaccine molecules [83].

Moreover, targeted delivery by biomaterials minimizes systematic exposure of antigenic materials and exhibits greatly enhanced safety, which is one of the major concerns for immunotherapy. Modern vaccines usually rely on an adjuvant, such as TLR agonists and proinflammatory cytokines, which promote a potent immune response through some level of controlled inflammation. However, these compounds tend to be extremely toxic when injected directly because they cause systematic inflammation as they quickly disperse into the bloodstream. Biomaterials can restrain antigenic materials stably before reaching targeted cells, which minimizes systemic exposure and improve both efficacy and safety of vaccines. For example, a study of TLR7/8 agonist demonstrated that loading of this compound into polymeric nanoparticles dramatically restricted its biodistribution and improved its pharmacokinetic profile, resulting in reduced morbidity and enhanced humoral and T-cell immunity [84]. In another study, water-soluble protein cytokine interleukin 2 (IL-2) was loaded into liposomal polymeric gels together with a small molecular inhibitor to achieve sustained local release. No systemic inflammatory toxicity was observed as opposed to direct administration of naked IL-2 [85].

Some biomaterials themselves have adjuvant effects, especially polymeric synthetic materials, simplifying the antigenic materials to be delivered and maximizing the chance of colocalization of all components in the vaccine. For example, polyethyleneimine (PEI) was simply absorbed to a mesoporous silica microrod (MSR) with a 10.9-nm pore size to serve as an adjuvant and to enhance antigen immunogenicity [86]. When loaded with peptide antigen pool, the MSR-PEI vaccine induced significantly stronger DC maturation and T-cell activation compared with existing vaccine formulations. Combination of the MSR-PEI vaccine with checkpoint blockade eradicated established syngeneic tumors and their lung metastases, making it a facile, powerful, and universal vaccine platform. In another approach, a pH-responsive polymeric material was exploited to carry peptide antigens in a micellar state in extracellular conditions [87]. Once inside the cells, the polymers became cationic and helped disruption of endosomal membranes and cytosolic delivery of the antigens. Interestingly, the polymers have this unexpected adjuvant effect by binding to and activating stimulator of interferon genes (STING), leading to upregulation of type I interferon-stimulated genes and cytokine release. As a result, this vaccine mounted strong CTL responses while the systemic cytokine expression was kept low, making it a simple, effective and safe strategy as an anticancer vaccine platform.

4.3 Anticancer vaccines based on biomaterial scaffolds

As mentioned above, although sipuleucel-T is the first and only FDA approved anticancer vaccine, as an autologous cell-based therapy, its implementation in clinics is quite complex. Therefore, experience from tissue engineering and regenerative medicine has inspired the idea of implantable scaffolds or hydrogels that slowly release antigens and danger signals in situ to enable the process of attracting, activating, and loading antigens on DCs and subsequent migration into LNs. For example, a porous biodegradable poly(lactic-*co*-glycolic acid) (PLGA)-based polymer disk loaded with tumor lysates, GM-CSF, and cytosine-phosphodiester-guanine oligodeoxynucleotide (CpG ODN) implanted subcutaneously has been shown to recruit and activate resident DCs, which greatly enhanced their homing to LNs and subsequent tumor-specific CTL response [88]. This vaccine platform is now tested under a first-in-human Phase I clinical trial for patients with stage IV melanoma. In contrast to implantable scaffolds, materials such as hydrogels and nanoparticles with high aspect ratio can form a vaccine depot in vivo through self-assembly. In one example, biodegradable mesoporous silica rods (MSRs) loaded with antigens, adjuvants, and GM-CSF were found to spontaneously form macroporous structures upon subcutaneous injection. The 3D cellular microenvironment successfully attracted host DCs and enhanced both CTL and humoral immune responses [89]. In addition to these injectable scaffolds, microneedles that penetrate the skin have also been explored for controlled and sustained release of antigens and adjuvants transdermally. The microneedle vaccines, usually composed of porous polymeric materials or even natural proteins like silk, are either coated or loaded with antigens and adjuvants on the surface or inside their tips, which mimic a local pathogen inflammation by targeting skin-resident DCs [90–92] (Fig. 9A).

4.4 Naturally inspired design of nanoparticles as an anticancer vaccine platform

Recent development of biomaterials as an anticancer vaccine platform has also been focused on naturally inspired designs that incorporate both antigens and adjuvants into the same compartment. For example, an interbilayer-crosslinked multilamellar vesicle (ICMV) liposome [93] has been developed, which stably entraps a model protein antigen ovalbumin in the core and incorporates monosphosphoryl lipid A (MPLA) as an adjuvant in the vesicle wall, a lipopolysaccharide molecule derived from fractions of the cell walls of gram-negative bacteria (Fig. 9B). It has been shown to stay intact before cellular uptake, while a fast release of antigenic materials is observed once inside endosomal compartment. The liposomal formulation also significantly increased the uptake of the protein antigen and the retention time of both antigens and adjuvants, resulting in a much stronger endogenous CTL and humoral response than the antigen or adjuvant alone. In another approach, cancer cell membranes were collected to functionalize a polymeric nanoparticle core [94]. When combined with adjuvants, the nanoparticle was able to efficiently induce DC maturation and promote T-cell activation. In a similar example, cancer cell membranes were extracted and sonicated to form nanovesicles with subsequent PEGylation (attachment of polyethylene glycol) [95]. Cholesterol-modified CpG ODN was further fused onto the surface to serve as an adjuvant. The resulting vesicles exhibited good serum stability and LN drainage, eliciting stronger antigen-specific CTL response and better survival rate than freeze-thawed cancer cell lysates. The key advantage of these cancer cell membrane-derived vaccine platform is that they offer a facile method for personalized tumor antigen extraction. At the same time, a lot of intracellular self-antigens that may dilute immune responses are excluded compared to whole-cell anticancer vaccines.

4.5 Summary and outlook

Within the past decade, immunotherapy has cemented itself as a paradigm shifting strategy to combat cancer. From the early successes of CTLA-4 and PD1 checkpoint blockades to the recently approved adoptive cell therapies, immunology has provided oncologists with an ever-expanding tool chest to combat the disease. Despite these exciting advancements, there is much to be learned and improved upon. Although cancer immunotherapy has proven invaluable to some patients, not everyone experiences

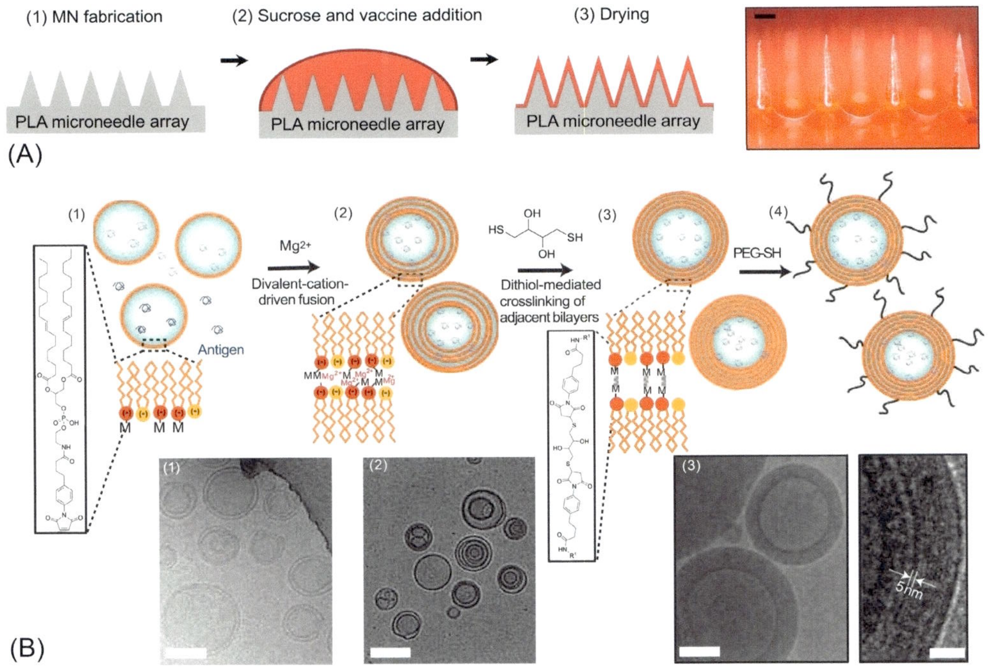

Fig. 9 (A) Schematic of the microneedle (MN) coating and optical micrograph of sucrose-coated microneedles (scale bar, 100 mm). Polylactic acid (PLA) microneedles were fabricated by poly(dimethyl siloxane) molding (1), which was followed by the application of an aqueous sucrose and replicon-incomplete adenoviral serotype 5 (Ad5, a viral antigen) vector solution (2) and drying under vacuum to solidify a conformal sucrose and Ad5 coating over the MN array (3). (B) Schematic illustration of ICMV synthesis and cryo-electron-microscope images: (1) anionic, maleimide-functionalized liposomes are prepared from dried lipid films, (2) divalent cations are added to induce fusion of liposomes and the formation of MLVs, (3) membrane-permeable dithiols are added, which crosslink maleimide lipids on apposed lipid bilayers in the vesicle walls, and (4) the resulting lipid particles are PEGylated with thiol-terminated PEG. Cryo-electron-microscopy images from each step of the synthesis show (1) initial liposomes, (2) MLVs and (3) ICMVs with thick lipid walls. Scale bars: D100 nm. The right-hand image of (3) shows a zoomed image of an ICMV wall, where stacked bilayers are resolved as electron-dense striations; scale bar D20 nm. *(From P.C. DeMuth, et al., Vaccine delivery with microneedle skin patches in nonhuman primates. Nat. Biotechnol. 31 (2013) 1082–1085, https://doi.org/10.1038/nbt.2759 and J.J. Moon, et al., Interbilayer-crosslinked multilamellar vesicles as synthetic vaccines for potent humoral and cellular immune responses. Nat. Mater. 10 (2011) 243–251, https://doi.org/10.1038/nmat2960.)*

benefits. Researchers are testing several promising avenues to try and expand the breath of patients that can ultimately benefit from immunotherapy. Areas of research such as identifying more efficacious checkpoint inhibitors, developing personalized cancer vaccines, blocking the effects of the inhibitory tumor microenvironment, engineering more efficacious CAR T cells, and combination immuno-chemotherapies are all being pursued. Advanced biomaterials will undoubtedly be part of many solutions for these perplexing problems.

References

[1] C.E. DeSantis, et al., Cancer treatment and survivorship statistics, 2014, CA Cancer J. Clin. 64 (2014) 252–271.

[2] K.D. Miller, et al., Cancer treatment and survivorship statistics, 2016, CA Cancer J. Clin. 66 (2016) 271–289.

[3] L.J. Van't Veer, R. Bernards, Enabling personalized cancer medicine through analysis of gene-expression patterns, Nature 452 (2008) 564.

[4] G.P. Dunn, A.T. Bruce, H. Ikeda, L.J. Old, R.D. Schreiber, Cancer immunoediting: from immunosurveillance to tumor escape, Nat. Immunol. 3 (2002) 991, https://doi.org/10.1038/ni1102-991.

[4a] R.D. Schreiber, L.J. Old, M.J. Smyth, Cancer immunoediting: integrating immunity's roles in cancer suppression and promotion, Science 331 (6024) (2011) 1565–1570, https://doi.org/10.1126/science.1203486.

[5] J.B. Swann, M.J. Smyth, Immune surveillance of tumors, J. Clin. Invest. 117 (2007) 1137–1146.

[6] D.R. Leach, M.F. Krummel, J.P. Allison, Enhancement of antitumor immunity by CTLA-4 blockade, Science 271 (1996) 1734–1736.

[7] P.C. Tumeh, et al., PD-1 blockade induces responses by inhibiting adaptive immune resistance, Nature 515 (2014) 568.

[8] T.L. Whiteside, Immune suppression in cancer: effects on immune cells, mechanisms and future therapeutic intervention, Semin. Cancer Biol. 16 (1) (2006) 3–15, https://doi.org/10.1016/j.semcancer.2005.07.008.

[9] I. Marigo, L. Dolcetti, P. Serafini, P. Zanovello, V. Bronte, Tumor-induced tolerance and immune suppression by myeloid derived suppressor cells, Immunol. Rev. 222 (2008) 162–179.

[10] C. Guo, et al., Therapeutic cancer vaccines: past, present, and future, Adv. Cancer Res. 119 (2013) 421–475 (Elsevier).

[11] Z. Hu, P.A. Ott, C.J. Wu, Towards personalized, tumour-specific, therapeutic vaccines for cancer, Nat. Rev. Immunol. 18 (2018) 168.

[12] I. Melero, et al., Therapeutic vaccines for cancer: an overview of clinical trials, Nat. Rev. Clin. Oncol. 11 (2014) 509.

[13] M.L. Davila, et al., Efficacy and toxicity management of 19-28z CAR T cell therapy in B cell acute lymphoblastic leukemia, Sci. Transl. Med. 6 (2014) 224ra225.

[13a] L. Ma, et al., Enhanced CAR-T cell activity against solid tumors by vaccine boosting through the chimeric receptor, Science 365 (6449) (2019) 162–168, https://doi.org/10.1126/science.aav8692.

[14] F. Teng, X. Meng, L. Kong, J. Yu, Progress and challenges of predictive biomarkers of anti PD-1/PD-L1 immunotherapy: a systematic review, Cancer Lett. 414 (2018) 166–173.

[15] F.S. Hodi, et al., Improved survival with ipilimumab in patients with metastatic melanoma, N. Engl. J. Med. 363 (2010) 711–723.

[16] J. Larkin, et al., Combined nivolumab and ipilimumab or monotherapy in untreated melanoma, N. Engl. J. Med. 373 (2015) 23–34.

[17] S.L. Topalian, et al., Safety, activity, and immune correlates of anti–PD-1 antibody in cancer, N. Engl. J. Med. 366 (2012) 2443–2454.

[18] W. Zou, J.D. Wolchok, L. Chen, PD-L1 (B7-H1) and PD-1 pathway blockade for cancer therapy: mechanisms, response biomarkers, and combinations, Sci. Transl. Med. 8 (2016) 328rv324.
[19] J.A. Owen, J. Punt, S.A. Stranford, P.P. Jones, Kuby Immunology, WH Freeman, New York, 2013.
[20] I. Hellström, K.E. Hellström, G.E. Pierce, J.P.S. Yang, Cellular and humoral immunity to different types of human neoplasms, Nature 220 (1968) 1352, https://doi.org/10.1038/2201352a0.
[20a] M.D. Hellmann, et al., Nivolumab plus ipilimumab in lung cancer with a high tumor mutational burden, N. Engl. J. Med. 378 (22) (2018) 2093–2104, https://doi.org/10.1056/NEJMoa1801946.
[21] J.D. Wolchok, et al., Nivolumab plus ipilimumab in advanced melanoma, N. Engl. J. Med. 369 (2013) 122–133, https://doi.org/10.1056/NEJMoa1302369.
[22] M.M. Gubin, et al., Checkpoint blockade cancer immunotherapy targets tumour-specific mutant antigens, Nature 515 (2014) 577–581, https://doi.org/10.1038/nature13988.
[23] N.A. Rizvi, et al., Mutational landscape determines sensitivity to PD-1 blockade in non–small cell lung cancer, Science 348 (2015) 124–128.
[24] A. Desrichard, A. Snyder, T.A. Chan, Cancer neoantigens and applications for immunotherapy, Clin. Cancer Res. 22 (2016) 807–812.
[25] S.M. Hatfield, et al., Immunological mechanisms of the antitumor effects of supplemental oxygenation, Sci. Transl. Med. 7 (2015) 277ra230, https://doi.org/10.1126/scitranslmed.aaa1260.
[26] S. Sethumadhavan, et al., Hypoxia and hypoxia-inducible factor (HIF) downregulate antigen-presenting MHC class I molecules limiting tumor cell recognition by T cells, PLoS ONE 12 (2017), https://doi.org/10.1371/journal.pone.0187314.
[27] S.A. Rosenberg, N.P. Restifo, Adoptive cell transfer as personalized immunotherapy for human cancer, Science 348 (2015) 62–68, https://doi.org/10.1126/science.aaa4967.
[28] C.H. June, R.S. O'Connor, O.U. Kawalekar, S. Ghassemi, M.C. Milone, CAR T cell immunotherapy for human cancer, Science 359 (2018) 1361–1365, https://doi.org/10.1126/science.aar6711.
[29] E.J. Delorme, P. Alexander, Treatment of primary fibrosarcoma in the rat with immune lymphocytes, Lancet 2 (1964) 117–120.
[30] A. Fefer, Immunotherapy and chemotherapy of Moloney sarcoma virus-induced tumors in mice, Cancer Res. 29 (1969) 2177–2183.
[31] E. Fernandez-Cruz, B.A. Woda, J.D. Feldman, Elimination of syngeneic sarcomas in rats by a subset of T lymphocytes, J. Exp. Med. 152 (1980) 823–841.
[32] D.A. Morgan, F.W. Ruscetti, R. Gallo, Selective in vitro growth of T lymphocytes from normal human bone marrows, Science 193 (1976) 1007–1008.
[33] S.A. Rosenberg, et al., Use of tumor-infiltrating lymphocytes and interleukin-2 in the immunotherapy of patients with metastatic melanoma. A preliminary report, N. Engl. J. Med. 319 (1988) 1676–1680, https://doi.org/10.1056/NEJM198812223192527.
[34] M.E. Dudley, et al., Cancer regression and autoimmunity in patients after clonal repopulation with antitumor lymphocytes, Science 298 (2002) 850–854, https://doi.org/10.1126/science.1076514.
[35] E. Tran, et al., Cancer immunotherapy based on mutation-specific CD4+ T cells in a patient with epithelial cancer, Science 344 (2014) 641–645, https://doi.org/10.1126/science.1251102.
[36] L.B. Alexandrov, et al., Signatures of mutational processes in human cancer, Nature 500 (2013) 415–421, https://doi.org/10.1038/nature12477.
[37] Y.C. Lu, et al., Efficient identification of mutated cancer antigens recognized by T cells associated with durable tumor regressions, Clin. Cancer Res. 20 (2014) 3401–3410, https://doi.org/10.1158/1078-0432.CCR-14-0433.
[38] P.F. Robbins, et al., Mining exomic sequencing data to identify mutated antigens recognized by adoptively transferred tumor-reactive T cells, Nat. Med. 19 (2013) 747–752, https://doi.org/10.1038/nm.3161.
[39] S.A. Rosenberg, N.P. Restifo, J.C. Yang, R.A. Morgan, M.E. Dudley, Adoptive cell transfer: a clinical path to effective cancer immunotherapy, Nat. Rev. Cancer 8 (2008) 299–308, https://doi.org/10.1038/nrc2355.
[40] J.D. Stone, A.S. Chervin, D.M. Kranz, T-cell receptor binding affinities and kinetics: impact on T-cell activity and specificity, Immunology 126 (2009) 165–176, https://doi.org/10.1111/j.1365-2567.2008.03015.x.

[41] M.S. Kuhns, M.M. Davis, K.C. Garcia, Deconstructing the form and function of the TCR/CD3 complex, Immunity 24 (2006) 133–139, https://doi.org/10.1016/j.immuni.2006.01.006.

[42] J. Hennecke, D.C. Wiley, T cell receptor-MHC interactions up close, Cell 104 (2001) 1–4.

[43] E.L. Reinherz, Alphabeta TCR-mediated recognition: relevance to tumor-antigen discovery and cancer immunotherapy, Cancer Immunol. Res. 3 (2015) 305–312, https://doi.org/10.1158/2326-6066.CIR-15-0042.

[44] D.J. Irvine, M.A. Purbhoo, M. Krogsgaard, M.M. Davis, Direct observation of ligand recognition by T cells, Nature 419 (2002) 845–849, https://doi.org/10.1038/nature01076.

[45] R.A. Morgan, et al., Cancer regression in patients after transfer of genetically engineered lymphocytes, Science 314 (2006) 126–129, https://doi.org/10.1126/science.1129003.

[46] L.A. Johnson, et al., Gene therapy with human and mouse T-cell receptors mediates cancer regression and targets normal tissues expressing cognate antigen, Blood 114 (2009) 535–546, https://doi.org/10.1182/blood-2009-03-211714.

[47] R.A. Morgan, et al., Cancer regression and neurological toxicity following anti-MAGE-A3 TCR gene therapy, J. Immunother. 36 (2013) 133–151, https://doi.org/10.1097/CJI.0b013e3182829903.

[48] A.P. Rapoport, et al., NY-ESO-1-specific TCR-engineered T cells mediate sustained antigen-specific antitumor effects in myeloma, Nat. Med. 21 (2015) 914–921, https://doi.org/10.1038/nm.3910.

[49] M.H. Kershaw, J.A. Westwood, P.K. Darcy, Gene-engineered T cells for cancer therapy, Nat. Rev. Cancer 13 (2013) 525–541, https://doi.org/10.1038/nrc3565.

[50] G. Bialer, M. Horovitz-Fried, S. Ya'acobi, R.A. Morgan, C.J. Cohen, Selected murine residues endow human TCR with enhanced tumor recognition, J. Immunol. 184 (2010) 6232–6241, https://doi.org/10.4049/jimmunol.0902047.

[51] C. Govers, Z. Sebestyen, M. Coccoris, R.A. Willemsen, R. Debets, T cell receptor gene therapy: strategies for optimizing transgenic TCR pairing, Trends Mol. Med. 16 (2010) 77–87, https://doi.org/10.1016/j.molmed.2009.12.004.

[52] M. Legut, G. Dolton, A.A. Mian, O.G. Ottmann, A.K. Sewell, CRISPR-mediated TCR replacement generates superior anticancer transgenic T cells, Blood 131 (2018) 311–322, https://doi.org/10.1182/blood-2017-05-787598.

[53] M. Sadelain, I. Riviere, S. Riddell, Therapeutic T cell engineering, Nature 545 (2017) 423–431, https://doi.org/10.1038/nature22395.

[54] G. Gross, T. Waks, Z. Eshhar, Expression of immunoglobulin-T-cell receptor chimeric molecules as functional receptors with antibody-type specificity, Proc. Natl. Acad. Sci. U. S. A. 86 (1989) 10024–10028.

[55] Y. Kuwana, et al., Expression of chimeric receptor composed of immunoglobulin-derived V regions and T-cell receptor-derived C regions, Biochem. Biophys. Res. Commun. 149 (1987) 960–968.

[56] F. Garrido, N. Aptsiauri, E.M. Doorduijn, A.M. Garcia Lora, T. van Hall, The urgent need to recover MHC class I in cancers for effective immunotherapy, Curr. Opin. Immunol. 39 (2016) 44–51, https://doi.org/10.1016/j.coi.2015.12.007.

[57] M.V. Maus, C.H. June, Making better chimeric antigen receptors for adoptive T-cell therapy, Clin. Cancer Res. 22 (2016) 1875–1884, https://doi.org/10.1158/1078-0432.CCR-15-1433.

[58] T. Brocker, K. Karjalainen, Signals through T cell receptor-zeta chain alone are insufficient to prime resting T lymphocytes, J. Exp. Med. 181 (1995) 1653–1659.

[59] M. Sadelain, R. Brentjens, I. Riviere, The promise and potential pitfalls of chimeric antigen receptors, Curr. Opin. Immunol. 21 (2009) 215–223, https://doi.org/10.1016/j.coi.2009.02.009.

[60] S.A. Grupp, et al., Chimeric antigen receptor-modified T cells for acute lymphoid leukemia, N. Engl. J. Med. 368 (2013) 1509–1518, https://doi.org/10.1056/NEJMoa1215134.

[61] D.L. Porter, B.L. Levine, M. Kalos, A. Bagg, C.H. June, Chimeric antigen receptor-modified T cells in chronic lymphoid leukemia, N. Engl. J. Med. 365 (2011) 725–733, https://doi.org/10.1056/NEJMoa1103849.

[62] S.L. Maude, D.T. Teachey, D.L. Porter, S.A. Grupp, CD19-targeted chimeric antigen receptor T-cell therapy for acute lymphoblastic leukemia, Blood 125 (2015) 4017–4023, https://doi.org/10.1182/blood-2014-12-580068.

[63] D.M. O'Rourke, et al., A single dose of peripherally infused EGFRvIII-directed CAR T cells mediates antigen loss and induces adaptive resistance in patients with recurrent glioblastoma, Sci. Transl. Med. 9 (2017), https://doi.org/10.1126/scitranslmed.aaa0984.

[64] C. Quintarelli, et al., Choice of costimulatory domains and of cytokines determines CAR T-cell activity in neuroblastoma, Oncoimmunology 7 (2018), https://doi.org/10.1080/2162402X.2018.1433518.

[65] M.C. Milone, et al., Chimeric receptors containing CD137 signal transduction domains mediate enhanced survival of T cells and increased antileukemic efficacy in vivo, Mol. Ther. 17 (2009) 1453–1464, https://doi.org/10.1038/mt.2009.83.

[66] J.H. Esensten, J.A. Bluestone, W.A. Lim, Engineering therapeutic T cells: from synthetic biology to clinical trials, Annu. Rev. Pathol. 12 (2017) 305–330, https://doi.org/10.1146/annurev-pathol-052016-100304.

[67] C.C. Kloss, M. Condomines, M. Cartellieri, M. Bachmann, M. Sadelain, Combinatorial antigen recognition with balanced signaling promotes selective tumor eradication by engineered T cells, Nat. Biotechnol. 31 (2013) 71–75, https://doi.org/10.1038/nbt.2459.

[68] C.Y. Wu, K.T. Roybal, E.M. Puchner, J. Onuffer, W.A. Lim, Remote control of therapeutic T cells through a small molecule-gated chimeric receptor, Science 350 (2015), https://doi.org/10.1126/science.aab4077.

[69] K.T. Roybal, et al., Engineering T cells with customized therapeutic response programs using synthetic notch receptors, Cell 167 (2016) 419–432. e416, https://doi.org/10.1016/j.cell.2016.09.011.

[70] B.L. Levine, J. Miskin, K. Wonnacott, C. Keir, Global manufacturing of CAR T cell therapy, Mol. Ther. Methods Clin. Dev. 4 (2017) 92–101, https://doi.org/10.1016/j.omtm.2016.12.006.

[71] J.A. Fraietta, et al., Disruption of TET2 promotes the therapeutic efficacy of CD19-targeted T cells, Nature 558 (2018) 307–312, https://doi.org/10.1038/s41586-018-0178-z.

[72] J. Eyquem, et al., Targeting a CAR to the TRAC locus with CRISPR/Cas9 enhances tumour rejection, Nature 543 (2017) 113–117, https://doi.org/10.1038/nature21405.

[73] M. Vyas, R. Muller, E. Pogge von Strandmann, Antigen loss variants: catching hold of escaping foes, Front. Immunol. 8 (2017), https://doi.org/10.3389/fimmu.2017.00175.

[74] D. Schneider, et al., A tandem CD19/CD20 CAR lentiviral vector drives on-target and off-target antigen modulation in leukemia cell lines, J. Immunother. Cancer 5 (2017) 42, https://doi.org/10.1186/s40425-017-0246-1.

[75] F.C. Thistlethwaite, et al., The clinical efficacy of first-generation carcinoembryonic antigen (CEACAM5)-specific CAR T cells is limited by poor persistence and transient pre-conditioning-dependent respiratory toxicity, Cancer Immunol. Immunother. (2017), https://doi.org/10.1007/s00262-017-2034-7.

[76] L.M. Whilding, J. Maher, CAR T-cell immunotherapy: the path from the by-road to the freeway? Mol Oncol. 9 (2015) 1994–2018, https://doi.org/10.1016/j.molonc.2015.10.012.

[77] B. Wiemann, C.O. Starnes, Coley's toxins, tumor necrosis factor and cancer research: a historical perspective, Pharmacol. Ther. 64 (1994) 529–564.

[78] P.W. Kantoff, et al., Sipuleucel-T immunotherapy for castration-resistant prostate cancer, N. Engl. J. Med. 363 (2010) 411–422.

[79] M.A. Cheever, C.S. Higano, PROVENGE (Sipuleucel-T) in prostate cancer: the first FDA-approved therapeutic cancer vaccine, Clin. Cancer Res. 17 (11) (2011) 3520–3526.

[80] D.N. McLennan, C.J. Porter, S.A. Charman, Subcutaneous drug delivery and the role of the lymphatics, Drug Discov. Today Technol. 2 (2005) 89–96.

[81] R. Kuai, L.J. Ochyl, K.S. Bahjat, A. Schwendeman, J.J. Moon, Designer vaccine nanodiscs for personalized cancer immunotherapy, Nat. Mater. 16 (2017) 489–496, https://doi.org/10.1038/nmat4822.

[82] J. Kim, et al., Injectable, spontaneously assembling, inorganic scaffolds modulate immune cells in vivo and increase vaccine efficacy, Nat. Biotechnol. 33 (2015) 64–72, https://doi.org/10.1038/nbt.3071.

[83] H. Liu, et al., Structure-based programming of lymph-node targeting in molecular vaccines, Nature 507 (2014) 519–522, https://doi.org/10.1038/nature12978.

[84] G.M. Lynn, et al., In vivo characterization of the physicochemical properties of polymer-linked TLR agonists that enhance vaccine immunogenicity, Nat. Biotechnol. 33 (2015) 1201–1210, https://doi.org/10.1038/nbt.3371.

[85] J. Park, et al., Combination delivery of TGF-beta inhibitor and IL-2 by nanoscale liposomal polymeric gels enhances tumour immunotherapy, Nat. Mater. 11 (2012) 895–905, https://doi.org/10.1038/nmat3355.

[86] A.W. Li, et al., A facile approach to enhance antigen response for personalized cancer vaccination, Nat. Mater. 17 (2018) 528–534, https://doi.org/10.1038/s41563-018-0028-2.
[87] M. Luo, et al., A STING-activating nanovaccine for cancer immunotherapy, Nat. Nanotechnol. 12 (2017) 648–654, https://doi.org/10.1038/nnano.2017.52.
[88] O.A. Ali, et al., The efficacy of intracranial PLG-based vaccines is dependent on direct implantation into brain tissue, J. Control. Release 154 (2011) 249–257.
[89] A.S. Cheung, D.K.Y. Zhang, S.T. Koshy, D.J. Mooney, Scaffolds that mimic antigen-presenting cells enable ex vivo expansion of primary T cells, Nat. Biotechnol. 36 (2018) 160–169, https://doi.org/10.1038/nbt.4047.
[90] P.C. DeMuth, et al., Polymer multilayer tattooing for enhanced DNA vaccination, Nat. Mater. 12 (2013) 367–376, https://doi.org/10.1038/nmat3550.
[91] P.C. DeMuth, et al., Vaccine delivery with microneedle skin patches in nonhuman primates, Nat. Biotechnol. 31 (2013) 1082–1085, https://doi.org/10.1038/nbt.2759.
[92] P.C. DeMuth, Y. Min, D.J. Irvine, P.T. Hammond, Implantable silk composite microneedles for programmable vaccine release kinetics and enhanced immunogenicity in transcutaneous immunization, Adv. Healthc. Mater. 3 (2014) 47–58, https://doi.org/10.1002/adhm.201300139.
[93] J.J. Moon, et al., Interbilayer-crosslinked multilamellar vesicles as synthetic vaccines for potent humoral and cellular immune responses, Nat. Mater. 10 (2011) 243–251, https://doi.org/10.1038/nmat2960.
[94] R.H. Fang, et al., Cancer cell membrane-coated nanoparticles for anticancer vaccination and drug delivery, Nano Lett. 14 (2014) 2181–2188, https://doi.org/10.1021/nl500618u.
[95] L.J. Ochyl, et al., PEGylated tumor cell membrane vesicles as a new vaccine platform for cancer immunotherapy, Biomaterials 182 (2018) 157–166, https://doi.org/10.1016/j.biomaterials.2018.08.016.

Further reading

[96] L.J. Rupp, et al., CRISPR/Cas9-mediated PD-1 disruption enhances anti-tumor efficacy of human chimeric antigen receptor T cells, Sci. Rep. 7 (2017) 737, https://doi.org/10.1038/s41598-017-00462-8.
[97] L. Cherkassky, et al., Human CAR T cells with cell-intrinsic PD-1 checkpoint blockade resist tumor-mediated inhibition, J. Clin. Invest. 126 (2016) 3130–3144, https://doi.org/10.1172/JCI83092.
[98] L. Tang, et al., Enhancing T cell therapy through TCR-signaling-responsive nanoparticle drug delivery, Nat. Biotechnol. 36 (2018) 707–716, https://doi.org/10.1038/nbt.4181.
[99] J.S. Ma, et al., Versatile strategy for controlling the specificity and activity of engineered T cells, Proc. Natl. Acad. Sci. U. S. A. 113 (2016) E450–E458, https://doi.org/10.1073/pnas.1524193113.
[100] D.T. Rodgers, et al., Switch-mediated activation and retargeting of CAR-T cells for B-cell malignancies, Proc. Natl. Acad. Sci. U. S. A. 113 (2016) E459–E468, https://doi.org/10.1073/pnas.1524155113.
[101] J. Barar, Y. Omidi, Translational approaches towards cancer gene therapy: hurdles and hopes, Bioimpacts 2 (2012) 127.
[102] L.M. Habenicht, S.B. Kirschbaum, M. Furuya, M.I. Harrell, A. Ruddell, Tumor regulation of lymph node lymphatic sinus growth and lymph flow in mice and in humans, Yale J. Biol. Med. 90 (2017) 403.
[103] A.D. Fesnak, C.H. June, B.L. Levine, Engineered T cells: the promise and challenges of cancer immunotherapy, Nat. Rev. Cancer 16 (9) (2016) 566–581.
[104] T. Shum, H.E. Heslop, A backpack revs up T-cell activity, Nat. Biotechnol. 36 (8) (2018) 702–703.
[105] G. Di Lorenzo, C. Buonerba, P.W. Kantoff, Immunotherapy for the treatment of prostate cancer, Nat. Rev. Clin. Oncol. 8 (9) (2011) 551.

CHAPTER 3.3

Robotic surgery in cancer

Timo Joric Corman Oude Vrielink[a], Valentina Vitiello[b], George P. Mylonas[a]
[a]Department of Surgery and Cancer, Imperial College London, London, United Kingdom
[b]School of Biomedical Engineering & Imaging Sciences, Kings College London, London, United Kingdom

1 Introduction

Since the advent of anesthesia and antisepsis one and a half centuries ago, cancer surgery has flourished driven by innovation and technological advances. These have allowed a radical change in the early perception that only radical cancer surgery can improve cure and survival rates. Today, preservation of form, function, and quality of life without compromising survival are of paramount importance [1]. The contemporary pursuance is not just to add years to cancer patient lives, but also to add life to those years, and robotic technology could play a pivotal role in this.

Oncology, the branch of medicine that encompasses prevention, diagnosis, and treatment of cancer, has achieved a 10-year age-standardized net survival for all cancers combined in England and Wales to be increased from 20% to 46% in men, and from 28% to 54% in women, over a period of 40 years between 1971 and 2011 [2]. It is argued that these improvements are most likely to be due to earlier detection and diagnosis, though advances in treatment may also have played a role [3].

There are several areas where robotic technology has been applied in cancer treatment. These areas include radiotherapy, focused energy delivery, needle placement, and organ-sparing or radical surgery. This chapter will be dedicated to the latter, which is currently the most common way of treating cancer, while other chapters in this book are dedicated to some of the other areas.

The remaining of this chapter introduces minimally invasive surgery (MIS) and how several types and generations of surgical robot technologies align with the quest for the ever-reducing trauma and better curative outcome for the benefit of the patient. Limitations of current technologies are discussed, and future opportunities are highlighted. Gastrointestinal cancer surgery is used as a representative example, where augmentation of conventional clinical instrumentation with robotics capabilities can demonstrate the true enabling nature of such synergy. Endoscopic submucosal dissection (ESD), for early-stage cancer removal from the bowel, is presented as a case study.

Bioengineering Innovative Solutions for Cancer
https://doi.org/10.1016/B978-0-12-813886-1.00012-7

2 Overview of robotic-assisted MIS for cancer

2.1 Scope of this review

The term MIS is used to refer to all surgical procedures and interventions aimed at minimization of the trauma to healthy or critical tissues. While laparoscopic surgery is the best-known type of MIS (and sometimes used as a synonym), the wider notion of MIS includes flexible endoscopy, neurosurgery, and percutaneous interventions, as well as noncontact therapeutic procedures involving the emission of energy into the patient. Cancer therapy has particularly benefitted from all these MIS techniques, especially since the introduction of robotic technology. In neurosurgery, for instance, MIS has obvious benefits: it minimizes trauma in the brain and avoids critical brain structures. Unsurprisingly, the first use of a robotic system for surgery was in 1985, when a UNIMATE PUMA 200 industrial arm was used to target a brain tumor for tissue biopsy [4]. Since then, over 30 robotic systems have been applied to neurosurgical procedures; however, due to the very specific nature of keyhole transcranial interventions, a comprehensive review of such systems is out of the scope of this chapter and can be found in [5]. On the other hand, robots designed for percutaneous procedures such as brachytherapy [6], cryotherapy, and radiofrequency needle ablation [7] are mostly used for surgical planning and accurate positioning of the needle under image guidance. As such, they are not required to perform extensive tissue manipulation and are subject to specific requirements such as compatibility with magnetic resonance imaging (MRI) scanners [8], leaving them outside of the scope of this chapter. Similarly, energy-emitting technologies such as those used in radiotherapy [9] or high-intensity focused ultrasound (HIFU), in which ultrasonic waves are used to achieve ablation of tissue [10], do not require any contact with tissue or any incision, thus are not discussed in this chapter. The focus of this section is instead on rigid robotic systems for laparoscopic and endoscopic surgery, as they are most commonly used for cancer treatment.

2.2 Multiport robotic systems for laparoscopic cancer surgery

As mentioned earlier, the best-known form of MIS is laparoscopy. A clear advantage of laparoscopic surgery is the use of smaller incisions for the insertion of surgical instruments, in contrast to the large incisions required for surgeons to be able to use their own hands when performing open surgery. The smaller incision size has many benefits for patients, such as faster recovery time, fewer postoperative complications, less pain, less analgesics, and improved cosmesis due to the reduction of visible scar tissue. Also, the shorter hospital stay—days rather than weeks, or even same-day discharge—reduces the strain on the healthcare system in terms of hospital bed occupation and care needs [11]. However, laparoscopic procedures are technically challenging to perform for the surgeon. The access points into the abdomen, via trocars, limit the laparoscopic instruments to 4 degrees of freedom (DoF) of motion. The entry points also result in the counterintuitive mirroring of the instruments motion, also referred to as the "fulcrum" effect (Fig. 1A).

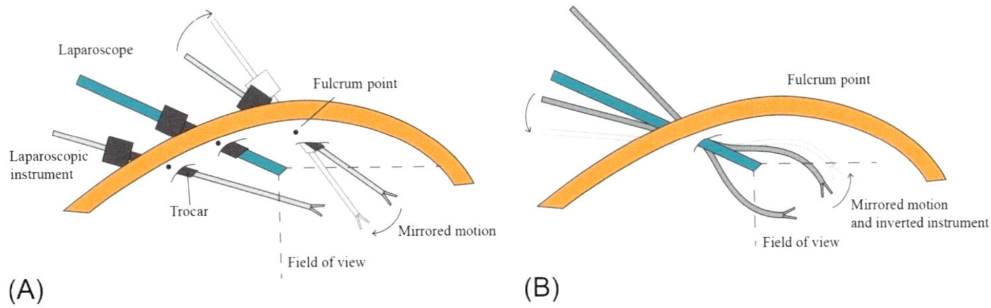

Fig. 1 (A) Multiport laparoscopic surgery. The fulcrum effect is shown for the instrument on the right, in which the motion outside the body is inverted on the inside. (B) Single-incision laparoscopic surgery. The "X"-configuration is shown for which the motion on the right instrument is visually displayed by the instrument on the left, and vice versa.

The added complexities that follow require additional training for laparoscopic skills and result in prolonged operation times. In addition, the suboptimal ergonomics, in which awkward positions must be maintained for a longer period of time, has caused an increase in musculoskeletal problems among surgeons. The introduction of robotic technology has allowed overcoming some of the limitations of laparoscopic surgery, which in turn has promoted its widespread adoption for many surgical procedures [12].

The range of oncological procedures for which robotic-assisted MIS is currently indicated is listed in Table 1 [13]. Most of these interventions are performed using the da Vinci Surgical System by Intuitive Surgical Inc., the market leader in robotic procedures, with 4986 surgical systems in use around the world [14] and over 6 million procedures performed at the end of 2018 [15]. Since becoming the first robotic surgery system approved by the FDA for general laparoscopic surgery in 2000, four generations of the system have been introduced with increasing versatility, modularity, visualization capabilities, and instrumentation functionality. However, the main design features remain largely the same as in the first generation (Fig. 2) [16].

The patient cart comprises multiple articulated arms for manipulating the laparoscope and two to three laparoscopic instruments. These are controlled in a master-slave fashion by the surgeon, who sits at an immersive console with 3D HD visualization and ergonomic handles. The main benefits of this robotic technology are clear: improved intuitiveness by eliminating the fulcrum effect for the operator, and better ergonomics, by creating a comfortable master console instead of requiring surgeons to spend hours standing looking at a screen while maneuvering laparoscopic instruments. In addition, the articulated end effectors of the surgical instruments are essential for the performance of more complex tasks, such as knot tying and suturing. Indeed, the EndoWrist, which is located at the tip of the rigid instrument shaft and mimics the function of the human wrist to provide additional dexterity, is considered to be one of the key factors contributing to the success of the da Vinci system.

Table 1 An overview of several minimally invasive surgical procedures in different areas of the body.

Anatomical location	Minimally invasive procedure or access methods
Head and neck	Transoral robotic surgery (TORS) Robotic supraglottic laryngectomy
Brain	Minimally invasive and robotics-assisted neurosurgical approaches: supraorbital, intraventricular, transsphenoidal, rectosigmoid, transparenchymal
Lung	Video-assisted thoracoscopic surgery (VATS) Robotic-assisted thoracoscopic lobectomy
Esophagus	Thoracoscopic esophagectomy Robotic-assisted esophagectomy
Gastric	Laparoscopic gastrectomy Robotic-assisted laparoscopic gastrectomy
Colorectal	Laparoscopic colectomy Robotic-assisted hemicolectomy Robotic-assisted low anterior resection (LAR) Robotic-assisted total mesorectal excision (TME)
Hepatobiliary	Laparoscopic distal pancreatectomy Robotic-assisted distal pancreatectomy Laparoscopic pancreatoduodenectomy Robotic-assisted pancreatoduodenectomy Laparoscopic hepatectomy Robotic-assisted hepatectomy
Gynecologic	Total laparoscopic radical hysterectomy (TLRH) Robotic-assisted radical hysterectomy (RRH)
Urologic	Laparoscopic radical prostatectomy (LRP) Robotic-assisted radical prostatectomy (RARP) Laparoscopic radical cystectomy (LRC) Robotic-assisted radical cystectomy (RARC) Laparoscopic radical/partial nephrectomy (LRN/LPN) Robotic-assisted radical/partial nephrectomy (RARN/RAPN)

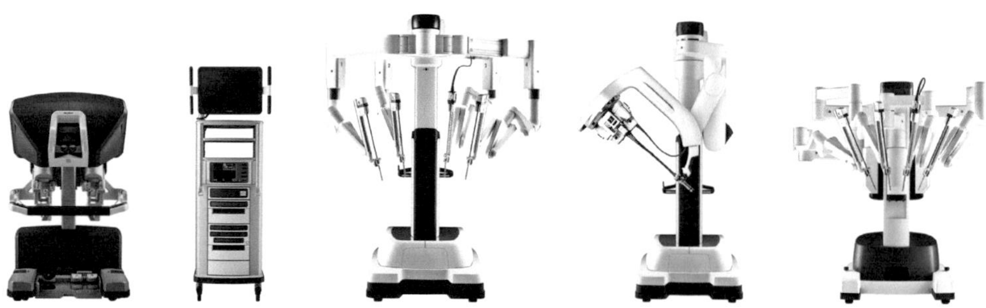

Fig. 2 The da Vinci robotic systems (Intuitive Surgical Inc). © 2019 Intuitive Surgical, Inc.

While the da Vinci system has remained dominant on the robotic surgery market over the last decade, several companies have recently launched new robotic platforms. The Senhance robotic system (TransEnterix Inc., Fig. 3A), formerly called the Telelap ALF-X, recently obtained FDA approval for laparoscopic colorectal, gynecological, inguinal hernia, and cholecystectomy surgery. Another company, Cambridge Medical Robotics (CMR Surgical Ltd., Fig. 3B), is developing a cost-effective, versatile, and portable system for laparoscopic surgery called Versius. Verb Surgical Inc., founded by Verily Life Sciences LLC and Ethicon Inc. (subsidiaries of Alphabet Inc. and Johnson & Johnson, respectively), is also developing a new robotic platform. While there is little public information available on this endeavor, expectations are high, due to the large corporations backing it. The term "digital surgery" used to describe the system suggests strong emphasis on the use of sophisticated data processing and machine-learning technology, which would be in line with the strengths of the mother company Alphabet. Medtronic Inc. is expected to launch its Hugo robotic system early in 2020, which is said to be based on the German MiroSurge technology introduced by DLR over a decade ago (Fig. 3C) [17]. The system used multiple 10-kg lightweight robotic arms (Miro) to manipulate 10-mm laparoscopic instruments with an integrated 7-DoF force sensor at the tip. The cable-driven instruments also featured a universal joint to provide two additional DoFs to the instrument. Other recent commercial developments are the Sina system (Sina Robotics and Medical Innovators Co. Ltd.) in Iran and the Mantra robot (SS Innovations Ltd.) in China. The latter has a similar approach to CMR Surgical as it aims at developing a cost-effective surgical system for a global population.

The Senhance is the only multiport surgical system, which was developed using conventional laparoscopic instruments as a control interface, including the fulcrum effect, but can measure the forces at the proximal end of the surgical instrument to provide haptic feedback to the surgeon. While the Versius system features a master console very

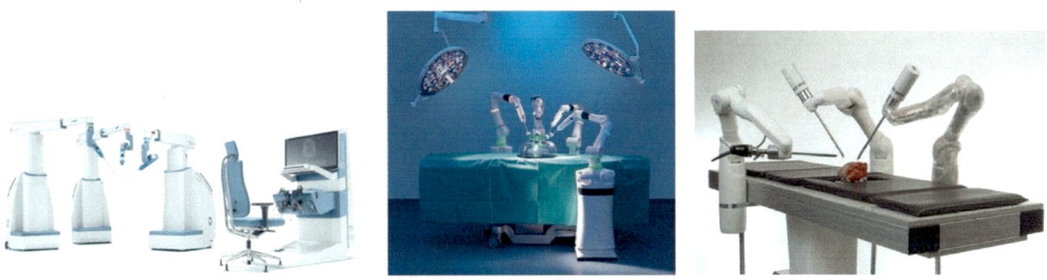

(A) (B) (C)

Fig. 3 (A) The Senhance™ robotic system, (B) The Versius® system, (C) DLR Mirosurge system. *(From (A) © 2019 TransEnterix Surgical, Inc.; (B) © 2019 CMR Surgical Ltd.; (C) Reprinted by permission from U. Hagn, R. Konietschke, A. Tobergte, M. Nickl, S. Jörg, B. Kübler, et al., DLR MiroSurge: a versatile system for research in endoscopic telesurgery, Int. J. Comput. Assist. Radiol. Surg. 5(2) (2010) 183–193, © 2010 Springer Nature Customer Service Centre GmbH: Springer Nature.)*

similar to the one of the da Vinci, the MiroSurge system used the sigma.7 haptic interface (developed in collaboration with Force Dimension Inc.) to control the 7-DoF instruments and provide haptic feedback on the translational, rotational, and grasping motions [18]. On the other hand, the mechanical design of the slave cart predominantly focuses on two aspects: the external mechanism providing tool positioning via a remote center-of-motion (RCM) around the entry point, and the development of articulated end effectors to increase the DoFs in the body. The primary function of the RCM is to safely control the rigid surgical instrument within the body, without violating the constraints imposed by the entry port on the skin. While many mechanical solutions exist, software-controlled "virtual" RCMs can be used for the same purpose. The use of virtual RCMs allows to decrease the size of the slave arms and the structure supporting them, which are largely responsible for a system's footprint in the operating room (OR). The da Vinci system uses a centralized support structure with multiple passive arms to position the RCM and instruments over the patient. The Senhance system has multiple overarching arms on large carts that can be positioned according to the specifications of the surgical procedure. Both the da Vinci and Senhance system use mechanical RCMs and take up a considerable amount of space inside the OR. Systems with smaller OR footprints are the CMR's Versius system, which uses virtual RCMs and much smaller carts supporting conventional robotic arms, and the DLR's MiroSurge system, which also uses virtual RCMs and bed-mounted robotic arms. Based on the patents filed by Verb Surgical, it seems that they are also considering the table-mounted approach for their new system [19]. In terms of end effectors, both the Versius and the MiroSurge feature cable-driven wristed instruments similar to the da Vinci EndoWrist, while the Senhance uses both wristed and straight laparoscopic tools.

While this short overview outlines the general background of the commercial activities in the area of surgical robotics, it represents but the tip of the iceberg. Many developments have taken place at institutions and companies throughout the world, especially in recent years. However, most of these endeavors are attempting to overcome the technical challenges that still limit the use of robotic systems for ultraminimally invasive surgical techniques such as single-access surgery. These have the potential to bring radical improvement of surgical outcomes through novel technology, as described in the following sections.

2.3 Single-port robotic platforms for laparoscopic cancer surgery

As described earlier, laparoscopic surgery typically uses three to four ports. Single-incision laparoscopic surgery (SILS)[a] reduces the number of ports to one in order to further decrease invasiveness. Using the umbilicus—a naturally occurring scar—as the

[a] In literature, other variations of the term SILS are used, such as laparoendoscopic single-site surgery (LESS), single-port access surgery (SPAS), and single-port surgery (SPS).

port access point further improves cosmesis resulting in virtually scarless surgical interventions. However, SILS is even more technically challenging than conventional laparoscopic surgery. In addition to the "fulcrum effect," SILS also involves instrument mirroring, in which the left hand controls the instrument that is visually displayed on the right, and vice versa (Fig. 1B). Another challenge is the effect of multiple instruments crowding around the port—also referred to as "sword-fighting"—which, among others, limits the surgeon's freedom of motion. In contrast to conventional laparoscopic surgery, SILS has not seen wide adoption in clinical practice, as its technical challenges outweigh the incremental clinical advantage it brings [20]. An extensive systematic review of SILS pelvic and abdominal surgery, including interventions for cancer therapy, showed an incremental improvement of postoperative recovery times for certain procedures [21]. Longer operative times have been reported, although this can be improved by training. Studies with regard to colorectal [22], gynecological [23], and urological [24] cancer surgery yielded similar findings. The most important conclusions of the aforementioned studies are that SILS procedures do not increase perioperative and postoperative complications, which shows feasibility and safety of the method. Overall, the main opportunity for improvement of SILS lies in the development of instruments and surgical techniques that reduce the technical challenges and make procedures more intuitive.

In robotics, there are two approaches to the performance of surgery through a single incision, depending on whether the instruments are integrated in a single shaft or not. The latter configuration (not integrated) resembles conventional SILS as it involves the use of multiple robotic instruments individually inserted and controlled through a single incision. The "X"-configuration, in which instruments cross, is the most common example of such approach. The use of the da Vinci system for single-incision cancer surgery has been explored for applications such as single-port nephrectomies [25], prostatectomies [26], and hysterectomies [27]. However, despite successes, instrument collision remains an issue. In an attempt to overcome this limitation, Intuitive Surgical introduced the single-port VeSPA instruments for the da Vinci Si [28]. These semirigid tools are inserted through a multichannel single-incision port that has prebent rigid cannulas, thereby providing instrument triangulation. The main limitation of the system is the lack of an EndoWrist to provide sufficient end-effector articulation, which results in reduced range of motion during surgery and makes complex procedures cumbersome [29]. Among the research platforms, notable is the work from Choi et al., who developed a robotic system exploiting a cone-type RCM mechanism to control two articulated instruments inserted through a single port. The instruments have six articulated joints, including a flexible linkage-driven elbow that allows triangulation and high end-effector payloads [30]. While the idea is promising, extensive preclinical validation data need to be collected to assess the system's clinical potential.

A second and more common approach to robotic single-port surgery is to integrate all instruments into a main shaft and deploy them individually after insertion [31]. The use of a rigid shaft requires the articulated instruments to have multiple segments to achieve sufficient triangulation when deployed. The specific mechanical topology and actuation methods vary between commercial and research systems, but with a few exceptions, these systems can be classified as redundantly actuated serial or continuum mechanisms. In both cases, there is motion redundancy in the system, i.e., the number of DoFs is higher than the number of individually controllable actuators. Intuitive Surgical developed a new generation da Vinci robot based on this design specifically for single-port surgery: the da Vinci SP system (Fig. 4A, FDA clearance obtained in June 2018). The main shaft contains the endoscope and two deployable surgical instruments with articulated tips [32]. The articulated tip has two sections: one "elbow" joint that provides instrument triangulation and a second snake-like mechanism, which can ostensibly be classified as a redundantly actuated cable-driven serial mechanism. While the system is currently clinically approved for urological procedures, integrating multiple instruments into a single shaft potentially allows the performance of different interventions such as transoral robotic surgery [33].

TransEnterix has also been developing a single-port system called the SurgiBot (after being denied FDA approval in 2016, the technology was transferred to a Chinese company for commercialization in China [34]). The main concept is based on the earlier

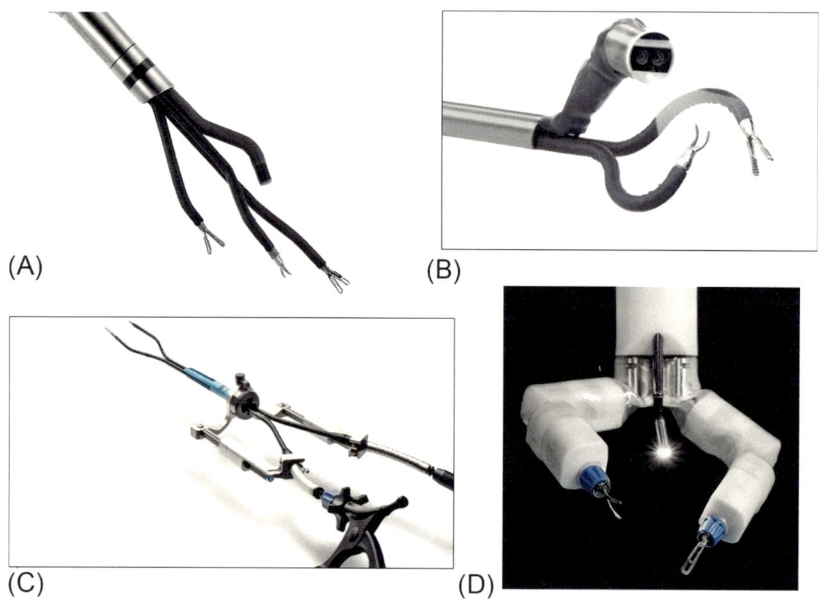

Fig. 4 (A) da Vinci SP, (B) SPORT Surgical System, (C) symphonX® Surgical Platform, (D) Virtual Incision Surgical Robot. *(From (A) © 2019 Intuitive Surgical Inc.; (B) © 2019 Titan Medical Inc.; (C) © Fortimedix Surgical 2019; (D) © 2019 Virtual Incision Corporation.)*

manually controlled SPIDER system, which uses a multichannel cannula to insert two flexible instruments, a scope and, if required, another rigid instrument. The tips of the two flexible instruments are controlled via hollow snake-like mechanisms, referred to as "instrument delivery tubes" [35]. While not specifically disclosed by the manufacturer, it is clear based on images that four cables are used for control of the end effector, indicating redundancy in actuation and hinting toward either a redundantly actuated serial mechanism or a continuum mechanism. An important addition to the mechanism is the use of a parallel structure, a bar mechanism, at the distal tip to offer stability and triangulation for the flexible instruments. A very similar design has been implemented in a recent manually controlled commercial product for SILS called the symphonX® (Fortimedix Surgical BV, Fig. 4C). The system is introduced via a 15-mm trocar port and features two 5-mm flexible instruments. Two additional channels (5 and 3 mm) are used for additional instruments and the endoscope. The system has received both FDA and CE marking; however, no extensive clinical study has been performed yet to validate its capabilities.

Titan Medical Inc. is also working toward obtaining FDA approval in 2019 for the single-port SPORT Surgical system (Fig. 4B), which originated from academic research under the name IREP and features a dual-arm system that can be inserted through a 15-mm port [36]. Both arms are snake-like continuum mechanisms with 4 DoF, each supported by a 2-DoF parallelogram. Push-pull nitinol rods are used for control and are actuated externally. Such continuum mechanisms have received attention from multiple research groups. However, this design has major limitations due to the inability to cope with high payloads and limited workspace [37]. One recent attempt to overcome these issues came from researchers at Tianjin University in China, who developed a manually controlled variable stiffness platform to increase the workspace and provide improved stability [38]. The system consists of one 2-DoF variable stiffness insertion tube designed using phase-change materials, used to deliver two 4-DoF flexible instruments and one 3-DoF stereo vision module. A plug at the distal end of the delivery tube diverges the two arms to provide triangulation. While the deployment of phase-change material in the OR might still prove challenging, variable stiffness is a promising concept to be further explored in the next generation of surgical robots.

Serial mechanisms, when not redundantly actuated, have been demonstrated to be able to exert higher forces than their redundantly actuated counterparts. One such example is the system developed by Virtual Incision Corp. which uses onboard motors with the aim to create a single-port technology in the form of a small portable device (Fig. 4D). The system originates from earlier academic research at the University of Nebraska [39]. The system shows the use of a geared mechanism for actuation of 2 serial links, with a total of 5 DoF per surgical instrument. The gripper is actuated with small DC motors placed in the last link of the end effector. The force exertion capability is high, with a theoretical end-effector force of 30 N in all directions. The footprint in the OR is impressively small, and the cross-sectional area of the shaft (4.13 cm^2) is comparable to other single-incision devices.

In conclusion, when designing single-port systems deployed through a single rigid shaft, there is a tradeoff between instrument diameter and instrument triangulation and payload capabilities. To achieve enough instrument triangulation, the most commonly applied technique is to incorporate additional joints (often an elbow joint) placed in a serial fashion. While this provides dexterity, these systems are redundantly actuated and therefore will inherently deform when external loads are applied. Also, as the articulated sections of the instruments are often long, thus forces applied on the tip result in large moments to compensate for. One mechanism that is frequently used to increase payload capacity takes the form of a parallelogram for extra support of the elbow joints. Typically, these mechanisms use rigid links to transmit forces and therefore cannot be easily adapted for flexible shafts. Similarly, the high payload capacities of direct transmission systems, such as the Virtual Incision, are attributable to the use of a rigid shaft. Therefore, such solutions are only applicable to procedures where the operative site is in a direct line-of-sight from the entry point. Flexible designs need to be introduced for most interventions where access is achieved via a lumen, as discussed in the next section.

2.4 Natural orifice endoscopic surgery and natural orifice transluminal endoscopic surgery

Alongside SILS, there are other minimally invasive techniques that can be performed through a single access point. For instance, natural orifice endoscopic surgery (NOES) uses existing anatomical points of entry to gain access into the body. In contrast with SILS, which is performed extraluminally, NOES requires instruments to navigate to the operative site inside a lumen. This is typically the gastrointestinal (GI) tract, as the mouth and the anus are the most commonly used access points, but other natural pathways such as the urethra or the nose are also used. Consequently, flexible endoscopy plays a major role in NOES to access the GI tract or the bronchi, for which there exists a variety of instruments in different sizes, with names referring to the anatomy of interest (e.g., colonoscope, sigmoidoscope, gastroscope, bronchoscope). For anatomies with a relatively short and straight access pathway, rigid endoscopes are also sometimes used, such as the rigid cystoscope for transurethral access. The number of surgical techniques that can be performed via NOES is still very limited and this is due to the lack of enabling technologies. However, it does play an important role in the removal and prevention of early-stage cancers. In the GI tract, the removal of lesions (e.g., polyps) with a snare is a common practice in many screening procedures. Although these are mostly performed using the standard flexible endoscope and simple instruments, more advanced therapeutic techniques have also been developed, e.g., ESD and endoscopic mucosal resection (EMR). A detailed review of flexible robotic systems specifically designed for these procedures is reported in Section 3 of this chapter.

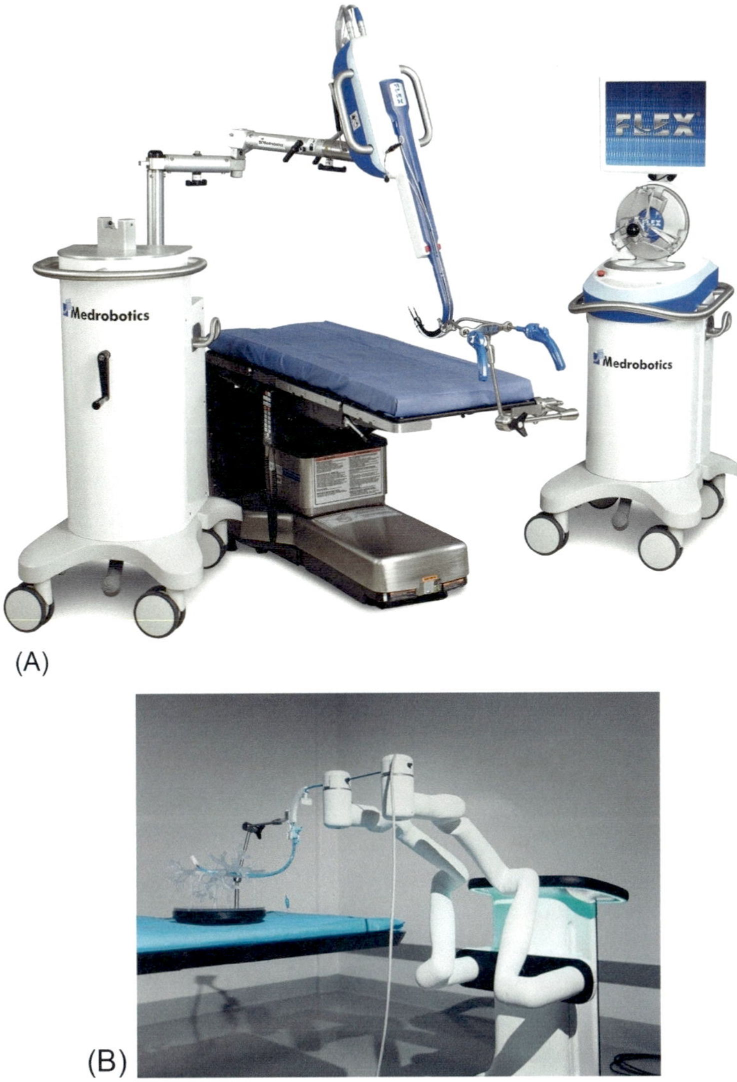

(A)
(B)

Fig. 5 (A) The Flex® Robotic System, (B) The Monarch™ Platform. *(From (A) Copyright © 2019 Medrobotics® Corporation; (B) Copyright 2019 Auris™ Health, Inc.)*

Recently developed commercial robotic systems for NOES include the Flex® Robotic System (Medrobotics® Corp., Fig. 5A) and the Monarch™ (Auris™ Health Inc., Fig. 5B). Auris™ Health received FDA clearance in March 2018 for their Monarch™ system, which is designed for the improved diagnosis and treatment of lung cancers. The system features a robotic arm driving a robotized flexible bronchoscope that can be controlled using a simple joystick. The navigation console combines traditional endoscopic views into the lung with computer-assisted navigation based on 3D models of the patient's own lung anatomy. The Flex® Robotic System uses a combination of a variable stiffness sheath with a flexible inner system to perform transoral robotic surgery and obtained FDA approval in 2015. In 2017 and 2018, additional FDA approval was obtained for colorectal, general, gynecological, thoracic, and urologic procedures. The main concept is based on the previously developed HARP technology [40]. HARP was originally designed for cardiac surgery and has also been referred to in the literature as CardioARM [41]. The system uses two concentric snake robots, which can change stiffness by cable tensioning. Three cables are used for steering and stiffening of the outer layer, whereas only one cable is used to change the stiffness of the inner tube. The change in stiffness is achieved by using spherical joints sliding over each other; when cables are tensioned, the friction between these joints increases, thereby stiffening the entire structure. While the HARP concept is mainly concerned with providing access methods for flexible endoscopy, two manually controlled flexible endoscopic instruments are attached to the tip of the Flex® system to perform tissue manipulation [42].

Another development in MIS techniques is natural orifice transluminal endoscopic surgery (NOTES). Unlike NOES, NOTES uses an existing anatomical entry into the body to pass through an anatomical border to access a second lumen [43]. Examples of NOTES access methods are transvaginal and transgastric, typically used to access the peritoneal cavity. The first transgastric NOTES procedure was published in 2004 by Kalloo et al., in which liver biopsies were taken in a porcine study [44]. Many other transgastric, transvaginal, and even transanal NOTES studies rapidly followed [45]. However, most procedures were performed using a hybrid approach, with the aid of one or more small laparoscopic instruments [46]. Despite this initial interest in NOTES, the technique remains experimental for most procedures and has not been adopted into clinical practice, as demonstrated by the decreasing number of publications and citations in recent years. This is due to a number of important challenges that have yet to be solved, especially the development of new endoscopic instruments, which could be used for suturing, tissue traction and counter traction, and endoscope stabilization [47]. Nonetheless, elected NOTES procedures to treat esophageal and low rectal cancer have become accepted, safe, and feasible approaches by trained surgeons for selected patients [48]. These clinical successes provide an important stimulus to further improve medical technology in order to make the technique more accessible.

3 Flexible robots for gastrointestinal cancer therapy

3.1 Colorectal cancer

Gastrointestinal (GI) cancers are one of the leading causes of mortality and morbidity in the world. In 2018, there were 685,800 new cases of esophageal, gastric, and colorectal cancers (CRC) and an estimated 389,800 gastrointestinal cancer-related deaths in Europe [49]. CRC was the second largest cause of death, with an estimated 242,000 deaths. In the

United Kingdom, bowel cancer was responsible for 41,804 new cancers cases and 16,384 deaths in 2016 [50]. In the United States, GI cancers together accounted for a total of 183,780 new cases and 77,280 deaths in 2018 [51] (gastric, esophageal, and colorectal). With an estimated 50,630 deaths, CRC mortality is the highest among GI cancers and after lung cancer the second highest when compared to all cancers.

3.2 Comparison of EMR/ESD

Endoscopic resection can be used for the minimally invasive removal of precancerous lesions (adenomas) and early-stage CRC. The endoscopic removal of polypoid lesions, polypectomy, is performed using a snare inserted into the biopsy channel of an endoscope. Nonpolypoid adenomas [52] (i.e., flat or depressed lesions) are more challenging to detect and remove, and are more likely to develop into adenocarcinomas. EMR and ESD are two endoscopic techniques (Fig. 6), which can be used to remove lesions, which have not yet invaded the deeper tissue layers (submucosal and muscular layers [53]). Both techniques typically use a submucosal injection to lift the mucosal layer from the underlying tissue. During EMR, the mucosa is resected using a snare placed over the elevated mucosal tissue.

Other techniques include the use of a suction cup to elevate the tissue before removal with the snare [54]. EMR is successful for *en bloc* removal of nonpolypoid lesions smaller than 20 mm, but requires piecemeal resection for larger lesions, referred to as piecemeal EMR. However, the assessment of tumor margins after resection is difficult and piecemeal EMR is often associated with a high recurrence rate, in particular for malignant lesions [55]. ESD, in contrast, uses a knife (typically in the form of electrocautery) to remove lesions by dissecting the submucosal layer. ESD can be used for *en bloc* dissection of lesions of all sizes, i.e., also for lesions larger than 20 mm. *En bloc* removal of lesions using ESD results in a lower local recurrence rate than piecemeal EMR [56,57]. In addition, *en bloc* resection is beneficial for the detection of negative tumor margins during

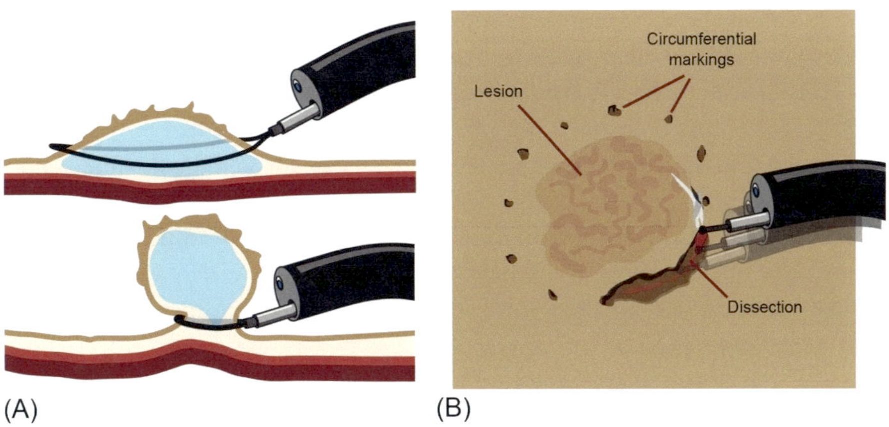

Fig. 6 Endoscopic mucosal resection (A) and endoscopic submucosal dissection (B).

pathological analysis and subsequent follow-up treatment [58]. However, ESD is associated with a higher bleeding and perforation rate than EMR [59]. ESD is also a challenging therapeutic procedure to perform, which is reflected in the long procedure times and long learning curves. The technical complexity of ESD arises from the limited control over the instruments motion. Flexible instruments are inserted into an endoscope biopsy channels and are controlled together with the endoscope's tip. However, to adequately perform ESD, one instrument is used to grasp and handle the tissue (providing traction), while a second instrument cuts the tissue, and therefore requires both instruments to be controlled independently from each other. Another aspect that makes ESD complex to perform is the mechanical compliance of flexible endoscopes, which is a major challenge in both NOES and NOTES procedures [60]. Compliance is required for navigation through the GI tract but decreases the positional stability of the endoscope leading to challenges in maintaining an adequate view and effective interacting with tissue.

Nevertheless, ESD is the preferred technique for lesions larger than 20 mm [61]. However, to realize the full potential of ESD, new techniques and technologies are required to make ESD easier to perform and thereby drastically reduce the procedure time and the high perforation rates.

3.3 Flexible systems for ESD

Novel endoscopes and endoscopy surgical platforms are aimed at providing the bimanual dexterity required for performing ESD. Olympus Corporation developed the R-scope with two articulated working channels to control endoscopic instruments, the first providing an up-down motion for grasping and lifting tissue, and a second a sweeping (left-right) motion (Fig. 7B). However, the R-scope still provides limited dexterity at the tip, and is complex to control manually with the numerous dials, leading to no overall reduction of procedure times [62]. A second system developed by Olympus is the EndoSamurai (Fig. 7A), in which has shown to be successful in bimanual tasks like suturing, ESD and full-thickness resection [63]. While the force exertion capability has not been explored, one study mentions that force is a limiting factor when using the system for traction and countertraction of larger maneuvers [64].

In 2010, Karl Storz developed the manually controlled Anubiscope system [65] in collaboration with IRCAD in Strasburg (Fig. 7C). They adapted the flexible endoscope to include two surgical graspers that are deployed at the end of the device. An additional elbow joint was added to provide additional triangulation to the instruments. Recently, a robotic version of the system was launched in the form of the STRAS system [66]. A second generation (STRAS v2) was successfully able to perform multiple ESDs on an in vivo pig model [67].

Phee et al. developed a robotic add-on to a conventional flexible endoscope that provides two instruments in order to enable the performance of complex surgical tasks [68]. The anthropomorphic design is based on the human arm and provides 5 DoF to the

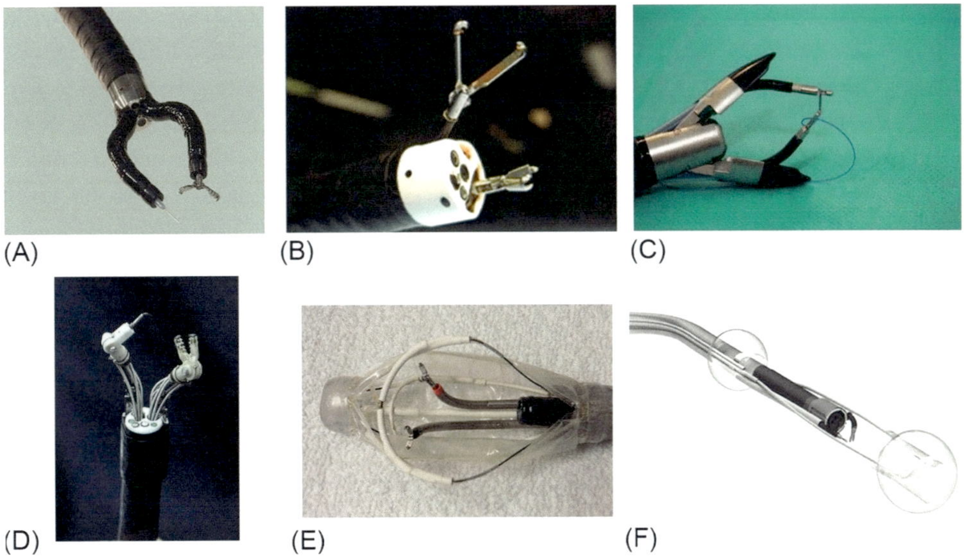

(A) (B) (C) (D) (E) (F)

Fig. 7 (A) Endosamurai system (Olympus Corp., Japan). (B) The R-scope (Olympus Corp., Japan). (C) The Anubiscope (IRCAD and Karl Storz Endoskope). (D) The MASTER system. (E) LumenR (Boston Scientific Corp.). (F) DiLumen C^2. *(Reprinted from (A) K. Ikeda, K. Sumiyama, H. Tajiri, K. Yasuda, S. Kitano, Evaluation of a new multitasking platform for endoscopic full-thickness resection, Gastrointest. Endosc. 73 (2011) 117–122, Copyright 2011, with permission from Elsevier. (B) M.T. Moyer, R.S. Haluck, J. Gopal, E.M. Pauli, A. Mathew, Transgastric organ resection solely with the prototype R-scope and the self-approximating transluminal access technique, Gastrointest. Endosc. 72 (2010) 170–176, Copyright 2010, with permission from Elsevier. (C) B. Dallemagne, J. Marescaux, The ANUBIS™ project, Minim. Invasive Ther. Allied Technol. 19 (2010) 257–261, by permission of Taylor & Francis Ltd. (D) W. Philip, Y. Chiu, S.J. Phee, Z. Wang, et al., Feasibility of full-thickness gastric resection using master and slave transluminal endoscopic robot and closure by overstitch: a preclinical study, Surg. End. 28 (2013) 319–324, Copyright 2013, by permission from Springer Nature Customer Service Centre GmbH. (E) S.V. Kantsevoy, M. Bitner, G. Piskun, New endoscopic platform for endoluminal enbloc tissue resection in the gastrointestinal tract (with videos), Surg. Endosc. 30 (2016) 3145–3151, Copyright 2016, by permission from Springer Nature Customer Service Centre GmbH. From (F) Copyright 2018 Lumendi Ltd.)*

surgical end effector. Actuation is performed by cables, guided through a flexible sheath (Bowden) along the endoscope. Two antagonistic cables are used to articulate a single DoF and are simultaneously actuated by a single DC motor placed at the proximal end of the endoscope. A more advanced version of the system, called MASTER, was used to perform advanced endoluminal and transluminal surgery during in vivo animal studies [69,70] (Fig. 7D). Force capacity tests of the system showed that the mechanism is able to exert forces up to 5.2 N at the end effector. The system is currently undergoing clinical trials and belongs to a Singaporean company called EndoMaster Pte Ltd.

Nakadate et al. developed an articulated 2.6-mm wire-driven instrument, which could be manually controlled by a joystick specifically developed for the system [71]. The system was tested with one articulated instrument inserted into the working channel, and a second mounted with a cap on the outside of the channel. The system was successfully used by endoscopists (both experts and novices in ESD) to perform ESD in ex vivo gastric porcine studies, proving itself to be a promising and cost-effective solution. Two experts used the system to perform in vivo gastric ESD on pigs. A 2019 study of 30 colonic ESDs performed on 10 pigs compared ESD performed using two of the articulated instruments ($n=15$) to conventional ESD ($n=15$) [72]. The same Kyushu University group developed the robotic-assisted flexible endoscope (RAFE) [73]. The RAFE is an add-on to the proximal dials of the endoscope to make any conventional endoscope robotic. The system showed success when performing several tasks, including ESD on ex vivo porcine stomach. The system did not use any additional articulated end effectors and the group remarked that future steps will be to integrate the articulated 2.6-mm instruments in the robotic system.

The LumenR Tissue Retraction System [74] is a flexible system specifically developed for ESD (Fig. 7E), which was acquired by Boston Scientific in 2016. The system uses a deployable operating chamber with two instrument guides, which can be controlled in 3 DoF, and a 4th DoF is created through retraction and extension of the flexible endoscopic instruments relative to the instrument guides. Both instruments are used as graspers to lift the submucosal layer, while a third electrocautery instrument is used via the endoscope's biopsy channel to dissect the tissue.

The DiLumen EIP oversheath (Lumendi Ltd., Fig. 7F) is a recent development that uses a double-balloon technique specifically for therapeutic interventions. By inflating both balloons, the DiLumen EIP device creates a working area around the lesion (the so-called therapeutic zone), which can be used to inflate and stretch the colonic folds. The endoscope tip is positioned between both balloons to perform ESD. Additionally, the distal balloon can be equipped with clips to provide mucosal traction. In 2018, this system with clips was compared to conventional cap-assisted ESD in ex vivo animal trials and one in vivo animal trial, demonstrating a significant decrease in operating time [75]. A similar result was found when performing the ESD in ex vivo tissue when the therapeutic zone was filled with water [76]. The combination of endoscope stabilization and tissue countertraction, as seen in the DiLumen system, are important to reduce the complexity of ESD procedures. However, the current system still uses conventional techniques for performing ESD, which is still challenging to perform due to less intuitive control and limited instrument dexterity. To overcome this, the company is currently working on the development of a bimanual system for ESD (DiLumen C2).

Despite these promising developments, it is not clear yet whether these systems will truly overcome the challenges seen in ESD and can be used for other advanced endoscopic therapeutic procedures. The authors of this chapter have been working on the CYCLOPS system, a novel system, which has several unique properties, which are beneficial for performing ESD and potentially for other NOES and NOTES procedures.

3.4 CYCLOPS for ESD and beyond

The CYCLOPS system is a novel robotic system developed for performing ESD. The system is mounted on a conventional endoscope and uses a deployable structure (or scaffold) to stabilize the endoscope and the tissue for performing surgery. The scaffold is used also provides the support for actuation cables that are guided to the endoscopic instruments and used for controlling the endoscopic instruments within the scaffold. The specific cable mechanism is referred to as cable-driven parallel mechanism (CDPM), and a first proof-of-concept of the CYCLOPS showed unique technical benefits, which can be used for performing ESD, including high payloads and large instrument workspace [77]. The actuation cables are guided through flexible hollow sheaths (i.e., Bowden cables) along the endoscope to the motor units. The surgeon intuitively controls both endoscopic instruments with two stylus-type joysticks, whose motions are scaled down and translated to the same motions of the instruments.

The first CYCLOPS prototype specifically developed for ESD (Fig. 8) has shown that it provides the required instrument dexterity without compromising in the amount of forces the instrument can deliver to manipulate the tissue [78]. The system was able to realize exceptionally high instrument payloads for a flexible endoscopy system with forces measured up to 46 N. Additionally, a high positional accuracy was shown during a user study, which resembled accuracies found with the da Vinci system. The scaffold in this prototype has been made from a deployable steel structure, covered by a soft silicone sleeve to prevent tissue damage. When the scaffold is undeployed, the cross-sectional width and height of is 30 mm and 18 mm, respectively. The small size allows the endoscope to be inserted and navigated into the colon. The scaffold can be deployed within a colon diameter of up to 50 cm, which corresponds to the mean diameter of the transverse colon in an American population [79]. The CDPMs are used to control two curved overtubes into which conventional endoscopic instruments are placed. The prototype was further evaluated for performing ESD and shown to be feasible in an ex vivo preclinical setting and preliminary results have been found for performing ESD in pigs.

The most recent developments are focused on the development of a deployable scaffold, which is inflatable. Such an inflatable scaffold structure is preferred for its compactness and safe interaction with tissue. Additionally, an inflatable scaffold can be manufactured inexpensively for single usage and thus eradicating the need for cleaning and sterilization after the procedure. A scaled-up inflatable scaffold can be seen in Fig. 9A. Currently, a manufacturing method is being explored in which such a scaffold can be created quickly and on the real scale, resulting to scaffolds shown in Fig. 9B.

In overall, the CYCLOPS offers endoscope and tissue stabilization while providing bimanual instrument dexterity and high payloads. While the CYCLOPS system has been developed for ESD specifically, these benefits are unique among flexible endoscopic systems developed for NOES and NOTES and thus the technology should be explored

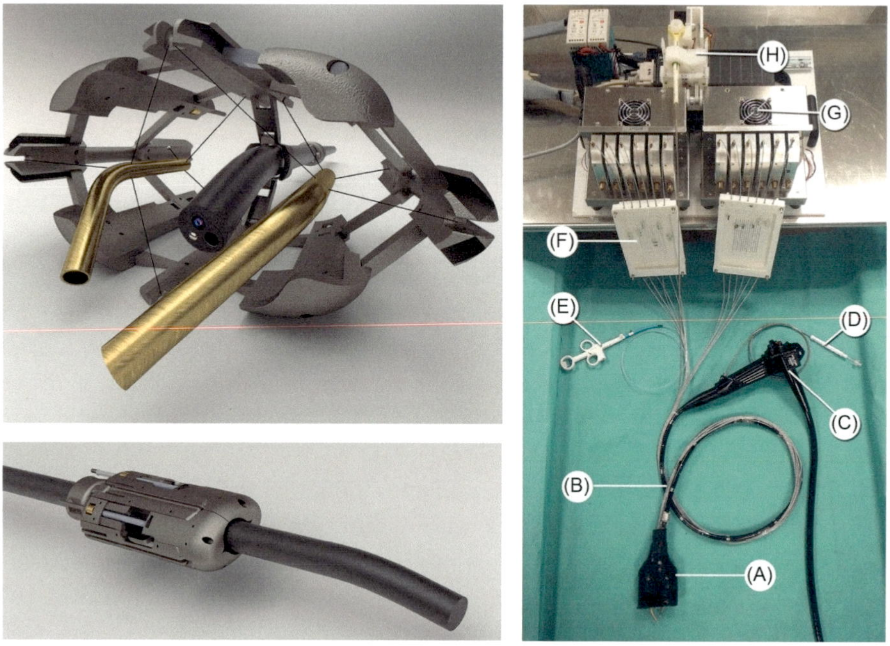

Fig. 8 *Left*: a render of the ESD CYCLOPS scaffold when deployed and undeployed. Two CDPMs are used to control the two curved brass overtubes containing endoscopic instruments. The renders do not include the flexible silicone sleeve used to prevent tissue damage. *Right*: an overview of the robotic system with (A) scaffold, (B) Bowden cables to guide the actuation cables, (C) flexible endoscope, (D) flexible needle placed into the remaining biopsy channel, (E) diathermy instrument, (F) cable splitters, and (G, H) mechatronics and actuators. *(From T.J. Oude Vrielink, M. Zhao, A. Darzi, G.P. Mylonas, ESD CYCLOPS: a new robotic surgical system for GI surgery, in: 2018 IEEE International Conference on Robotics and Automation (ICRA), 2018, pp. 150–157.)*

further for other MIS surgeries. For CRC specifically, full-thickness resection (FTR) is an even more complex procedure as entire sections of the colon wall are dissected to remove tumors that have invaded into tissue layers below the mucosa. In FTR traction and cutting remain two essential tasks, but additionally requires a method to close the colon wall after resection. Nevertheless, the shown advantages of the CYCLOPS system provide an important starting point toward the development of a new surgical platform to perform FTR.

4 Conclusion

MIS has brought unprecedented benefits to surgery by reducing recovery time, pain, analgesics, and postoperative complications. New access methods through increasingly smaller incisions and natural orifices can further improve MIS and enable advanced therapeutic cancer procedures in remote areas of the body. However, the small incisions pose

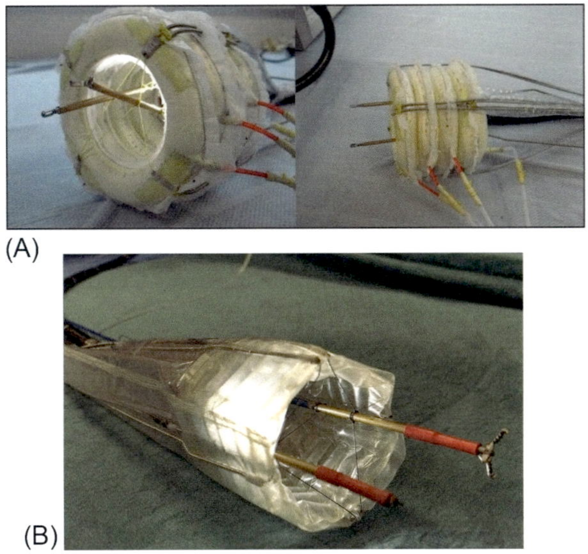

Fig. 9 (A) A scaled-up version of the inflatable scaffold for the CYCLOPS system. (B) A recent development for a real-sized inflatable scaffold. *(From (A) T.J. Oude Vrielink, M. Zhao, A. Darzi, G.P. Mylonas, ESD CYCLOPS: a new robotic surgical system for GI surgery, in: 2018 IEEE International Conference on Robotics and Automation (ICRA), 2018, pp. 150–157.)*

a technical challenge for the development of new surgical platforms with sufficient dexterity and high payloads.

In this context, robotic technology has the potential to show benefits whenever it can demonstrate its enabling nature. On one hand is the surgeon, for whom robotics have undeniably introduced ergonomic improvements, such as better control over the surgical instruments, a better view of the surgical site, and a more comfortable working posture. On the other hand, there is the most important stakeholder: the patient. Urologic procedures for excision of prostate cancer have provided evidence, but with considerable uncertainty, that robotic approaches can offer more complete removal of cancer and less side-effects from conventional laparoscopic prostatectomy [80]. However, this study also concluded that robotic prostatectomy for healthcare systems such as England's National Healthcare System (NHS) will always be more costly because of the fixed capital and maintenance charges for the robotic system. The study showed that this excess cost could be reduced if the cost of the robot is minimized and by maintaining a high case volume for each robotic system of at least 100–150 procedures per year. Even so, these findings are primarily based on limited evidence on reduced positive margin rates, as a result of robotic surgery. The study concludes that there is a need for further research to establish how these reduced positive margin rates, impact on long-term patient outcomes.

Currently, a very active area of research in surgical robotics is this of flexible endoscopy, where robotic augmentation can have an enabling effect. Complex MIS techniques that could be greatly facilitated are ESD and FTR, which are used for the *en bloc* removal of cancers in the gastrointestinal tract. Flexible robotic platforms are expected to make such procedures more accessible and safer. The CYCLOPS, a bimanual endoscopic surgical system that provides a high instrument dexterity and payloads, while stabilizing the endoscope and tissue through a scaffold structure, is one such example.

In general, no conclusive evidence has yet emerged to support superior curative outcomes from robotic surgery compared to conventional approaches [81]. The high cost of current robotic systems and a market monopolized for nearly 20 years by the da Vinci robot (Intuitive Surgical Inc.), have played a significant role in suppressing wider adoption of robotics in clinical settings. A more competitive market would have certainly encouraged more investment in relevant research and development (R&D) and translation of technologies into clinical practice, with a consequent effect on new and lower-cost robotic technologies emerging. Wider adoption and lower costs would naturally encourage a less conservative attitude toward innovation in the field and could eventually provide evidence toward superior curative outcomes offered by surgical robotics. As Intuitive Surgical Inc. early patents have recently started expiring, there has been an increasing mobility on the market and on R&D in surgical robotics. However, recent acquisitions of smaller surgical robotics companies by larger companies, such as Mazor Robotics acquisition by Medtronic, and Auris™ Health Inc. acquisition by Johnson & Johnson, raise further concerns of a market shifting from a monopoly to an oligopoly. It is the authors' opinion that this will further suppress wider adoption of surgical robotics and further delay the long-awaited clear evidence supporting improved curative outcomes.

Notwithstanding the commercial messages emanating from large surgical robotics companies, the consensus of the medical community is that the high cost of robotic technologies cannot justify the marginal gains these offer in clinical outcomes. It is the authors' view that the true strength of robotics, for benign or cancer treatment, is yet to emerge. This is expected to happen through the synergy of surgical robotic technologies with other modalities. These modalities include engineering and materials sciences, computer vision, intraoperative image guidance and real-time diagnosis, artificial intelligence and machine learning, data sciences, just to name a few. Recently introduced concepts and research areas, such as soft and variable stiffness robotics [82–84], perceptually enabled smart operating theaters [85], digital surgery [86], and task automation [87], are expected to drive innovation in robotic surgery in the years to come. In conclusion, just like any healthcare technology, robotic surgery is going to continue to improve as devices get smaller, more intelligent, and more accessible.

References

[1] L. Wyld, R.A. Audisio, G.J. Poston, The evolution of cancer surgery and future perspectives, Nat. Rev. Clin. Oncol. 12 (2014) 115.

[2] M. Quaresma, M.P. Coleman, B. Rachet, 40 year trends in an index of survival for all cancers combined and survival adjusted for age and sex for each cancer in England and Wales, 1971–2011: a population-based study, Lancet 385 (2015) 1206–1218.
[3] C. R. UK, Cancer Survival Statistics for all Cancers Combined, Available from: www.cancerresearchuk.org/health-professional/cancer-statistics/survival/all-cancers-combined, 2018.
[4] Y.S. Kwoh, J. Hou, E.A. Jonckheere, S. Hayati, A robot with improved absolute positioning accuracy for CT guided stereotactic brain surgery, IEEE Trans. Biomed. Eng. 35 (1988) 153–160.
[5] H.J. Marcus, C.A. Seneci, C.J. Payne, D. Nandi, A. Darzi, G.-Z. Yang, Robotics in keyhole transcranial endoscope-assisted microsurgery: a critical review of existing systems and proposed specifications for new robotic platforms, Oper. Neurosurg. 10 (2014) 84–96.
[6] A. Mukherji, An introduction to brachytherapy, in: Basics of Planning and Management of Patients during Radiation Therapy: A Guide for Students and Practitioners, Springer Singapore, Singapore, 2018, pp. 171–182.
[7] B.J. Wood, J.R. Ramkaransingh, T. Fojo, M.M. Walther, S.K. Libutti, Percutaneous tumor ablation with radiofrequency, Cancer 94 (2002) 443–451.
[8] R. Monfaredi, K. Cleary, K. Sharma, MRI robots for needle-based interventions: systems and technology, Ann. Biomed. Eng. 46 (2018) 1479–1497.
[9] S.S. Lo, T. Biswas, R.J. Ellis, P.C. Gerszten, Radiosurgery, in: A. Gulati, V. Puttanniah, B.M. Bruel, W.S. Rosenberg, J.C. Hung (Eds.), Essentials of Interventional Cancer Pain Management, Springer International Publishing, Cham, 2019, pp. 235–240.
[10] J.E. Kennedy, High-intensity focused ultrasound in the treatment of solid tumours, Nat. Rev. Cancer 5 (2005) 321.
[11] M.J. Mack, Minimally invasive and robotic surgery, JAMA 285 (2001) 568–572.
[12] V. Vitiello, K.W. Kwok, G.Z. Yang, (Chapter 1). Introduction to robot-assisted minimally invasive surgery (MIS), in: P. Gomes (Ed.), Medical Robotics, Woodhead Publishing, 2012, p. 1-P1.
[13] V. Sun, Y. Fong, Minimally invasive cancer surgery: indications and outcomes, Semin. Oncol. Nurs. 33 (2017) 23–36.
[14] Intuitive Surgical Inc, Q4 2018 Preliminary Financial Data Tables, Available from: https://isrg.gcs-web.com/static-files/3a12c816-6637-4a3a-bfef-247a60c1fbd7, 2018.
[15] Intuitive Surgical Inc, Intuitive Surgical Announces Preliminary Fourth Quarter and Full Year 2018 Results, Available from: https://isrg.gcs-web.com/node/16601/pdf, 2019.
[16] G.S. Guthart, J.K. Salisbury, The Intuitive telesurgery system: overview and application, in: Proceedings of the IEEE International Conference on Robotics and Automation—ICRA 2000, 2000, pp. 618–621.
[17] U. Hagn, R. Konietschke, A. Tobergte, M. Nickl, S. Jörg, B. Kübler, et al., DLR MiroSurge: a versatile system for research in endoscopic telesurgery, Int. J. Comput. Assist. Radiol. Surg. 5 (2010) 183–193.
[18] A. Tobergte, P. Helmer, U. Hagn, P. Rouiller, S. Thielmann, S. Grange, et al., The sigma.7 haptic interface for MiroSurge: a new bi-manual surgical console, in: 2011 IEEE/RSJ International Conference on Intelligent Robots and Systems, 2011, pp. 3023–3030.
[19] D. Cagle, P. E. G. Kilroy, K. S. Koenig, J. S. Gee, W. Grout, and M. P. Schaller, Table Adapter for Mounting Robotic Arms to a Surgical Table, US20180078439A1, 2018.
[20] S. Morales-Conde, A. Peeters, Y.M. Meyer, S.A. Antoniou, I.A. del Agua, A. Arezzo, et al., European association for endoscopic surgery (EAES) consensus statement on single-incision endoscopic surgery, Surg. Endosc. 33 (2019) 996–1019.
[21] K. Ahmed, T.T. Wang, V.M. Patel, K. Nagpal, J. Clark, M. Ali, et al., The role of single-incision laparoscopic surgery in abdominal and pelvic surgery: a systematic review, Surg. Endosc. 25 (2011) 378–396.
[22] H.-J. Li, L. Huang, T.-J. Li, J. Su, L.-R. Peng, W. Liu, Short-term outcomes of single-incision versus conventional laparoscopic surgery for colorectal diseases: meta-analysis of randomized and prospective evidence, J. Gastrointest. Surg. 21 (2017) 1931–1945.
[23] L. Yang, J. Gao, L. Zeng, Z. Weng, S. Luo, Systematic review and meta-analysis of single-port versus conventional laparoscopic hysterectomy, Int. J. Gynecol. Obstet. 133 (2016) 9–16.

[24] X. Fan, T. Lin, K. Xu, Z. Yin, H. Huang, W. Dong, et al., Laparoendoscopic single-site nephrectomy compared with conventional laparoscopic nephrectomy: a systematic review and meta-analysis of comparative studies, Eur. Urol. 62 (2012) 601–612.
[25] J.H. Kaouk, R.K. Goel, G.-P. Haber, S. Crouzet, R.J. Stein, Robotic single-port transumbilical surgery in humans: initial report, BJU Int. 103 (2009) 366–369.
[26] M.A. White, G.-P. Haber, R. Autorino, R. Khanna, S. Forest, B. Yang, et al., Robotic laparoendoscopic single-site radical prostatectomy: technique and early outcomes, Eur. Urol. 58 (2010) 544–550.
[27] E.J. Nam, S.W. Kim, M. Lee, G.W. Yim, J.H. Paek, S.H. Lee, et al., Robotic single-port transumbilical total hysterectomy: a pilot study, J. Gynecol. Oncol. 22 (2011) 120–126.
[28] G.-P. Haber, M.A. White, R. Autorino, P.F. Escobar, M.D. Kroh, S. Chalikonda, et al., Novel robotic da Vinci instruments for laparoendoscopic single-site surgery, Urology 76 (2010) 1279–1282.
[29] L. Morelli, S. Guadagni, G. Di Franco, M. Palmeri, G. Di Candio, F. Mosca, Da Vinci single site© surgical platform in clinical practice: a systematic review, Int. J. Med. Robot. 12 (2016) 724–734.
[30] H. Choi, H. Kwak, Y. Lim, H. Kim, Surgical robot for single-incision laparoscopic surgery, IEEE Trans. Biomed. Eng. 61 (2014) 2458–2466.
[31] V. Vitiello, S. Lee, T.P. Cundy, G. Yang, Emerging robotic platforms for minimally invasive surgery, IEEE Rev. Biomed. Eng. 6 (2013) 111–126.
[32] J.H. Kaouk, G.-P. Haber, R. Autorino, S. Crouzet, A. Ouzzane, V. Flamand, et al., A novel robotic system for single-port urologic surgery: first clinical investigation, Eur. Urol. 66 (2014) 1033–1043.
[33] F.C. Holsinger, A flexible, single-arm robotic surgical system for transoral resection of the tonsil and lateral pharyngeal wall: next-generation robotic head and neck surgery, Laryngoscope 126 (2016) 864–869.
[34] Business Wire, TransEnterix Announces Global SurgiBot System Agreement, Available from: https://www.businesswire.com/news/home/20171218005245/en/TransEnterix-Announces-Global-SurgiBot-System-Agreement, 2017.
[35] A.D. Pryor, J.R. Tushar, L.R. DiBernardo, Single-port cholecystectomy with the TransEnterix SPIDER: simple and safe, Surg. Endosc. 24 (2010) 917–923.
[36] K. Xu, R.E. Goldman, J. Ding, P.K. Allen, D.L. Fowler, N. Simaan, System design of an insertable robotic effector platform for single port access (SPA) surgery, in: 2009 IEEE/RSJ International Conference on Intelligent Robots and Systems, 2009, pp. 5546–5552.
[37] J. Shang, V. Vitiello, G.-Z. Yang, Single-port access robots for MIS, in: R. Patel (Ed.), The Encyclopedia of Medical Robotics, 2018, pp. 143–179.
[38] J. Li, X. Li, J. Wang, Y. Xing, S. Wang, X. Ren, Design and evaluation of a variable stiffness manual operating platform for laparoendoscopic single site surgery (LESS), Int. J. Med. Robot. 13 (2017).
[39] T.D. Wortman, J.M. Mondry, S.M. Farritor, D. Oleynikov, Single-site colectomy with miniature in vivo robotic platform, IEEE Trans. Biomed. Eng. 60 (2013) 926–929.
[40] A. Degani, H. Choset, A. Wolf, M.A. Zenati, Highly articulated robotic probe for minimally invasive surgery, in: Proceedings 2006 IEEE International Conference on Robotics and Automation, 2006. ICRA 2006, 2006, pp. 4167–4172.
[41] T. Ota, A. Degani, D. Schwartzman, B. Zubiate, J. McGarvey, H. Choset, et al., A highly articulated robotic surgical system for minimally invasive surgery, Ann. Thorac. Surg. 87 (2009) 1253–1256.
[42] S. Lang, S. Mattheis, P. Hasskamp, G. Lawson, C. Güldner, M. Mandapathil, et al., A european multicenter study evaluating the flex robotic system in transoral robotic surgery, Laryngoscope 127 (2017) 391–395.
[43] D. Rattner, A. Kalloo, ASGE/SAGES working group on natural orifice translumenal endoscopic surgery, Surg. Endosc. Other Interv. Tech. 20 (2006) 329–333.
[44] A.N. Kalloo, V.K. Singh, S.B. Jagannath, H. Niiyama, S.L. Hill, C.A. Vaughn, et al., Flexible transgastric peritoneoscopy: a novel approach to diagnostic and therapeutic interventions in the peritoneal cavity, Gastrointest. Endosc. 60 (2004) 114–117.
[45] E. Della Flora, T.G. Wilson, I.J. Martin, N.A. O'Rourke, G.J. Maddern, A review of natural orifice translumenal endoscopic surgery (NOTES) for intra-abdominal surgery: experimental models, techniques, and applicability to the clinical setting, Ann. Surg. 247 (2008) 583–602.

[46] M.H. Sodergren, J. Clark, T. Athanasiou, J. Teare, G.-Z. Yang, A. Darzi, Natural orifice translumenal endoscopic surgery: critical appraisal of applications in clinical practice, Surg. Endosc. 23 (2009) 680–687.
[47] D.W. Rattner, R. Hawes, S. Schwaitzberg, M. Kochman, L. Swanstrom, The second SAGES/ASGE white paper on natural orifice transluminal endoscopic surgery: 5 years of progress, Surg. Endosc. 25 (2011) 2441–2448.
[48] A.M. Lacy, F. Borja De Lacy, S. Valverde, Transluminal cancer surgery, Surg. Oncol. Clin. 28 (2019) 101–113.
[49] J. Ferlay, M. Colombet, I. Soerjomataram, T. Dyba, G. Randi, M. Bettio, et al., Cancer incidence and mortality patterns in Europe: estimates for 40 countries and 25 major cancers in 2018, Eur. J. Cancer 103 (2018) 356–387.
[50] C. R. UK, Bowel Cancer Statistics, Available from: https://www.cancerresearchuk.org/health-professional/cancer-statistics/statistics-by-cancer-type/bowel-cancer, 2018.
[51] N. C. Institute, Cancer Stat Facts, Available from: https://seer.cancer.gov/statfacts/, 2018.
[52] A. Facciorusso, M. Antonino, M. Di Maso, M. Barone, N. Muscatiello, Non-polypoid colorectal neoplasms: classification, therapy and follow-up, World J. Gastroenterol. 21 (2015) 5149.
[53] N. Yoshida, Y. Naito, N. Yagi, A. Yanagisawa, Importance of histological evaluation in endoscopic resection of early colorectal cancer, World J. Gastrointest. Pathophysiol. 3 (2012) 51.
[54] J.H. Hwang, V. Konda, B.K.A. Dayyeh, S.S. Chauhan, B.K. Enestvedt, L.L. Fujii-Lau, et al., Endoscopic mucosal resection, Gastrointest. Endosc. 82 (2015) 215–226.
[55] G.J. Seo, D.K. Sohn, K.S. Han, C.W. Hong, B.C. Kim, J.W. Park, et al., Recurrence after endoscopic piecemeal mucosal resection for large sessile colorectal polyps, World J. Gastroenterol. 16 (2010) 2806.
[56] Y. Saito, M. Fukuzawa, T. Matsuda, S. Fukunaga, T. Sakamoto, T. Uraoka, et al., Clinical outcome of endoscopic submucosal dissection versus endoscopic mucosal resection of large colorectal tumors as determined by curative resection, Surg. Endosc. 24 (2010) 343–352.
[57] S. Oka, S. Tanaka, I. Kaneko, R. Mouri, M. Hirata, T. Kawamura, et al., Advantage of endoscopic submucosal dissection compared with EMR for early gastric cancer, Gastrointest. Endosc. 64 (2006) 877–883.
[58] J.T. Maple, B.K.A. Dayyeh, S.S. Chauhan, J.H. Hwang, S. Komanduri, M. Manfredi, et al., Endoscopic submucosal dissection, Gastrointest. Endosc. 81 (2015) 1311–1325.
[59] Y. Cao, C. Liao, A. Tan, Y. Gao, Z. Mo, F.J.E. Gao, Meta-analysis of endoscopic submucosal dissection versus endoscopic mucosal resection for tumors of the gastrointestinal tract, Endoscopy 41 (2009) 751–757.
[60] D. Rattner, A. Kalloo, ASGE/SAGES Working Group, ASGE/SAGES working group on natural orifice translumenal endoscopic surgery, Surg. Endosc. 20 (2006) 329–333.
[61] B.-H. Min, J. Lee, J. Kim, S. Shim, D. Chang, Y.-H. Kim, et al., Clinical outcomes of endoscopic submucosal dissection (ESD) for treating early gastric cancer: comparison with endoscopic mucosal resection after circumferential precutting (EMR-P), Dig. Liver Dis. 41 (2009) 201–209.
[62] H. Neuhaus, G. Costamagna, J. Devière, P. Fockens, T. Ponchon, T. Rösch, ARCADE Group, Endoscopic submucosal dissection (ESD) of early neoplastic gastric lesions using a new double-channel endoscope (the "R-scope"), Endoscopy 38 (2006) 1016–1023.
[63] K. Ikeda, K. Sumiyama, H. Tajiri, K. Yasuda, S. Kitano, Evaluation of a new multitasking platform for endoscopic full-thickness resection, Gastrointest. Endosc. 73 (2011) 117–122.
[64] K.-H. Fuchs, W. Breithaupt, Transgastric small bowel resection with the new multitasking platform EndoSAMURAI™ for natural orifice transluminal endoscopic surgery, Surg. Endosc. 26 (2012) 2281–2287.
[65] B. Dallemagne, J. Marescaux, The ANUBIS™ project, Minim. Invasive Ther. Allied Technol. 19 (2010) 257–261.
[66] A. De Donno, L. Zorn, P. Zanne, F. Nageotte, M. de Mathelin, Introducing STRAS: a new flexible robotic system for minimally invasive surgery, in: 2013 IEEE International Conference on Robotics and Automation, 2013, pp. 1213–1220.

[67] L. Zorn, F. Nageotte, P. Zanne, A. Legner, B. Dallemagne, J. Marescaux, et al., A novel telemanipulated robotic assistant for surgical endoscopy: preclinical application to ESD, IEEE Trans. Biomed. Eng. 65 (2018) 797–808.
[68] S. Phee, S. Low, Z. Sun, K. Ho, W. Huang, Z.M. Thant, et al., Robotic system for no-scar gastrointestinal surgery, Int. J. Med. Robot. 4 (2008) 15–22.
[69] S.J. Phee, S.C. Low, V. Huynh, A.P. Kencana, Z. Sun, K. Yang, Master and slave transluminal endoscopic robot (MASTER) for natural orifice transluminal endoscopic surgery, in: 2009 Annual International Conference of the IEEE Engineering in Medicine and Biology Society, 2009, pp. 1192–1195.
[70] Z. Wang, S.J. Phee, D. Lomanto, R. Goel, P. Rebala, Z. Sun, et al., Endoscopic submucosal dissection of gastric lesions by using a master and slave transluminal endoscopic robot: an animal survival study, Endoscopy 44 (2012) 690–694.
[71] R. Nakadate, S. Nakamura, T. Moriyama, H. Kenmotsu, S. Oguri, J. Arata, et al., Gastric endoscopic submucosal dissection using novel 2.6-mm articulating devices: an ex vivo comparative and in vivo feasibility study, Endoscopy 47 (2015) 820–824.
[72] Y. Okamoto, R. Nakadate, S. Nakamura, J. Arata, S. Oguri, T. Moriyama, et al., Colorectal endoscopic submucosal dissection using novel articulating devices: a comparative study in a live porcine model, Surg. Endosc. 33 (2019) 651–657.
[73] T. Iwasa, R. Nakadate, S. Onogi, Y. Okamoto, J. Arata, S. Oguri, et al., A new robotic-assisted flexible endoscope with single-hand control: endoscopic submucosal dissection in the ex vivo porcine stomach, Surg. Endosc. 32 (2018) 3386–3392.
[74] S.V. Kantsevoy, M. Bitner, G. Piskun, New endoscopic platform for endoluminal enbloc tissue resection in the gastrointestinal tract (with videos), Surg. Endosc. 30 (2016) 3145–3151.
[75] S. Sharma, K. Momose, H. Hara, J. East, K. Sumiyama, K. Nakajima, et al., Facilitating endoscopic submucosal dissection: double balloon endolumenal platform significantly improves dissection time compared with conventional technique (with video), Surg. Endosc. 33 (2019) 315–321.
[76] S.K. Sharma, T. Hiratsuka, H. Hara, J.W. Milsom, Antigravity ESD–double-balloon-assisted underwater with traction hybrid technique, Endosc. Int. Open 6 (2018) E739–E744.
[77] G.P. Mylonas, V. Vitiello, T.P. Cundy, A. Darzi, G.-Z. Yang, CYCLOPS: a versatile robotic tool for bimanual single-access and natural-orifice endoscopic surgery, in: 2014 IEEE International Conference on Robotics and Automation (ICRA), 2014, pp. 2436–2442.
[78] T.J. Oude Vrielink, M. Zhao, A. Darzi, G.P. Mylonas, ESD CYCLOPS: a new robotic surgical system for GI surgery, in: 2018 IEEE International Conference on Robotics and Automation (ICRA), 2018, pp. 150–157.
[79] M. Khashab, P. Pickhardt, D. Kim, D. Rex, Colorectal anatomy in adults at computed tomography colonography: normal distribution and the effect of age, sex, and body mass index, Endoscopy 41 (2009) 674–678.
[80] C. Ramsay, R. Pickard, C. Robertson, A. Close, L. Vale, N. Armstrong, et al., Systematic review and economic modelling of the relative clinical benefit and cost-effectiveness of laparoscopic surgery and robotic surgery for removal of the prostate in men with localised prostate cancer, Health Technol. Assess. 16 (2012) 1–313.
[81] A. Tan, H. Ashrafian, A.J. Scott, S.E. Mason, L. Harling, T. Athanasiou, et al., Robotic surgery: disruptive innovation or unfulfilled promise? A systematic review and meta-analysis of the first 30 years, Surg. Endosc. 30 (2016) 4330–4352.
[82] M. Runciman, A. Darzi, G.P. Mylonas, Soft robotics in minimally invasive surgery, Soft Robot. Epub ahead of print. (2019), https://doi.org/10.1089/soro.2018.0136.
[83] M. Cianchetti, T. Ranzani, G. Gerboni, I.D. Falco, C. Laschi, A. Menciassi, STIFF-FLOP surgical manipulator: mechanical design and experimental characterization of the single module, in: 2013 IEEE/RSJ International Conference on Intelligent Robots and Systems, 2013, pp. 3576–3581.
[84] M. Runciman, J. Avery, M. Zhao, A. Darzi and G.P. Mylonas, Deployable, variable stiffness, cable driven robot for minimally invasive surgery, Front. Biomed. Eng. (in press).
[85] A.A. Kogkas, A. Darzi, G.P. Mylonas, Gaze-contingent perceptually enabled interactions in the operating theatre, Int. J. Comput. Assist. Radiol. Surg. 12 (2017) 1131–1140.

[86] O. Zisimopoulos, E. Flouty, M. Stacey, S. Muscroft, P. Giataganas, J. Nehme, et al., Can surgical simulation be used to train detection and classification of neural networks? Healthc. Technol. Lett. 4 (2017) 216–222.
[87] F. Ficuciello, G. Tamburrini, A. Arezzo, L. Villani, B. Siciliano, Autonomy in surgical robots and its meaningful human control, Paladyn 10 (2019) 30–43.

Further reading

[88] M.T. Moyer, R.S. Haluck, J. Gopal, E.M. Pauli, A. Mathew, Transgastric organ resection solely with the prototype R-scope and the self-approximating transluminal access technique, Gastrointest. Endosc. 72 (2010) 170–176.

PART 4

In vitro models

CHAPTER 4.1

In vitro models of cancer

Alexandra Boussommier-Calleja
Department of Mechanical Engineering, Massachusetts Institute of Technology, Cambridge, MA, United States

1 Introduction

In Latin, in vitro, means "in the glass." Hence, in vitro experiments are experiments done in, or rather on, the glass. This refers to the process by which cells are isolated from living tissues onto a dish to study them outside their normal environment. Cells are studied as they survive and proliferate in the laboratory; in other words, they are grown in vitro, a technique called cell culture. Originally, glass slides were used as a support for cell culture, explaining the use of the term "in vitro." Nowadays, there exist many different types of supports that are used for cell culture, not necessarily made of glass. The term in vitro is however widely used to denote more generally studies of cells performed outside of their normal habitat, in a laboratory setting. Today, in vitro research has become a field of its own in biomedical research.

While in vitro studies have advanced knowledge of all diseases, the development of in vitro research is particularly intertwined with that of cancer research. Burrows, one of the pioneers of cell culture, was convinced that "if one was to solve the formation of [..] the growth of cells in cancer, [then] one must find a means to study the functions of body cells in vitro" [1]. Among other things, that is because cancer is fundamentally characterized by abnormal and aggressive cell proliferation. It follows that this exacerbated proliferation has the potential to facilitate the survival, replication, and study of cells in nonphysiological laboratory settings. Cancer cells are immortalized cells that have evaded their normal finite life cycle: they can in theory grow indefinitely. This became of vital use for in vitro research, as it enabled the study of cells in the laboratory for prolonged periods of time. For example, some of the first cell lines used in vitro were isolated from cancer patients, such as the now famous HeLa cells, isolated in 1961 from Henrietta Lacks. Eventually, the understanding of the mechanisms underlying cancer cell immortality also shaped the development of other cell lines [2], now commonly used for in vitro research. Thus, progress in cancer research has helped the establishment of cell culture.

Conversely, in vitro models have brought invaluable insight to cancer research and promise to deliver much more. Indeed, in vitro models are rapidly evolving to meet the ever-growing demand of more sophisticated models to gain for a deeper understanding of the disease and discover new drugs. All eyes are now set on being able to recreate personalized "avatars" of patients in vitro to tailor treatment to the unique needs of each

Bioengineering Innovative Solutions for Cancer
https://doi.org/10.1016/B978-0-12-813886-1.00013-9

individual. These personalized models could be particularly powerful for cancer patients as we now grasp how unique and diverse each cancer and patient is. While we are not there yet, recent progress in the development of in vitro models and the field of stem cells suggests it is within the reach of the field, as discussed in this chapter.

More specifically, we review the progress made in the development of in vitro models in cancer research. We first briefly discuss their importance in the biomedical sciences. We define in vitro models and situate them in the broader landscape of medical and cancer research. Third, we discuss how in vitro research evolved from a 2D to a 3D in vivo like configuration. We expose the current formats of 3D cancer models and describe examples of in vitro studies that model individual steps of cancer development. Fourth, we analyze the emerging trends of in vitro modeling for cancer. Finally, we discuss the challenges associated with the use of in vitro models for cancer research, as well as progress made toward implementing in vitro personalized avatars.

2 Why in vitro models matter

Before delving into the details of in vitro models for cancer research, we will briefly explain why they have become crucial to research and why they deserve special attention from future biomedical scientists. These seemingly separate but related applications warrant the use of in vitro models today, namely, fundamental biological research, drug discovery, diagnostics, and regenerative medicine.

First, in vitro models are helping a great deal to advance fundamental research in understanding the biological mechanisms underlying normal and pathological processes. By culturing and studying cells "on dishes," scientists can design simple but powerful experiments that help them gain new insights into how cells behave in vivo. These experiments typically generate results that might have been otherwise hard to obtain in animal studies, let alone humans. In addition, in vitro studies can be cheaper and faster than studies performed in animals. Thus, scientists use in vitro studies to test numerous hypotheses about the biology underlying the development of diseases. The selected hypotheses that yield interesting results in vitro are subsequently tested in more involved experimental animal models.

Second, in vitro models are being used for drug discovery. This process generally starts with fundamental research, which generally helps to identify pharmacological targets involved in the onset of a disease. In vitro models can subsequently be used as tools for screening through libraries of drugs that can modulate the pharmacological target of interest. In parallel, in vitro models will be used to assess, study, and optimize specific parameters such as the dosage of the drug to predict its efficacy or toxicity in humans. The concept remains the same, scientists test numerous drugs on cells in vitro, to filter out more quickly those that will not work in mice or humans.

Third, in vitro models can be powerful tools to advance the field of diagnostics. Scientists have recreated simplified models of the disease in vitro to discover biomarkers that characterize the development the disease. These biomarkers can be powerful signs to look for in patients to achieve an early and specific diagnosis of the disease. In addition, these biomarkers can be used to assess the effects of the treatment: that is, their presence can be tracked as a proof that the disease is not fading away during the therapy, suggesting that other modalities of treatment should be chosen.

Fourth, in vitro models are becoming key in regenerative medicine, which aims at replacing, restoring, or regenerating a tissue. In the field of cancer, this has recently taken the form of engineering immune cells in vitro so that they can better fight cancer upon reimplantation into the patient. To evaluate the efficacy of the engineered immune cells prior to their introduction into the patient, scientists have tested them inside in vitro models of cancerous tissues [3].

While in vitro models have already advanced these fields, there is an important bottleneck today that warrants urgent additional progress, especially with regard to drug discovery. It is estimated that it currently takes up to 2.5 billion USD and 10–12 years to develop a drug [4]. These facts suggest that the current discovery process is not efficient and must be revisited. Generally, studies have consisted in using simplistic in vitro models as preliminary drug screening tests, prior to testing these drugs in animals, such as mice. Shockingly, only 1 in 10 drugs that were successful in mice are successful in human clinical trials [5]. This shows that, as of today, both in vitro and animal models are failing at predicting the human drug response. This gap in predictive preclinical in vitro models is one of the main challenges faced today by biomedical scientists.

There are several reasons for the mismatch between results observed in animals and humans: the most important reason is likely the inherent biologic differences between species [6]. Studies suggest that the bulk of the failure of drugs in humans is explained by a lack of efficacy of the drug or unforeseen toxicity effects [7]. Other factors could be at stake, such as flaws in the design of the methodology used in animal studies [5]. Yet, animal models are not the only ones to blame in this failure as they are generally accompanied by in vitro studies. This implies that current in vitro models are also failing to filter out successful drugs prior to animal testing. It is believed that current in vitro models fail to predict human response to drugs because they are too simplistic and lack physiological relevance. Current efforts are therefore focusing on improving in vitro models such that they become surrogates that are more reliable than current in vitro assays and animal models.

3 In vitro models in the biomedical research landscape

3.1 In vitro vs. in vivo vs. ex vivo

In vitro research is one of several types of biological research. In vitro is contrasted with in vivo, which means "inside the living" in latin. In vivo experiments consist of research done directly in living organisms. This encompasses research done on animals, as well as human clinical trials, and is typically considered to be the gold standard in biological research. For cancer research, numerous powerful mouse models have been established [8], and more recently, groups have been using zebrafish as an interesting model to visualize cancer development [9].

An alternative type of biological research exists in the form of ex vivo experiments. In latin, ex vivo means outside of the organism. Similar to in vitro research, the tissue of interest is not studied in its normal environment, inside the body. Rather, ex vivo research typically relies on the study of tissue explants or whole organ cultures isolated from the body. Ex vivo studies have for example consisted in imaging whole organs dissected from mice such as the lungs to visualize cancer cell extravasation through intact vessels [10]. These studies offer the advantage of studying cells in a microenvironment that is somewhat conserved. In contrast, in vitro studies are typically based on the use of cells dissociated from tissues, which have therefore lost the macroscopic structural integrity of the original tissue. Ex vivo studies allow for easier and higher resolution imaging compared to in vivo studies, particularly for visceral organs. They can however be limited in time due to tissue degradation. They also fail to replicate systemic signaling events that happen across different organs because tissues have been isolated from the organism [11].

The main advantages and disadvantages of each research format are summarized in Table 1. The most important advantage of in vivo studies is that they enable the study of cellular processes in a complete and physiologically relevant microenvironment. Equally important is the fact that the phenomenon of interest can be studied across different organs, at the scale of the entire body. This is particularly important for studying the effect of drugs, since drugs are metabolized by different organs. In addition, drugs might be affecting cells anywhere in the body other than those intentionally targeted; therefore, studying drugs in a living organism helps identify any potential side effects in any part of the body. On the other hand, in vivo studies can raise ethical issues, which are becoming increasingly recognized: in 2013 NIH decided to downscale facilities for research on nonhuman primates to minimize animal research. Finally, studies performed in vivo can exhibit species-related differences from human physiology. This is assumed to partly explain the important failure rate of drugs on humans that were previously shown to work in mice [6]. In vivo cancer studies have predominantly been performed in mice [8] and, more recently, zebrafish [9]. Both can be readily manipulated genetically and allow for long-term studies. There are numerous established murine metastasis models, but these studies can be time consuming and expensive. Zebrafish allow for improved imaging, have shorter generation times, and a large number of offsprings. However, evolutionarily, they are not as similar to humans as mice and lack some organs present in humans, e.g., mammary tissue, which limits the options for orthotropic transplantation. For cancer research, studies done in animals pose one important problem: if one wants to study human cancer cells, these will be interacting with a microenvironment made of animal cells, introducing potential interspecies discrepancies. This is crucial as it is now known that cancer cells heavily interact with their microenvironment [12]; thus,

Table 1 Pros and cons of in vitro, ex vivo, and in vivo models.

	In vivo	In vitro	Ex vivo
Pros	1. Complete and truly physiologically relevant models	1. All human cells	1. True and complex local cellular and extracellular microenvironment
	2. Interconnectivity of organs: phenomenon observed at the scale of entire organism	2. Control of cellular, biochemical, and biophysical content	2. Conserved local microarchitecture
	3. Long term studies	3. High throughput	3. Easier for imaging than in vivo
	4. Numerous well-established mouse models for cancer	4. High resolution imaging	
Cons	1. Ethical issues	1. Isolated system: no systemic analysis	1. Isolated from rest or organism
	2. Expensive	2. Not long term	2. Limited in time
	3. Low throughput	3. Limited physiological relevance, incomplete microenvironment	3. Requires access to fresh living tissue
	4. Nonhuman microenvironment	4. Use of genetically homogeneous cell lines	
	5. Imaging is hard and expensive		
	6. Limited control of microenvironment		

it is of paramount importance to replicate the human microenvironment as faithfully as possible. Recent efforts have moved into humanized mice to make part of the cellular microenvironment inside mouse models more human-like [13]. Briefly, this implies using immune-deficient mice (i.e., that lack part of their immune system) so that one can introduce human cells into the mouse and minimize their rejection. The human cells to be introduced into the mouse have typically been either cancerous human cells or immune human cells, to study their interactions as happens in patients. Alternatively, one can modify genetically mouse embryos with human DNA using CRISPR technology to obtain human-like cells inside mice. This however still fails to represent all human cellular microenvironment, which normally surround cancer cells in patients, and which is now known to play a role in cancer development.

The main advantage of in vitro studies is that they offer a higher control of the cellular and biochemical content of the assay. The user controls exactly what is placed inside the assay, and this modularity is unique for studying the effects of individual variables. In contrast, in vivo models are subject to numerous macroenvironmental factors that are hard to account for; there is also a myriad of microenvironmental factors at play at the cellular level that is similarly hard to control, let alone be aware of. In addition, in vitro assays facilitate the visualization of processes that would be otherwise hard to stumble upon or image in vivo. While in vivo imaging has enabled the visualization of cancer progression in animals [14], it requires specialized and expensive equipment. In contrast, imaging cells in vitro is easier, and typically allows for high-resolution imaging because it offers a short distance between the cells of interest and the objective. Importantly, in vitro studies can rely on the use of cells that are exclusively human: this means one can study human cancer cells as they interact with an all-human microenvironment, in contrast with animal studies. While in vitro models also present some limitations listed in Fig. 1, many have the potential of being overcome, unlike those associated with animal models that are inherent to working with living nonhuman organisms.

All in all, in vitro systems offer a unique control of the biochemical and cellular content of the assay, but still fall short of the physiological relevance provided by in vivo studies. The end goal is for the physiological relevance of in vitro models to increase, so that they can reduce if not replace the use of animals in biomedical sciences.

3.2 The market of in vitro modeling

In vitro research is performed both in academic and pharmaceutical laboratories. Typically, the basic science is initiated in academia, while pharmaceutical companies are the ones who often link scientific findings with medicinal and process chemistry, pharmacokinetics, and safety sciences. It is estimated that between 67% and 97% of drug development is performed by private companies [4]. Since much of the basic science has been gradually developed in academic labs over decades, cutting-edge in vitro models often come out of academic labs, which either spin-off into companies or form collaborations with pharmaceutical companies for a combined effort to accelerate drug discovery.

In vitro research is used in all biomedical fields—it is hard to have an overall estimate of the share of in vitro research dedicated to cancer, but Market Allied Research has reported that in 2014, 32.4% of the 3D cell culture market was dedicated solely to cancer research [15]. Since much of the field is moving toward 3D cell culture, this can be considered as a good estimate of how much of in vitro research is dedicated to studying cancer. The market of 3D cell culture for cancer research is estimated to reach 1.4$ billion by 2022, with a compound annual growth rate of 28.4%. These numbers support the idea that there is much enthusiasm for in vitro models and what they can accomplish for cancer research.

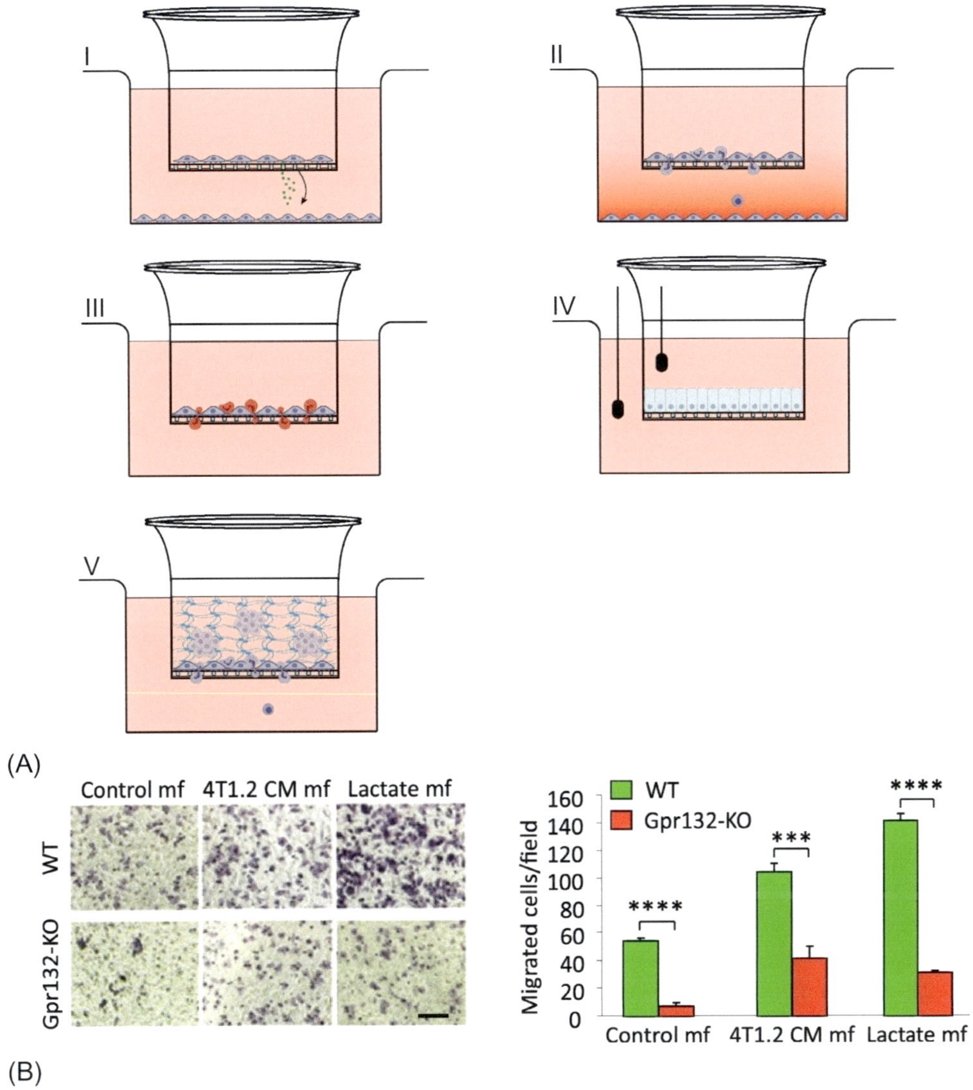

Fig. 1 (A) Schematic of the Boyden chamber and its different applications. I. Different cell types can be placed in each compartment to isolate physical vs. paracrine interactions. II. The invasion of cells through the filter membrane can be studied with or without a chemotactic gradient. III. The invasion of cells can also be studied through a monolayer of cells formed on top of the filter membrane. IV. The transepithelial electrical resistance of a cellular monolayer formed on top of the filter membrane can be measured using electrodes. V. Cells can be placed inside a 3D hydrogel placed on top of the filter insert to mimic 3D like invasion. (B) Example of a study that used the Boyden Chamber to quantify the number of tumor cells that crossed through the porous membrane during coculture with macrophages. 4T1.2 cells were plated on the upper chamber inserts with macrophages (wild type of Gpr132-KO) that were either left untreated (control), activated with media conditioned by 4T1.2 tumor cell or activated with lactate-activated spleen-derived macrophages. After migration for 6 the migrated cells were stained with *crystal violet* and counted as cells per field of view under the microscope ($n = 3$–4). $^{***}P < 0.005$, $^{****}P < 0.001$. Scale bars, 500 μm. *(Reproduced from P. Chen, et al., Gpr132 sensing of lactate mediates tumor–macrophage interplay to promote breast cancer metastasis, Proc. Natl. Acad. Sci. U. S. A. 114 (2017) 580–585 with permission from PNAS.)*

3.3 What can be studied in vitro?

A myriad of in vitro studies exists, each with their own set of experimental design. However, often, in vitro research relies on the analysis of key metrics found repeatedly in most studies, some of which are discussed here. This is not an exhaustive list of metrics extracted from in vitro assays, but rather a sample of those most commonly found in cancer in vitro research.

First, because in vitro studies enable scientists to visualize cells at high resolution, they tend to generate precious phenotypic data, extracted from images of cells. One can investigate their shape, which can be very informative about their health or migration mode. Second, the migratory properties of cells are often investigated in vitro. Scientists image cells in vitro over time to determine how fast they move or to identify what mode of migration they use. This is of paramount importance to cancer research, as migration is a key component of cancer development and tumor metastasis. Third, in vitro assays can reveal information about the genes and proteins expressed in cells. This can be done either in situ inside the in vitro assay, using immunostaining or by postprocessing the cells with conventional biochemical assays, such as PCR or Western blots. Immunostaining consists in staining your assay with a biomarker specific to a protein or gene. This biomarker is generally fluorescent and can therefore be visualized inside your in vitro assay to test whether it is present in the cells. This technique provides the advantage of being able to correlate the presence of the biomarker with its localization in the assay. In other words, one can see where in your assay the gene or protein of interest is found. In contrast, if cells are taken out of the assay for further biochemical testing this spatial information is lost. At any rate, this enables scientists to correlate genotypic and phenotypic information, which can be very powerful to unravel the role of specific genes and proteins in the development of diseases.

Another important parameter often measured in vitro is the permeability of endothelial of epithelial barriers. Such experiments generally consist in introducing a dye of known size and imaging its diffusion across the cellular barrier. Scientists can extract the permeability of the barrier by measuring the quantity of the dye that has managed to cross the barrier, and its timing. Other techniques such as transepithelial electrical resistance (TEER) also exist, as discussed in Section 4.1. This is of utmost importance in cancer research, as many key processes of the disease require the passage of cells across barriers. In fact, the endothelium is often found to be leakier in cancerous tissues to facilitate the progression of the disease [16]. Normal and pathological permeability is therefore a key parameter to mimic and measure in vitro.

4 The evolution from 2D to 3D in vitro cancer models

4.1 In vitro models of cancer in 2D

The first cell culture experiments are dated back to the early 1900s, when scientists first thought of isolating cells from tissues, placing them on dishes, and observing them as they evolved and even multiplicated [1]. The first experiments reported in the literature consisted in observing nerve cells elongate and survive for a few weeks on a coverslip or visualizing the growth of chicken embryos tissue extracts. Originally, these scientists imagined cell or tissue culture as a step for organ transplantation. While this is still an application of active interest, cell culture ended up opening many other avenues for biomedical research. Ironically, Thomas Burrows, one of the pioneers of cell culture, wrote in 1942 that "tissue culture has not changed biology as it should have done" [1]. It would only take a few more years for cell culture to become key in many areas of biological research, such as virus isolation, antibody production, in vitro fertilization, and fundamental research.

We are now a long way from these first in vitro experiments, characterized by the culture of cells in two dimensions (2D), composed of only one cell type. Typically, cancer cells would be grown as 2D monolayers inside wells and studied following exposure to a drug or genetic manipulation. These simple but powerful experiments have enabled scientists to dissect many important pathways in cancer development and are still being used to this day. Here, we do not provide an exhaustive list of all types of 2D in vitro assays; rather, we describe only a few formats of 2D experiments, which are commonly used for cancer research.

The Boyden Chamber assays has become one of the most common 2D in vitro assay in cancer research, as it can be used to study several facets of cellular migration, invasion, and interactions (Fig. 1A). It consists of placing a filter made of a porous membrane inside of a well. The filter sits inside the well, effectively separating the well into two compartments. A major reason to use the Boyden Chamber is to coculture different types of cells and study their paracrine interactions. In this case, one subpopulation of cells is cultured at the bottom of the well while the other is placed inside the filter, on top of the membrane. This allows both cell subpopulations to bathe in the same media, thus being able to interact via the secretion of factors, without interacting physically. This powerful experimental design can help test whether the effect of interest resulting from the interactions of cells depends on their physical interactions or paracrine signaling.

Second, the Boyden Chamber has often been used to study cancer cell migration through the membrane pores [17] of the filter. In this case, the pore size must be large enough to let cells pass through; these pores are meant to mimic small gaps that cells have to navigate inside tissues, such as intercellular junctions, or the extracellular matrix. For example, Chen et al. have used a Boyden chamber to study the effect of macrophages placed at the bottom of the well on the invasion of tumor cells placed on top of the transwell (Fig. 1B) [18]. Briefly, they show that macrophages previously activated with lactate increase the number of cancer cells that cross the transwell filter compared to samples with nonactivated macrophages. This simple but powerful setup allowed them to identify the G protein receptor 132 in macrophages to be at stake, which was subsequently confirmed in mice.

Third, scientists have used the Boyden Chamber to study the effect of chemical gradients on cells, by introducing the chemical of interest in only one of the compartments and letting it diffuse, thus creating a transient gradient across the filter. This can be very interesting to study gradients in vitro, an important phenomenon to mimic as gradients often act as cues to guide cells. For example, chemokines gradients (or chemotaxis) are known to attract cancer cells to specific tissues.

While the Boyden chamber has been a powerful tool in cancer research, it does present limitations: first, it is not easy to obtain a precise count of cells to estimate how many have crossed the filter because it is hard to be sure that all cells that did cross were collected. In addition, cells can generally not be imaged during their migration as the filter tends to be far from the objective. Not being able to image cells during migration means that important phenotypic information is being missed out on. In addition, the chemotactic gradients tend to only be transient as they eventually equilibrate over time. Finally, cells must cross a nonphysiologic polycarbonate or polypropylene filter, and it is often hard to achieve statistically significant results if only a limited number of cells were collected [19].

Using the Boyden chamber, scientists have also performed transepithelial electrical resistance (TEER). This assay is used to measure the permeability of a 2D monolayer of cells formed on top of the filter. This is done by placing an electrode in the upper compartment and another on in the lower compartment, thus separated by the cellular barrier. A voltage is applied to the electrodes, and the resulting current is measured; the ratio of the voltage and the current indicates the ohmic resistance of the cellular barrier [20]. This can help measure the permeability of cellular barriers, or the invasiveness of cancer cells through the cellular barrier [21].

Note that the Boyden chamber can be used to mimic 3D-like invasion, by placing cells inside or on top of a 3D layer of ECM, itself placed on top of the filter insert. This enables cells to invade through a 3D scaffold prior to crossing the filter membrane. While this type of assay does enable 3D-like invasion through the scaffold, it still suffers from the limitations previously discussed inherent to the Boyden chamber.

Another typical 2D experiment is the scratch assay. This experimental design consists in growing a tight monolayer of cells in a well. A wound is created across the monolayer by scratching the monolayer and cells are imaged as they invade the void created by this wound to study their migratory properties. This assay has the advantage of being technically easy to prepare and image; it therefore lends itself to high-throughput analysis. Scientists have used it in 96-well plates to screen over 1000 siRNA compounds and their effect on breast cancer cell migration [22]. However, there are questions as to the reproducibility of these experiments, as methods to manually perform the scratches vary between labs and can even vary within the same experiment between wells. Recently, more sophisticated methods have been developed to automate the scratch process, which however requires specialized equipment. It is also hard to guarantee that the confluency of cells is the same between wells. Finally, creating a wound kills cells, which can release factors that affect the overall experiment [19].

4.2 The jump from 2D to 3D models

One of the turning points in the field came when cells in 3D were shown to respond to drugs differently than cells in 2D; perhaps more importantly was the finding that drugs tested on cells in 3D were giving results closer to those observed in vivo [23]. In as early as 1984, Erlichman and Vodgen compared the cytotoxic activity of adriamycin in the human bladder carcinoma cell line MGH-U1 grown as monolayer cultures, spheroids, or xenografts in immune-deprived mice [24]. They showed that tumor cells near the surface of the spheroids exhibited the same cytotoxic response as tumor cells grown as 2D monolayers, while tumor cells inside the spheroids were more resistant, as observed in vivo, presumably because of poorer infiltration of adriamycin inside the spheroid. These pioneer studies strongly suggested that spheroids and more generally 3D cell culture are more appropriate models for drug screening than 2D monolayers of cells.

There are many factors at play that can explain why cells respond to drugs differently in 2D vs. 3D, which motivates researchers to move toward 3D cell culture. Culturing cells in 3D affects them in numerous ways that we are still trying to fully understand, summarized in Fig. 2 and detailed here.

First, culturing cells in 3D allows the cells to adopt a morphology similar to that found in vivo. Recapitulating the normal morphology of cells is important as their shape can directly affect their biological activity [25]. In 2D, cells are lying flat and often adopt a cobblestone morphology, which forces them to adopt an apical-basal polarity. While this polarity is relevant for epithelial cells, it is not appropriate for mesenchymal cells that tend to adopt a stellate morphology or polarize from rear to back during migration [26]. Forcing the cells to adopt an unnatural shape is problematic as cell shape directly affects certain processes: for example, flattening cells has been shown to affect signaling [27], and apical-basal polarity can affect apoptosis [28]. This 2D phenotype can be far from representative of their 3D configuration, where cells are much more voluminous. Studies have

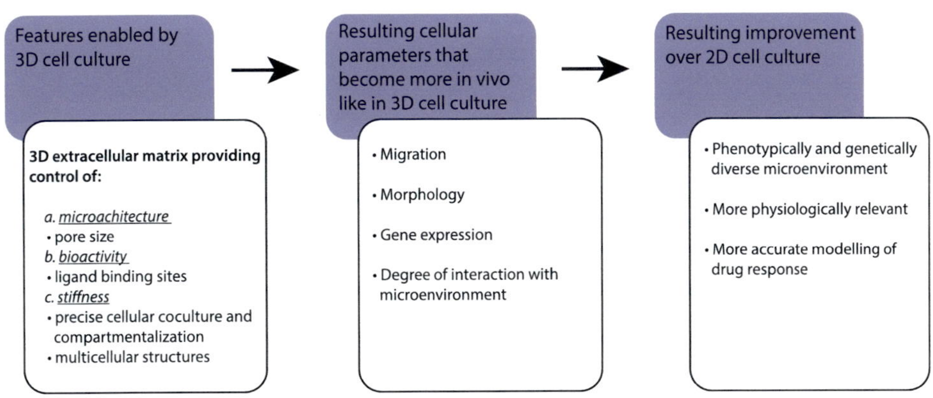

Fig. 2 Features enabled by 3D cell culture and resulting improvements over 2D cell culture.

shown that cancer cells displayed in vivo-like histological features when cultured in 3D that could not be detected in 2D culture specimens (Fig. 3) [29]. In their study, Lee et al. show that cells cultured in 2D often remain isolated, whereas in 3D they form multicellular clusters that resemble more structures found in vivo.

Second, cells in 3D assays resort to drastically different modes of migrations compared to 2D situations; these 3D migration modes resemble those found in vivo more closely than those observed in 2D. For cancer research, it is particularly fundamental to replicate migration modes faithfully in vitro, as cell migration is a central process both in primary

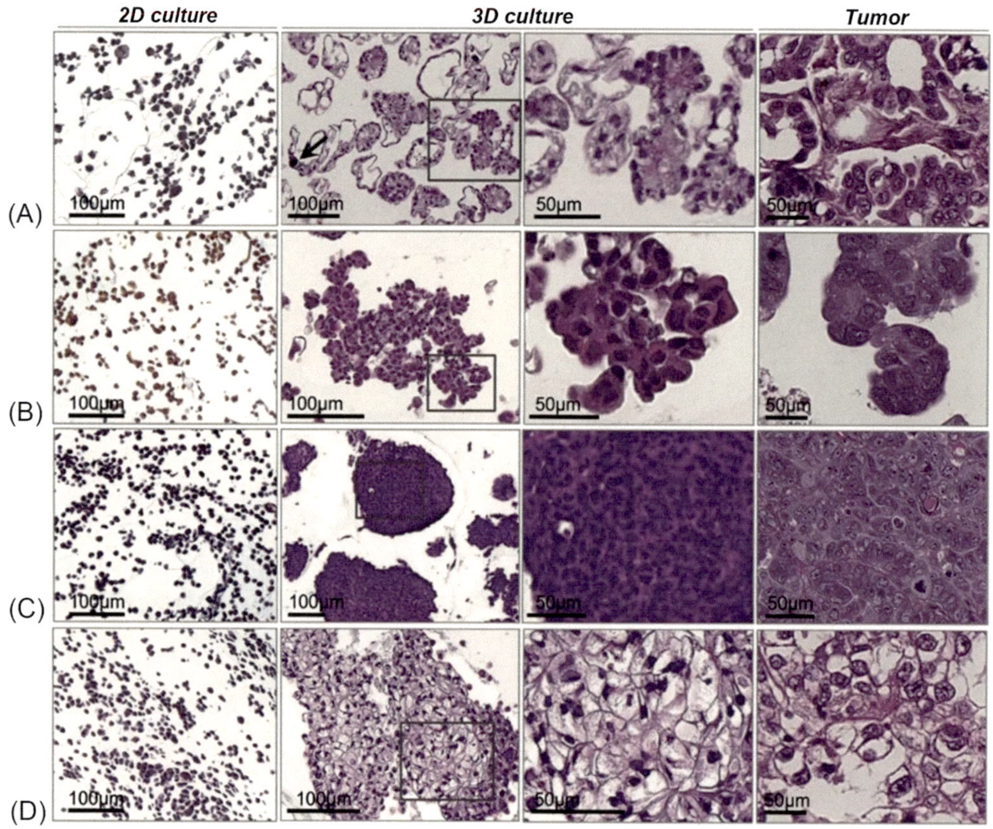

Fig. 3 Histological comparison of epithelial ovarian cancer cells lines cultured in 2D (first column, "2D culture"), 3D (second and third columns, "3D culture"), or taken from human ovarian carcinoma specimens representative of each tumor histological subtype (fourth column, "Tumor") [29]. (A) OAW42, (B) UWB1.289+BRCA1, (C) HEY.A8, and (D) SKOV3.ip cell lines. Each row corresponds to a different type of cell line supposed to represent a given subtype of ovarian cancer. Cells were stained histologically with hematoxylin and eosin, which stains the nucleus and ribosomes in *purple* and the proteins in the cytoplasm in *pink*, and imaged bright-field microscopy. *(Reproduced with permission from J.M. Lee, et al., A three-dimensional microenvironment alters protein expression and chemosensitivity of epithelial ovarian cancer cells in vitro, Lab. Investig. 93 (2013) 528–542.)*

and metastatic cancer. 2D cell migration has taught us immensely but is inherently limited to one plane; in contrast, within 3D in vitro assays, cells can evolve in all directions, as they can do in vivo. In addition, cells in 2D have been shown to use the mesenchymal mode of migration based on the formation of lamellipodial protrusions [30]. In reality, this type of migration is only one of others that cells can adopt, all of which can be replicated in 3D cell culture. For example, amoeboid migration is another important type of migration that cancer cells can adopt in 3D [30]. Cancer cells are known to switch between those two modes, which has been shown to contribute to more widespread invasion in vivo [31]. Each mode is important for migrating through different microenvironments, and therefore needs to be recapitulated in vitro to study them and target them differentially [32].

Third, placing cells in 3D effectively increases their degree of interaction with their immediate surroundings, as they do in vivo. Cells cultured in 3D interact more heavily through their entire surface area with their microenvironment (whether it is with a scaffold, a hydrogel, or other cells). The surface of cells is lined with a myriad of receptors that sense signals from the extracellular environment and transduces them into intracellular signals to respond and adapt to their microenvironment. Cellular interactions with their ECM and microenvironment through these receptors are essential to many cellular functions and steps in cancer [12,33,34]. During 2D cell culture only half of the cell's membrane is exposed to media, and cells only interact physically with each other through their junctions. In contrast, cells in 3D cell culture interact through their entire surface area either with ECM or other cells, as they do in vivo, and are bathed in media (Fig. 4).

Fourth, cells in 3D have a gene expression profile that is different to that found in 2D [29]. It is generally thought that this effect observed in 3D cell culture is mediated through ECM interactions that trigger unique intracellular signals [35], or via signaling with other cells found in their 3D environment, such as stromal cells. Differences in gene expression can naturally have an important impact on cellular physiology but also on the response of cells to drugs. For example, cells in 3D have been shown to upregulate genes that increase drug resistance [36], compared to cells cultured in 2D.

Fifth, cells in 3D make use of a 3D support system that can be engineered to replicate in vivo conditions. This 3D environment provides physical and biochemical anchorage. Both the stiffness, pore size, or ligands that make up the extracellular matrix can be finetuned in 3D culture to resemble the tissue of interest. Because these parameters can vary between organs, having the ability to finetune them provides additional flexibility to study cells in more specific microenvironments in vitro. When cells are grown in 2D, they often lie on glass or plastic substrates that have a nonphysiologic and high stiffness. While cells can be placed in 2D on top of biomaterials with varying stiffness, this still only exposes part of the cellular membrane to a physiologically relevant stiffness. Stiffness can modulate many crucial cellular processes such as motility of tumor cells [37] and should therefore be close to that found in vivo. The 3D support can also be controlled to have

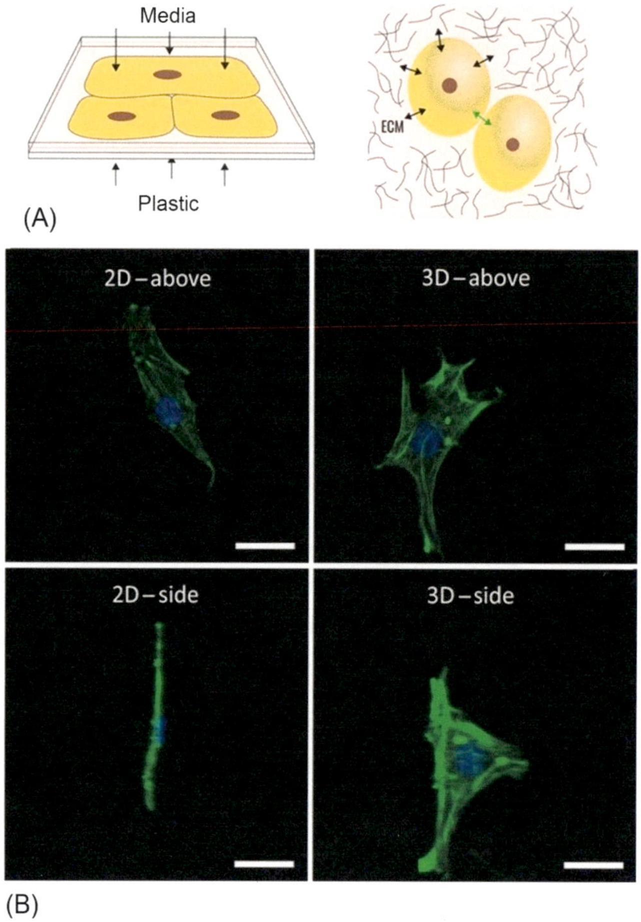

Fig. 4 (A) Schematic of cells cultured in 2D and in a 3D extracellular matrix. In 2D, only half of the cell's surface is exposed to media *(pink)*, while the other half is exposed to a high stiffness, and cells only interact minimally at their junctions. In contrast, in 3D, cells are more spherical interact through their entire surface area wither with other cells *(green arrow)* or the extracellular matrix *(black arrows)* and bathe in media. (B) Confocal images of a single fibroblast grown in 2D or 3D culture. The cell has been stained with phalloidin to visualize the primary structural elements of the F-actin cytoskeleton and 4′,6-diamidino-2-phenylindole (DAPI) for the nucleus. The images show the shape of a typical cell when visualized from above *(top panels)* or from the side *(bottom panels)*. Note how thin a cell can become when cultured on a flat substrate as in conventional 2D culture *(left)* compared with the more normal structure of a cell in a 3D culture model *(right)*. Scale bars: 10 μm. *((A) Drawing by Michel Boussommier-Calleja, reproduced from A. Boussommier-Calleja, 3D cell culture: market and industrial needs, an Elveflow Tutorial, 2018 website with permission. (B) Images courtesy of Dr. F. Tholozan, Durham University. Reproduced from E. Knight, S. Przyborski, Advances in 3D cell culture technologies enabling tissue-like structures to be created in vitro, J. Anat. 227 (2015) 746–756 with permission granted through the creative commons license (https://creativecommons.org/licenses/by/3.0/legalcode).)*

the desired pore size through which cells navigate, an important feature that can also affect cell motility [38]. This intricate tunable microarchitecture is clearly not present in 2D assays, where cells are placed on a flat glass or plastic. Finally, the 3D support system can be made of proteins that offer ligand sites to which cell receptors bind to achieve migration and other cellular processes. Thus, 3D in vitro assays are also more physiologically relevant because they let cells use these sites around them to evolve as they do in vivo in three dimensions.

On the other hand, 3D cell culture allows the formation and maintenance of more complex multicellular structures, such as spheroids, organoids or a microvasculature. Inherently, 2D cell culture was limited to structures that remained in direct contact with the dish, thus lacking any depth or thickness. The clearest example of a multicellular structure is the spheroid, defined as a spherical clump of cancer cells. Multicellular structures such as the spheroid force cells inside the clump to interact heavily with each other. Another example is the microvasculature: groups have now formed and cultured a 3D vasculature in vitro inside hydrogels or in microfluidic chips [39,40], that can recreate blood flow and other processes that are key to cancer development [41]. This would not be possible without a 3D support, although spheroids on the other hand can be grown without a 3D support. Groups had studied angiogenesis by culturing endothelial cells which formed networks in 2D, but these 2D vascularized patterns were not perfusable as they did not have an open 3D lumen.

Finally, culturing cells in 3D significantly enhances the possibilities for coculturing different cell types. Mixing different cell types in 2D is possible but limited to one plane. In contrast, when cells are cultured in 3D, they tend to be freer to interact in all three dimensions as they would in tissues, migrating toward each other through the 3D support at physiological distances. Progress has also seen organs on chips develop whereby different cells are organized in 3D by compartmentalizing them at physiological distances as they would be in organs. The ability to coculture different cell types is very powerful, especially for cancer research, as we now know that tumor cells make constant use of other specific cell types in their surroundings to progress in their endeavor. In addition, it has been shown that the cellular microenvironment (such as immune cells) can affect the tumor cells response to drugs [42]. Hence, replicating the cellular microenvironment in 3D is crucial to predicting drug efficiency.

Another important feature of 3D cell culture is that it can recapitulate physical barriers that 2D cell culture cannot mimic as faithfully. For example, drugs will not penetrate cells in 3D as easily as those cultured in 2D, thus partly explaining discrepancies in drug responses between 2D and 3D cell cultures [43]. More specifically, cells in 2D are directly in contact with the drug, while a drug tested on cells in 3D has to surpass barriers like an ECM, or other cells before reaching the target. Thus, cells in 3D might not all encounter the drug in same quantity or timing, depending on their location in the

3-dimensional assay, whereas cells in 2D are all exposed to the same amount of drug. This physical barrier to drug diffusion exists in the body and must therefore be recapitulated in vitro. Recapitulating the 3D structure of tumor is also important to mimic the response to treatment in other ways: for example, studies have also shown that spheroids subjected to radiotherapy or chemotherapy sometimes undergo regrowth kinetics [23]. This consists in spheroids initially shrinking because the well-oxygenated cells on the edge of the spheroid die in response to the treatment. Yet this causes the underlying quiescent cells to become situated at the edge of the spheroid, and thus switching to a proliferative state and resume growth. These cell kinetics could presumably not be recapitulated in 2D cell culture composed of only one layer of cells, but have been recreated in 3D spheroids [44].

Overall, the ability to finetune these numerous parameters in 3D cell culture promotes cellular heterogeneity, as found in vivo. In contrast, cells grown in 2D tend to be homogeneous, as they are all exposed similarly to nutrients, oxygen, or physical factors. A clear example of heterogeneity is found in 3D spheroids. 3D tumor spheroids have cells in the center that form a necrotic core because they do not have as much oxygen or nutrients; as a result of this gradient, cells within the spheroid have varying proliferation activity [45] or increased resistance to selected treatments [23]. This better replicates the widely different characteristics of tumor cells found in patients in vivo [46]. Importantly, the heterogeneity that is better reproduced in 3D cell culture might also better mimic the heterogeneity of clinical tumor response [23]. While 2D assays often fail to recapitulate as much cellular heterogeneity, they can still be used to study selected aspects of subpopulations of cells: for example, monolayer cultures have been adapted to model populations of quiescent tumor cells [47], as those found inside tumor spheroids. However, monolayers do not pose the barrier to drug penetration, nor provide many of the microenvironmental influences found in solid tumors and 3D cultures [48]. Thus, 3D multicellular heterogeneous structures such as spheroids ensure that several aspects are recapitulated at once in the same assay, thus resembling more in vivo tumors.

4.3 2D dimensions still matter!

While we have discussed numerous reasons for using 3D cell culture to recreate more faithful in vitro models, there is no reason to systematically discard 2D assays. A lot has been learned from them, as discussed previously, and they are often a good stepping stone to start experiments. Moving toward 3D in vitro systems should be a gradual process and is often conditional on having access to additional equipment. One major downside to switching to 3D cell culture is financial: it is more expensive to do 3D cell culture, as it can be more time consuming and can require additional reagents and plastic ware. It is also more complicated: protocols can be tricky, but more importantly, these assays can be harder to image as one has to image through a scaffold, and cells can be found in the scaffold too far away from the objective. Imaging 3D assays generally requires the use of 3D microscopy tools, which can be expensive and time consuming.

On the other hand, chances are that 2D cell culture will remain crucial in the foreseeable future. For example, cells are generally grown in 2D for their expansion, characterization, and quality control, as they tend to proliferate faster and are easier to handle in 2D compared to 3D. In other words, 2D cell culture will remain an important part of in vitro experiments and can still be used as a powerful type of reductionist assay. On the other hand, even though cells are found in 3D tissues in vivo, the behavior of interest is sometimes governed at a smaller 2D or even 1D scale [49]. For example, cells often migrate along microfibers; in these cases, it can be interesting to study cells in vitro as they migrate in 1D along microfibers, with no need to use 3D scaffolds [50]. Such reductionistic migration assays can provide a clearer understanding of the roles of each dimensions in migration, and dissect the role of dimensionality in the variable being studied [49].

5 Existing 3D models of primary and metastatic cancer

5.1 Formats of 3D cancer in vitro models

There are numerous formats of 3D in vitro models, many of them used for cancer research. These formats all allow cancer cells to evolve in 3D but differ in their configuration or in the technology they rely on. Here we discuss some of the formats that are the most used as of today: spheroids, scaffolds, microfluidic devices, and bioreactors. Note that they can be combined, for example, scaffolds are often placed inside microfluidic devices or spheroids can be placed inside scaffolds.

5.1.1 Spheroids

Spheroids are self-organized clusters of cells that can be made with different types of cells, among others, tumor cells. Tumor spheroids recapitulate important aspects of the tumor, which is itself a cluster of tumor cells. The aggregation of cells into spheroids can be promoted through different well-established protocols that minimize the adherence of cells to the well. Tumor spheroids can be formed either by using nonadherent surfaces, suspending cells in a drop of fluid that hangs from an upside down well plate, or by agitating the cells with spinner flasks or rotary systems [51]. Not all tumor cell lines form spheroids, but many do and have been characterized in the literature. Tumor cells inside spheroids secrete their own extracellular matrix. Because of reduced diffusion of nutrients and oxygen, the core of the spheroids has been shown to become hypoxic and necrotic, as found in large tumors in vivo [51]. This also results in areas of varying proliferation rate, with cells found at the center of the spheroid becoming more quiescent that those at the surface (Fig. 5A).

Spheroids have become a predominant tool for high-throughput drug screening in cancer research. One current limitation of spheroids grown in vitro is their lack

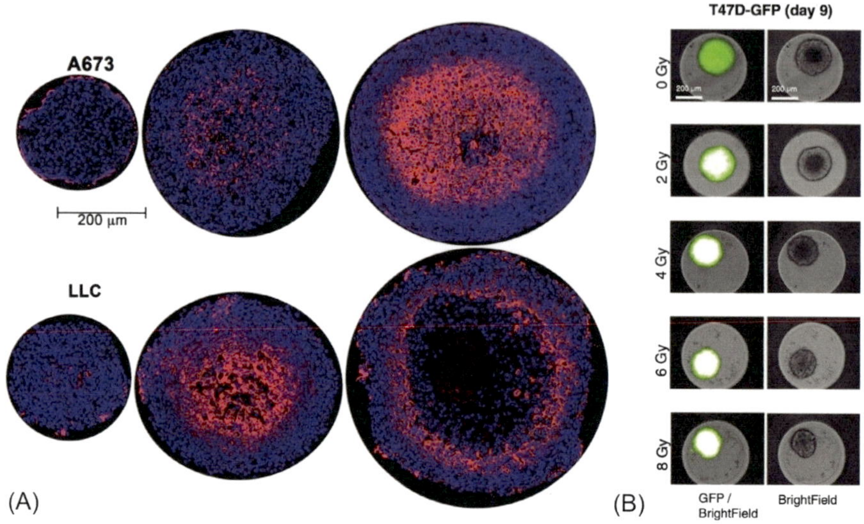

Fig. 5 Spheroids in vitro. (A) Immunostaining of A673 and LLC spheroid cryosections for the hypoxia marker pimonidazole *(red)* shows hypoxia development at different sizes in the central core of A673 and LLC spheroids. Spheroids sections counterstained with Hoechst. (B) Fluorescent *(left column)* and bright-field *(right column)* images of T47D spheroids, 9 days after radiation therapy with varying intensity of radiation (0–8Gy). Increasing radiation cause spheroids to shrink. *((A) Reproduced from S. Riffle, R.N. Pandey, M. Albert, R.S. Hegde, Linking hypoxia, DNA damage and proliferation in multicellular tumor spheroids, BMC Cancer 17 (2017) 1–12 with permission granted through the creative commons license (https://creativecommons.org/licenses/by/3.0/legalcode). (B) Reproduced from N. Anastasov, et al., A 3D-microtissue-based phenotypic screening of radiation resistant tumor cells with synchronized chemotherapeutic treatment, BMC Cancer 15 (2015) 466 with permission granted through the creative commons license (https://creativecommons.org/licenses/by/3.0/legalcode).)*

of vascularization. Thus, spheroids are not necessarily relevant for studying vascularized tumors, and are rather used to mimic avascular primary tumors, some micrometastases, or avascular microregions of tumor [51]. Note that spheroids are established models of primary tumors, but they do not mimic all types of metastatic tumor formation: for example, some metastatic tumors do not initially consist in spherical clusters, but rather, in tumor cells co-opting blood vessels and proliferating along the walls of the blood vessels [52]. Spheroids have been cocultured with other cell types, typically stromal cells such as immune cells [53] or fibroblasts [54], to recreate a more faithful and diverse microenvironment. For example, culturing tumor spheroids with fibroblasts was shown to promote epithelial-to-mesenchymal transition of tumor cells at the edge of the tumor mass [54]. Spheroids in cancer research have also been used to study cancer stem cells (CSCs), as CSCs have been shown to grow as nonadherent 3D cluster in vitro and maintain their differentiation state inside spheroids. These cells have particular self-renewal properties thought to be key in promoting tumor and metastasis initiation [55].

Spheroids present several advantages that have transformed them into one of the most important 3D cell culture models for drug discovery. They are relatively easy to form, use and suitable for high-throughput drug screening [56], as individual spheroids can be formed and studied independently inside plates that contain numerous wells and thus, can test numerous drugs. Fluorescence imaging is particularly powerful to study spheroids composed of fluorescent tumor cells, as fluorescent images can provide measurements of the size of the spheroids (Fig. 5B), their proliferation, or drug response. High content imaging platforms have now been developed to automatically extract such metrics from well plates for increased throughput [57].

5.1.2 Scaffolds

Scaffolds are structures used in 3D in vitro assays to provide biochemical and mechanical anchorage to cells in three dimensions [58]. They can be made from natural or synthetic material. Natural scaffolds generally are proteins that make up the extracellular matrix in tissues, such as collagen, fibrin, or hyaluronic acid, or other naturally occurring biomaterials such as gelatin. These biomaterials offer adhesion sites that can differ between proteins. Therefore, it is important to select a scaffold with the appropriate ligands to mimic the tissue of interest, as organs can be composed of different ECMs. Different ECM components can also favor different processes: for example, it is known that fibrin promotes angiogenesis, and is therefore often the preferred scaffold used for growing vasculature in vitro [59]. The obvious advantage of naturally derived biomaterials is that they are found in vivo and therefore are physiologically relevant (Fig. 6B). However, naturally derived biomaterials can

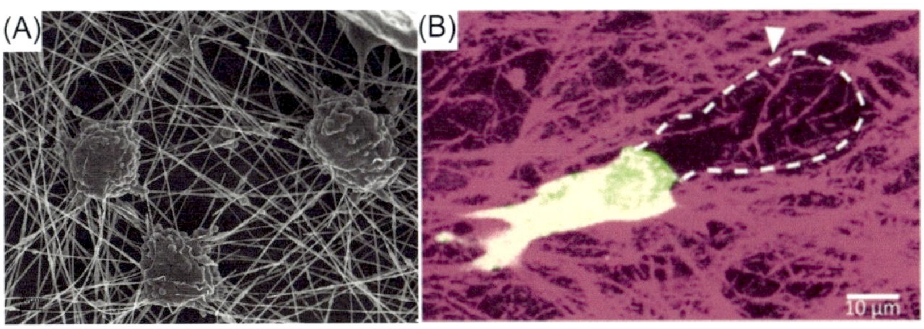

Fig. 6 (A) Scanning electron microscopy image of HeLa cells on the nanostructured grid fibers (fabricated using electro-spinning method) after 48 h of incubation. Original magnification ×100. (B) Representative confocal image showing MDA231 cells *(green)* migrating through dense collagen I ECM *(magenta)* by degrading the matrix, leaving behind a microtrack *(arrow)*. *((A) Reproduced from M. Wolun-cholewa, et al., An efficient 3D cell culture method on biomimetic nanostructured grids, PLoS One. 8 (2013) 4–9 with permission granted through the creative commons license (https://creativecommons.org/licenses/by/3.0/legalcode). (B) R. Li, J.D. Hebert, T.A. Lee, H. Xing, A. Boussommier-Calleja, R.O. Hynes, D.A. Lauffenburger, R.D. Kamm, Macrophage-secreted TNFα and TGFβ1 influence migration speed and persistence of cancer cells in 3D tissue culture via independent pathways, Cancer Res. 77(2) (2017) 279–290. https://doi.org/10.1158/0008-5472.CAN-16-0442.)*

yield highly variable results because they are extracted each time freshly from a different animal. Besides, their exact constituents are not always clearly defined. This is the case for Matrigel, a popular hydrogel composed of a mixture of ECM proteins such as laminin, collagen, secreted by cultured Engelbreth-Holm-Swarm mouse sarcoma cells. Finally, naturally derived scaffolds can be prone to biodegradability. In studies where scaffolds are implanted in the body, biodegradability is powerful and needed. However, for in vitro studies, this biodegradability can limit the duration of the study and cause unwanted and uncontrolled effects due to by-products of the scaffold degradation.

In contrast, scaffolds made from synthetic biomaterials are made of polymers, ceramic materials such as bioactive glasses, or self-assembled peptides. The advantage of these scaffolds compared to naturally derived products lies in their reproducibility: they have defined chemical composition that can be more precisely finetuned and consistently reproduced. Synthetic scaffolds can be inert, with no biodegradation, which ensures longer viability and less toxic effects. They are also generally free of animal-based components unlike natural biomaterials that tend to be extracted some animal tissues, which can generate interspecies discrepancies when used with human cells. However, some synthetic biomaterials might lack adhesion sites that cells require to migrate. Groups are overcoming this limitation by biofunctionalizing synthetic scaffolds by adding for example cell-binding motif arginine-glycine-aspartic acid (RGD) [60].

Another interesting way of considering scaffolds is to distinguish them into hydrogels and solid scaffolds [58], either of which can be obtained using naturally extracted or engineered scaffolds. A hydrogel is a loose scaffold framework of a crosslinked natural base material such as agarose, fibrin, collagen, or hyaluronic acid with high water content. Cells can be mixed and encapsulated in these hydrogels. In contrast, solid scaffolds can be either fibrous or porous. Fibrous scaffolds are generally formed via electrospinning, which produces a mesh of fibers by passing a polymer jet through an electric field and collecting the material on a grounded surface (Fig. 6A). An advantage of electrospinning lies with its ability to align fibers, which induces cell alignment and directionality in cultures. On the other hand, porous scaffolds are sponge-like polymers, formed with voids or pores. Different techniques exist to create them, one of them being particulate leaching, where a polymer is cast around soluble beads that can be dissolved to open up pores inside the material—sometimes with limited connectivity between voids, which can be a problem. Fibrous or porous scaffolds can contain cells either between the fibers or inside the pores where they can interact in 3D. It has been noted that it can be hard to induce cells to infiltrate these scaffolds. Some studies have pointed out that fibrous scaffolds might not truly allow cells to evolve in 3D, because cells tend to migrate along the fibers and bridge the gap between fibers only occasionally [58].

An emerging technique to create naturally derived scaffolds consists in decellularizing ECM from tissues. This approach removes the cellular components of tissues but maintains the ECM. This has several advantages: first, it maintains the native microarchitecture of the tissues, and second it removes the allogenic or xenogenic cellular antigens from the tissues as they are the sources for immunogenicity upon implantation. The ECM in contrast is conserved among species and therefore well tolerated immunologically [61]. Decellularizing is generally done by lysing cell membranes inside tissue extracts, prior to solubilizing and removing the cellular debris to only conserve the ECM scaffold of the tissue. Once decellularization is complete, the scaffold offers native biological and mechanical properties, and might even preserve growth factors inside the matrix, which helps cell growth. The seeding of cells inside the decellularized scaffold can however be hard and inhomogeneous [61].

5.1.3 Microfluidic devices

Microfluidic models have emerged in the past decade as important tools for in vitro cancer research. Microfluidic systems contain small channels, ranging from tens to hundreds of micrometers in height or width for handling small fluid volumes [62]. Overall, microfluidic assays represent a dramatic improvement in physiological relevance over other in vitro models because they allow for precise control of the cellular, physical, and biochemical microenvironment [63]. They can be easily designed according to the experimental requirements and display multiple compartments and channels. Each microfluidic compartment can contain a specific cell population separately, allowing for well-defined cocultures. This compartmentalization enables excellent spatial control of cell distribution at physiological length scales. Uniquely, the presence of microfluidic channels permits precise control of flow, allowing the study of many important biomechanical processes such as shear stress. The channels also allow for the guided delivery of cells or nanoparticles into the assay. In addition, valves can be added to direct and control flow. These channels can also be used to precisely establish physical (e.g., interstitial pressure) or chemical (e.g., cytokines) complex gradients in the microfluidic assay in a more precise and sustained manner than in macroscale in vitro systems [64]. The channels permit local control of other environmental physical factors, such as pressure and strain or hypoxia, by connecting the microchannels to sources of gas or to vacuum chambers. For example, a microfluidic device consisting of a porous membrane covered with cells placed between two apposed microchannels was developed to mimic the alveolar-capillary interface in the lungs (Fig. 7) [65]. Two larger flanking channels were connected to a vacuum pump that exerts a cyclical pressure on the central channel containing the porous membrane, thus cyclically stretching the cells as would occur during normal breathing. Microfluidic devices can contain cells cultured in 2D or in 3D embedded in hydrogels. Since these platforms are small, they require only minute amounts of cells and reagents that are often expensive, rare, or difficult to obtain. More importantly, the distance from the biological samples in the device to the microscope objective is short and, therefore, it is easier to image all cells at high resolution, as opposed to imaging larger traditional 3D in vitro models [11].

There are, however, several limitations to microfluidic technology. For example, retrieving cells from microfluidic devices for subsequent biochemical studies is possible

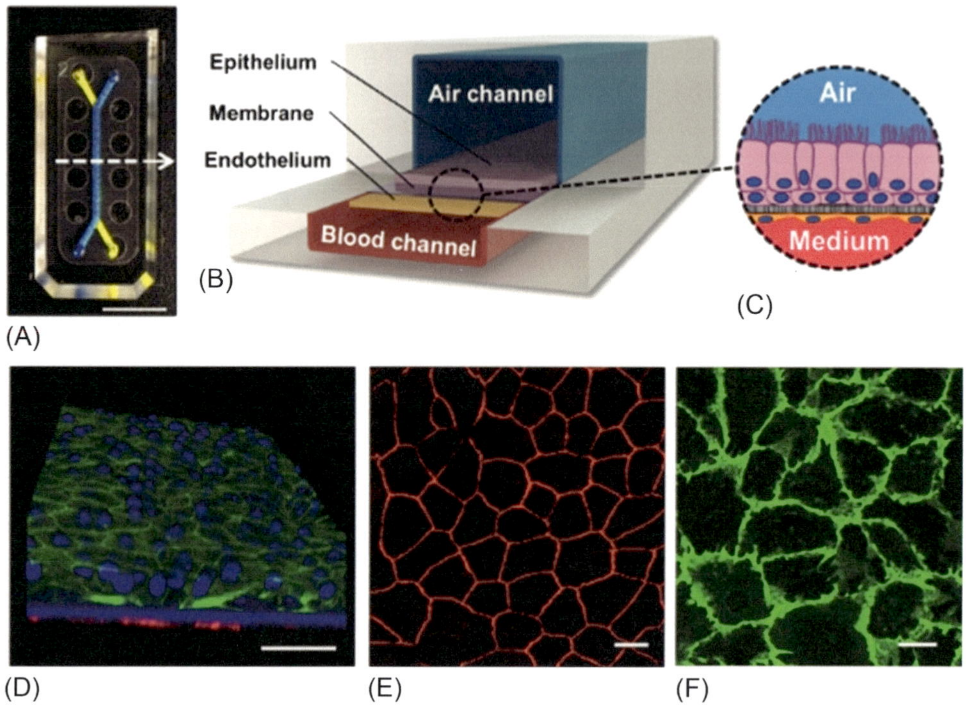

Fig. 7 The lung-on-a-chip microfluidic system. (A) Photograph of the microfluidic small airway-on-a-chip device (airway channel is filled with *blue dye*; microvascular channel is filled in *yellow*; bar, 1 cm). (B) Schematic diagram of a cross section through the small airway-on-a-chip shown in (A), visualizing its two hollow linear channels separated by a thin, porous, collagen-coated membrane which supports growth and differentiation of human primary airway epithelial cells on its upper surface and human pulmonary microvascular endothelial cells on its lower surface. (C) Diagram of the tissue-tissue interface that forms on-chip, with a differentiated airway epithelium cultured on top of the porous collagen-coated membrane at an air-liquid interface in the upper channel, and the endothelium below with medium flowing beneath it that feeds both tissue layers. (D–H) Confocal immunofluorescence views of cells grown for 3–5 weeks at air liquid interface in the small airway chip (bar, 20 μm). (D) The airway epithelium (*green*, F-actin) underlined by human pulmonary microvascular endothelial cells (*red*, F-actin) on the opposite side of the membrane (*blue*, DAPI-stained nuclei). (E) The differentiated human airway epithelium exhibits continuous tight junctional connections on-chip, as evidenced by ZO1 staining *(red)*. (F) The human (lung blood microvascular) endothelial monolayer formed on chip also contains continuous adherens junctions between adjacent cells, as indicated by PECAM-1 staining *(green)*. *(Reproduced from K.H. Benam, et al., Small airway-on-a-chip enables analysis of human lung inflammation and drug responses in vitro, Nat. Methods 13 (2016) 151–157.)*

but not straightforward, especially when cells are embedded in 3D scaffold formed inside the microfluidic chip [66]. The low cell number can also be a drawback if a large number of cells are needed for downstream biochemical studies (e.g., by western blotting) or if secreted proteins need to be quantified (e.g., by ELISA), although microfluidic platforms that include such miniaturized assays on-chip are currently being developed [67,68].

Microfluidic devices are generally made with polydimethylsiloxane (PDMS), which has several great properties such as being transparent and thus allowing imaging. Its use has been uniquely facilitated by photolithography, which enables scientists to fabricate molds with intricate designs for complex cellular assays. PDMS is poured onto the mold, left to polymerize and subsequently peeled off. This enables the features of the mold to be etched into the surface of the PDMS slab. These indentations in the PDMS can subsequently become channels after being sealed with a glass cover. However, PDMS has been shown to adsorb small hydrophobic molecules [69] which can be problematic for drug screening. To resolve this issue, thermoplastic microfluidic devices are being developed along with new materials [70].

Microfluidic technologies have clear potential to advance cancer research. They are particularly relevant for studying the metastatic cascade, since metastatic spread occurs through several steps that are often difficult to resolve in vivo. While intravital imaging has proven extremely useful in the study of metastasis [14], it requires specialized expertise and equipment, and high image resolution is harder to attain in some visceral organs due to the depth of their location. Therefore, many microfluidic assays have been designed to replicate certain aspects of several of these steps as described in Section 5.2.

5.1.4 Bioreactors

Bioreactors are closed systems that locally control factors such as perfusion, temperature, humidity, and gas exchange [71]. Typically, a bioreactor contains a main chamber (containing the cells), actuators components, sensors, and software to control and monitor the experiment. Cells can be placed in 3D scaffolds inside those bioreactors to facilitate 3D cell culture. Bioreactors provide a tight control of the culture environment and enable physical conditioning of the tissue to mimic the forces the cells experience inside the body. Sensors help to build computation models and obtain precise experimental data [71]. Bioreactors have been used in studies for cancer in several unique ways [72]. First, bioreactors have been used to expand cells. Second, they have been used to facilitate a uniform colonization of scaffolds by cells: this is more efficiently done under dynamic flow conditions inside a bioreactor than under static conditions, when cells are dropped onto the scaffold. Third, bioreactors have been used to control important environmental factors in cancer development such as hypoxia or tissue pressure [72]. Most importantly, perfusion inside bioreactors facilitates the long-term survival of cells, by continuously delivering fresh media and facilitating the removal of metabolic products of cells, especially in thick 3D scaffolds. In contrast, typical in vitro experiments rely on discrete periodical replacement of media, which diffuses into the 3D scaffold only to a limited distance.

There is a myriad of different types of bioreactors, often customized in the lab for specific needs. Different types of bioreactors exist to ensure optimal mass transfer to the cells inside the scaffolds, such as rotary, stirring, perfusion, or hollow-fiber bioreactors [72]. Other bioreactors exist to control and study the effects of pressure or strain on tumor cells. For example, a study has used a bioreactor to apply pressure using flat rollers on 3D gels containing tumor cells, showing that compression altered gene expression and metastatic potential. These are interesting variables to study in the context of cancer development seeing that nonphysiological mechanobiological stimuli typically occur in tumors [73]. Recently, a bioreactor was developed for long-term maintenance of 3D liver tissue under tight perfusion and oxygen control [74]. This LiverChip bioreactor has been exploited in several studies to study the poorly understood process of metastasis formation in the liver. For example, Chao et al. showed that the liver microenvironment induces the re-expression of E-cadherin in breast cancer cells [75], and that this confers cancer cells with enhanced survival following extravasation by increasing the attachment of cancer cells to hepatocytes and promoting the resistance of tumor cells to chemotherapy [76]. Note that microfluidic devices can be considered as a type of bioreactor seeing that they can be connected to external actuators such as pumps to similarly control local microenvironment factors.

This section was not meant to be an exhaustive list of all current available 3D cell culture assays. Rather, it details the most important ones for cancer research, and highlights how they can each in their own way advance our understanding of the disease. Other methods are emerging, such as cell-sheet technology [77], which have however not yet been used in the context of cancer research to the best of our knowledge. This consists of layering confluent monolayers of cells along with the underlying ECM they secreted to recreate 3D tissues rich in ECM resembling the cornea, the kidney, or liver. In any case, most of the 3D cell culture formats discussed are often used in combination with each other and can be hard to distinguish. New methods are bound to borrow aspects from existing methods we have described to maximize the benefits.

5.2 Individual models of cancer development

Cancer is a complex disease, composed of different stages that can take place in different microenvironments. Each step is characterized by a unique cellular and biochemical microenvironment. Thus, in vitro models have been created to model individual steps of the entire process of cancer development, to understand them better and design drugs that can target specific steps of the disease. Briefly, cancer can be distinguished into two phases: primary and metastatic cancer. Metastasis is composed of several steps that include an epithelial-to-mesenchymal transition (EMT), invasion, intravasation, transport of circulating tumor cells (CTCs) in the bloodstream, extravasation in the distant organ, and recolonization at the metastatic site [78]. This process can vary in nature and timing; for example, in some instances, cancer cells stay dormant at the secondary site, with the possibility of becoming active again at a later stage. Metastasis is responsible for 90% of cancer-related deaths: it is therefore vital to gain a better understanding of this stage of the disease, which we will therefore detail in more depth here than primary cancer. Note that most of the assays discussed here are microfluidic assays (Fig. 8) because they offer a unique control over the microenvironment as explained previously. This is

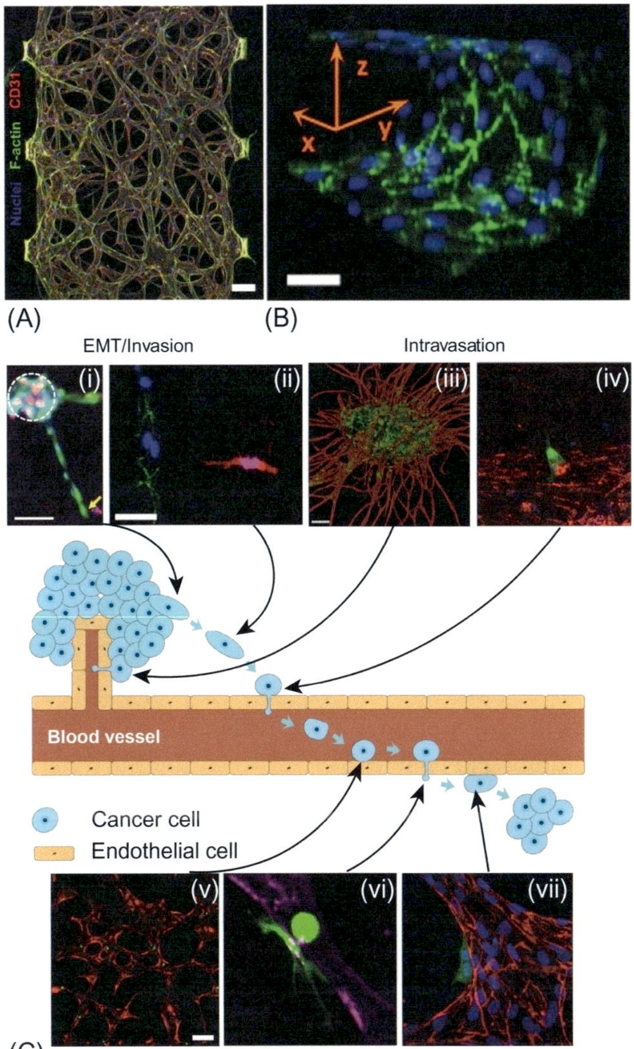

Fig. 8 Examples of a microvascular (A) network or (B) monolayer developed in microfluidic assays [26, 116]. (C) Schematic showing the principal stages of the metastatic cascade, accompanied by confocal images of the corresponding steps reproduced in different microfluidic devices. (i) Spheroid of GFP-MCF-7 labeled cells to model epithelial to mesenchymal transition (EMT), where the dashed circle shows the body of the spheroid and the arrow points to cells migrating away from it [13]; (ii) Invasion of a single RFP- HT1080 cell toward a monolayer of GFP-HUVECs [116]; (iii) Intravasation of a single GFP-MDA-MB-231 cell through a monolayer of RFP-HUVECs [19]; (iv) adhesion of GFP-MDA-MB-231 in a vascular network of RFP-HUVECs [21]; (v) extravasation of a GFP-MDA-MB-231 through a channel of a vascular network stained for phalloidin in pink [21]; and (vi) GFP-MDA-MB-231 in a perivascular position in the ECM after extravasation through the vascular network of RFP-HUVECs (with blue DAPI nuclear staining) [21]. *(Reproduced with permission from A. Boussommier-Calleja, R. Li, M.B. Chen, S.C. Wong, R.D. Kamm, Microfluidics: a new tool for modeling cancer-immune interactions, Trends Cancer 2 (2016) 6–19.)*

however not an exhaustive list and is only meant to provide a general scope of the state of the art in the field.

5.2.1 Primary tumor

Primary cancer starts with the neoplastic transformation of epithelial cells, which is generally followed by their abnormal multiplication. This process is often modeled in vitro with the use of tumor spheroids, previously described in Section 5.1.1. Studies have seeded spheroids with numerous different types of cell lines and studied the kinetics of their growth. They have also used these models to recapitulate key characteristics of primary tumor previously described such as the hypoxic and necrotic core (Fig. 5A). More recently, tumor organoids have been developed to study the primary tumors of individual patients, which we describe in more detail in Section 6.4. Spheroids have been used to mimic the response of the primary tumor to treatments. For example, Anastasov et al. imaged tumor spheroids comprising cancer cells mixed with stromal cells over 14 days to quantify their growth under different combinations of radiation and chemotherapy. They demonstrated in vitro, the shrinking effect of radiation on tumor spheroids as they would during treatment in patients [79] (Fig. 5B).

5.2.2 Angiogenesis

Several approaches for recapitulating tumor angiogenesis in vitro exist. 2D studies were helpful in deciphering the effect of tumor cells on endothelial cells, showing that direct contact between the tumor and endothelial cells, as well as the release of proangiogenic factors could promote angiogenic behavior [80]. Numerous 3D studies have studied tumor angiogenesis, typically by embedding tumor cells in a gel covered by an endothelial monolayer, thus allowing to study the formation of endothelial sprouts into the underlying gel toward the cancer cells [81]. Chen et al. have embedded microcarrier beads coated with endothelial cells inside a collagen gel that was covered with glioblastoma cells [82]. Interestingly, they show that in presence of tumor cells, the endothelial cells form sprouts, which are blocked in presence of anti-VEGF. The sprouting was enhanced by 50%–100% under hypoxic conditions. Similar experiments have been reproduced in microfluidic devices for enhanced control of flow, and have shown that angiogenesis induced by cancer cells could be inhibited by antivascular endothelial growth factor (bevacizumab) [83].

5.2.3 EMT and invasion

EMT and invasion has been modeled in microfluidic models by analyzing the migration of cancer cells from spheroids into an underlying collagen gel [84,85]. For example, Kuo et al. showed the expression of the epithelial marker EpCAM decreased in migratory cells compared to those remaining in the spheroid, indicating that cells had undergone EMT, as occurs in vivo. In other studies, human mammary fibroblasts were shown to promote the transition of breast cancer cells to an invasive phenotype, which lead to collagen remodeling [86,87]. Hypoxia was also shown to increase invasion in vitro [88], as well as the presence of C-X-C motif chemokine 12 (CXCL12) or matrix metalloproteinases (MMP) inhibitors [89].

5.2.4 Intravasation

Intravasation has been studied in vitro by recreating an endothelial monolayer along a microfluidic channel in direct contact with a collagen gel containing tumor cells [90]. Individual tumor cells were imaged as they crossed the endothelial monolayer from the collagen gel into the microfluidic channel. This study showed that TNF-α [83,90] and macrophages increase tumor intravasation rates [90], as observed in vivo [91]. Alternatively, intravasation has also been modeled using a Boyden chamber coupled with a microchannel and a controlled fluidic environment, in this case, to model tumor cell intravasation through a monolayer of lymphatic endothelial cells. This study showed that luminal and transmural flow increased intravasation of individual tumor cells [92]. Using tumor spheroids formed in well plates, Ehsan et al. studied tumor cell intravasation as they directly left the tumor mass [93] in presence of a microvascular network. This interesting study showed that intravasation of tumor cells from the spheroid into the microvasculature was enhanced under hypoxia and depended on the transcription factor Slug, consistent with in vivo results [94].

5.2.5 Adhesion

The simplest adhesion assay measures the number of tumor cells that adhere to an endothelial monolayer, typically after washing away nonadherent cells. This means these assays overlook differences in adhesion kinetics beyond the single time point examined [95]; if the washing away is done manually, it is also hard to guarantee it is exactly the same between conditions and experiments repeats. Parallel flow experiments have studied tumor cell adhesion on an endothelial monolayer under controlled flow, which did enable measuring real-time adhesion kinetics and rolling, but in most cases require observation with a microscope, dramatically reducing throughput. These powerful studies have brought invaluable insight into tumor cell adhesion kinetics, and the role of immune cells in tumor cell adhesion: for example, tumor cells were shown to adhere to the endothelium by tethering to neutrophils previously adhered to the endothelium, which was affected by shear rate [96]. Microfluidic adhesion assays of either single cancer cells or aggregates of cells onto an endothelial monolayer showed that E-selectin expression in human umbilical vein endothelial cells (HUVECs), CXCL12 [97], and shear stress [98] are all capable of affecting adhesion. Interestingly, studies replicating tumor cell extravasation inside microvascular networks did report entrapment in small capillary-like vessels as an alternative means for tumor cells to stop prior to their extravasation [99]. A recent study has increased the throughput of adhesion assays, by forming endothelial monolayers inside well plates, introducing cancer cells inside the wells, imaging the fluorescently labeled cancer cells before and after washing of the nonadherent cells with the use of shaft that generate shear stress inside each well. The shaft can also be used to generate shear stress and mimic adhesion under physiological flow. They use this system to screen through libraries of kinases and integrins in the cancer cell adhesion. Thus, this system is a promising tool for the preclinical discovery of new pathways and pharmacological inhibitors of cancer metastasis [95].

5.2.6 Extravasation

Extravasation is the process by which cells, in this context tumor cells, cross the vasculature from the inside of the vessel to the outside, into a new tissue. Conventionally, extravasation was studied in Boyden chambers: an endothelial monolayer was grown on top of the insert placed into the well, tumor cells were placed on top of the monolayer and the number of cells found in the lower compartment was considered to have extravasated through the monolayer. However, this technique suffered limitations explained in Section 4.1. Alternatively, planar endothelial monolayers have been grown on a thin layer of basement membrane; tumor cells were placed on top of the monolayer and observed as they crossed the monolayer downwards into the basement membrane gel. This enabled higher-resolution imaging than that allowed by Boyden chambers, and showed that extravasation is reliant on proteolysis of the basement membrane underlying endothelial cells [100].

Imaging downward extravasation of tumor cells can however be difficult as the resolution in the z-direction is generally not as high as that in the x- or y-direction. Thus, recent studies have used microfluidic models to image at higher resolution the extravasation of tumor cells through a vertical endothelial monolayer formed along the microfluidic channel [101], or through microvessels formed inside 3D hydrogels [41]. Microfluidic assays have shown that extravasation of breast cancer cells generally occurs within 24 h [101,102] paracellularly through transient gaps between the endothelial cells of a microvascular network. In contrast, extravasation of aggregates of tumor cells was shown to irreversibly disrupt an endothelial monolayer [103]. Extravasation rates were increased in the presence of osteo-like cells [104], replicating the in vivo organ selectivity of breast cancer cells to bones. Extravasation also increased in the presence of CXCL12 [98] and was blocked by AMD3100, a CXCR4 receptor antagonist currently in clinical trials [98].

Other microfluidic experiments have studied extravasation by analyzing the deformability of cells through narrow 2D channels that mimic the restricted space tumor cells have to cross between endothelial cells during transendothelial migration (in 2D or channels filled with hydrogels [105]). A recent study has managed to isolate the tumor cells that did extravasate through a porous membrane by collecting them through microfluidic channels to analyze differences between tumor that did or did not extravasate. Interestingly they show that tumor cells that did migrate through the membrane seemed to express more palladin compared to the bulk population of tumor cells cultured in 2D. This suggests that palladin may reflect the extravasation potential of tumor cells, a hypothesis that is supported by studies that have shown that palladin is correlated with cancer metastasis [106].

5.2.7 Dormancy and recolonization

Less recolonization assays have been reported to analyze events beyond extravasation, likely because it is hard to maintain in vitro assays viable for long periods of time. A few extravasation studies quantified cancer cell proliferation or invasion in the ECM shortly after extravasation [101], showing that CXCL12 exacerbated invasion [103]. A microfluidic study has tracked tumor cell proliferation 1–6 days post extravasation through a microvascular network [41], showing interesting trends of tumor cell growth along the microvasculature as observed in some in vivo studies [52]. A separate study studied the growth of cells after they extravasated with or without the presence of macrophages around the vasculature, seeing little difference in the first 5 days following extravasation [99]. Platforms such as these have only been recently characterized and are very promising tools to progressively add different stromal or immune cell types to dissect their effect on micrometastasis formation. Note that micrometastases have been sometimes observed in vivo to initiate intravascularly, prior to extravasating "en masse" [107]; this should be further studied in vitro to better understand the role of the intravascular microenvironment in the development of metastasis. This could be particularly powerful as micrometastases are difficult to identify in vivo due to their small size.

An interesting study has studied tumor cell dormancy in a 3D hydrogel in presence of a microvascular network (Fig. 9) [108]. First, the authors formed a microvasculature inside a 3D fibrin gel inside a well; subsequently, they placed tumor cells on top of the gel, and studied the tumor cells once they had migrated down into the gel toward the microvasculature. Using high-resolution imaging, they show that endothelial cells in a stable microvasculature induced breast tumor cell dormancy through the secretion of thrombospondin-1. In contrast, neovascular sprouting did not, thus favoring tumor cell growth: this suggests that sprouting neovasculature could be responsible for sparking micrometastasis formation. This study demonstrates the potential of in vitro model to further understand micrometastasis formation, which is crucial as they are generally associated with poor prognosis [109].

A different group has modeled micrometastases formation in a bioreactor that mimics the liver microenvironment. Their results show that the presence of hepatocytes induces the re-expression of E-cadherin in breast cancer cells [75], which confers a survival advantage by promoting attachment to hepatocytes [76]. This E-cadherin re-expression also favored chemoresistance of cancer cells. This study demonstrates the potential of in vitro models for investigating the organ-specific basis for successful colonization of secondary metastatic sites.

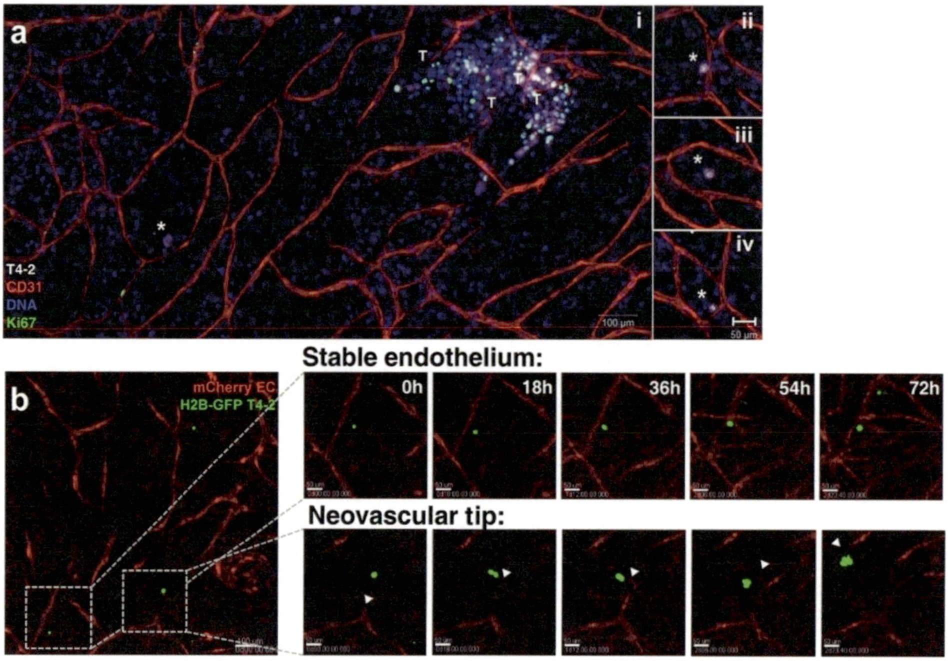

Fig. 9 In vitro model of tumor cell dormancy regulated by endothelial subniches (Ai) Immunofluorescence of YFP T4-2 tumor cells in microvascular niche after 1 days *: Ki67-negative tumor cluster; T: neovascular tips surrounding proliferative tumor. Scale bar = 100 μm. Insets ii–iv: additional examples of Ki67-negative tumor cells residing on/near microvasculature from that same culture shown in (Ai). Scale bar = 50 μm. (B) Stills from live tracking of histone H2B-GFP T4-2 tumor cells in microvascular subniches (stable endothelium and neovascular tip) from 0 to 72 h. White arrowheads denote approaching tip and subsequent T4-2 division. Scale bars: left-most panel = 100 μm, right panels = 50 μm. *(Reproduced from C.M. Ghajar, et al., The perivascular niche regulates breast tumour dormancy, Nat. Cell Biol. 15 (2013) 807–817.)*

5.2.8 Multistage models

Few studies have combined several metastatic steps in the same in vitro assay. This is likely because mimicking several stages adds complexity to the assay. Yet, these multistage assays are bound to advance our understanding of the disease for several reasons. First, it would increase throughput as much more data could be collected from one chip that mimics a larger part of the metastatic process. But more interestingly, it could help recapitulate phenomena that arise from the synergy of different tissues at play during metastasis. Indeed, while metastasis is often dissected into different stages (as we have done in this chapter), these stages can often progress at the same time and influence each other. For example, researchers have shown that a premetastatic niche seems to arise before the arrival of cancer cells to the metastatic site to facilitate their adhesion and growth [110]. Findings suggest that the presence of the primary tumor conditions at a distance the metastatic tissue, such that circulating tumor cells are more able to adhere and colonize the metastatic tissue. Much more needs to be understood about this process, as treatments could be designed to prevent the initiation of the premetastatic niche to prevent metastasis. So far, few studies have studied this process in vitro; however, studies have cocultured tumor cells with cells characteristic of the metastatic tissue to investigate how organ-specific cells condition tumor cells. For example, Sieh et al. have shown that the coculture of osteoblasts in a tissue-engineered bone with prostate cancer cells promotes osteomimicry and modulate the expression of androgen-responsive genes in prostate cancer cells, thus demonstrating that the bone microenvironment can prepare tumor cells to better adapt and survive in the bone [111]. Similar studies could be performed to study, conversely, the role that tumor cells have on the organ-specific microenvironment to prepare the premetastatic niche. Future work could model in the same assay the primary tumor and the metastatic site to directly observe how the primary tumor conditions the endothelial cells and the ECM of the metastatic site and as a result affect tumor cell adhesion or extravasation.

Multistep metastatic assays could benefit research on other synergic processes, such as primary tumor self-seeding [112]. It was generally thought that metastasis was a unidirectional process, whereby the primary tumor gave rise to the metastatic tumors. Recent research has however unveiled an additional possibility, wherein circulating tumor cells colonize their tumor of origin. This "tumor self-seeding" could explain recurrence of the primary tumor after their apparent removal. In vitro assays that recapitulate both the primary and metastatic site could potentially help gain further insight into this process so as to design drugs that can block it.

One of the few multisteps metastatic models was designed by Shin et al. who combined in one microfluidic device invasion of cancer cells out of a hydrogel to mimic intravasation and their subsequent adhesion to a monolayer of HUVECs [89]. Future studies could combine more complex models of individual metastatic steps, as described earlier in this chapter for a more comprehensive in vitro metastatic model.

6 Emerging paradigms for in vitro cancer models

It is clear that in vitro models have fueled an unprecedented interest and have already brought much insight to cancer research. In vitro assays are however still evolving to meet the ever-growing demand of scientists for more complex and physiologically relevant models. Here we discuss four main emerging trends that are, at varying degrees, still in their infancy and are bound to become ubiquitous to in vitro assays for cancer research.

6.1 Immunocompetent in vitro models

While conventional therapies seem to fail at containing metastasis, cancer immunotherapy has recently shown promising results for several types of cancer [113], such as metastatic melanoma [114]. This has generated acute interest to develop additional immunotherapeutic drugs, and thus has created a need to study immune-cancer cell interactions both in vivo and in vitro. Powerful in vivo models exist to study the role of immune cells in cancer, such as mouse models [115], and more recently, humanized mouse models. In vivo models are unique because they allow scientists to study immune-cancer interactions at a systemic level across the entire organism. This is particularly important for immune cells studies, as immune cells are part of a highly complex and hierarchized system spread across different parts of the body which is, as of today, hard to recapitulate in vitro. In vivo systems do however present some limitations for immune-cancer studies that warrant the use of in vitro models. First, scientists often must use immune-suppressed mice: this is because they inject human tumor cells inside the animals, which are typically rejected by the murine immune system. To prevent the rejection of human tumor cells, part of the immune system of mice is suppressed. Intrinsically, the lack a complete functioning immune system could fail to replicate all immune-cancer cell interactions at play during cancer. Animal models used to study immune-cancer interactions also suffer from another limitation: the microenvironment inside the mouse is murine, which means one is studying interactions between human tumor cells and immune, stromal, or other organ-specific cells of mouse origin. This could cause species-related discrepancies that can lead to erroneous predictions of the effects of drugs in humans.

In contrast, in vitro studies can contain tumor, immune, and stromal cells that are exclusively of human origin, which circumvents interspecies discrepancies. Microfluidic models show particular potential to study immune cells for several reasons [11]: first, immune cells are often guided by chemokine gradients, which can be recreated in microfluidic devices in a more controlled fashion than in other in vitro assays. Second, immune cells often evolve under flow conditions inside the vasculature, which can be precisely recreated inside microfluidic channels. Finally, immune cells can exist in different compartments inside the body, ranging from the bone marrow, the vasculature to the target tissue, which can be captured in microfluidic models through spatial compartmentalization.

Numerous in vitro studies have investigated the role of immune cells in the primary tumor using tumor spheroids cocultured with immune cells in 3D hydrogels. However, in vitro immunocompetent models to study cancer metastasis are still in their infancy, with a small but growing number of them reported in the literature. This gap needs to be filled, as it is now known that immune cells also play an important but still poorly understood role in metastasis [116]. In one study, Vacchelli et al. identified a loss-of-function mutation in the allele of the gene coding for formyl peptide receptor 1 (Fpr1) that correlated with poor survival in breast and colorectal cancer patients receiving adjuvant chemotherapy [117]. Their data showed that cancer cells growing in $Fpr1^{-/-}$ mice were resistant to anthracyclines and suggested that Fpr1-deficient immune cells failed to migrate toward dying cancer cells. Using a microfluidic platform, they tested this hypothesis by seeding cancer cells and immune cells in separate compartments and imaged over time the migration of immune cells toward the cancer cells in 2D through microchannels. This confirmed that Fpr1-deficient immune cells interacted much less with dying cancer cells treated with anthracyclines than immune cells expressing functional Fpr1, explaining why Fpr1 deficiency leads to defective anticancer immune response and less efficient chemotherapy. These results show great potential to further understand how immune cells affect cancer progression and treatment. While this study was done in 2D, others have emerged wherein cancer and immune cells are seeded in 3D. In a 3D vascularized microfluidic model, we showed that monocytes seem to decrease breast cancer cell extravasation through a 3D microvascular network [99].

Pavesi et al. developed a microfluidic device where tumor cell aggregates were seeded inside 3D collagen gels and imaged as they were lysed by T cells. These in vitro models hold great promise for being able to test in vitro different types of immune therapies prior to treating patients [118]. In a separate study, Lee et al. study the interactions of two immune cell types with tumor cells. More specifically, they investigate the immune suppressive role that monocytes have on the tumor-killing capacity of T-cell receptor-redirected T cells [119]. Their study shows that monocytes suppress the T-cell killing of tumor cells via the PD-1/PDL-1 axis. This demonstrates the potential and need of in vitro models to study complex immune-cancer interactions, by including several types of immune cells and showing that these can have synergistic effect on cancer development (Fig. 10). Importantly, recent work has shown that immune cells can induce resistance to treatment [120] or cause relapse after chemotherapy [42]. This further demonstrates that it is crucial to include immune cells inside in vitro models for cancer research, so that drugs can be designed as a function of their true immune microenvironment inside the body.

6.2 Vascularization

Scientists have been interested in being able to vascularize cellular models for decades. One the major reason driving this line of research was the need to engineer tissues viable for implantation in patients. While thin or avascular tissues have been successfully implanted in vivo, implanting thicker tissues has proven more difficult. That is because they require a functional vasculature to provide gas and nutrient exchange and waste removal [16], and forming a functional vasculature in vitro has proven to be difficult. While learning to form a vasculature in vitro is crucial for tissue engineering, it became clear that vascularizing cellular models could more generally make them more physiologically relevant models. For example, introducing a vasculature inside in vitro assays can help better mimic the effect of drugs which are often delivered through the bloodstream. Recent study have even introduced whole blood inside microfluidic endothelium-lined channels to study therapeutics in even more physiologically relevant conditions, during thrombosis [121].

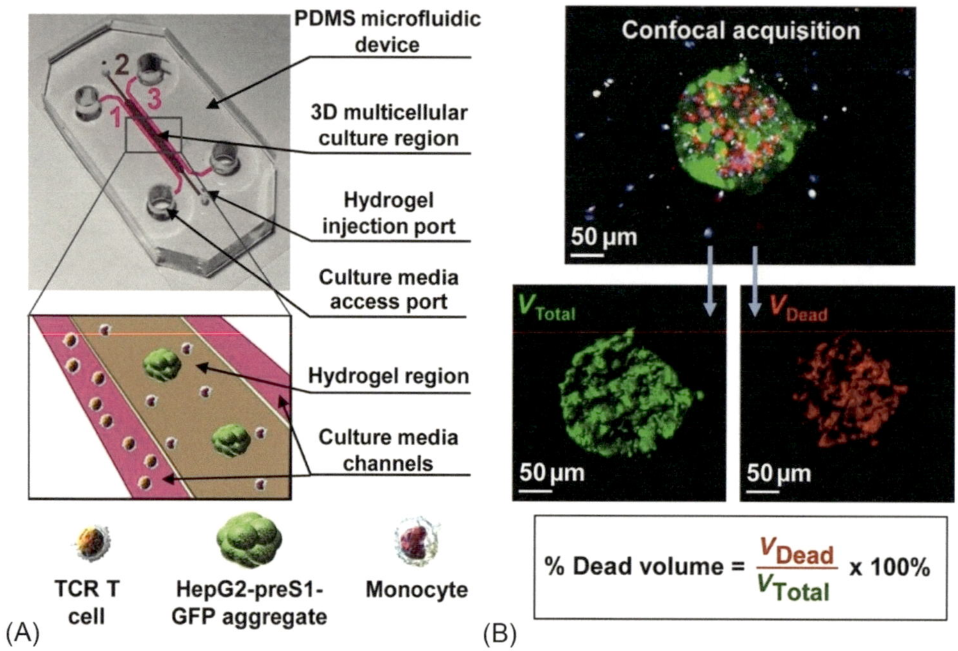

Fig. 10 (A) A 3D multicellular tumor microenvironment microfluidic model consisting of a middle hydrogel channel (2) flanked by two media channels (1, 3) for the mechanistic study of the effect of monocytes on T cell receptor-redirected T cell (TCR T cell) killing of tumor cell aggregates. Human monocytes were inserted together with target HepG2-preS1-GFP cell aggregates in collagen gel in the central hydrogel region (2), while hepatitis B virus (HBV)-specific TCR T cells were added into one fluidic channel (1) to mimic the intrahepatic carcinoma environment. (B) Representative confocal image of a target cell aggregate *(in green)* surrounded by monocytes *(in blue)* and HBV-specific TCR T cells *(in white)*, in which the presence of dead target cells is DRAQ7+ *(in red)*. Target cell death is quantified as shown based on the DRAQ7+ volumetric portion in the total volume of each GFP-labeled aggregate. *(Reproduced from S.W.L. Lee, et al., Characterizing the role of monocytes in T cell cancer immunotherapy using a 3D microfluidic model, Front. Immunol. 9 (2018) 416.)*

Vascularizing in vitro models will be particularly important for cancer research, especially for modeling cancer metastasis. That is because an essential feature of the metastatic cascade is the interaction between cancer cells and blood or lymphatic vessels. Both intravasation, adhesion and extravasation require for the tumor cells to interact with the endothelium. The endothelium has also been shown to potentially regulate tumor cell dormancy [108]. Clearly, the endothelium is a crucial feature of the microenvironment to be included inside in vitro models of metastasis.

There are several ways of recreating a vasculature in vitro. The most basic vascular assays for cancer research consist in forming a 2D monolayer of endothelial cells; once formed, tumor cells are placed on top of the endothelial monolayer and studied as they cross the vascular wall. However, these 2D assays do not recapitulate the 3D geometry of vessels in vivo and are not always amenable to perfusion. Recently, methods for recreating more physiological 3D like vessels have emerged: briefly, they consist in either endothelial-lined patterned channels or self-assembled microvascular networks [16]. The former assays consist of depositing endothelial cells along the walls of a channel; this can be a microfluidic channel, or a hollow region cast inside a hydrogel. The latter assays consist in mixing endothelial cells in a hydrogel with or without stromal cells and letting them assemble over time into a microvascular network. Microfluidic tools lend themselves particularly well for both assay formats, as they enable the precise control of flow through the vessel-like structure. For example, while one can form self-assembled microvascular networks in a hydrogel in a well plate, it will not be perfusable [93]; however, placing that hydrogel inside a microfluidic device enables its perfusion. Several groups have formed microvascular networks inside hydrogels flanked by channels in microfluidic platforms [39,122], where the lumens of the microvascular networks open to the channels allowing the microvessels to be perfused.

Being able to perfuse the vessels is key as it enables direct access inside the vessels, and thus the intravascular delivery of media, cells, or reagents to mimic their presence inside the vessels as it happens in vivo. For cancer research, this means that tumor cells can be perfused into the microvascular networks, and studied as they flow, stop, or adhere to the endothelial wall, cross the endothelium, and behave following extravasation, as they would in vivo [41]. The vasculature is also key to studying other metastatic steps, such as dormancy. As previously discussed in Section 5.2.7, studies have used microvascular networks formed in a hydrogel in vitro to study how they induced or interrupted dormancy in tumor cells [108]. Finally, vascularization is not only important to in vitro models mimicking metastasis: it is also an important area of interest for in vitro models of primary tumors. As discussed earlier, spheroids are important tools for screening drugs but still lack vascularization. Groups are now focusing on vascularizing spheroids to replicate a more faithful response to drugs of large vascularized tumors with encouraging success (Fig. 11) [93].

6.3 Organs on chip

In the race toward ever more physiologically relevant in vitro models, scientists have focused on engineering tissues that mimic more closely organ-specific functions. Recreating organ-specific in vitro models is particularly important for cancer research, because cancer progression can be organ-specific: that is, tumor cells metastasize in selected organs depending on the origin of the primary tumor. For example, it is known that breast cancer tends to develop secondary tumor in the bones and lungs. The basis for this organ selectivity has been long investigated, dating back to Stephen Paget, who first hypothesized in 1889 that CTCs (or "seeds") only metastasize in organs (or "soils") with which

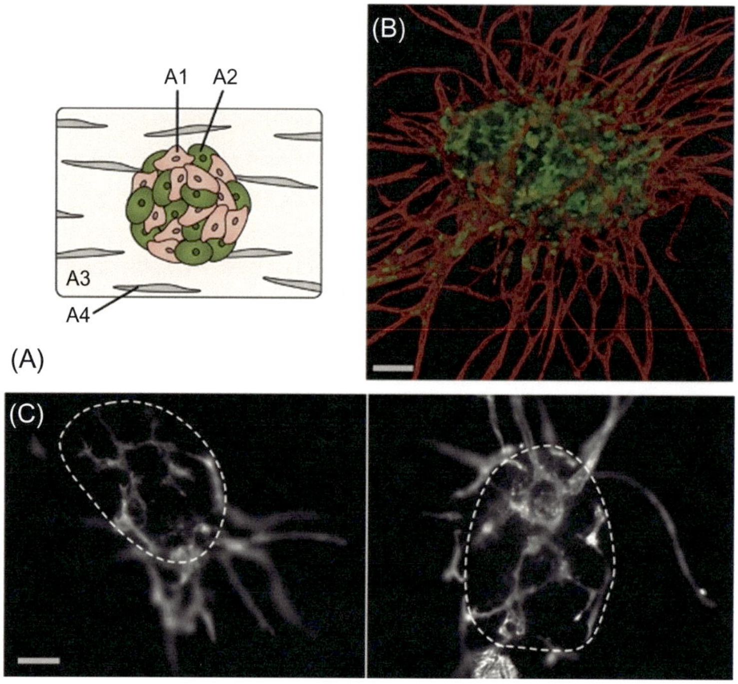

Fig. 11 Prevascularized tumor (PVT) spheroid model. (A) Schematic of model shows coculture spheroids composed of endothelial (A1) and tumor cells (A2) embedded in a fibrin gel (A3) distributed with fibroblasts (A4). (B) Representative fluorescent image of PVT spheroid demonstrates robust radial sprouting of lumenized capillaries. EC are labeled with CD31 antibodies *(red)*, and tumor cells (here, SW620) are transduced with EGFP *(green)*. Additionally, the PVT model features a contiguous vessel network that vascularizes the spheroid itself. Scale bar represents 100 mm. (C) Fluorescent images of vascular network reveal inner capillaries are characteristically shorter, jagged, and more branched compared to radial sprouting capillaries. Boundaries of spheroids are outlined with dashed lines. Scale bar represents 100 mm. *(Reproduced from S.M. Ehsan, K.M. Welch-Reardon, M.L. Waterman, C.C.W. Hughes, S.C. George, A three-dimensional in vitro model of tumor cell intravasation, Integr. Biol. (Camb.) 6 (2014) 603–610.)*

they have binding affinity. It was later speculated tumor cells simply colonize the first organs they encounter during their transport in the blood circulation [123]. However, autopsy and animal studies have demonstrated that hemodynamics cannot fully explain the patterns of secondary tumors [124–126]. Thus, at present, it is thought that mechanical factors dictate the distribution of CTCs to preferential organs, but tumor cell extravasation and colonization is favored in organs, which can interact specifically with the tumor cells. In light of these findings, efforts are concentrating on including organ-specific cells inside in vitro assays to better replicate the local microenvironment that tumor cells encounter in different organs and identify drugs that can pharmacologically block organ-specific signaling.

Groups have resorted to several strategies to recapitulate organ specific characteristics in vitro. The most basic assay consists in coculturing tumor cells with organ-specific cell types in 2D or 3D or exposing tumor cells to chemokine gradients present in the organ of interest. Another strategy to replicate organ specificity in vitro consists in tuning the ECM to resemble that of the desired organ. For example, Kostic et al. seeded breast cancer cells on top of 3D gels with varying rigidity to mimic different organs that have corresponding stiffness [127]. Briefly, they used distinct subpopulations of breast cancer cells that were isolated in mice by the Massague group [128] for their propensity to metastasize to the bones or the lungs. When Kostic et al. used the breast cancer cells prone to metastasizing in bones, they detected a higher proliferation and invasion on top of stiffer 3D gels compared to softer gels. This supports the idea that the bone-metastasizing tumor cells favor the gel that mimic the rigidity of bones. On the other hand, the tumor cells that caused metastasis primarily in the lungs, but also, at lower frequency, to the bone, proliferated faster on soft substrates. This shows that replicating organ-specific stiffness in vitro can reflect the metastatic potential of tumor cells. While these have been powerful studies to start understanding organ-specific cues that direct metastasis, the field has since leapt forward, by focusing much effort on recreating more physiologically relevant organs on chips.

Organs on chips are miniaturized cellular models that recapitulate key functions of human organs. Organs on chips have been uniquely potentiated by microfluidic technology, which as described earlier, facilitates an enhanced control of the microenvironment at cellular length scales. In addition to replicating the biochemical and cellular microenvironment, organs on chips have also been modeled to mimic mechanical cues experienced in specific organs, such as the cyclical stretching of the lungs that occurs during breathing [65]. As of today, simplified models of many organs exist, such as the lungs or the gut [129]. There are several parameters that can vary between organs, which are known to affect the development of cancer and should therefore be replicated in vitro. Among them are the presence of organ-specific cells, mechanical cues, or physical barriers, as described later.

Organs on chips can contain organ-specific cells to recreate the biochemical microenvironment of unique organ. This is important for studying metastasis, as organ-specific cells have been shown to promote tumor cell colonization. For example, Jeon et al. mimicked a bone-like or muscle-like environment in a 3D microfluidic model by culturing microvascular networks with bone marrow-derived mesenchymal stem cells or myoblasts. Interestingly, they show that breast tumor cells had a higher tendency to extravasate in the in vitro bone-like microenvironment compared to the muscle-like microenvironment. This replicates the organ specificity of breast cancer in patients, which generally metastasizes to the bones and rarely to the muscles [104].

In addition, organs on chips can mimic organ specific mechanical cues; interestingly, these can also affect cancer progression and should therefore be considered in future in vitro models of cancer. In a very interesting recent study, Hassell et al. [130] used the well characterized lung-on-a-chip model to replicate and study non-small-cell lung cancer as it progresses around the alveolar epithelium during normal breathing (Fig. 12). Interestingly, they show that cyclic stretching of the lung epithelium suppresses cancer cell proliferation and invasion. This novel finding raises the possibility that loss of lung motion due to filling of the alveolar spaces by growing cancer cells in patients may further help the disease progression. On the other hand, they show that tumor cells become resistant to a drug because of the cyclical stretching. This highlights the importance of testing drugs in models that mimic organ-specific mechanical cues.

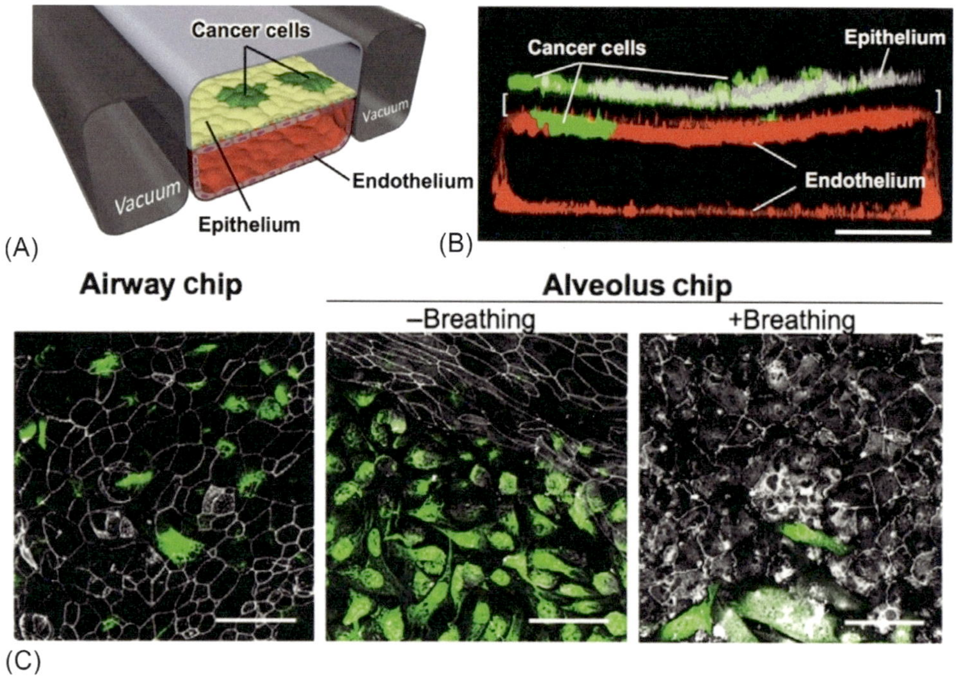

Fig. 12 (A) Schematic diagram of a cross-section through the 2-channel microfluidic chip containing human lung epithelial cells and a tumor cells cultured on the upper surface of a porous ECM-coated membrane with human lung microvascular endothelial cells cultured on all four walls of the lower channel, forming a hollow vascular lumen. (B) Confocal fluorescence micrograph of a cross-section of the two central cell-lined channels of an alveolus chip, showing GFP-labeled lung cancer cells (*green*, anti-GFP) cocultured with primary lung alveolar epithelial cells labeled with antibodies against the tight junction protein ZO-1 *(white)*. Primary lung microvascular endothelial cells labeled with anti-VE-cadherin *(red)* form a continuous monolayer that covers all four sides of the lower microchannel, creating a vascular (scale bar, 200 μm). (C) GFP-labeled lung cancer cells *(green)* exhibit distinct growth patterns within the normal epithelial monolayer containing cells delineated by ZO1-containing tight junctions *(white)* after 2 weeks of culture in the airway chip *(left)* versus the alveolus chip in the absence *(middle)* or presence of physiological breathing motions *(right)* (scale bar, 50 μm). *(Reproduced from B.A. Hassell, et al., Human organ chip models recapitulate orthotopic lung cancer growth, therapeutic responses, and tumor dormancy in vitro, Cell Rep. 21 (2017) 508–516.)*

Organs on chips often recreate cellular or physical barriers, which can vary immensely between organs. For example, it is known that the blood brain barrier has a particularly low permeability compared to other endothelium walls in the body, which can make it harder for tumor cells or drugs to enter the brain. Thus, it is important to replicate a low endothelial permeability in vitro to study tumor cells inside the brain. To this end, Xu et al. [131] cultured brain endothelial cells along a microfluidic channel juxtaposed to a 3D gel containing astrocytes. In this study, Xu et al. show that both the presence of astrocytes and that of flow decrease permeability to values similar to those found in the brain in vivo. This model was then used to screen the effect of several drugs used against glioma, by placing glioma cells inside the 3D gel with astrocytes. The drugs were perfused inside the microfluidic channel and had to cross the endothelium to be able to kill the tumor cells. Their results show that, while all drugs were able to kill cancer cells upon direct exposure, only one was able to cross the in vitro blood brain barrier and kill tumor cells in an in vivo-like setting. This study warrants the use of organs on chips that model the unique microenvironmental conditions of each organ for enhanced drug screening.

Recently, groups have combined several organs on the same chip for enhanced systemic modeling of drug effects. For example, in vitro models of both the gut and liver have been combined on chip because drugs are first absorbed through the gut and then metabolized by the liver [132]. Their combination has been shown to affect drug fate differently than chips where liver of guts are cultured separately [133]. Similarly, body on chips that include several organ-specific-environment could be modeled on chip for a more comprehensive model of cancer.

6.4 Organoids

Finally, organoids must be mentioned as an emerging and promising technique for studying cancer in vitro. Organoids are three-dimensional (3D) constructs that can be developed from embryonic stem cells (ESCs), induced pluripotent stem cells (iPSCs), somatic SCs [134]. These cells can differentiate into multiple cell types that organize themselves in a similar fashion to that observed in vivo and exhibit functions of the organ of interest [135]. The initial breakthrough came in 2009, when Sato et al. cultured in 3D an organoid of the intestines [136]. Briefly, they managed to culture 3D mouse crypt structures for 8 months generating villus-like epithelial domains in which all differentiated cell types found in vivo were present. Since groups have successfully recreated organoids of several organs, such as the liver or stomach. These organoids have the ability to recapitulate the genotype, phenotype, and cellular behaviors of parent tissues [134]; importantly, their normal genome remains stable over time [137]. In addition to being able to recreate organoids representative of healthy tissues, tumor organoids can also be formed with cancer cells to study diseased tissues.

Both normal and tumor organoids are already proving invaluable for in vitro cancer research. For example, Drost et al. formed normal colon organoids as a model to identify the origin of mutational signatures in cancer [138]. Using CRISPR/Casp9, they knocked out a mismatch DNA repair gene known to predispose to colorectal cancer into the organoids. This allowed them to study the effect of disrupting a single DNA repair gene on an otherwise normal genetic background. Following subcloning and passaging of the organoids for 2 months, they find that disrupting this gene was enough to lead to a mutational landscape in the organoids similar to that found in mismatch repair-deficient colorectal cancer. Using this novel strategy to track the origin of mutations, they investigate the less known role of base excision repair genes in causing cancer. Their results show that disrupting one of these genes leads to a mutational signature found in breast cancer. Thus, organoids can be used as a means to decipher the development of cancer and systematically dissect the role of individual genes in the initiation of the disease. Organoids have also been used to understand how bacteria contribute to cancer initiation in a predisposed subpopulation of patients in India. First, Scanu et al. formed gallbladder organoids with a mutation to mimic the genetic predisposition of these patients [139]. Second, they infected gallbladder organoids with Salmonella and showed that it led to neoplastic cellular transformation, as well as abnormal growth in conditions with restricted of growth factors (Fig. 13A and B).

While these studies were studying the onset of cancer in normal tissues, others have used organoids to model and characterize already-diseased tissues. For example, Gao et al. [140] characterized a protocol for developing prostate organoids with biopsies or circulating tumor cells obtained from prostate cancer patients. These organoids recapitulated distinct mutations characteristic of prostate cancer and were shown to be suitable for drug testing. Similarly, Seidlitz et al. [141] develop human gastric organoids using cells isolated from gastric cancer patients. They also show that these organoids showed a similar mutational spectrum to that found in vivo, and interestingly, that organoids from different patients responded distinctly to chemotherapy. Therefore, organoids hold great potential for creating biobanks that can be used to identify biomarkers to predict therapy response and resistance in individual patients.

On the other hand, organoids can also be inserted into microfluidic chips for even more complex and physiologically relevant models. Recently, Kasendra et al. developed an intestine chip containing epithelial cells isolated from healthy regions of intestinal biopsies (Fig. 13C) [142]. The primary epithelial cells were first expanded into 3D organoids prior to culturing them on a porous membrane, placed between two microfluidic channels (in a design similar to the lung-on-a-chip described earlier). This caused the epithelium in the intestine chip to develop villi-like structure, as happens in vivo. Yet, unlike previous studies, organoids placed in this chip were exposed on their apical surface to an endothelium, which could be perfused. In addition, the porous membrane could be stretched cyclically to mimic the peristaltic movement. Importantly, they

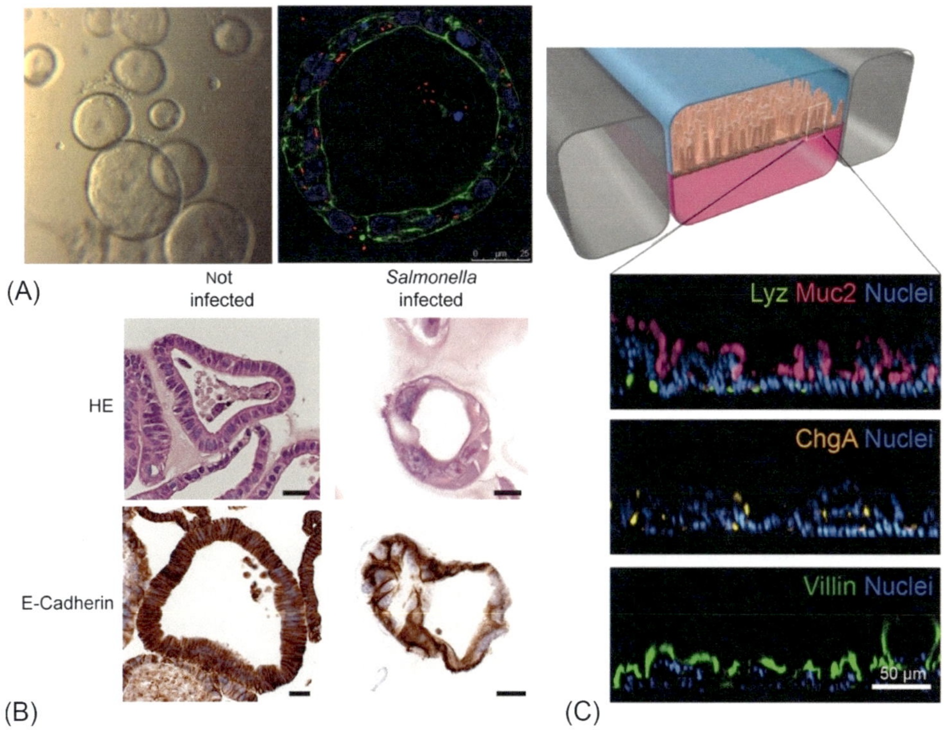

Fig. 13 (A) Representative wide-field microscopy image of a gallbladder organoid culture. *Bottom panel*, confocal images of one representative gallbladder organoid infected with *S. typhimurium (in red)*. Nuclei are stained with DAPI *(blue)*, and actin is shown in *green* (scale bar, 25 mm). Z stack projections of gallbladder organoids. (B) H&E staining *(top)* and E-cadherin IHC *(bottom)* of gallbladder organoids grown from cells infected *(right panels)* or not *(left panels)* with *S. typhimurium*. The sections are counterstained by hematoxylin. Scale bar, 20 μm. (C) A schematic cross-sectional representation of the 3D intestinal epithelial tissue architecture developed on chip *(top)* and confocal immunofluorescence micrographs *(bottom)* showing vertical cross-sections of the differentiated epithelium in intestine chip stained for lysozyme (Lyz, *green*), mucin 2 (Muc2, *magenta*), chromogranin A (ChgA, *yellow*), and villin *(green)*. Cell nuclei were counterstained with DAPI *(blue)*. *((A,B) Reproduced from T. Scanu, et al., Salmonella manipulation of host signaling pathways provokes cellular transformation associated with gallbladder carcinoma, Cell Host Microbe 17 (2015) 763–774. (C) Reproduced from M. Kasendra, et al., Development of a primary human Small Intestine-on-a-Chip using biopsy-derived organoids, Sci. Rep. 8 (2018) 1–14).*

compare the gene profiles of the epithelial cells taken from the intestine chip (following 12 days of continuous perfusion and cyclical stretching) to that of epithelial cells taken from patients and cells taken from standard organoids cultured for the same amount of time. Their results demonstrate that the transcriptome of the cells in the intestine chip better matched that of cells in vivo than standard organoids. This lends enormous weight to the added value brought by organs on chips and the huge benefits that can be derived from combining organoids with microfluidic technology.

Lastly, organoids also represent great models of healthy tissues, which can be used to predict whether vital organs will suffer unwanted toxic side effects from drugs used for treating cancer. As previously mentioned, many drugs fail in clinical trials due to toxicity linked with the liver, kidneys, or heart; thus, organoids of either of these three organs can now be used for toxicity screening, thus massively helping with predicting which drugs have a risk of causing toxic side effects among patients [134].

7 Challenges and future of in vitro models for cancer research

While it is undeniable that in vitro assays have already had a massive impact on cancer research, they promise to deliver even more in the coming years, as they potentially become companion clinical tools for personalized medicine. Yet before reaching that point, the field will have to face some challenges, of which a few are discussed here.

7.1 Cell lines: Validity and quality control

Cell lines are often used for in vitro cancer research. Typically, they represent transformed and aggressive cancer cells from advanced stage cancers. They can be derived either spontaneously through selection for rapid growth in culture or through exogenous expression of viral or mammalian oncogenes [2]. Numerous types of cancer cell lines exist and have been incredibly powerful and instrumental in facilitating in vitro studies of cancer cells. However, several issues exist with respect to the use of cell lines, which might explain partly why they fail to recapitulate the human response to drugs. First, they do not retain the tumor heterogeneity present in cancer, second, they accumulate mutations during culture over time, and third they do not represent all relevant components of the tumor microenvironment [143]. Yet, studies have shown that failure of cell lines to predict the human response to drugs could simply be caused by the insufficient number of cell lines tested, as it limits the number of subsets of cancers represented. Accordingly, a large-scale study describing the Cancer Cell Line Encyclopedia using numerous numbers of cell lines per cancer correctly identified multiple known features, including tissue type (lineage), as top predictors of sensitivity to several agents [144]. This suggests that cell lines can be relevant models but only if used in large enough numbers, which is often hard to do in preliminary and exploratory experiments. If large number of cell lines cannot be tested, scientists might have to choose carefully the ones selected for their study. For example, Domcke et al. analyzed the genome and transcriptome of 47 ovarian cell lines and compared it to more than 300 primary cancers. Their results suggest that the most commonly used ovarian cell lines in vitro are probably not good models of high-grade serous ovarian carcinoma, the most common human form of ovarian cancer [145].

Alternatively, patient-derived cells can be used, but their expansion is still variable and low [143], even though that field is currently leaping forward, as discussed in the following section [146]. At any rate, using patient-derived cells can also suffer from similar limitations such as using them in too small numbers, thus failing to represent many subsets of patients. They can also, just as cell lines, suffer from a lack of heterogeneity over time, due to the selective pressure of culture in the lab that will favour some sub-populations of cells over others. To prevent this, scientists are now developing patient-derived xenografts, which are based on the transfer of primary tumors directly from the patient into an immunodeficient mouse. These cells are maintained by passaging cells directly from mouse to mouse once the tumor burden becomes too high, thus by-passing their culture in the lab [146a].

Therefore, cell lines are bound to still being used in the near future while other protocols for extracting, growing, and expanding patient-specific cells in vitro are being optimized. Despite the limitations that cell lines present, they can be powerful if properly used. In particular, care should be taken toward authenticating cell lines. According to Horbach et al., more than 30,000 studies have reported research with misidentified cell lines [147]. This warrants the introduction of regular validation and screening procedures (e.g., using short tandem repeats [148], which are not necessarily performed as of today). This is probably because it takes too much time and money that scientists cannot afford to put in additional experiments. Hopefully progress in lab automation could help facilitate these tasks to ensure proper cell culture quality control. In addition to cell line misidentification, cells used in vitro can also become contaminated with microbes, which can affect results. Thus, improved practices and screens are needed to ensure scientists generate valid data in vitro. This quality control is bound to become increasingly important as in vitro assays become clinical companion tools.

7.2 High content assays and cell culture automation

A separate issue faced by scientists that use in vitro assay is their limited throughput. Many of the in vitro assays described in this chapter can be hard to set up. For example, while microfluidic device are meant to be "lab-on-a-chip," intended to facilitate time-consuming bench-top procedures, it has been said that they are instead often "chips-in-a-lab" [149] that require cumbersome external equipment (e.g., pumps, microscope) and much human intervention. Thus, there should be continuing effort toward the development of efficient workflows that can ensure higher throughput. This is particularly important for drug discovery, which generally requires the screening of large number of drugs. These numbers are becoming larger with the onset of combinatorial treatments that aim to combine several drugs together to prevent the development of cancer resistance toward one particular drug.

The throughput of in vitro assays can be limited by the need for imaging, which can be time consuming, especially when studying 3D in vitro assays. Yet, progress is being made in high content imaging assays that automate the imaging procedure and analysis. For example, Anastasov et al. grew numerous tumor spheroids, comprising cancer cells mixed with stromal cells, and imaged them over 14 days to quantify their growth under different combinations of radiation and chemotherapy. Using an automated high content imaging platform, they observed, in vitro, the shrinking effect of radiation on tumor spheroids as they would during treatment in patients (Fig. 5B). More interestingly, they identified vinblastine from a large panel of chemotherapeutic agents as a radio sensitizing agent, which, in combination with radiation was even more efficient at decreasing the size of spheroids compared to assays where vinblastine was used alone. The high content imaging platform uniquely facilitated the screening of numerous combinations of treatments. Scientists should make an effort to develop assays that are amenable to high-throughput workflows, while companies should develop instruments that facilitate the high-throughput analysis of samples. The automation of many laboratory tasks not only promises to increase throughput but also reproducibility, all of which will accelerate drug discovery.

7.3 In vitro cancer models: Future personalized avatars?

Even though one might dwell on these limitations, ongoing academic breakthroughs promise to provide enough incentives to overcome them. Currently, the field is focusing much of its energy on developing 3D personalized cellular models using cells derived from patients. One of the long-term goals is for in vitro assays to become miniaturized avatars of patients to test the unique effect of drugs on individual patients. Currently, there is growing interest in combining the use of patient-derived cells with microfluidic devices to maximize physiological relevance: that is, using the patient's very own cells in a controlled cellular, biophysical, and biochemical microenvironment.

To that end, scientists are learning to extract and expand cells from patients, which is proving to be hard. One avenue consists in obtaining induced pluripotent stem cells (iPSC) cells from patients by reprogramming already differentiated adult cells. This has the advantage of by-passing the need to isolate adult stem cells. iPSCs are cells that can be maintained in a self-sustaining pluripotent state equivalent to that of embryonic stem cells. They possess two unique properties: they can be maintained indefinitely in culture in an undifferentiated pluripotent state, and they can be directed to differentiate into any cell type of the human body [146]. This means that one could use iPSCs from a patient to create a disease model in vitro that captures the genome of the patient, and thus its unique response to drugs. iPSCs have been derived from immortalized human cells lines, and to a lesser extent from primary human cells. Importantly, iPSCs can be derived from normal or malignant primary cells. Briefly, cells are manipulated genetically or exposed to growth factors to be reprogrammed into iPSCs. Once they are reprogrammed, they can in theory be maintained in culture indefinitely and cryopreserved without any drift in their genetic, epigenetic, or phenotypic properties. At that point, they can in principle be differentiated into any cell type of interest.

These novel workflows are however suffering from experimental issues that are still being worked out. iPSCs offer tremendous potential for cancer research, from basic research, drug discovery, and precision oncology. They are bound to be of particular interest to drug discovery as they could be used to generate different cell types with the same genetic background corresponding to one patient, such that drugs can be tested not only on an in vitro model of the diseased organ but also on models of other healthy organs that could suffer from toxic side effects. At any rate, one of their more important appeals remains the fact that they can pave the way for personalized medicine. Specifically, in vitro models composed of iPSC derived cells could be used as companion tools for patients to assess and optimize the treatment according to the patient's own response to drugs. Many challenges remain before these iPSC-based models can be routinely used in the clinic, but they have already demonstrated their potential in recent publications. For example, iPSC cells derived from patients have already been used to draw associations between genotype and drug responses and to identify biomarkers to inform patient selection for enrollment in clinical trials [146]. One ground-breaking study has used iPSCs derived from a patient suffering from cardiac arrythmia to identify the mutation at stake and find the appropriate treatment. More specifically, they generated cardiomyocytes with iPSCs from the patients and identified a mutation in a sodium channel as responsible for the arrythmia. Mechanistic investigations allowed to pinpoint the precise origin of the arrythmia and improve the management of the disease in a patient-specific manner [150]. Thus, one could imagine having in vitro models with iPSCs derived from a patient differentiated into several different cell types to recreate a personalized diseased organ-on-chip in vitro model. Recent studies show that iPSCs are amenable to being used in microfluidic devices. A recent study has managed to recreate a perfusable microvascular network inside a microfluidic device with iPSCs. A separate group recreated a blood brain barrier inside a microfluidic device using iPSC-derived neurons [151,152].

On the other hand, circulating tumor cells (CTCs) are also believed to play an important part in the future of personalized medicine for cancer. In the last decades, numerous groups have developed tools to isolate them directly from the blood of patients using microfluidic chips. This has enabled the molecular characterization of CTCs and could provide opportunities for biomarker identification and dynamic tailoring of treatment. CTCs isolation has been studied in the context of several cancers, such as prostate cancer. Androgen receptor (AR) is a transcription factor that drives growth and progression of prostate cancer cells; drugs that target AR are standard care for prostate cancer patients, who respond well initially to anti-AR drugs but often develop resistance. CTCs might be key to understanding how this resistance develops but more urgently, to predict which patients will develop resistance to the treatment [153]. In an effort to identify biomarkers for predicting response to treatment, Sperger et al. evaluated AR expression and subcellular localization in CTCs isolated from patients with prostate cancer. They showed that CTCs from patients responding to treatment generally had lower AR expression and lower percentage of AR in the nucleus [154]. CTCs are particularly powerful sources of biomarkers as they can in theory be isolated repeatedly from patients through blood samples to dynamically assess the effect of drugs on individual patients. While there has been enormous progress on developing tools to isolate them from blood, scientists are still having trouble expanding them in vitro. Once that becomes feasible, these CTCs could be integrated inside in vitro models of metastasis such as those described in Section 5.2 to study the unique response of patient's CTCs inside physiologically relevant microenvironment.

Similarly, organoids generated from patient's cells promise to become key toward the development of personalized medicine, as previously discussed in Section 6.4. In a recent study, Hill et al. generated organoids from ovarian cancer patients and performed genomic analyses to screen for defects in DNA repair [155]. Interestingly, they find that defects in homologous recombination and replication fork protection corresponded to sensitivities to different treatments. This study further demonstrates the potential of using patient-derived cells to predict and tailor treatment to patients according to their unique genetic profiles. In future studies, such organoids could be introduced inside in vitro models of cancer for a more complete assessment of the patient's response to drugs as a function of its cellular microenvironment.

8 Conclusion

In this chapter, we have reviewed the importance and development of in vitro models for cancer research. We show that many types of in vitro models exist, with microfluidic devices showing particular potential for recreating physiologically relevant models as they enable a unique control of the microenvironment of cells at physiological length scales. We summarize the reasons explaining why culturing cells in 3D is important to recapitulate more faithfully numerous phenotypical and genetic features of cancer cells in vivo, even though 2D reductionist assays should still be considered to study selected aspects of cell motility. We review numerous studies that have brought forward in vitro models of distinct steps of cancer. We discussed the emerging trends in the field of in vitro modeling, suggesting that immune-competent models will become crucial to develop new cancer immunotherapies, or that organ on chips are already becoming invaluable models to better understand organ-specific cues that shape metastatic dissemination. All in all, these in vitro studies have not only helped visualize and understand cells as they complete key steps in the development of cancer but have also formed the basis for the creation of in vitro models that could be used as tools to screen drugs in realistic environments. We have discussed some limitations that will need to be overcome for such in vitro models to become reliable tools for drug discovery, such as their low throughput and lack of cell authentication. Finally, examples of recent ground-breaking studies were described to provide proof that patient-derived cells such as IPSC, CTCs, or organoids are giving unprecedented momentum to precision medicine. Now that we have learned to create complex tissue-like in vitro models with cell lines, eyes are set on upgrading these models with the use of patient-derived cells to turn them into personalized avatars of patients.

While there has been incredible progress in the ability to recreate complex cellular models of cancer, this should not make scientists forget that in vitro assays remain laboratory surrogates for humans. In other words, in vitro findings must always be validated against data obtained in humans. This might be one of the most difficult tasks for future biomedical scientists, as it is often difficult to obtain tissue from patients, let alone in large enough numbers. Therefore, advances in in vitro modeling will go hand in hand with progress made in developing reliable protocols for extracting and expanding patient-derived cells. On the other hand, the identification of valid biomarkers will prove invaluable to match findings of in vitro studies to patients. While it may sound like much remains to be done, small advances in each of these directions are closing the gap at an unprecedented speed. This progress is all the more acute now that these in vitro models do not pertain only to the academic realm and are also becoming attractive tools for pharmaceutical companies. The current landscape of active academic and pharmaceutical collaboration is bound to prioritize the optimization of workflows for the development of reproducible and efficient protocols that can have a more immediate and reliable impact on patients.

References

[1] C.T. Ambrose, An amended history of tissue culture: concerning harrison, burrows, mall, and carrel, J. Med. Biogr. 27 (2) (2016) 95–102, https://doi.org/10.1177/0967772016685033.

[2] M.I. Maqsood, M.M. Matin, A.R. Bahrami, M.M. Ghasroldasht, Immortality of cell lines: challenges and advantages of establishment, Cell Biol. Int. 37 (2013) 1038–1045.

[3] Y. Ando, et al., Evaluating CAR-T cell therapy in a hypoxic 3D tumor model, Adv. Healthc. Mater. 8 (2019).

[4] PHRMA, Biopharmaceutical Research & Development: The Process Behind New Medicines, 2015, pp. 1–21.

[5] D.G. Hackam, D.A. Redelmeier, Translation of research evidence from animals to humans, JAMA 296 (2006) 1731–1732.

[6] M. de Jong, T. Maina, Of mice and humans: are they the same?—implications in cancer translational research, J. Nucl. Med. 51 (2010) 501–504.

[7] D. Cook, D. Brown, R. Alexander, R. March, P. Morgan, Lessons learned from the fate of AstraZeneca's drug pipeline: a five-dimensional framework, Nat. Publ. Group 13 (2014) 419–431.

[8] D.-J. Cheon, S. Orsulic, Mouse models of cancer, Annu. Rev. Pathol. 6 (2011) 95–119.

[9] R. White, K. Rose, L. Zon, Zebrafish cancer: the state of the art and the path forward, Nat. Rev. Cancer 13 (2013) 624–636.

[10] B. Qian, et al., A distinct macrophage population mediates metastatic breast cancer cell extravasation, establishment and growth, PLoS One 4 (2009).

[11] A. Boussommier-calleja, R. Li, M.B. Chen, S.C. Wong, R.D. Kamm, Microfluidics: a new tool for modeling cancer-immune interactions, Trends Cancer 2 (2016) 6–19.

[12] J.A. Joyce, J.W. Pollard, Microenvironmental regulation of metastasis, Nat. Rev. Cancer 9 (2009) 239–252.

[13] P. De La Rochere, et al., Humanized mice for the study of immuno-oncology, Trends Immunol. 39 (2018) 748–763.
[14] E. Sahai, Illuminating the metastatic process, Nat. Rev. Cancer 7 (2007) 737–749.
[15] P. Jaiswal, Allied Market Research, 3D Cell Culture Market (Report Code LI171474), 2017.
[16] K. Haase, R.D. Kamm, Advances in on-chip vascularization, Regen. Med. 12 (2017) 285–302.
[17] M. Ogasawara, T. Matsubara, H. Suzuki, Screening of natural compounds for inhibitory activity on colon cancer cell migration, Biol. Pharm. Bull. 24 (2001) 720–723.
[18] P. Chen, et al., Gpr132 sensing of lactate mediates tumor–macrophage interplay to promote breast cancer metastasis, Proc. Natl. Acad. Sci. U. S. A. 114 (2017) 580–585.
[19] K.I. Hulkower, R.L. Herber, Cell migration and invasion assays as tools for drug discovery, Pharmaceutics 3 (2011) 107–124.
[20] B. Srinivasan, et al., TEER measurement techniques for in vitro barrier model systems, J. Lab. Autom. 20 (2015) 107–126.
[21] R. Mandic, T. Ludwig, H. Oberleithner, J.A. Werner, Evaluation of head and neck squamous cell carcinoma invasiveness by the electrical resistance breakdown assay, Clin. Exp. Metastasis 21 (2004) 699–704.
[22] K.J. Simpson, et al., Identification of genes that regulate epithelial cell migration using an siRNA screening approach, Nat. Cell Biol. 10 (2008) 1027–1038.
[23] B. Desoize, J.C. Jardillier, Multicellular resistance: a paradigm for clinical resistance? Crit. Rev. Oncol. Hematol. 36 (2000) 193–207.
[24] C. Erlichman, D. Vidgen, Cytotoxicity off adriamycin in MGH-U1 cells grown as monolayer cultures, spheroids, and xenografts in immune-deprived mice, Cancer Res. 44 (1984) 5369–5375.
[25] B.M. Baker, C.S. Chen, Deconstructing the third dimension—how 3D culture microenvironments alter cellular cues, J. Cell Sci. 125 (2012) 3015–3024.
[26] T. Mseka, J.R. Bamburg, L.P. Cramer, ADF/cofilin family proteins control formation of oriented actin-filament bundles in the cell body to trigger fibroblast polarization, J. Cell Sci. 120 (2007) 4332–4344.
[27] J. Meyers, J. Craig, D.J. Odde, Potential for control of signaling pathways via cell size and shape, Curr. Biol. 16 (2006) 1685–1693.
[28] V.M. Weaver, et al., β4 integrin-dependent formation of polarized three-dimensional architecture confers resistance to apoptosis in normal and malignant mammary epithelium, Cancer Cell 2 (2002) 205–216.
[29] J.M. Lee, et al., A three-dimensional microenvironment alters protein expression and chemosensitivity of epithelial ovarian cancer cells in vitro, Lab. Investig. 93 (2013) 528–542.
[30] R.J. Petrie, K.M. Yamada, At the leading edge of three-dimensional cell migration, J. Cell Sci. 125 (2012) 5917–5926.
[31] K. Talkenberger, E.A. Cavalcanti-Adam, A. Voss-Böhme, A. Deutsch, Amoeboid-mesenchymal migration plasticity promotes invasion only in complex heterogeneous microenvironments, Sci. Rep. 7 (2017) 9237.
[32] K. Paňková, D. Rösel, M. Novotný, J. Brábek, The molecular mechanisms of transition between mesenchymal and amoeboid invasiveness in tumor cells, Cell. Mol. Life Sci. 67 (2010) 63–71.
[33] M.J. Bissell, A. Rizki, I.S. Mian, Tissue architecture: the ultimate regulator of breast epithelial function, Curr. Opin. Cell Biol. 15 (2003) 753–762.
[34] M.W. Pickup, J.K. Mouw, V.M. Weaver, The extracellular matrix modulates the hallmarks of cancer, EMBO Rep. 15 (2014) 1243–1253.
[35] S.-H. Kim, J. Turnbull, S. Guimond, Extracellular matrix and cell signalling: the dynamic cooperation of integrin, proteoglycan and growth factor receptor, J. Endocrinol. 209 (2011) 139–151.
[36] P. Longati, et al., 3D pancreatic carcinoma spheroids induce a matrix-rich, chemoresistant phenotype offering a better model for drug testing, BMC Cancer 13 (2013) 95.
[37] M.H. Zaman, et al., Migration of tumor cells in 3D matrices is governed by matrix stiffness along with cell-matrix adhesion and proteolysis, Proc. Natl. Acad. Sci. U. S. A. 103 (2006) 10889–10894.
[38] E.L. Baker, R.T. Bonnecaze, M.H. Zaman, Extracellular matrix stiffness and architecture govern intracellular rheology in cancer, Biophys. J. 97 (2009) 1013–1021.

[39] J.A. Whisler, M.B. Chen, R.D. Kamm, Control of perfusable microvascular network morphology using a multiculture microfluidic system, Tissue Eng. Part C Methods 20 (2014) 543–552.
[40] J.H. Yeon, H.R. Ryu, M. Chung, Q.P. Hu, N.L. Jeon, In vitro formation and characterization of a perfusable three-dimensional tubular capillary network in microfluidic devices, Lab Chip 12 (2012) 2815–2822.
[41] M.B. Chen, et al., On-chip human microvasculature assay for visualization and quantification of tumor cell extravasation dynamics, Nat. Protoc. 12 (2017) 865–880.
[42] R. Hughes, et al., Perivascular M2 macrophages stimulate tumor relapse after chemotherapy, Cancer Res. (2015), https://doi.org/10.1158/0008-5472.CAN-14-3587.
[43] F. Perche, V.P. Torchilin, Cancer cell spheroids as a model to evaluate chemotherapy protocols, Cancer Biol. Ther. 13 (2012) 1205–1213.
[44] R.E. Durand, S.L. Vanderbyl, Tumor resistance to therapy: a genetic or kinetic problem? Cancer Commun. 1 (1989) 277–283.
[45] R.-Z. Lin, R.-Z. Lin, H.-Y. Chang, Recent advances in three-dimensional multicellular spheroid culture for biomedical research, Biotechnol. J. 3 (2008) 1172–1184.
[46] N.D. Marjanovic, R.A. Weinberg, C.L. Chaffer, Cell plasticity and heterogeneity in cancer, Clin. Chem. 59 (2013) 168–179.
[47] D. Horiatis, Q. Wang, J. Pinski, A new screening system for proliferation-independent anti-cancer agents, Cancer Lett. 210 (2004) 119–124.
[48] H.R. Mellor, D.J.P. Ferguson, R. Callaghan, A model of quiescent tumour microregions for evaluating multicellular resistance to chemotherapeutic drugs, Br. J. Cancer 93 (2005) 302–309.
[49] A.D. Doyle, R.J. Petrie, M.L. Kutys, K.M. Yamada, Dimensions in cell migration, Curr. Opin. Cell Biol. 25 (2013) 642–649.
[50] H.M. Estabridis, A. Jana, A. Nain, D.J. Odde, Cell migration in 1D and 2D nanofiber microenvironments, Ann. Biomed. Eng. 46 (2018) 392–403.
[51] J.A. Hickman, et al., Three-dimensional models of cancer for pharmacology and cancer cell biology: capturing tumor complexity in vitro/ex vivo, Biotechnol. J. 9 (2014) 1115–1128.
[52] Y. Kienast, et al., Real-time imaging reveals the single steps of brain metastasis formation, Nat. Med. 16 (2010) 116–122.
[53] A. Giannattasio, et al., Cytotoxicity and infiltration of human NK cells in in vivo-like tumor spheroids, BMC Cancer 15 (2015) 351.
[54] S.-A. Kim, E.K. Lee, H.-J. Kuh, Co-culture of 3D tumor spheroids with fibroblasts as a model for epithelial-mesenchymal transition in vitro, Exp. Cell Res. 335 (2015) 187–196.
[55] D.D. Fang, et al., Expansion of CD133(+) colon cancer cultures retaining stem cell properties to enable cancer stem cell target discovery, Br. J. Cancer 102 (2010) 1265–1275.
[56] Y.-C. Tung, et al., High-throughput 3D spheroid culture and drug testing using a 384 hanging drop array, Analyst 136 (2011) 473–478.
[57] C. Wenzel, et al., 3D high-content screening for the identification of compounds that target cells in dormant tumor spheroid regions, Exp. Cell Res. 323 (2014) 131–143.
[58] E. Knight, S. Przyborski, Advances in 3D cell culture technologies enabling tissue-like structures to be created in vitro, J. Anat. 227 (2015) 746–756.
[59] K.T. Morin, R.T. Tranquillo, In vitro models of angiogenesis and vasculogenesis in fibrin gel, Exp. Cell Res. 319 (2013) 2409–2417.
[60] A.K. Kudva, F.P. Luyten, J. Patterson, RGD-functionalized polyethylene glycol hydrogels support proliferation and in vitro chondrogenesis of human periosteum-derived cells, J. Biomed. Mater. Res. A 106 (2018) 33–42.
[61] B.P. Chan, K.W. Leong, Scaffolding in tissue engineering: general approaches and tissue-specific considerations, Eur. Spine J. 17 (2008).
[62] G.M. Whitesides, The origins and the future of microfluidics, Nature 442 (2006) 368–373.
[63] W.J. Polacheck, R. Li, S.G.M. Uzel, R.D. Kamm, Microfluidic platforms for mechanobiology, Lab Chip 13 (2013) 2252–2267.
[64] S. Kim, H.J. Kim, N.L. Jeon, Biological applications of microfluidic gradient devices, Integr. Biol. 2 (2010) 584.

[65] D. Huh, et al., Reconstituting organ-level lung functions on a chip, Science 328 (2010) 1662–1668.
[66] B. Patra, et al., A microfluidic device for uniform-sized cell spheroids formation, culture, harvesting and flow cytometry analysis, Biomicrofluidics 7 (2013).
[67] N. Thaitrong, R. Charlermroj, O. Himananto, C. Seepiban, N. Karoonuthaisiri, Implementation of microfluidic sandwich ELISA for superior detection of plant pathogens, PLoS One 8 (2013).
[68] A.J. Hughes, A.E. Herr, Microfluidic Western blotting, Proc. Natl. Acad. Sci. U. S. A. 109 (2012) 21450–21455.
[69] K.J. Regehr, et al., Biological implications of polydimethylsiloxane-based microfluidic cell culture, Lab Chip 9 (2009) 2132–2139.
[70] J. Lachaux, et al., Thermoplastic elastomer with advanced hydrophilization and bonding performances for rapid (30 s) and easy molding of microfluidic devices, Lab Chip 17 (2017) 2581–2594.
[71] D. Wendt, S.A. Riboldi, M. Cioffi, I. Martin, Potential and bottlenecks of bioreactors in 3D cell culture and tissue manufacturing, Adv. Mater. 21 (2009) 3352–3367.
[72] A.E. Guller, P.N. Grebenyuk, A.B. Shekhter, A.V. Zvyagin, S.M. Deyev, Bioreactor-based tumor tissue engineering, Acta Nat. 8 (2016) 44–58.
[73] Z.N. Demou, Gene expression profiles in 3D tumor analogs indicate compressive strain differentially enhances metastatic potential, Ann. Biomed. Eng. 38 (2010) 3509–3520.
[74] K. Domansky, W. Inman, J. Serdy, A. Dash, Perfused multiwell plate for 3D liver tissue engineering, Lab Chip 10 (2010) 51–58.
[75] Y.L. Chao, C.R. Shepard, A. Wells, Breast carcinoma cells re-express E-cadherin during mesenchymal to epithelial reverting transition, Mol. Cancer 9 (2010) 179.
[76] Y. Chao, Q. Wu, C. Shepard, A. Wells, Hepatocyte induced re-expression of E-cadherin in breast and prostate cancer cells increases chemoresistance, Clin. Exp. Metastasis 29 (2012) 39–50.
[77] K. Matsuura, R. Utoh, K. Nagase, T. Okano, Cell sheet approach for tissue engineering and regenerative medicine, J. Control. Release 190 (2014) 228–239.
[78] A.F. Chambers, A.C. Groom, I.C. Mac Donald, Dissemination and growth of cancer cells in metastatic sites, Nat. Rev. Cancer 2 (2002) 563–572.
[79] N. Anastasov, et al., A 3D-microtissue-based phenotypic screening of radiation resistant tumor cells with synchronized chemotherapeutic treatment, BMC Cancer 15 (2015) 466.
[80] L.C. Roudsari, J.L. West, Studying the influence of angiogenesis in in vitro cancer model systems, Adv. Drug Deliv. Rev. (2015), https://doi.org/10.1016/j.addr.2015.11.004.
[81] S.S. Verbridge, et al., Oxygen-controlled three-dimensional cultures to analyze tumor angiogenesis, Tissue Eng. Part A 16 (2010) 2133–2141.
[82] Z. Chen, et al., In vitro angiogenesis by human umbilical vein endothelial cells (HUVEC) induced by three-dimensional co-culture with glioblastoma cells, J. Neurooncol. (2009) 121–128, https://doi.org/10.1007/s11060-008-9742-y.
[83] H. Lee, W. Park, H. Ryu, N.L. Jeon, A microfluidic platform for quantitative analysis of cancer angiogenesis and intravasationa, Biomicrofluidics 8 (2014).
[84] C.-T. Kuo, et al., A spatiotemporally defined in vitro microenvironment for controllable signal delivery and drug screening, Analyst (2014) 15–23, https://doi.org/10.1039/C4AN00936C.
[85] C. Te Kuo, et al., Modeling of cancer metastasis and drug resistance via biomimetic nano-cilia and microfluidics, Biomaterials 35 (2014) 1562–1571.
[86] L.L. Bischel, D.J. Beebe, K.E. Sung, Microfluidic model of ductal carcinoma in situ with 3D, organotypic structure, BMC Cancer 15 (2015) 1–10.
[87] K.E. Sung, et al., Transition to invasion in breast cancer: a microfluidic in vitro model enables examination of spatial and temporal effects, Integr. Biol. (Camb.) 3 (439–450) (2011).
[88] M.A. Acosta, et al., A microfluidic device to study cancer metastasis under chronic and intermittent hypoxia, Biomicrofluidics 8 (5) (2014).
[89] M.K. Shin, S.K. Kim, H. Jung, Integration of intra- and extravasation in one cell-based microfluidic chip for the study of cancer metastasis, Lab Chip 11 (2011) 3880–3887.
[90] I.K. Zervantonakis, et al., Three-dimensional microfluidic model for tumor cell intravasation and endothelial barrier function, Proc. Natl. Acad. Sci. U. S. A. 109 (2012) 13515–13520.

[91] J.B. Wyckoff, et al., Direct visualization of macrophage-assisted tumor cell intravasation in mammary tumors, Cancer Res. 67 (2007) 2649–2656.
[92] M. Pisano, V. Triacca, K.A. Barbee, M.A. Swartz, An in vitro model of the tumor–lymphatic microenvironment with simultaneous transendothelial and luminal flows reveals mechanisms of flow enhanced invasion, Integr. Biol. (2015) 1–5, https://doi.org/10.1039/C5IB00085H.
[93] S.M. Ehsan, K.M. Welch-Reardon, M.L. Waterman, C.C.W. Hughes, S.C. George, A three-dimensional in vitro model of tumor cell intravasation, Integr. Biol. (Camb.) 6 (2014) 603–610.
[94] J. Kim, et al., Cooperative actions of p21WAF1 and p53 induce slug protein degradation and suppress cell invasion, EMBO Rep. 15 (10) (2014) 1062–1068.
[95] A. Spencer, et al., A high-throughput mechanofluidic screening platform for investigating tumor cell adhesion during metastasis, Lab Chip 16 (2016) 142–152.
[96] S. Liang, M.J. Slattery, D. Wagner, S.I. Simon, C. Dong, Hydrodynamic shear rate regulates melanoma-leukocyte aggregation, melanoma adhesion to the endothelium, and subsequent extravasation, Ann. Biomed. Eng. 36 (2008) 661–671.
[97] J.W. Song, et al., Microfluidic endothelium for studying the intravascular adhesion of metastatic breast cancer cells, PLoS One 4 (2009).
[98] R. Riahi, et al., A microfluidic model for organ-specific extravasation of circulating tumor cells, Biomicrofluidics 8 (2014).
[99] A. Boussommier-Calleja, et al., The effects of monocytes on tumor cell extravasation in a 3D vascularized microfluidic model, Biomaterials (2018), https://doi.org/10.1016/j.biomaterials.2018.03.005.
[100] E.B. Voura, et al., Proteolysis during tumor cell extravasation in vitro: metalloproteinase involvement across tumor cell types, PLoS One 8 (2013) 1–15.
[101] J.S. Jeon, I.K. Zervantonakis, S. Chung, R.D. Kamm, J.L. Charest, In vitro model of tumor cell extravasation, PLoS One 8 (2013).
[102] M.B. Chen, J.A. Whisler, J.S. Jeon, R.D. Kamm, Mechanisms of tumor cell extravasation in an in vitro microvascular network platform, Integr. Biol. (Camb.) 5 (2013) 1262–1271.
[103] Q. Zhang, T. Liu, J. Qin, A microfluidic-based device for study of transendothelial invasion of tumor aggregates in realtime, Lab Chip 12 (2012) 2837.
[104] S. Bersini, J.S. Jeon, G. Dubini, C. Arrigoni, S. Chung, J.L. Charest, M. Moretti, R. D. Kamm, A microfluidic 3D in vitro model for specificity of breast cancer metastasis to bone, Biomaterials 35 (2014) 2454–2461.
[105] K.C. Chaw, M. Manimaran, F.E.H. Tay, S. Swaminathan, A quantitative observation and imaging of single tumor cell migration and deformation using a multi-gap microfluidic device representing the blood vessel, Microvasc. Res. 72 (2006) 153–160.
[106] X. Cui, et al., A microfluidic device for isolation and characterization of transendothelial migrating cancer cells, Biomicrofluidics 11 (2017) 1–12.
[107] S. Ito, et al., Real-time observation of micrometastasis formation in the living mouse liver using a green fluorescent protein gene-tagged rat tongue carcinoma cell line, Int. J. Cancer 93 (2001) 212–217.
[108] C.M. Ghajar, et al., The perivascular niche regulates breast tumour dormancy, Nat. Cell Biol. 15 (2013) 807–817.
[109] S. Braun, et al., A pooled analysis of bone marrow micrometastasis in breast cancer, N. Engl. J. Med. 353 (2005) 793–802.
[110] R.N. Kaplan, et al., VEGFR1-positive haematopoietic bone marrow progenitors initiate the pre-metastatic niche, Nature 438 (2005) 820–827.
[111] S. Sieh, et al., Paracrine interactions between LNCaP prostate cancer cells and bioengineered bone in 3D in vitro culture reflect molecular changes during bone metastasis, Bone 63 (2014) 121–131.
[112] M.-Y. Kim, et al., Tumor self-seeding by circulating cancer cells, Cell 139 (2009) 1315–1326.
[113] D.M. Pardoll, The blockade of immune checkpoints in cancer immunotherapy, Nat. Rev. Cancer 12 (2012) 252–264.
[114] F.S. Hodi, et al., Improved survival with ipilimumab in patients with metastatic melanoma, N. Engl. J. Med. 363 (2010) 711–723.

[115] G. Dranoff, Experimental mouse tumour models: what can be learnt about human cancer immunology? Nat. Rev. Immunol. 12 (2011) 61–66.
[116] T. Kitamura, B.-Z. Qian, J.W. Pollard, Immune cell promotion of metastasis, Nat. Rev. Immunol. 15 (2015) 73–86.
[117] E. Vacchelli, et al., Chemotherapy-induced antitumor immunity requires formyl peptide receptor 1, Science (2015) 1–11, https://doi.org/10.1126/science.aad0779.
[118] A. Pavesi, et al., A 3D microfluidic model for preclinical evaluation of TCR-engineered T cells against solid tumors, JCI Insight 2 (2017) 89762.
[119] S.W.L. Lee, et al., Characterizing the role of monocytes in T cell cancer immunotherapy using a 3D microfluidic model, Front. Immunol. 9 (2018) 416.
[120] L.B. Rivera, et al., Intratumoral myeloid cells regulate responsiveness and resistance to antiangiogenic therapy, Cell Rep. 11 (2015) 577–591.
[121] A. Jain, et al., Primary human lung alveolus-on-a-chip model of intravascular thrombosis for assessment of therapeutics, Clin. Pharmacol. Ther. 103 (2018) 332–340.
[122] S. Kim, H. Lee, M. Chung, N.L. Jeon, Engineering of functional, perfusable 3D microvascular networks on a chip, Lab Chip 13 (2013) 1489–1500.
[123] I.J. Fidler, The pathogenesis of cancer metastasis: the 'seed and soil' hypothesis revisited, Nat. Rev. Cancer 3 (2003) 453–458.
[124] L. Weiss, Metastasis of cancer: a conceptual history from antiquity to the 1990s, Cancer Metastasis Rev. 19 (I–XI) (2000) 193–383.
[125] E.V. Sugarbaker, Cancer metastasis: a product of tumor-host interactions, Curr. Probl. Cancer 3 (1979) 1–59.
[126] I.R. Hart, I.J. Fidler, Role of organ selectivity in the determination of metastatic patterns of B16 melanoma, Cancer Res. 40 (1980) 2281–2287.
[127] A. Kostic, C.D. Lynch, M.P. Sheetz, Differential matrix rigidity response in breast cancer cell lines correlates with the tissue tropism, PLoS One 4 (2009).
[128] G.P. Gupta, et al., Identifying site-specific metastasis genes and functions, Cold Spring Harb. Symp. Quant. Biol. 70 (2005) 149–158.
[129] H.J. Kim, D. Huh, G. Hamilton, D.E. Ingber, Human gut-on-a-chip inhabited by microbial flora that experiences intestinal peristalsis-like motions and flow, Lab Chip 12 (2012) 2165.
[130] B.A. Hassell, et al., Human organ chip models recapitulate orthotopic lung cancer growth, therapeutic responses, and tumor dormancy in vitro, Cell Rep. 21 (2017) 508–516.
[131] H. Xu, et al., A dynamic in vivo-like organotypic blood-brain barrier model to probe metastatic brain tumors, Sci. Rep. 6 (2016) 1–12.
[132] D.W. Lee, S.K. Ha, I. Choi, J.H. Sung, 3D gut-liver chip with a PK model for prediction of first-pass metabolism, Biomed. Microdevices 19 (2017) 100.
[133] J.M. Prot, et al., First pass intestinal and liver metabolism of paracetamol in a microfluidic platform coupled with a mathematical modeling as a means of evaluating ADME processes in humans, Biotechnol. Bioeng. 111 (2014) 2027–2040.
[134] H. Xu, et al., Organoid technology and applications in cancer research, J. Hematol. Oncol. (2018) 1–15, https://doi.org/10.1186/s13045-018-0662-9.
[135] M.A. Lancaster, J.A. Knoblich, Organogenesis in a dish: modeling development and disease using organoid technologies, Science 345 (2014) 1247125.
[136] T. Sato, et al., Single Lgr5 stem cells build crypt-villus structures in vitro without a mesenchymal niche, Nature 459 (2009) 262–265.
[137] F. Weeber, S.N. Ooft, K.K. Dijkstra, E.E. Voest, Tumor organoids as a pre-clinical cancer model for drug discovery, Cell Chem. Biol. 24 (2017) 1092–1100.
[138] J. Drost, et al., Use of CRISPR-modified human stem cell organoids to study the origin of mutational signatures in cancer, Science 358 (2017) 234–238.
[139] T. Scanu, et al., Salmonella manipulation of host signaling pathways provokes cellular transformation associated with gallbladder carcinoma, Cell Host Microbe 17 (2015) 763–774.
[140] D. Gao, et al., Organoid cultures derived from patients with advanced prostate cancer, Cell 159 (2014) 176–187.

[141] T. Seidlitz, et al., Human gastric cancer modelling using organoids, Gut (2018) 1–11, https://doi.org/10.1136/gutjnl-2017-314549.
[142] M. Kasendra, et al., Development of a primary human Small Intestine-on-a-Chip using biopsy-derived organoids, Sci. Rep. 8 (2018) 1–14.
[143] J.L. Wilding, W.F. Bodmer, Cancer cell lines for drug discovery and development, Cancer Res. 74 (2014) 2377–2384.
[144] J. Barretina, et al., The cancer cell line encyclopedia enables predictive modelling of anticancer drug sensitivity, Nature 483 (2012) 603–607.
[145] S. Domcke, R. Sinha, D.A. Levine, C. Sander, N. Schultz, Evaluating cell lines as tumour models by comparison of genomic profiles, Nat. Commun. 4 (2013) 2126.
[146] E.P. Papapetrou, Patient-derived induced pluripotent stem cells in cancer research and precision oncology, Nat. Med. 22 (2016) 1392–1401.
[146a] D. Siolas, G.J. Hannon, Patient derived tumor xenografts: transforming clinical samples into mouse models, Cancer Res. 73 (17) (2013) 5315–5319.
[147] S.P.J.M. Horbach, W. Halffman, The ghosts of HeLa: how cell line misidentification contaminates the scientific literature, PLoS One 12 (2017) 1–16.
[148] J.R. Masters, et al., Short tandem repeat profiling provides an international reference standard for human cell lines, Proc. Natl. Acad. Sci. U. S. A. 98 (2001) 8012–8017.
[149] A.M. Streets, Y. Huang, Chip in a lab: microfluidics for next generation life science research, Biomicrofluidics 7 (2013).
[150] C. Terrenoire, et al., Induced pluripotent stem cells used to reveal drug actions in a long QT syndrome family with complex genetics, J. Gen. Physiol. 141 (2013) 61–72.
[151] J.A. Brown, et al., Recreating blood-brain barrier physiology and structure on chip: a novel neurovascular microfluidic bioreactor, Biomicrofluidics 9 (2015).
[152] D.G. Belair, et al., Human vascular tissue models formed from human induced pluripotent stem cell derived endothelial cells, Stem Cell Rev. Rep. 11 (2015) 511–525.
[153] G. Galletti, D. Worroll, D.M. Nanus, P. Giannakakou, Using circulating tumor cells to advance precision medicine in prostate cancer, J. Cancer Metastasis Treat. (2017) 190–205, https://doi.org/10.20517/2394-4722.2017.45.
[154] J.M. Sperger, et al., Integrated analysis of multiple biomarkers from circulating tumor cells enabled by exclusion-based analyte Isolation, Clin. Cancer Res. 23 (2017) 746–756.
[155] S.J. Hill, et al., Prediction of DNA repair inhibitor response in short term patient-derived ovarian cancer organoids, Cancer Discov. 8 (11) (2018) 1404–1421, https://doi.org/10.1158/2159-8290.CD-18-0474.

Further reading

[156] A. Boussommier-Calleja, 3D cell culture: market and industrial needs, an Elveflow Tutorial, 2018, https://www.elveflow.com/organs-on-chip/3d-cell-culture-market-industrial-needs/.
[157] S. Riffle, R.N. Pandey, M. Albert, R.S. Hegde, Linking hypoxia, DNA damage and proliferation in multicellular tumor spheroids, BMC Cancer 17 (2017) 1–12.
[158] M. Wolun-cholewa, et al., An efficient 3D cell culture method on biomimetic nanostructured grids, PLoS One 8 (2013) 4–9.
[159] R. Li, et al., Macrophage-secreted TNFα and TGFβ1 influence migration speed and persistence of cancer cells in 3D tissue culture via independent pathways, Cancer Res. 77 (2017) 279–290.
[160] K.H. Benam, et al., Small airway-on-a-chip enables analysis of human lung inflammation and drug responses in vitro, Nat. Methods 13 (2016) 151–157.

CHAPTER 4.2

Biomechanics of cancer cells

Alistair Rice, Armando del Rio Hernandez
Cellular and Molecular Biomechanics Laboratory, Department of Bioengineering, Imperial College London, London, United Kingdom

1 Introduction

Whether squeezing through a small gap in a basement membrane, migrating through a dense stroma, or becoming malignant in response to extracellular rigidity, cancer cells survive in an environment full of physical forces. Cells experience the forces of compression, shear stress and tension, and in response, cells deform viscoelastically and initiate intracellular signaling pathways. Furthermore, cells maintain a physical equilibrium with their surroundings by generating tension within their cytoskeleton, this force generation being also essential for cell migration.

Cancer cells are known to differ in their mechanical attributes in relation to their healthy counterparts, and hence an appreciation of oncological cell mechanics is necessary to understand the deeply complex field of cancer progression. Tumor development is associated with a rigid extracellular matrix (ECM) due to increased ECM protein secretion and remodeling, which promotes the acquisition of malignant properties in healthy cells through mechanosensing. Cancer cells are generally softer but also more contractile than healthy cells, allowing them to squeeze and push through the dense stroma that develops alongside the tumor [1,2]. Recent research using novel biomechanical techniques has elucidated many key ideas about cancer cell mechanics and their relevance to tumor progression, though many complexities have not been fully determined. For example, there is still debate about whether cell softening occurs for all cancer types [1], and how the optimal mechanical properties for various steps in metastasis are likely to differ [2] (Fig. 1).

Here we describe the current methods for mechanical analysis of cancer and cancer-related cells, highlighting the specific information that each assay provides and how studies using these methods have begun to unravel the details of cancer mechanics.

1.1 The cytoskeleton in cancer cells

The mechanical properties of cancer cells are determined by the cytoskeleton, an interconnected network of proteins within cells that defines their shape and movement. The

Bioengineering Innovative Solutions for Cancer
https://doi.org/10.1016/B978-0-12-813886-1.00014-0

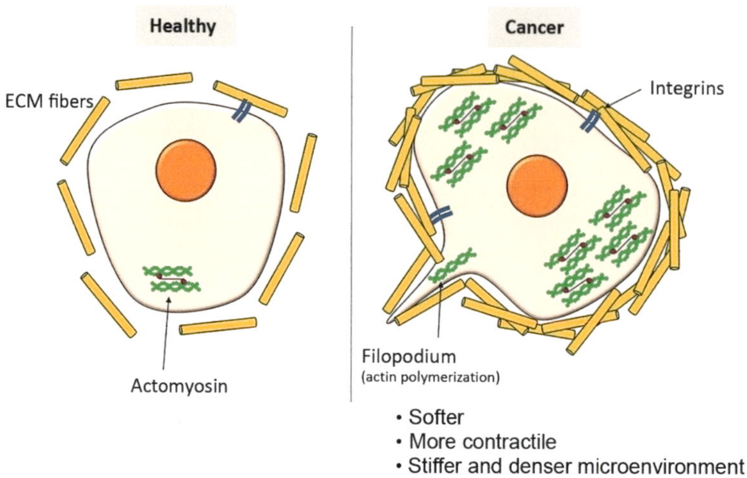

Fig. 1 Mechanical differences between healthy and cancer cells. Diagram of healthy and cancer cell. Cancer cells are surrounded by a higher density of ECM proteins, e.g., collagen and fibronectin. The behavior of cancer cells is influenced by integrin-mediated mechanotransduction of a stiff and dense microenvironment. Cancer cells are generally softer than healthy cells, and show increased contractility produced by actomyosin tension.

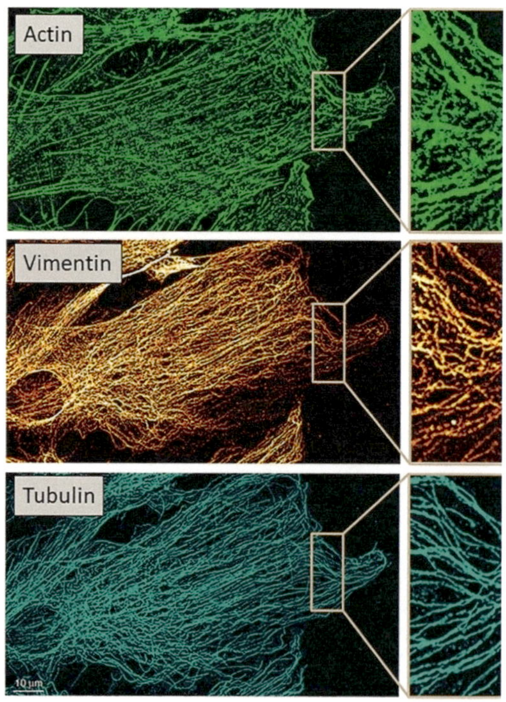

Fig. 2 Cytoskeletal proteins and their cellular organization. (Left) Actin filaments *(green)*, intermediate filaments (vimentin, *yellow*), and microtubules (tubulin, *blue*) and (Right) their organization within cells. *(Adapted from C. Leduc, S. Etienne-Manneville, Intermediate filaments in cell migration and invasion: the unusual suspects, Curr. Opin. Cell Biol. 32 (2015) 102–112.)*

cytoskeleton is made of actin, intermediate filaments, and microtubules (Fig. 2), all of which contribute to overall cell mechanics through their abundance, organization, and dynamic remodeling.

1.1.1 Actin filaments

The actin network consists of F-actin fibers, formed from the dynamic process of polymerization/depolymerization of G-actin monomers. Actin polymerization generates polarized actin filaments, with assembly faster at the plus end and disassembly faster at the minus end, generating force at the growing end in a process known as treadmilling. Additionally, actin filaments can assemble into a variety of structures such as stress fibers, filopodia, lamellipodia, or the cell cortex, with branching initiated by the Arp2/3 complex. Actin also binds myosin filaments to form actomyosin, which drives cell contractility (Fig. 3A).

Actin structures can be force generating and/or structural. One important actin structure is the stress fiber, formed from large bundles of actin fibers, which connects to large protein complexes called focal adhesions at the cell membrane. Ventral stress fibers link to focal adhesions at both ends, providing a link between the external environment and internal cytoskeleton. Additionally, actin in ventral stress fibers binds to myosin filaments, where relative movement of actin and myosin leads to contraction. Dorsal stress fibers on the other hand, which link transverse arcs to focal adhesions, do not bind myosin, and do not generate force [3]. Force is also generated by actin polymerization and is essential in the extension of membrane protrusions by lamellipodia or filopodia in cell migration.

The dynamics of the actin cytoskeleton are primarily determined by members of the Rho GTPase family: Rho, Rac, and Cdc42. These proteins are activated when guanine nucleotide exchange factors (GEFs) swap bound guanosine diphosphate (GDP) for guanosine triphosphate (GTP), and then intrinsic GTPase activity and/or GTPase activating proteins (GAPs) facilitate the hydrolysis of GTP back to GDP allowing transient activation (Fig. 4). When activated following GTP binding, Rho activates ROCK, which phosphorylates myosin light chain (MLC), an accessory protein that promotes actomyosin contraction. Rac and Cdc42 promote actin polymerization through Arp2/3 or mDia. Additionally, Cdc42 regulates cell polarity through Par6. Intrinsic GTPase activity, as well as GTPase activity by other proteins, leads to inactivation of these Rho GTPase family members. Abundance of either Rho or Rac drives different mechanisms of motility, facilitated by the mutual antagonism between the two GTPases [4,5]. Rho GTPases are known to be activated downstream of mechanosensing proteins such as integrins, linking the external environment to actin dynamics.

1.1.2 Microtubules

Microtubules are also polymers, made up of hollow tubes formed from the polymerization of α-β tubulin heterodimers, and are the stiffest of the cytoskeletal polymers (Fig. 3B) [6]. According to the tensegrity model of cell mechanics, microtubules carry compressive

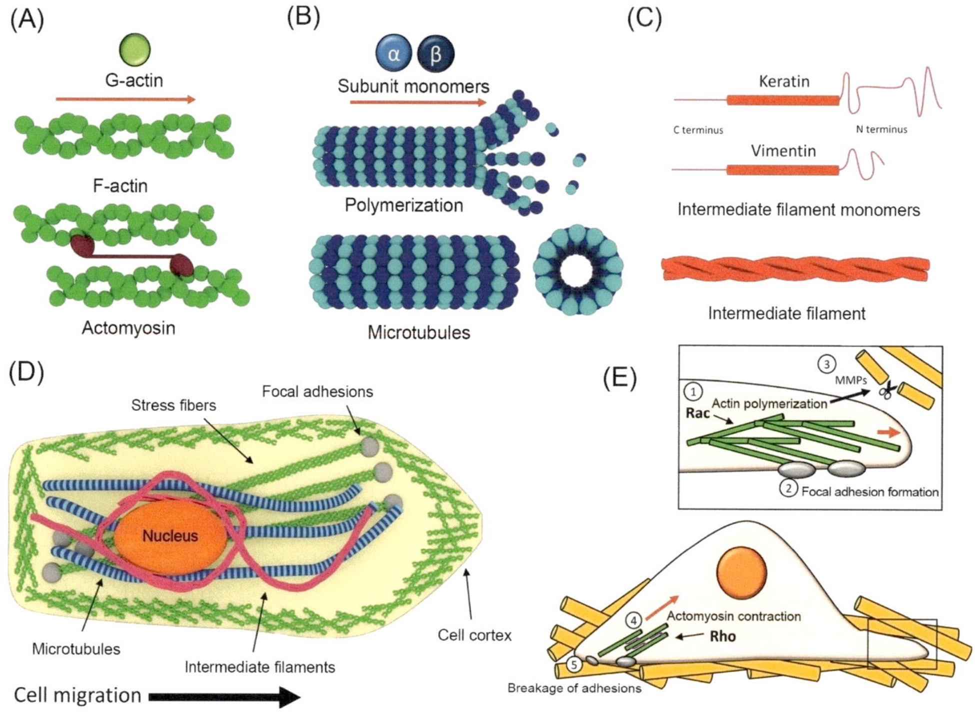

Fig. 3 Cytoskeletal filaments and mesenchymal motility. (A) Structure of actin. G-actin monomers *(green circles)* polymerize into polarized F-actin fibers. These fibers *(green cylinders)* can associate with myosin *(purple)* to form actomyosin, a contractile actin structure. *Red arrow* indicates polarity from − to +. (B) Structure of microtubules. α and β subunits *(light blue and dark blue)* polymerize to form a polarized hollow tube. *Red arrow* indicates polarity from − to +. (C) Structure of intermediate filaments. Individual monomers, keratin and vimentin, are made of α helical rod and different N and C termini. Monomers combine into dimers, then tetramers, eventually formed intertwined filaments with high tensile strength. (D) Structure of cytoskeleton during mesenchymal migration. From earlier, cytoskeletal structures are highlighted. Actin makes up the cell cortex, as well as contractile stress fibers linked to focal adhesions. Microtubules and intermediate filaments are also present. (E) Mechanism of mesenchymal motility. The five steps of mesenchymal motility are highlighted. (1) Rac-mediated actin polymerization at the leading edge, (2) focal adhesion formation, (3) secretion of proteases, (4) actomyosin contraction at the cell rear, (5) breakage of focal adhesions. *Red arrows* indicate force application.

force as part of balancing mechanical stress within living cells [7]. Dynein and kinesin motors use the track provided by microtubules for transport of intracellular cargo, and in lamellipodia formation during migration, microtubules deliver membrane vesicles for membrane extension. The polymerization/depolymerization of microtubules is suggested to be important in force generation, and microtubules can additionally regulate focal adhesion dynamics and cell polarization [8].

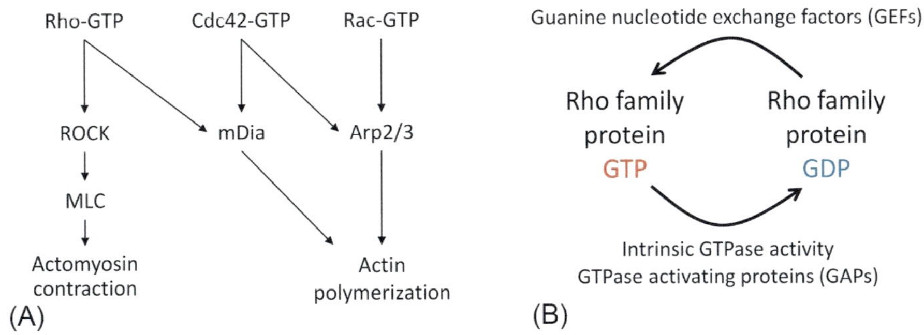

Fig. 4 Rho GTPase family signaling. (A) Downstream effects of Rho family proteins. Rho-GTP activates both ROCK and mDia, with ROCK promoting activation of myosin light chain (MLC) and therefore actomyosin contraction. Cdc42-GTP activates both mDia and Arp2/3, with both effectors promoting actin polymerization. Rac-GTP also activates Arp2/3. (B) G-protein cycle. Members of the Rho family are activated by guanine nucleotide exchange factors (GEFs), which swap bound GDP for GTP. GTP is then hydrolyzed by intrinsic GTPase activity and/or facilitated by GTPase activating proteins (GAPs).

1.1.3 Intermediate filaments

Intermediate filaments are a group of α-helix rich proteins found in both the cytoplasm and around the nuclear envelope. Monomers are formed from an α-helical rod domain, with N and C terminal domains that differ between different types of intermediate filament. These monomers polymerize into dimers, tetramers, and then finally into a helical array, which lacks the polarity of either actin filaments or microtubules (Fig. 3C). The rope-like structure of intermediate filaments lends them high tensile stress, which is important when cells are stretched or experiencing shear force, though resistance to compressive stress is limited. Intermediate filaments crosslink to each other, as well as to actin and microtubules, and therefore are essential in cytoskeletal interconnectivity [6].

Of the types of intermediate filaments, keratin and vimentin filaments have a known role in cancer. In epithelial sheets, keratin filaments span across the entire cell and link to keratin filaments in adjacent cells via desmosomes, providing tensile strength across the entire sheet and preventing rupture. Keratin filaments contribute to cell stiffness, where their absence significantly increases the deformability of cells [9]. The epithelial-mesenchymal transition (EMT), a phenotypic change associated with cancer progression that can be induced by external stiffness, involves the switch from a keratin to a vimentin-based cytoskeleton [10], with vimentin increasing cell stiffness and directional motility [11].

1.2 Quantifying cancer cells biomechanics

We have categorized cell mechanics into three mechanical attributes: viscoelasticity, force generation, and mechanosensing. Each of these can be quantified in a variety of ways, and differences in these attributes are associated with oncogenic transformation and/or malignancy.

- *Viscoelasticity* represents the ability of a cell to resist both instantaneous and gradual deformation following force application. This occurs since cells exhibit both elastic and viscous behavior in response to stress. In the context of cancer, the deformability of a cell is important in squeezing through gaps in processes like intravasation or amoeboid migration. Furthermore, due to reported differences between cancer and healthy cells, cell elasticity has been proposed as a cancer cell biomarker [12].
- *Force generation* represents the ability of a cell to generate mechanical tension. This is required for the tensional homeostasis of a cell with its physical environment, as well as surface adhesion and traction force generation, which are required for cell migration.
- *Mechanosensing* represents the ability of a cell to alter its physical and/or chemical phenotype in response to mechanical stimulation. Cancer cells are known to mechanosense their environment, where physical changes surrounding cells can promote phenotypic changes required for invasion and metastasis. Additionally, cancer-associated cells can augment the tumor microenvironment by also responding to its rigidity, leading to cancer-associated cell activation and a fibrotic positive feedback loop.

The cytoskeleton influences all aspect of cell mechanics and the three mechanical properties of viscoelasticity, force generation, and mechanosensing do not work in isolation. Hence, a key aspect of mechanical analysis is studying the interplay between these different attributes. For example, the complex metastatic cascade, from initial tumor growth to reaching the secondary site, involves many steps in which the mechanical properties of cells contribute to metastatic efficiency.

1.3 Mechanics of cancer cell dissemination during metastasis

The initial growth of a tumor in a confined space leads to compression, and the hypoxia and increasing concentration of growth factors that occurs alongside this compression can further stimulate cell growth [13]. Compression, as well as the increased rigidity of the ECM as promoted by cancer associated fibroblasts, leads to phenotypic changes such as the epithelial-mesenchymal transition, which leads to metastatic properties [10].

Cells then escape through the basement membrane and into the surrounding stroma. Degradation of the basement membrane is achieved through secretion of proteolytic enzymes by actin-rich invadopodia, which are known to be activated by mechanosensing of ECM stiffness and are essential in the initial breach of the basement membrane. The extension of protrusions into the stroma is controlled by Rac-mediated actin polymerization. Rac also suppresses actomyosin contractility that would force the cell into a rounded morphology by inhibition of Rho. Adhesion of protrusions to the ECM activates Rho, inhibiting protrusion by inhibiting Rac, and promoting contractility for forward migration through ROCK [14].

Cells then migrate through the dense stroma and can use a variety of migration modes. In amoeboid migration, cells move in a rounded shape, with propulsion achieved through rapid cycles of expansion and contraction of the cell body. Filamentous actin in the cell cortex generates tension downstream of Rho and ROCK signaling, leading to cortex stiffening and contraction, promoting propulsion of the cell with an ellipsoid morphology. Rac and Cdc42 signaling is largely absent from this migratory mechanism as dynamic protrusions requiring actin polymerization are not present. This method of migration requires a high deformability to squeeze through small gaps in the ECM. Weak interactions between the cell and the ECM, i.e., not requiring integrins, make amoeboid migration a very fast method for movement [15,16].

Mesenchymal migration involves a polarized cell, with a spindle-shaped morphology, much like that of fibroblasts, and is made up of five main stages (Fig. 3D and E).

(1) Extension of lamellipodia/filopodia through Rac- and Cdc42-mediated actin polymerization.

(2) Attachment of the protrusion to the surface via focal adhesions. Integrins form at the leading edge and serve as sites where traction force is applied, additionally acting as foci where ECM properties can be sensed through mechanosensing.

(3) Secretion of proteases, leading to proteolysis of the ECM to allow cell transit.

(4) Movement of the cell body forwards through actomyosin contractility. At the cell rear, Rho signaling is uninhibited by Rac and is involved in myosin-dependent contraction to push the cell body forwards.

(5) Breakage of adhesions at the cell rear, facilitated by internal cell contractility and in some cases, proteases [5]. Migration through the stroma also requires contractility for force-mediated reorientation of collagen fibers to facilitate movement [10].

Finally, collective migration involves cells maintaining cell-cell contacts via cadherins, with leader cells generating traction forces and determining the migration path. These leader cells show a mesenchymal-type motility, with focal adhesion formation and focalized proteolysis. The phenotype differs from mesenchymal motility in that leader cells must maintain cell-cell contacts [5].

Following migration through the stroma, cells move from the primary site to the secondary site, traveling through the blood stream or lymphatic system. Cells migrate through the endothelium through the process of intravasation, where physical restrictions force cells to either degrade their surroundings or squeeze through them. Cytoskeletal contractility is used to deform the cell body, in particular the nucleus, the largest and stiffest of all cellular structures [14]. A soft deformable cell is thought to be necessary for the changes in cell shape required for migration through small gaps and the dense ECM [1]. After intravasation, within the vasculature or lymphatic system, cancer cells must then resist shear forces. Sensing of these forces by mechanotransduction can additionally alter the behavior of the metastasizing cancer cells. The exit of cancer cells from the vasculature or lymphatic system into the secondary site involves the process of diapedesis, the extension of pseudopodia to penetrate cell-cell junctions in the endothelium of distant site, which requires cytoskeletal remodeling [10].

1.4 Cancer-associated cells

Tumors and their microenvironments involve a plethora of cells in a complex interaction network, and therefore studying the relevance of cell mechanics to cancer requires analysis of a wide range of cell types in addition to cancer cells themselves. For example, cancer-associated fibroblasts play a key role in remodeling the extracellular matrix surrounding cancer cells to facilitate invasion, and this process requires fibroblasts to be mechanosensitive and generate forces. Positive feedback loops allow cancer-associated fibroblasts to respond to external rigidity and remodel and stiffen the ECM, leading to an increase in external rigidity. Furthermore, durotaxis, the directed movement of a cell along a rigidity gradient, involves a finely tuned mechanical signaling network involving mechanosensing and force generation. This is essential in the profibrotic positive feedback loop in which cancer-associated fibroblasts promote ECM stiffening and subsequently recruit additional cancer-associated fibroblasts. Recent studies have analyzed the effect of drugs on deactivating these cells to prevent their mechanical influence on the environment [17,18].

2 Viscoelasticity

In general, cancer cells appear to be softer than healthy cells, with many studies demonstrating that cancer cells tend to have reduced elasticity and viscosity in vitro. This increased deformability is suggested to facilitate processes in metastasis, including intravasation and extravasation, as cells more easily squeeze through small gaps in the endothelium to enter or exit the vasculature or lymphatic system. A few studies, however, link a higher stiffness of cancer cells to increased malignancy, and demonstrate the uncertainty that still exists within the field [1].

Cells exhibit both viscous and elastic behavior. Viscous behavior, where deformation in response to an applied force is gradual, is primarily controlled by the water content of the cell, determined by the rate at which water can redistribute itself within the cell body following deformation. Elastic behavior, where deformation in response to an applied force is instantaneous, is mediated principally through the actin cytoskeleton, with (i) the bending stiffness of individual fibers, (ii) overall fiber organization and (iii) intracellular tension from contraction, all affecting stiffness. The interconnectivity of the cytoskeletal network also allows forces to be easily distributed across the cell in the resistance of deformation. However, a highly dense network also inhibits the flow of fluid (i.e., viscosity) and limits the rate of stress propagation [6,19]. Hence, cell rheology is not as straightforward as a simple elastic model obeying Hooke's law.

To add further complexity to cell rheology, the cytoskeleton is also highly dynamic, with polymerization and depolymerization of fibers in response to biochemical and mechanical signals, and this rearrangement alters the viscoelastic behavior of the cell. For example, shear stresses applied to actin networks can induce filament entanglement and subsequent stiffening of the network [6]. Accordingly, viscoelastic parameters are often dependent on deformation frequency, where at low frequencies, cells mechanically respond to applied forces by cytoskeletal reorganization, whereas they cannot do this at high frequencies [1].

Common analysis methods can be separated based on whether they analyze cells adhered onto a surface or in suspension. When adherent, the cell cytoplasm is structurally different, since actin fibers are involved in attachments to focal adhesions. Nonadherent assays measure whole-cell mechanics, reporting parameters that represent the contribution of all elastic components within the cell. Adherent cell assays are therefore generally more localized in their measurements, as probes can be positioned more accurately, and as such, report more localized parameters [1].

2.1 Definition of viscoelastic parameters

Viscoelastic analysis involves applying force to cells, and either visualizing the deformation of the cell, or measuring the force required to move the force application probe in contact with the cell a set distance. Though some methods are only semiquantitative, most methods can be calibrated and be used to determine absolute values for parameters, allowing comparison between studies. In vitro mechanical studies quantify cell deformability through a variety of parameters that describe different aspects of the viscoelastic response (Fig. 5A).

Purely elastic materials are described by their spring constant (K), also known as Young's modulus (E), where strain (ε) is instantaneous and directly proportional to applied stress (σ). These parameters are related by $E = \frac{\sigma}{\varepsilon}$. Similarly, the shear modulus (G) is determined from shear stress applied perpendicular to the cross section of the cell and the consequent shear strain, with these parameters similarly related by $G = \frac{\sigma}{\varepsilon}$ [1]. Purely viscous materials are described by their viscosity (η), where the time derivative of strain is proportional to applied stress, $\sigma = \eta \frac{d\varepsilon}{dt}$.

Viscoelasticity can also be quantified through complex moduli, representative of deformation frequency-dependent rheology. The complex elastic modulus, the ratio of stress to strain under oscillatory conditions, is defined as $E^* = E' + iE''$, where E' is the storage modulus, E'' is the loss modulus, and i is the imaginary unit ($\sqrt{-1}$). The use of the imaginary unit derives from Euler's formula, with $E' = \frac{\sigma_0}{\varepsilon_0} \cos\delta$ and $E'' = \frac{\sigma_0}{\varepsilon_0} \sin\delta$, where δ is the phase lag between stress and strain. The complex shear

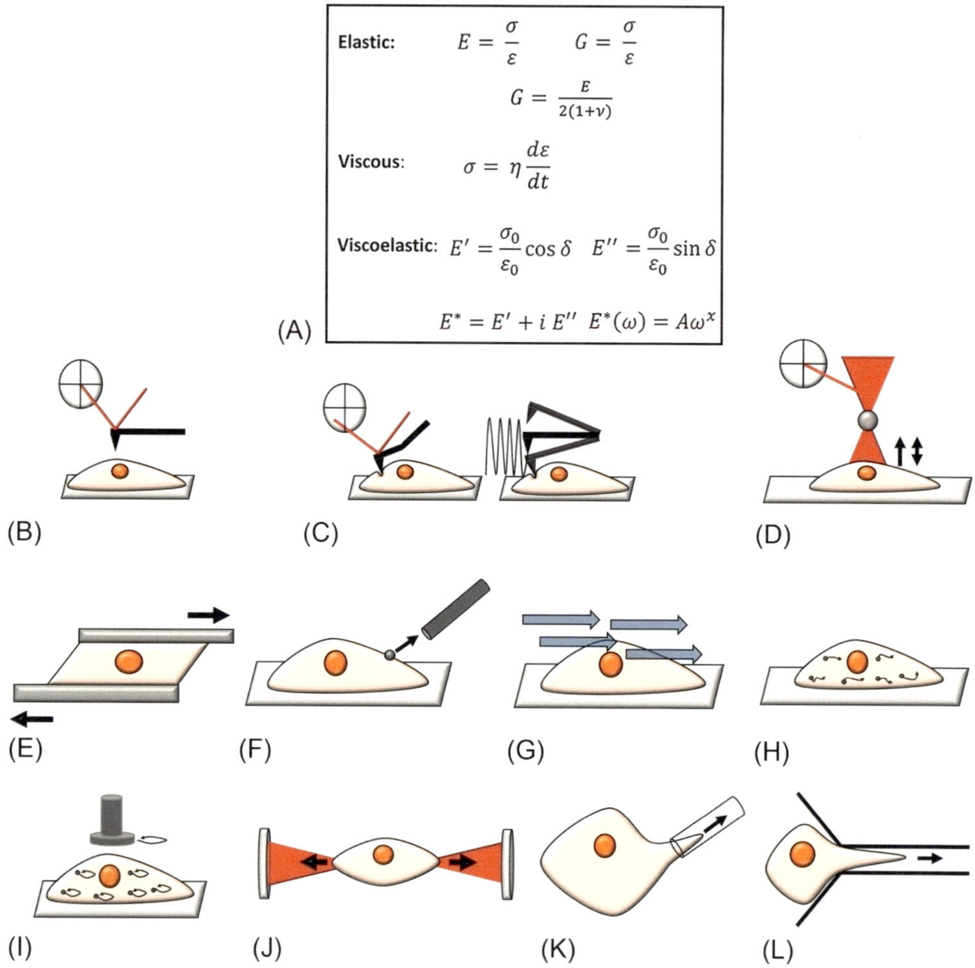

Fig. 5 Viscoelasticity analysis techniques. (A) Parameters. Equations relating stress and strain with elastic, viscous, and viscoelastic parameters. E = Young's modulus, σ = stress, ε = strain, G = shear modulus, ν = Poisson's ratio, η = viscosity, t = time, E' = storage modulus, δ = phase lag, E'' = loss modulus, ω = oscillation frequency, A and x = power law coefficients. (B) Contact AFM. A cantilever *(black)* with a laser *(red)* reflected onto a photodiode is lowered onto a cell. Deflection of the cantilever, in a manner dependent on cell elasticity, is detected by movement of the laser on the photodiode. This laser displacement is then converted into Young's modulus. (C) Intermittent contact AFM. A cantilever is oscillated, with the frequency-dependent elastic modulus used to determine the complex modulus of the cell. (D) Optical tweezer nanoindentation. A bead is suspended in an optical trap *(red)*, and a cell is moved toward it. Depending on the cell elasticity, the bead is displaced and this is detected by a change in the laser's reflection onto a photodiode. *Arrows* represent movement of stage, either in one direction or oscillatory. (E) Microplate stretcher. Cells are constrained between two parallel fibronectin-coated plates *(gray)* and gradually pulled apart. The elasticity of the cell affects the force required for stretching. *Arrows* indicate motor movement of plates. (F) Magnetic tweezers. A magnetic bead is attached to the surface of a cell

(Continued)

modulus is similarly defined as $G^* = G' + iG''$, and G and E are related by Poisson's ratio (ν), $G = \frac{E}{2(1+\nu)}$, where ν is 0.5 for cells. This value for ν, which in materials can range from −1.0 to 0.5, comes from an assumption that cells are incompressible [20]. The storage modulus (E' or G') represents the elastic component and the loss modulus (E'' or G'') represents the viscous component.

Many viscoelastic models of the cell predict that the complex elastic modulus (E^*) is a function of oscillation frequency (ω) and obeys a weak power law, $E^*(\omega) = A\omega^x$, and many studies report and/or compare the values of A and x. A purely elastic material, which responds instantaneously to stress, has the exponent x value $= 0$, as elastic responses are frequency independent. A purely viscous material has an exponent x value of 1, as response is directly proportional to frequency [1].

2.2 Adherent cell methods

2.2.1 Atomic force microscopy

The most common analysis method for determining the elasticity of adherent cells is atomic force microscopy (AFM). A semielastic cantilever with a sharp tip is brought into contact with the cell by a piezoelectric transducer, and the physical interaction with the cell is used to assess viscoelastic properties such as Young's modulus, viscosity, and complex elastic and shear moduli.

To obtain elastic data in contact mode, the cantilever is moved toward the cell until it interacts and is deformed. A laser is reflected off the back of the cantilever onto a photodiode. Upon indentation, the cantilever is deflected as the cell undergoes strain, causing the reflected laser to also be deflected and change its position on the photodiode (Fig. 5B). Force curves of "distance from cell" vs. "force applied to cantilever" are plotted from the data obtained (Fig. 6). Multiple models, such as the Hertz contact model, can be fitted to

---

Fig 5, cont'd and pulled by a magnetic probe *(gray)* with a defined force. By Hooke's law, the elasticity can be determined from the bead's displacement. Arrow represents movement of bead due to magnetic field applied. (G) Shear flow. By flowing a liquid over cells, their movement due to shear force is representative of elasticity. *Blue arrows* represent flow of liquid. (H) Passive microrheology. Markers are added to cells and their movement over time due to thermal motion and active cytoplasmic structures is visualized to semiquantitatively analyze cell compliance. *Arrows* represent marker movements. (I) Magnetic twisting cytometry. Markers in cells are actively moved by an external force. In this example, magnetic twisting cytometry, a magnetic probe *(gray)* twists embedded beads, with their movement dependent on cell elasticity. *Arrows* represent magnetic field and marker movements. (J) Optical stretcher. A cell is trapped by two divergent laser beams *(red)*. Due to the transparency of the cell membrane, this stretches the cell in the optical trap, with deformation dependent on elasticity. *Arrows* represent stretching force. (K) Micropipette aspiration. A micropipette is used to suck up a cell with a defined force, with the speed and displacement of the cell edge used to calculate elasticity. *Arrow* represents force induced by suction pressure. (L) Microfluidics. By forcing a cell to enter a smaller channel, the elasticity of the cell is revealed, with softer cells entering channels faster. *Arrow* represents direction of cell movement.

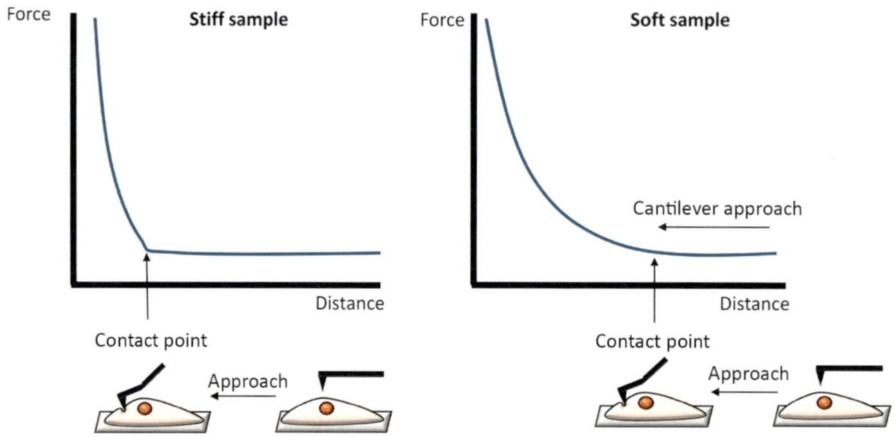

Fig. 6 Example AFM curves for stiff and soft samples. (Left) Example curve for stiff sample. (Right) Example curve for soft sample. The cantilever approaches the samples (from right to left on *x*-axis) and makes contact, applying a force. For a stiff sample (left), the curve rises sharply as the cantilever is deflected greatly over a short indentation distance. For a soft sample (right), the curve rises gently as the cantilever is deflected to the same degree over a longer indentation distance.

this curve to calculate the viscoelastic properties of the cell, most commonly Young's modulus.

AFM can be performed in a wide variety of ways, with differently shaped and functionalized cantilevers, providing a wealth of mechanical results. Local force curves can be obtained for one discrete point per cell and values averaged across multiple cells (localized force spectroscopy) or force curves can be obtained for multiple points on a single cell, creating a topographical and mechanical profile for a cell (force mapping). Sharp pyramidal tips are generally used for cellular imaging and generation of high-resolution force maps, though the small contact area applies a very high stress even at very low forces, which can disrupt the membrane [21], as well as lead to an overestimation of elasticity constants [22]. To alleviate these effects, spherical beads are often attached to cantilever tips to more widely distribute force to the cell, allowing for a better representation of whole cell rheology using a localized technique. Force mapping with these cantilevers is necessarily at a lower resolution, but can still provide important topographical data regarding cell elasticity [23].

Local force spectroscopy has been used to analyze a variety of cancer cell types, using either bare tips with a radius <20 nm [24,25], but more commonly with large spherical beads (diameters upwards of 20 μm) [26–29], with all studies demonstrating that Young's modulus decreased in parallel with oncogenic transformation. Alternatively, force mapping with a sharp tip (radius = 20 nm) has been used to show that healthy lung epithelial cells were stiffer than lung adenocarcinoma cells, and that treatment of the adenocarcinoma cells with the anticancer drug doxorubicin leads to cell stiffening [30]. Using a small spherical probe (radius = 1 μm), force mapping has been used to determine that the cytokine LeftyA decreases the stiffness of human endometrial cancer cells by inhibiting Rac1 and therefore promoting actin depolymerization [31].

AFM is often combined with confocal or epifluorescence imaging, and force maps can be correlated with visualization of key cytoskeletal proteins for elucidation of their contribution to cell mechanics. In Fig. 7, melanoma cells, with or without the actin depolymerizing agent cytochalasin D, were stained for actin and measured using AFM, indicating that depolymerization leads to cell softening [32].

In addition, a recent study used AFM combined with confocal fluorescence microscopy to show that metastatic breast adenocarcinoma cells invading through a 3D extracellular matrix were stiffer than noninvading cells, using confocal microscopy to optically locate the cell membrane and collagen within the matrix, allowing for precise measurements of cell elasticity [33].

As an alternative to contact mode, tapping mode involves oscillating the cantilever tip at a fixed frequency, giving intermittent contact between the cantilever and the cell (Fig. 5C). The oscillatory nature of tapping AFM allows the extraction of the complex elastic and shear moduli, which provide additional information on the viscoelastic nature of the cell. The storage and loss moduli obtained from this analysis form the complex modulus, and this allows the fitting of cell behavior to viscoelastic models. Intermittent contact AFM is also useful for topography analysis, but the high frequency, and therefore high forces, required for adequate throughput make this unsuitable for mechanical

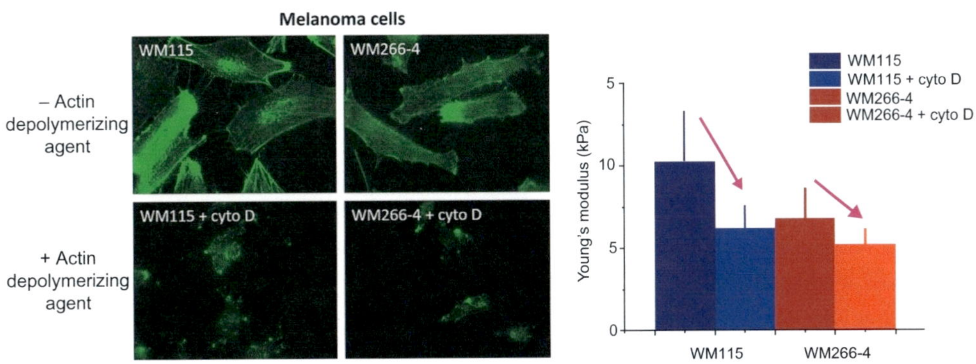

Fig. 7 Correlating the cytoskeleton with cellular mechanical properties. (Left) Melanoma cells (WM115 and WM266-4) stained for actin *(green)*, with or without treatment with the actin depolymerizing agent cytochalasin D (cyto D). Nontreated cells show aligned actin fibers, whereas with cytochalasin D promotes a disorganized actin structure. (Right) Young's modulus for melanoma cells as determined by AFM, indicating that actin depolymerization reduces stiffness. *(Adapted from M. Lekka, Discrimination between normal and cancerous cells using AFM, BioNanoSci 6 (2016) 65–80.)*

analysis. Low forces must be used in the mechanical analysis of cells in order to (i) ensure that the response observed is an elastic response, and (ii) prevent rupture of the cell membrane [34].

For example, a study on bladder cancer cells demonstrated changes in the elastic storage modulus (E') and elastic loss modulus (E'') as a function of cantilever oscillation frequency (1–300 Hz). By running this experiment on different regions of the cells, combined with immunofluorescence imaging, the importance of elastic behavior at the nuclear periphery due to actin fibers was demonstrated [35]. Intermittent contact AFM, using frequencies from 0.2 to 100 Hz, has also been used to track the mechanical influence of anticancer drugs on a prostate cancer cell line, indicating that they all lead to cell stiffening, with more stiffening at higher-frequency force application [36].

The low frequency required for the application of low force in intermittent contact AFM makes it a time-consuming experiment and therefore unsuitable for force mapping. A recent development, peak force tapping AFM, has improved the ability of low-frequency cantilevers to detect their position relative to the sample, and has therefore improved the speed at which force curves can be obtained. This has allowed the generation of force maps of breast cancer cells, using a maximum force of 1 nN at 250 Hz [37].

Other developments in intermittent contact AFM involve the use of harmonic data to extract further mechanical data from cancer cells. Multiharmonic AFM uses a higher drive frequency, around 8 kHz, and measures changes in the amplitude, frequency, and/or phase of the harmonics of the drive frequency, i.e., a frequency that is a positive integer multiple of the drive frequency. These are used to generate cell maps with the parameters of the elastic modulus, and the complex elastic modulus and its components. Multiharmonic AFM has been used to demonstrate that expression of the tyrosine kinase Syk increased elasticity in breast cancer cells by promoting a more substantial microtubule network [38].

2.2.2 Optical tweezer vertical indentation

Though AFM is the most common form of nanoindentation in viscoelasticity assays, elasticity can also be measured through optical tweezer vertical indentation. In this method, a bead is trapped by a highly focused laser beam. A cell is slowly moved upwards towards the bead, and upon contact, the bead is displaced in a manner dependent on cell elasticity, which can be detected by a photodiode (Fig. 5D).

Optical tweezer vertical indentation allows the application of smaller forces than in AFM due to the low stiffness of the optical trap. Additionally, the use of beads instead of sharp tips further reduces the applied force and evens out heterogeneity in the cell, allowing for more representative mechanical quantification. The low frequency of optical tweezer vertical indentation allows specific measurement of the elastic properties of the cell as high frequency stress application often leads to viscous behavior by cells. The elastic properties of a set of breast cancer cell lines, from healthy to malignant, have been assessed using optical tweezer vertical indentation, indicating that the most aggressive cell line had a reduced Young's modulus [34,39].

2.2.3 Shear force application

The foregoing methods both involve the use of compressive forces. Another type of force that can be applied is shear force, where forces are applied parallel to the planar cross section of the cell.

One simple method of applying shear forces involves physically constraining cells between two fibronectin-coated glass microplates, which are then gradually moved apart at a set rate to stretch the cell. By using a feedback mechanism, the forces required to move the plates at a set speed can be used to determine the elasticity of the trapped cell (Fig. 5E). This method has been used to show that sphingosylphosphorylcholine (SPC) reduces the elastic modulus of the pancreatic cancer cell line Panc-1 [40].

Magnetic tweezers are also used in the application of shear force. Fibronectin-coated magnetic beads are incubated with cells, which allow them to attach to the cell surface. A tweezer produces a magnetic field, which tugs on the beads with a defined force. Based on the displacement of the bead in response to this force, viscoelastic parameters can be determined using Hooke's law (Fig. 5F). This method has been used in multiple studies; to show that BMP-2, a cell regulatory protein, promotes the softening of pancreatic cancer cells [41], and also to correlate invasiveness with cell compliance in ovarian cancer cell lines [42]. Since movement of the bead is parallel to the cross section of the cell and substrate, this method can be classified as providing shear force.

The controlled flow of culture media over adherent cells as a shear force creates shear stress on the cell. This faithfully recapitulates the shear stresses that cells would experience in blood vessels in vivo (Fig. 5G). Skin cancer cells have been analyzed in response to shear flow and it was shown that cytochalasin D, a toxin, which disrupts actin polymerization, promoted cell softening [43].

2.2.4 Microrheology

The movement of particles within a cell, either endogenous or injected, can indicate the viscoelasticity of the cytoplasm. Passive microrheology visualizes the movement of these particles without applying any external force, with particle movement due to thermal motion and active cytoplasmic structures, and frictional forces dependent on viscoelasticity (Fig. 5H). Though direct calculation of the elastic modulus is impossible, increased bead movement indicates a more pliable cytoplasm. In breast cancer cell lines, it has been shown, using passive microrheology, that invasiveness and metastatic potential are associated with a more pliable cytoplasm [44].

In contrast to passive microrheology, the application of external forces to intracellular probes in active microrheology allows absolute quantification of elastic parameters. Much like intermittent contact AFM, the oscillatory nature of force application allows determination of complex elastic moduli. For example, optical tweezers have been used to oscillate 500-nm colloidal particles in melanoma cells to determine the complex shear modulus, demonstrating that these cells were more elastic than viscous in the frequency range studied [45].

The use of magnetic probes and an oscillating magnetic field is a common form of active microrheology. In magnetic twisting cytometry (MTC), magnetic beads attached to cells are used to determine complex moduli (Fig. 5I). For example, MTC has been used to determine the complex shear modulus of liver cancer cells to show that TMBS1, a natural compound from traditional Chinese herbal medicine, decreases cytoskeletal stiffness [46]. Furthermore, the correlation between the metastatic potential of fibroblast, breast, skin, kidney, prostate, and bladder cancer cells and complex shear modulus and power law exponent has been explored using MTC [47]. A similar method, rotational magnetic spectroscopy, uses superparamagnetic wires as intracellular magnetic probes and has been used to determine the shear elasticity and viscosity of the cytoplasm of human cancer HeLa cells [48].

2.3 Nonadherent methods

Nonadherent methods involve cells in suspension, and therefore a key aspect of each assay is in how cells are trapped in a position where forces can be applied in a controlled manner.

Optical stretchers can be used to trap and stretch cells for viscoelastic analysis. Much like the bead in optical tweezers nanoindentation, forces from lasers guide cells into a specific position within the beam. With two divergent laser beams, the trapped cell is stretched with a defined force. Changes in cell shape are used to quantify deformation, which is then used to calculate elasticity (Fig. 5J). This has been used to show differences in the elasticity of breast cancer cell lines, demonstrating that MDA-MB-231 breast cancer cells became stiffer and lost their aggressive nature when treated with all-trans retinoic acid [49]. Cells can also be trapped and stretched using sound waves. This deformation has been quantified to demonstrate that the invasiveness of breast cancer cells correlated with cell softening [50].

The ability of cancer cells to squeeze through small gaps is an important aspect of malignancy, and techniques such as micropipette aspiration quantify deformability by forcing cells into narrow channels. A micropipette is used with a constant pressure to aspirate the surface of a cell, and movement of the cell edge is tracked. A softer cell will be aspirated into the micropipette at a faster rate (Fig. 5K). This method has been used to demonstrate that high-grade colon cancer cells are softer than primary grade colon cancer cells [51], as well as to demonstrate that nontransformed prostate epithelial cells were stiffer than transformed prostate cancer cells [52].

Viscoelastic analysis of cells in suspension can be made high throughput with the application of microfluidics. Microfluidics describes systems with small channels through which the flow of liquids can be precisely controlled, allowing the precise application of force as a cell suspension flows through. For example, acoustic waves have also been used in combination with microfluidic chambers to determine that breast, liver, and colon cancer cell lines were more compressible than normal breast cells [53]. Alternatively, a channel that narrows can be used, with analysis of entry velocity used to quantify the viscoelastic properties of cells, i.e., faster entry indicates softer cells (Fig. 5L). One study visualized breast cancer cells entering into narrow channels, where cancerous MCF-7 cells showed a faster entry time than benign MCF-10A cells, indicating their increased deformability [54].

3 Force generation

The ability of a cancer cell to generate traction forces is essential in cell migration and metastasis but is also important in processes such as cell adhesion and mechanosensing. In response to genetic changes and/or environmental changes, cancer cells generate higher forces, and subsequently, force-dependent processes promote malignant transformation and metastasis. The modes of migration discussed in section 1.3 all require cells to generate force, for example cortical actomyosin contraction during amoeboid migration and lamellipodia extension through actin polymerization in mesenchymal migration. The facilitation of cancer progression is further augmented by cancer associated cells, most commonly cancer-associated fibroblasts (CAFs). These cells facilitate the migration and metastasis of cancer cells by force-mediated remodeling of the extracellular matrix, and therefore their ability to generate forces is essential in understanding cancer mechanics [55].

By determining the distribution of forces, their magnitudes, and how this correlates with the organization of intracellular proteins, therapeutics may be able to be developed to target specific aspects of migration and metastasis. For example, ROCK inhibitors showed promise for cancer treatment but were unsuccessful in later clinical trials, and an understanding of the reasons behind this will likely generate improved therapeutics [56].

Quantification of force generation is most commonly achieved with forms of either 2D or 3D traction force microscopy, in which the forces generated by cells deform materials or objects which cells are in contact with. This deformation is often converted in absolute values of force generation, in systems where calibration is possible. Traction force microscopy is also often presented alongside assays which visualize the cytoskeleton to look in detail at the mechanisms present. For example, quantification assays for the abundance of actomyosin-associated proteins or adhesion strength are often performed concurrently.

3.1 Traction force microscopy

3.1.1 Two-dimensional TFM

The forces that cells apply to their underlying substrates are traction forces, with 2D-traction force microscopy (2D-TFM) the most common way of quantitatively analyzing these. Cells are seeded onto a compliant substrate, such as a polyacrylamide gel, which contains fiduciary markers such as small fluorescent beads. As the cell applies forces to the underlying substrate, the positions of these markers are deformed from their original position and this is visualized by microscopy. Chemically removing the cell, or allowing it to migrate away, reveals the initial position of the beads, and therefore the bead displacements induced by the cell. Computational analysis then works out the most likely traction field to produce the observed displacements (Fig. 8A) [57].

The traction forces of human metastatic breast, prostate, and lung cancer cell lines have been shown through TFM to be increased in comparison to their nonmetastatic cell lines, and also indicated the role of matrix stiffness and collagen density in promoting force generation [58]. CAFs have also been analyzed by TFM and have been shown to exert larger traction forces than normal fibroblasts, with this force required to form an aligned fibronectin-rich microenvironment that promotes directional cancer cell migration [59]. Furthermore, TFM of clusters of colon cancer cell has been used to show the transmission of force within the group of clustered cells [60].

Time-dependent TFM, where TFM data are collected at multiple time points, has been used to compare the forces generated by two bladder cancer cell lines of different invasiveness. It was shown that the more invasive cell line exhibited periodic peaks of traction force of high magnitude, whereas the less invasive cell line exhibited more constant traction forces without the high force peaks. These forces and their time-dependence correlated with migration velocity and actin structure, indicating the difference in migration methods between the cell lines [61]. Furthermore, holographic traction force microscopy has improved the resolution of the method by using micron-sized beads as fiduciary markers, and has been used in the analysis of colorectal cancer cells [62].

3.1.2 Elastic pillars

Another form of 2D-TFM involves an array of semiflexible pillars, most commonly made of polydimethylsiloxane (PDMS), onto which ECM proteins are adsorbed. Cells are then seeded onto the pillars, and pillar deflection over time, along with pillar rigidity, is used to determine the forces applied to the surface (Fig. 8B). These pillars generally have a diameter around 1 μm and height around 6 μm. This set up can be modified with different heights of pillars to alter the rigidity of the surface, and a lower limit of substrate stiffness can be used than in polyacrylamide gel TFM. However, some controversy exists about the suitability of these pillars, due to the noncontinuous topography of the contact surface, which may affect the behavior of cells.

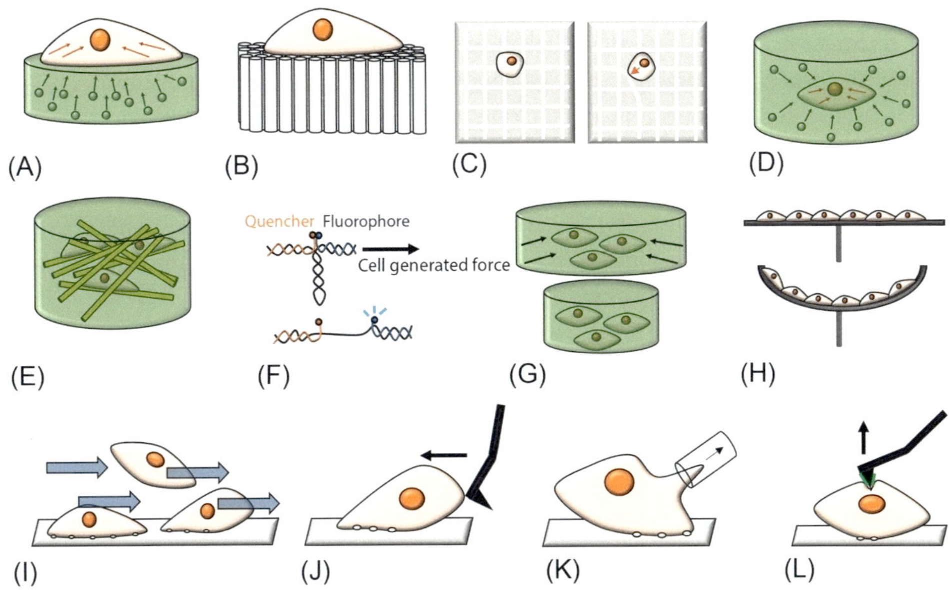

Fig. 8 Force generation analysis techniques. (A) 2D traction force microscopy. Cells are seeded on gels *(green)* containing fiduciary markers *(gray)*. As the cell generates force *(red arrows)*, this promotes movement of the fiduciary markers, which is visualized through microscopy. (B) Elastic pillars. Cells are seeded on fibronectin-coated elastic pillars *(gray)*. Traction forces lead to displacement of pillars, which are visualized through microscopy. (C) Photolithography gels. Cells are seeded onto gels *(dark gray)* with a regular pattern *(light gray)*. Disruption of the regular pattern by cell-generated force *(red arrow)* is visualized through microscopy. (D) 3D Traction force microscopy. Cells are embedded within gels *(green)* containing fiduciary markers *(gray)*. As the cell generates force *(red arrows)*, this promotes movement of the fiduciary markers, which is visualized through microscopy. (E) Collagen gel embedding. Cells are embedded within gels *(green)* containing collagen *(yellow)*. Remodeling of collagen allows a semiquantitative analysis of force generation. (F) Fluorescent force sensor. A DNA probe is used with a quencher near a fluorophore, such that absorption of light by the fluorophore does not lead to light emission. When cells generate force and pull the sensor apart, the quencher and fluorophore are separated. Absorption of light by the fluorophore now does lead to light emission, which allows a semiquantitative analysis of force generation. (G) Gel contraction. Cells are embedded within gels *(green)*. As the cell generates force, this contracts the gel *(black arrows)*, which is visualized through microscopy, which allows a semiquantitative analysis of force generation. (H) Cantilever bending. Cells are seeded on a soft PDMS cantilever *(gray)*. With traction force generation, the cantilever is bent and then visualized through microscopy, which allows a semiquantitative analysis of force generation. (I) Shear flow detachment. This adhesion assay involves flowing liquid over adherent cells and analyzing the kinetics of their detachment. Furthermore, molecular biology methods can be used to characterize the attached/detached cells. *Blue arrows* indicate flowing liquid. *Small gray ovals* indicate points of adhesion to substrate. (J) AFM lateral detachment. This adhesion assay involves pushing a cell with an AFM cantilever *(black)* with a defined force, assessing the force required for detachment. *Small gray ovals* indicate points of adhesion to substrate. (K) Micropipette detachment. A micropipette is used to suck up a cell with a defined force until detachment. *Arrow* represents force induced by suction pressure. *Small gray ovals* indicate points of adhesion to substrate. (L) AFM vertical detachment. Using a cantilever *(black)* functionalized with ECM proteins *(green)*, a cell is pulled upwards away from its substrate until detachment. *Small gray ovals* indicate points of adhesion to substrate.

One study used microfabricated pillars to assess the effect of the drug emodin on the ability of breast cancer cells to apply traction forces, showing that low concentrations increased traction forces due to cell rounding, but that higher concentrations decreased traction forces due to apoptosis [63]. The traction forces of other cancer related cells such as pancreatic stellate cells [18] and hepatic stellate cells [17] have also been analyzed, with these studies demonstrating how certain drugs reduce traction force generation. A further development has been the use of nanowires of less than 100-nm diameter as pillars, leading to an increased resolution in the traction force map. This method has been used to demonstrate that the traction forces of breast cancer cells are higher than healthy breast cells [64].

The regular pattern of elastic pillars is beneficial, as cells do not need to be trypsinized in order to determine the original position of the pillars with which the deformed pillars are compared. Soft photolithography has been used to create regular patterns on PDMS sheets, either pits or fluorescent photoresist dots, creating a surface for cells to deform in a manner that can be quantified (Fig. 8C). A grid pattern on PDMS has been used to quantify the forces applied by rat cardiac fibroblasts, demonstrating that the average force at an individual focal adhesion was around 10 nN [65].

These 2D force generation assay systems can additionally be combined with the application of mechanical stimuli such as stretching of the substrate, shear flow, or dynamically tuning the rigidity of gels (e.g., increasing rigidity of methacrylated hyaluronic acid gels by UV light–induced crosslinking). Regulation of gel rigidity can be used to optimize assays, as the optimum stiffness gel is compliant enough that cells can apply forces leading to observable marker displacements, but stiff enough that differences in force generation can be observed. Furthermore, tuning the system allows assessment of the role of substrate rigidity on force generation, allowing the assessment of mechanosensing [66].

3.1.3 3D traction forces

The physiological traction forces of cancer cells and CAFs are applied in three dimensions in vivo, and therefore the development of in vitro 3D TFM methods has improved our understanding of processes such as cell migration and ECM remodeling. In these methods, cells are either embedded within, or allowed to invade into, hydrogel matrices. The movement of either exogenously added beads or endogenous ECM components are visualized and analyzed to determine 3D traction forces (Fig. 8D and E) [67].

For 3D gels with embedded beads, 3D strain maps from the observed movement of the beads shed light on the forces applied by the cells, and also indicate the location within the cell where forces are being exerted onto the substrate. For example, fibroblasts were shown to generate force at the tips of pseudopodia in PEG hydrogels [68]. More complex setups, such as those containing ECM components, are more representative of the in vivo environment for cells, and using embedded microbeads, have been used to determine the traction forces of MDA-MB-231 breast cancer cells during invasion, showing that accumulation of F-actin and myosin-IIa at the rear of the cell promote contractility during invasion [69]. Additionally, visualization of the movement of the ECM components themselves has been used for traction force analysis. For example, the alignment of ECM fibers around pseudopodia has been shown in fibroblasts, indicating force application in the alignment process [70].

3.2 Other force probes

While traction force microscopy with gels may be the most common way of measuring forces applied to a substrate, other probes have been successfully used in quantifying traction forces. One example is the use of fluorescent DNA probes. A nucleic acid hairpin is used where in an unfolded state, a fluorophore is located close to a quencher, and therefore no fluorescence is observed. Part of the probe is able to bind to cell surface receptors, and force application by cells pulls open the DNA probe and separates the fluorophore and quencher, leading to fluorescence (Fig. 8F). A specific benefit of this method is that the probe can be modified to only bind to specific surface receptors such as integrins. Fluorescent DNA probes have been used to analyze the forces associated with integrin dynamics in breast cancer cells. The study showed that integrin tension increased along with integrin density during the formation of cell-substrate adhesions, indicating the requirement for molecular recognition of the substrate for mechanosensing [71].

3.3 Population contractile forces

Not all force analysis methods directly calculate the forces applied by individual cells to their surroundings but use assays where a population of cells is assessed. The readout is dependent on the assay, though does not necessarily report exact values of force generation, unless complex analysis is applied. For example, cells embedded within hydrogels can remodel and contract the gel, with large changes in gel size or component organization used as a metric for contractility (Fig. 8G). This has been used to demonstrate the importance of $\beta 1$ integrin in force-mediated ECM remodeling in pancreatic stellate cells [72]. Similarly, the contractile ability of a group of cells can be measured by seeding them on a PDMS cantilever that bends in response to contraction (Fig. 8H). This method has been used to assess the contractile ability of breast cancer cell lines, and compare the results to previously reported values [73].

To further assess force applied in 3D, invasion assays have been modified to allow determination of traction forces during the invasion process. By binding 5-nm gold beads onto rabbit corneas, researchers created a basement membrane mimic with properties similar to traction force microscopy gels. By visualizing the clustering of beads, representative of compression of the basement membrane, a semiquantitative analysis of force application by a group of cells invading through the cornea was achieved. This study showed the importance of the $\alpha_6\beta_4$ integrin in epithelial cells in generating forces during invasion [74].

3.4 Complementary force generation assays

3.4.1 Protein quantification

Multiple proteins interact with the cytoskeletal components that are involved in force generation, and changes in protein or mRNA abundance and/or cytoskeletal organization of these accessory proteins can detail the method by which changes in force generation occur. For example, Western blots or immunofluorescence staining can be performed to assess the relative abundance of pMLC-2, the active form of MLC-2. Changes in pMLC-2 abundance between conditions can indicate an increased contractility, and if no difference in MLC-2 abundance is concurrently shown, this indicates that posttranscriptional phosphorylation is the method by which pMLC-2 has increased [17]. The abundance of other actomyosin-interacting proteins are known to be altered in cancers at either the mRNA of protein level, including cofilin, Arp2/3, formins, WASP, WAVE, cortactin, ena/VASP, and ERM [56].

Many of the force generation assays previously mentioned can be easily modified to include immunofluorescent microscopy analysis. For example, cells attached to pliable surfaces such as traction force microscopy gels or elastic pillars can be fixed in place with a fixation agent such as paraformaldehyde, making them amenable for immunofluorescent staining [75]. Alternatively, cells can be modified to express proteins of interest with fluorescent tags, so protein analysis can be performed alongside the live cell imaging aspect of the force generation assay [76].

3.4.2 Cell adhesion studies

Adhesion to a substrate is a relevant phenotype that requires force generation by cancer cells. Although the quantification of adhesion strength does not directly assess force generation, it is often presented alongside force generation experiments for a more comprehensive analysis of cell mechanics. Many assays have been developed for analysis of the complex process of cell adhesion and have been well reviewed previously [77].

The kinetics of cell attachment to a surface provide a semiquantitative marker for adhesion strength and force generation. Setups that analyze these parameters include piezoelectric acoustic wave resonator biosensors and microfluidic devices. The washing of cells already attached to a substrate provides information on the strength of adhesion, and therefore internal force generation, as weakly adhered cells will become detached with the shear stress (Fig. 8I). Furthermore, cells that remain attached, or become detached, during this assay can be analyzed for protein or mRNA content to compare the phenotype of attached/detached cells, giving insight into the biochemical background of attachment. For example, the essential role of three kinases, PAK2, GRK6, and DBF4, in cancer cell adhesion has been demonstrated by combing the washing assay with mass spectrometry on the adhered cells [78].

Forced cell detachment of individual cells is also used. Single cells can be forced to detach by an inclined AFM probe, which approaches and then generates shear force by pushing the cell laterally until it detaches (Fig. 8J). This has been used to show that cervical carcinoma cells attach more strongly to protein-coated hydrophilic surfaces than protein-coated hydrophobic surfaces, demonstrating that multiple properties of the surface must be considered when analyzing cell adhesion [79]. Micropipette aspiration has also been used for forced cell detachment (Fig. 8K), and has been used to show that prostate and breast cancer cells become less adherent with increasing metastatic potential, and that the blocking of voltage-gated Na^+ channels promoted adhesion [80].

AFM or optical tweezer probes can be coated with ECM proteins to facilitate the attachment of cells to the cantilever, and then the force required to drag the cell away from the surface is measured. The force required for each of these detachment events to occur therefore correlates with adhesion strength (Fig. 8L). This method has been used to determine that human osteosarcoma cells are much more loosely attached to the substrate during the M phase of the cell cycle compared to during interphase [81].

At the cell population level, a group of cells attached to a surface can be forced to detach to measure adhesion strength. In centrifugation, multiwell cell culture plates with cells attached in each well are spun at high speed, such that centrifugal force acts away from the substrate leading to detachment. The adhesiveness of human fibrosarcoma cells have been analyzed with this method, demonstrating that adhesion strength correlated with ECM protein density [82]. Alternatively, cells can also be seeded on disks that are spun such that shear stress from the rotation leads to detachment. This method has been used to show that magnesium and calcium decrease the adhesiveness of metastatic breast cancer cells [83].

Microfluidic devices with controlled flow can be used to apply shear stress that promotes detachment. With this method, using fabricated surfaces with different patterns for cells to attach to in microfluidic channels, breast cancer cells have been shown to have stronger adhesion independent of surface topography [84]. Flow-induced detachment has also been combined with electric cell-substrate impedance sensing to assess the attachment status of cells, and therefore determine the kinetic constants of detachment. This has been used to assess the role of small molecule inhibitors on the adhesiveness of breast cancer cells [85].

4 Mechanosensing

Mechanosensing describes the process in which cells respond to force application, either physically or biochemically. Experimental methods in mechanosensing therefore involve the application of a force, and the subsequent visualization of the cellular response. Tumor tissues increase in rigidity with cancer development, and this altered mechanical environment can induce signaling in cancer cells that promotes their growth. For example, matrix stiffness can induce the malignant mesenchymal phenotype in pancreatic cells [86], but can also activate pancreatic stellate cells, which secrete collagen, and remodel and maintain the rigid microenvironment [87].

When cells bind to a surface, focal adhesions form as contact points. Both the actin cytoskeleton and the extracellular matrix, interacting through focal adhesions, form a tensional equilibrium as actomyosin-generated contractility tugs on the surface. For a stiffer substrate, increased actomyosin force generation is required to maintain an equilibrium, leading to force-induced conformational changes of intracellular signaling effector proteins. In this way, cells can respond to a stiff external environment through mechanosensing. For example, talin, one of the most studied mechanosensitive molecules, contains multiple consecutive alpha helix bundles, and external tension promoted by a stiff microenvironment leads to the unraveling of some of these domains, leading to exposure of previously hidden binding sites for multiple effector proteins, such as DLC-1, which alter cell mechanics [88]. Talin activation also targets vinculin and paxillin to focal adhesions and therefore promotes focal adhesion maturation in response to a stiff environment [89].

Many of the actin structures in the cell also respond to mechanical stimulation. Tensile forces increase the pitch length of actin filaments, i.e., the height of one turn of the helix, leading to an increase in the affinity of the motor protein myosin II and therefore increased actomyosin contractility [90]. Transverse arcs convey contractile force to the surrounding environment through their connection to dorsal stress fibers, and the perinuclear actin cap conveys force from the cell environment to the nucleus. Additionally, actin polymerization and/or contractility in response to mechanical force promotes nuclear localization of the transcriptional regulator YAP, a known mechanoregulator, which regulates gene expression of cell proliferative processes [91].

Methods for force application used in mechanosensing assays are often the same as those used in viscoelasticity assays, which look at the deformation resistance of a cell to applied force. The difference in mechanosensing assays is that other parameters, such as changes in protein expression, are visualized in response to force. Many mechanical parameters are often used as a mechanosensing readout such as traction force generation and migration, highlighting the interconnectivity of mechanical analysis techniques. Much like viscoelastic assays, force application can be performed at a very localized level, the whole cell level, or even at a cell population level.

4.1 Nonlocalized forces

4.1.1 Internal force

To assess the effect of external rigidity, which promotes an increase in internal force via tensional homeostasis, a variety of gels with tunable mechanical properties are used (Fig. 9A). These gels are formed by altering the concentration of one of the gel

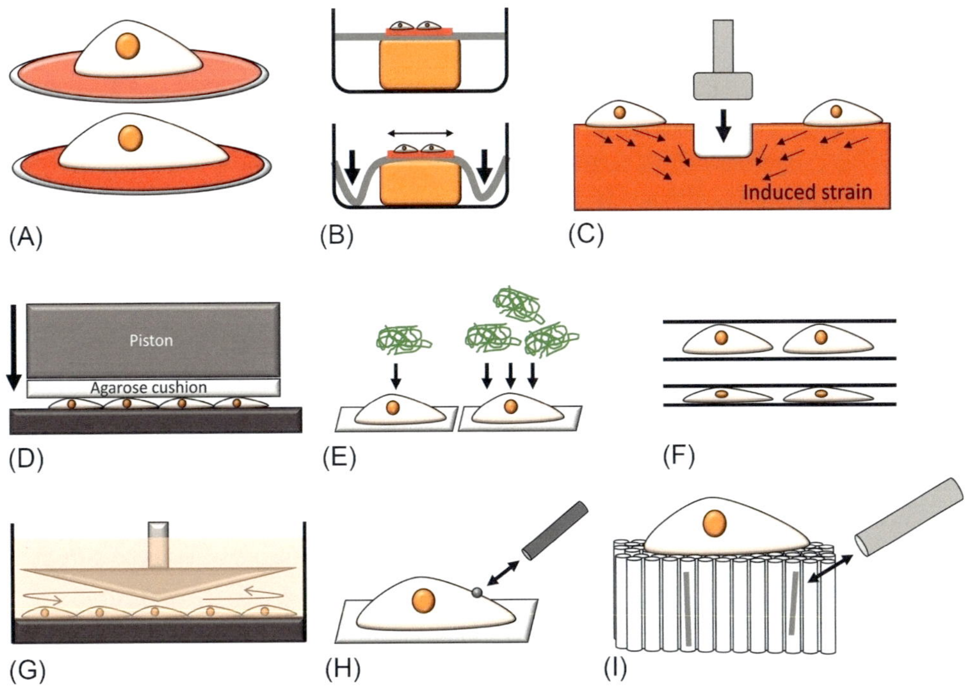

Fig. 9 Mechanosensing analysis techniques. (A) Tunable rigidity gels. Cells are seeded on gels *(red)* on coverslips *(gray)*. *Light red gel*, soft; *dark red gel*, stiff. On these gels, cells can be fixed and stained for protein analysis. (B) Vacuum pressure stretching. Cells are seeded on polyacrylamide gels *(red)* on a silicon membrane *(gray)* on a loading post *(orange)*. Vacuum pressure is applied beneath to stretch the silicon membrane, stretching the polyacrylamide gels and cells. (C) Cell-mapping rheometry. Cells are seeded onto a gel *(red)* and the gel is indented by an punch *(gray)*. This induces a stain field in the gel leading to cell stretching. (D) Piston compression. Cells seeded on a surface and then compressed with a piston attached to an agarose cushion. (E) Dextran concentration. By increasing the concentration of the sugar dextran *(green)* in cell culture medium, increasing compressive force can be applied to adherent cells. (F) Confined growth. By confining the growth of cells, a constant compressive force can be applied. (G) Cone-and-plate viscometer. Cells are seeded on a surface and a cone *(gray)* is lowered into the cell culture medium *(pink)*. The rotation of the cone generates flow and therefore shear stress. (H) Magnetic tweezers. Fibronectin beads are attached to the surface of a cell and pulled with a magnetic tweezer in an oscillatory manner. The reduction of bead displacement over time represents cell reinforcement from mechanotransduction. (I) Magnetic nanowire elastic pillars. By embedding magnetic nanowires into PDMS pillars, which are then manipulated with a magnetic tweezer, forces can be applied to individual regions of a cell for localized mechanotransduction assays.

components, e.g., a crosslinking agent, which affects the final rigidity of the material. Additionally, these materials must be able to be functionalized with extracellular matrix proteins to allow cell attachment, and also in some cases, to provide a consistent surface composition that is independent of the gel rigidity. This allows specific analysis of the effect of external rigidity on cells only. 2D substrates involve the seeding of cells on top of a functionalized material, whereas 3D substrates involve embedding cells inside a material, i.e., by mixing a cell suspension with a nonsolid form of the gel and then allowing solidification of the material to its tuned rigidity [92].

For 2D substrates, the most commonly used gel is polyacrylamide, formed from acrylamide monomers, bis-acrylamide as a crosslinking molecule, and other components, which promote the crosslinking process. By altering the ratio of acrylamide to bis-acrylamide, a wide range of rigidities can be achieved. Fibronectin-coated polyacrylamide gels have recently been used to assess the role of matrix stiffness on the mesenchymal phenotype and chemoresistance properties of pancreatic cancer cells [86]. Polydimethylsiloxane, with its rigidity tuned by the concentration of its curing agent, is also used to assess mechanosensing in 2D [93]. PDMS is versatile as it can be used to make elastic pillar arrays of tunable rigidity, which can then be used in a force generation assay with their deflection dependent on traction force application as previously described. This has been used in a pillar array containing pillars of multiple rigidities to observe force application by fibroblasts during the mechanosensory process of durotaxis, the process by which cells migrate based on stiffness gradients [94].

For 3D tunable rigidity assays, the individual components of the gels must be cytocompatible, as cell are embedded within the gel before it is fully set. Polyacrylamide is unsuitable, as the acrylamide monomer is toxic to cells. Conversely, collagen gels are suitable for the embedding of cells, and have been used to demonstrate the effect of ECM rigidity on Rho and FAK activity in breast cancer cells [95]. Polyethylene glycol (PEG) gels are also cytocompatible, with rigidity tuned by PEG concentration. They have been used in a wide variety of cancer mechanosensing studies indicating that breast cancer cells become more malignant with 3D external rigidity [96].

Gels have alternatively been made from alginate, derived from seaweed, and have been used for both 2D and 3D environments for cells, where stiffness is tuned by diffusion of calcium ions, which promote crosslink formation. Sugars like alginate contain multiple functional groups allowing chemical modifications to the gel; hence, assays can analyze both biophysical and biochemical signaling [97]. Alginate gels have been used to study the cell viability of breast cancer cells in response to external stiffness, with stiffer gels reducing cell proliferation [98]. The viscoelasticity of prostate cancer cells, as determined through particle tracking microrheology, has been measured in response to changes in stiffness in 2D polyacrylamide gels, as well as 2D and 3D collagen gels, where the architecture of the 3D gel was suggested to be the main contributor to changes in cell stiffness [99].

4.1.2 External force

The active application of force to cells can be in the form of tension, compression, or shear stress. Tension involves two forces pulling on a cell from opposite directions, compression involves pushing down onto the cell, and shear involves forces applied parallel to the cross section of the cell. Shear force application is only applicable for adherent cells. Different mechanosensing behaviors may arise from different types of force application, as signaling molecules are activated in a variety of ways. For example, the focal adhesion protein talin is activated through tension, but not compression.

Tension can be actively applied to adherent cells by physically stretching their substrate. For example, a polyacrylamide gel can be stretched using vacuum pressure, thereby stretching the cells attached to its surface (Fig. 9B). Combined with the tunable rigidity characteristic of polyacrylamide gels, this has been used to show that human mesenchymal stem cells are affected by both matrix stiffness and frequency-dependent external stretching, with spreading behavior increased by rigidity and stretching [100]. Similarly, cyclical mechanical strain using vacuum pressure has been used in a lung "organ-on-a-chip" microdevice to mechanically recapitulate breathing, demonstrating that the toxic and inflammatory response of lung epithelial and endothelial cells to silica nanoparticles is increased by cyclic mechanical strain [101].

Another method used for gel stretching is cell mapping rheometry, which involves indenting a gel with an annular punch, inducing a stretch field. Removing the punch allows the gel to elastically recoil and permits cyclical stretch application (Fig. 9C). Combined with traction force microscopy and optical magnetic twisting cytometry, this has been used to show that stretch leads to a reduction in traction forces and cell stiffness in human airway smooth muscle cells [102].

Compressive forces can also be applied to individual cells and/or groups of cells. For example, breast cancer cells in a monolayer have been compressed with a piston of adjustable weight (Fig. 9D), and this compression has been shown to promote migration and fibronectin deposition [103]. Another study used the polysaccharide dextran in cell culture medium to exert mechanical stress on multicellular tumor spheroids (Fig. 9E), where increasing concentrations of dextran reduce the volume available to spheroids and therefore apply stress [104]. This study showed that compression induced the proliferation inhibitor $p27^{Kip1}$ [105]. Another alternative method of compression involves confining the growth or movement of cells in fabricated channels (Fig. 9F), and has been used to demonstrate that confinement impairs the migration of breast cancer cells [106]. Compression has also been achieved with the use of ultrasound, showing that invasive prostate and bladder cancer cell lines show calcium elevation in response to force [107].

Shear stress entails force application parallel to the cell and is most commonly experienced in vivo as fluid shear stress. By controlling the flow of fluid over cells, and analyzing the biochemical events that follow, mechanosensing can be demonstrated. A cone-and-plate viscometer uses a rotating inverted cone above a group of cells to provide fluid flow (Fig. 8G), and has been used to show that shear stress sensitizes colorectal and prostate cancer cell lines to apoptosis-inducing agents [108]. In a more complex assay, glioma cells inside a 3D collagen gel were subjected to shear stress as hydrostatic pressure was used to drive media through the gel, leading to a reduction in motility and expression of ECM-degrading proteins [109].

Flow chambers are also common, having been to assess the role of shear stress on cancer cells. In one study, elastic pillars were placed inside the flow chamber, such that traction forces of cells adhered onto the pillar array could be measured under shear stress. By placing pillars at different positions within the chamber, both laminar and disturbed flow of liquid could be achieved and compared. Cells under laminar flow exhibited higher traction forces than those cells under disturbed flow [110]. Another study combined an adhesion study with a flow chamber, indicating that shear stress increased adhesiveness for nonmetastatic breast cancer cells but decreased adhesiveness for metastatic breast cancer cells [111]. The controlled flow of cell suspensions through needles has also been used to demonstrate that cancer cells of epithelial origin are more resistant to fluid shear stress than nontransformed cells [112].

4.2 Localized force application

The localized application of stress leads to local strain, which can induce biochemical signaling in specific regions of mechanotransductive cells. For example, stress applied to focal adhesions can lead to the unraveling of talin, revealing binding sites for downstream effector proteins [88].

One technique for assessing local mechanosensing uses magnetic tweezers, which applies a lateral force to a ligand-coated magnetic bead attached to the cell of interest. One method of mechanosensing that can be assessed is cell reinforcement in response to applied force. Oscillatory pulses are applied, and reinforcement of the cell is visualized through a decrease in bead displacement for each pulse (Fig. 9H). This has been used to show that a vitamin A metabolite reduces the mechanosensory properties of pancreatic stellate cells [18].

Another lateral force application method uses a modified form of elastic pillars. Magnetic nanowires are embedded into PDMS pillar arrays, such that pillars containing the nanowires can be moved with a magnetic field. This movement applies nanonewton forces to specific areas of cells, such as focal adhesions (Fig. 9I). When combined with immunofluorescence, it has been shown that in NIH 3T3 fibroblasts, localized force application leads to increased focal adhesion size at the point of force generation, but also a reduction in traction force [113].

AFM is a popular method for vertical force application. The controlled vertical movement of a cantilever is used to apply defined forces to distinct areas of a cell. Combined with biochemical visualization techniques, mechanosensing in response to this force can be observed. For example, compression of mouse embryonic fibroblasts using AFM was seen to promote nuclear localization of the transcriptional regulator YAP. Visualization was achieved by transfecting these cells with a green fluorescent protein form of YAP [114]. Another study combined traction force microscopy and tunable rigidity substrates with AFM to determine that muscle myoblasts generate traction forces in response to nanonewton forces, and that this force generation is dependent on substrate rigidity [115]. Optical tweezers have also been used to apply vertical localized force by immobilizing, and then controlling the movement of, a microbead in an optical trap. This has been used to demonstrate the induction of calcium ion transients, using fluorescent calcium dyes, in neuroblastoma cells with piconewton forces applied [116].

5 Conclusion

There exist a wide range of in vitro assays that evaluate a multitude of mechanical parameters, and these assays have been used in cancer cell studies to begin to comprehend the way in which cancer cells are mechanically different from healthy cells. This will improve our understanding of cancer initiation and propagation and behaviors from a mechanical perspective and may even lead to the development of therapeutics, which seek to alter the mechanical signaling that maintains malignancy. Additionally, an understanding of the mechanical difference between health and disease may be useful in diagnostic procedures, with different mechanical attributes as possible biomarkers for cancer.

Though many of these assays dominate their particular field (i.e., AFM for viscoelasticity, TFM for force generation), less common analysis techniques are also very useful. The fully quantitative nature of these assays, which emerges from their ability to be calibrated, is of particular benefit for comparison across different studies. While semi-quantitative assays are useful, the fact that absolute parameter values are not reported affects the replicability of their conclusions across independent in vitro studies, though patient-to-patient variability limits the use of absolute values in vivo. Furthermore, the fact that multiple assays exist to analyze the same parameters is beneficial, as it allows comparison between studies, where different results from different assays may be representative of phenotypic variations that one assay alone would not detect.

There are other important aspects, which must be considered when choosing an analysis technique. The type of in vitro analysis chosen should represent in vivo events as faithfully as possible. For example, viscoelasticity as analyzed through cell stretching is not directly applicable to the compression that cells experience in intravasation. Also, the speed at which an assay can be carried out is important. For example, force mapping of cells is lower throughput than localized force analysis but does provide more details on viscoelasticity. The information required for reaching a conclusion should therefore be considered before embarking on a complex analysis technique.

Many studies tend to focus on one mechanical technique, comparing values between cancer and healthy, or different stages of cancer, and often combining these data with immunofluorescence imaging to identify the important molecules. Comparison of the different mechanical attributes of viscoelasticity, force generation, and mechanosensing is rarely performed, and hence the interplay between mechanical behaviors is understudied. One key gap in our knowledge is how to combine the seemingly incompatible, yet commonly referred to, observations that cancer cells generate more force than healthy cells but are also softer. Many cell models predict that intracellular tension from force generation correlates with a stiffer cell. This contradiction neatly illustrates how far our mechanical understanding of cancer needs to progress, and the use of a variety of in vitro mechanical assays is our toolkit for seeking out the answer.

The full comprehension of the mechanical phenotype of cancer in vitro must follow analysis of all aspects of cell mechanics. While individual assays for viscoelasticity, force generation and mechanosensing give important data, combinatorial studies, which assess multiple mechanical attributes, will push our understanding of the interdependency of the elements of cell mechanics. The overlap of the field of physics and biology, as well as the modeling of complex mechanical phenomena in silico, has already made great strides in cancer cell characterization and highlights the importance of interdisciplinary approaches to analysis.

The therapeutics development field has been inspired by the beginnings of mechanical comprehension in cancer, and drugs have been developed that inhibit cell contractility, interstitial fluid pressure, and external stiffness. Specific inhibition of important molecules is a promising approach, though side effects and tumor heterogeneity may limit the applications. Our increased understanding of cancer cell mechanics, as achieved through in vitro assays, will be instrumental in determining the usefulness of mechanically driven therapeutics.

References

[1] C. Alibert, B. Goud, J.B. Manneville, Are cancer cells really softer than normal cells? Biol. Cell. 109 (2017) 167–189, https://doi.org/10.1111/boc.201600078.

[2] D. Wirtz, K. Konstantopoulos, P.C. Searson, The physics of cancer: the role of physical interactions and mechanical forces in metastasis, Nat. Rev. Cancer 11 (2011) 512–522, https://doi.org/10.1038/nrc3080.

[3] S. Tojkander, G. Gateva, P. Lappalainen, Actin stress fibers—assembly, dynamics and biological roles, J. Cell Sci. 125 (2012) 1855–1864, https://doi.org/10.1242/jcs.098087.

[4] T. Doerks, R.R. Copley, J. Schultz, C.P. Ponting, P. Bork, Systematic identification of novel protein domain families associated with nuclear functions, Genome Res 12 (2002) 47–56, https://doi.org/10.1101/gr.203201.

[5] M. Parri, P. Chiarugi, Rac and Rho GTPases in cancer cell motility control, Cell Commun. Signal. 8 (2010) 23, https://doi.org/10.1186/1478-811X-8-23.

[6] D.A. Fletcher, R.D. Mullins, Cell mechanics and the cytoskeleton, Nature 463 (2010) 485–492, https://doi.org/10.1038/nature08908.

[7] D. Stamenovic, S.M. Mijailovich, I.M. Tolic-Norrelykke, J. Chen, N. Wang, Cell prestress. II. Contribution of microtubules, Am. J. Physiol. Cell Physiol. 282 (2002) C617–C624.

[8] D.D. Tang, B.D. Gerlach, The roles and regulation of the actin cytoskeleton, intermediate filaments and microtubules in smooth muscle cell migration, Respir. Res. 18 (2017) 54, https://doi.org/10.1186/s12931-017-0544-7.

[9] K. Seltmann, A.W. Fritsch, J.A. Kas, T.M. Magin, Keratins significantly contribute to cell stiffness and impact invasive behavior, Proc. Natl. Acad. Sci. U. S. A. 110 (2013) 18507–18512, https://doi.org/10.1073/pnas.1310493110.

[10] S. Kumar, V.M. Weaver, Mechanics, malignancy, and metastasis: the force journey of a tumor cell, Cancer Metastasis Rev. 28 (2009) 113–127, https://doi.org/10.1007/s10555-008-9173-4.
[11] C.Y. Liu, H.H. Lin, M.J. Tang, Y.K. Wang, Vimentin contributes to epithelial-mesenchymal transition cancer cell mechanics by mediating cytoskeletal organization and focal adhesion maturation, Oncotarget 6 (2015) 15966–15983.
[12] W. Xu, et al., Cell stiffness is a biomarker of the metastatic potential of ovarian cancer cells, PLoS One 7 (2012) e46609, https://doi.org/10.1371/journal.pone.0046609.
[13] F. Broders-Bondon, T.H. Nguyen Ho-Bouldoires, M.E. Fernandez-Sanchez, E. Farge, Mechanotransduction in tumor progression: the dark side of the force, J. Cell Biol. 217 (2018) 1571–1587, https://doi.org/10.1083/jcb.201701039.
[14] S.P. Carey, T.M. D'Alfonso, S.J. Shin, C.A. Reinhart-King, Mechanobiology of tumor invasion: engineering meets oncology, Crit. Rev. Oncol. Hematol. 83 (2012) 170–183, https://doi.org/10.1016/j.critrevonc.2011.11.005.
[15] P. Friedl, Prespecification and plasticity: shifting mechanisms of cell migration, Curr. Opin. Cell Biol. 16 (2004) 14–23, https://doi.org/10.1016/j.ceb.2003.11.001.
[16] N.V. Krakhmal, M.V. Zavyalova, E.V. Denisov, S.V. Vtorushin, V.M. Perelmuter, Cancer invasions: patterns and mechanisms, Acta Nat. 7 (2014) 17–28.
[17] E. Cortes, et al., RAR-beta is downregulated in HCC & cirrhosis and its expression inhibits myosin-driven activation and durotaxis in hepatic stellate cells, Hepatology (2018), https://doi.org/10.1002/hep.30193.
[18] A. Chronopoulos, et al., ATRA mechanically reprograms pancreatic stellate cells to suppress matrix remodelling and inhibit cancer cell invasion, Nat, Commun. 7 (2016) 12630, https://doi.org/10.1038/ncomms12630.
[19] E. Moeendarbary, A.R. Harris, Cell mechanics: principles, practices, and prospects, Wiley Interdiscip. Rev. Syst. Biol. Med. 6 (2014) 371–388, https://doi.org/10.1002/wsbm.1275.
[20] N. Nijenhuis, X. Zhao, A. Carisey, C. Ballestrem, B. Derby, Combining AFM and acoustic probes to reveal changes in the elastic stiffness tensor of living cells, Biophys. J. 107 (2014) 1502–1512, https://doi.org/10.1016/j.bpj.2014.07.073.
[21] K. Haase, A.E. Pelling, Investigating cell mechanics with atomic force microscopy, J. R. Soc. Interface 12 (2015) 20140970, https://doi.org/10.1098/rsif.2014.0970.
[22] A.R. Harris, G.T. Charras, Experimental validation of atomic force microscopy-based cell elasticity measurements, Nanotechnology 22 (2011) 345102, https://doi,org/10.1088/0957-4484/22/34/345102.
[23] L. Puricelli, M. Galluzzi, C. Schulte, A. Podesta, P. Milani, Nanomechanical and topographical imaging of living cells by atomic force microscopy with colloidal probes, Rev. Sci. Instrum. 86 (2015) 033705, https://doi.org/10.1063/1.4915896.
[24] S.E. Cross, et al., AFM-based analysis of human metastatic cancer cells, Nanotechnology 19 (2008) 384003, https://doi,org/10.1088/0957-4484/19/38/384003.
[25] S.E. Cross, Y.S. Jin, J. Rao, J.K. Gimzewski, Nanomechanical analysis of cells from cancer patients, Nat. Nanotechnol. 2 (2007) 780–783, https://doi.org/10.1038/nnano.2007.388.
[26] S. Park, D. Koch, R. Cardenas, J. Kas, C.K. Shih, Cell motility and local viscoelasticity of fibroblasts, Biophys. J. 89 (2005) 4330–4342, https://doi.org/10.1529/biophysj.104.053462.
[27] A.N. Ketene, E.M. Schmelz, P.C. Roberts, M. Agah, The effects of cancer progression on the viscoelasticity of ovarian cell cytoskeleton structures, Nanomedicine 8 (2012) 93–102, https://doi,org/10.1016/j.nano.2011.05.012.
[28] E.M. Darling, S. Zauscher, J.A. Block, F. Guilak, A thin-layer model for viscoelastic, stress-relaxation testing of cells using atomic force microscopy: do cell properties reflect metastatic potential? Biophys. J. 92 (2007) 1784–1791, https://doi.org/10.1529/biophysj.106.083097.
[29] M.E. Grady, R.J. Composto, D.M. Eckmann, Cell elasticity with altered cytoskeletal architectures across multiple cell types, J. Mech. Behav. Biomed. Mater. 61 (2016) 197–207, https://doi.org/10.1016/j.jmbbm.2016.01.022.
[30] L. Xiao, M. Tang, Q. Li, A. Zhou, Non-invasive detection of biomechanical and biochemical responses of human lung cells to short time chemotherapy exposure using AFM and confocal Raman spectroscopy, Anal. Methods 5 (2013) 874, https://doi.org/10.1039/c2ay25951f.

[31] M.S. Salker, et al., LeftyA decreases actin polymerization and stiffness in human endometrial cancer cells, Sci. Rep. 6 (2016) 29370, https://doi.org/10.1038/srep29370.
[32] M. Lekka, Discrimination between normal and cancerous cells using AFM, Bionanoscience 6 (2016) 65–80, https://doi.org/10.1007/s12668-016-0191-3.
[33] J.R. Staunton, B.L. Doss, S. Lindsay, R. Ros, Correlating confocal microscopy and atomic force indentation reveals metastatic cancer cells stiffen during invasion into collagen I matrices, Sci. Rep. 6 (2016) 19686, https://doi.org/10.1038/srep19686.
[34] G. Coceano, et al., Investigation into local cell mechanics by atomic force microscopy mapping and optical tweezer vertical indentation, Nanotechnology 27 (2016) 065102, https://doi.org/10.1088/0957-4484/27/6/065102.
[35] Y. Abidine, V.M. Laurent, R. Michel, A. Duperray, C. Verdier, Local mechanical properties of bladder cancer cells measured by AFM as a signature of metastatic potential, Eur. Phys. J. Plus 130 (2015), https://doi.org/10.1140/epjp/i2015-15202-6.
[36] J. Ren, H. Huang, Y. Liu, X. Zheng, Q. Zou, An atomic force microscope study revealed two mechanisms in the effect of anticancer drugs on rate-dependent Young's modulus of human prostate cancer cells, PLoS One 10 (2015), https://doi.org/10.1371/journal.pone.0126107.
[37] A. Calzado-Martin, M. Encinar, J. Tamayo, M. Calleja, A. San Paulo, Effect of actin organization on the stiffness of living breast cancer cells revealed by peak-force modulation atomic force microscopy, ACS Nano 10 (2016) 3365–3374, https://doi.org/10.1021/acsnano.5b07162.
[38] M.O. Krisenko, A. Cartagena, A. Raman, R.L. Geahlen, Nanomechanical property maps of breast cancer cells as determined by multiharmonic atomic force microscopy reveal Syk-dependent changes in microtubule stability mediated by MAP1B, Biochemistry 54 (2015) 60–68, https://doi.org/10.1021/bi500325n.
[39] F. Ndoye, et al., The influence of lateral forces on the cell stiffness measurement by optical tweezers vertical indentation, Int. J. Optomechatronics 10 (2016) 53–62, https://doi.org/10.1080/15599612.2016.1149896.
[40] S. Suresh, et al., Connections between single-cell biomechanics and human disease states: gastrointestinal cancer and malaria, Acta Biomater. 1 (2005) 15–30, https://doi.org/10.1016/j.actbio.2004.09.001.
[41] L.D. Osborne, et al., TGF-β regulates LARG and GEF-H1 during EMT to affect stiffening response to force and cell invasion, Mol. Biol. Cell 25 (2014) 3528–3540, https://doi.org/10.1091/mbc.E14-05-1015.
[42] V. Swaminathan, et al., Mechanical stiffness grades metastatic potential in patient tumor cells and in cancer cell lines, Cancer Res. 71 (2011) 5075–5080, https://doi.org/10.1158/0008-5472.CAN-11-0247.
[43] W.J. Eldridge, A. Sheinfeld, M.T. Rinehart, A. Wax, Imaging deformation of adherent cells due to shear stress using quantitative phase imaging, Opt. Lett. 41 (2016) 352–355, https://doi.org/10.1364/OL.41.000352.
[44] N. Gal, D. Weihs, Intracellular mechanics and activity of breast cancer cells correlate with metastatic potential, Cell Biochem. Biophys. 63 (2012) 199–209, https://doi.org/10.1007/s12013-012-9356-z).
[45] M. Guo, et al., Probing the stochastic, motor-driven properties of the cytoplasm using force spectrum microscopy, Cell 158 (2014) 822–832, https://doi.org/10.1016/j.cell.2014.06.051.
[46] H. Zhao, et al., In vitro assay of cytoskeleton nanomechanics as a tool for screening potential anticancer effects of natural plant extract, tubeimoside I on human hepatoma (HepG2) cells, Chinese Sci. Bull. 58 (2013) 2576–2583, https://doi.org/10.1007/s11434-013-5757-7.
[47] M.F. Coughlin, et al., Cytoskeletal stiffness, friction, and fluidity of cancer cell lines with different metastatic potential, Clin. Exp. Metastasis 30 (2013) 237–250, https://doi.org/10.1007/s10585-012-9531-z.
[48] J.F. Berret, Local viscoelasticity of living cells measured by rotational magnetic spectroscopy, Nat. Commun. 7 (2016) 10134, https://doi.org/10.1038/ncomms10134.
[49] J. Guck, et al., Optical deformability as an inherent cell marker for testing malignant transformation and metastatic competence, Biophys. J. 88 (2005) 3689–3698, https://doi.org/10.1529/biophysj.104.045476.

[50] J.Y. Hwang, et al., Cell deformation by single-beam acoustic trapping: a promising tool for measurements of cell mechanics, Sci. Rep. 6 (2016) 27238, https://doi.org/10.1038/srep27238.
[51] M. Pachenari, et al., Mechanical properties of cancer cytoskeleton depend on actin filaments to microtubules content: investigating different grades of colon cancer cell lines, J. Biomech. 47 (2014) 373–379, https://doi.org/10.1016/j.jbiomech.2013.11.020.
[52] V.K. Chivukula, B.L. Krog, J.T. Nauseef, M.D. Henry, S.C. Vigmostad, Alterations in cancer cell mechanical properties after fluid shear stress exposure: a micropipette aspiration study, Cell Health Cytoskelet. 7 (2015) 25–35, https://doi.org/10.2147/CHC.S71852.
[53] D. Hartono, et al., On-chip measurements of cell compressibility via acoustic radiation, Lab Chip 11 (2011) 4072–4080, https://doi.org/10.1039/c1lc20687g.
[54] H.W. Hou, et al., Deformability study of breast cancer cells using microfluidics, Biomed. Microdevices 11 (2009) 557–564, https://doi.org/10.1007/s10544-008-9262-8.
[55] B. Hinz, et al., Recent developments in myofibroblast biology: paradigms for connective tissue remodeling, Am. J. Pathol. 180 (2012) 1340–1355, https://doi.org/10.1016/j.ajpath.2012.02.004.
[56] C.M. Fife, J.A. McCarroll, M. Kavallaris, Movers and shakers: cell cytoskeleton in cancer metastasis, Br. J. Pharmacol. 171 (2014), https://doi.org/10.1111/bph.2014.171.issue-24.
[57] C.M. Kraning-Rush, S.P. Carey, J.P. Califano, C.A. Reinhart-King, Quantifying traction stresses in adherent cells, Methods Cell Biol. 110 (2012) 139–178, https://doi.org/10.1016/B978-0-12-388403-9.00006-0.
[58] C.M. Kraning-Rush, J.P. Califano, C.A. Reinhart-King, Cellular traction stresses increase with increasing metastatic potential, PLoS One 7 (2012), https://doi.org/10.1371/journal.pone.0032572.
[59] B. Erdogan, et al., Cancer-associated fibroblasts promote directional cancer cell migration by aligning fibronectin, J. Cell Biol. 216 (2017) 3799–3816, https://doi.org/10.1083/jcb.201704053.
[60] X. Tang, A. Tofangchi, S.V. Anand, T.A. Saif, A novel cell traction force microscopy to study multicellular system, PLoS Comput. Biol. 10 (2014), https://doi.org/10.1371/journal.pcbi.1003631.
[61] V. Peschetola, et al., Time-dependent traction force microscopy for cancer cells as a measure of invasiveness, Cytoskeleton 70 (2013) 201–214, https://doi.org/10.1002/cm.21100.
[62] S. Makarchuk, N. Beyer, C. Gaiddon, W. Grange, P. Hebraud, Holographic traction force microscopy, Sci. Rep. 8 (2018) 3038, https://doi.org/10.1038/s41598-018-21206-2.
[63] S.C. Wuang, B. Ladoux, C.T. Lim, Probing the chemo-mechanical effects of an anti-cancer drug emodin on breast cancer cells, Cell. Mol. Bioeng. 4 (2011) 466–475, https://doi.org/10.1007/s12195-011-0163-1.
[64] Z. Li, et al., Cellular traction forces: a useful parameter in cancer research, Nanoscale 9 (2017) 19039–19044, https://doi.org/10.1039/c7nr06284b.
[65] N.Q. Balaban, et al., Force and focal adhesion assembly: a close relationship studies using elastic micropatterned substrates, Nat. Cell Biol. 3 (2001) 466–472.
[66] A.J. Ribeiro, A.K. Denisin, R.E. Wilson, B.L. Pruitt, For whom the cells pull: hydrogel and micropost devices for measuring traction forces, Methods 94 (2016) 51–64, https://doi.org/10.1016/j.ymeth.2015.08.005.
[67] T.I. Yin, et al., IEEE, 16th International Solid-State Sensors, Actuators and Microsystems Conference, 2011.
[68] W.R. Legant, et al., Measurement of mechanical tractions exerted by cells in three-dimensional matrices, Nat. Methods 7 (2010) 969–971, https://doi.org/10.1038/nmeth.1531.
[69] R. Poincloux, et al., Contractility of the cell rear drives invasion of breast tumor cells in 3D Matrigel, Proc. Natl. Acad. Sci. U. S. A 108 (2011) 1943–1948, https://doi.org/10.1073/pnas.1010396108.
[70] A. Kim, N. Lakshman, W.M. Petroll, Quantitative assessment of local collagen matrix remodeling in 3-D culture: the role of rho kinase, Exp. Cell Res. 312 (2006) 3683–3692.
[71] Y. Zhang, C. Ge, C. Zhu, K. Salaita, DNA-based digital tension probes reveal integrin forces during early cell adhesion, Nat. Commun. 5 (2014) 5167, https://doi.org/10.1038/ncomms6167.
[72] B.K. Robinson, E. Cortes, A.J. Rice, M. Sarper, A. Del Rio Hernandez, Quantitative analysis of 3D extracellular matrix remodelling by pancreatic stellate cells, Biol. Open 5 (2016) 875–882, https://doi.org/10.1242/bio.017632.

[73] M.T. Holley, A. YekrangSafakar, M. Maziveyi, S.K. Alahari, K. Park, Measurement of cell traction force with a thin film PDMS cantilever, Biomed. Microdevices 19 (2017) 97, https://doi.org/10.1007/s10544-017-0239-3.

[74] I. Rabinovitz, I.K. Gipson, A.M. Mercurio, Traction forces mediated by a6b4 integrin: implications for basement membrane organization and tumor invasion, Mol. Biol. Cell 12 (2001) 4030–4043.

[75] J.L. Tan, et al., Cells lying on a bed of microneedles: an approach to isolate mechanical force, Proc. Natl. Acad. Sci. U. S. A 100 (2003) 1484–1489, https://doi.org/10.1073/pnas.0235407100.

[76] C. Mohl, N. Kirchgessner, C. Schafer, B. Hoffmann, R. Merkel, Quantitative mapping of averaged focal adhesion dynamics in migrating cells by shape normalization, J. Cell Sci. 125 (2012) 155–165, https://doi.org/10.1242/jcs.090746.

[77] A.A. Khalili, M.R. Ahmad, A review of cell adhesion studies for biomedical and biological applications, Int. J. Mol. Sci. 16 (2015) 18149–18184, https://doi.org/10.3390/ijms160818149.

[78] Y. Chen, et al., Combined integrin phosphoproteomic analyses and small interfering RNA-based functional screening identify key regulators for cancer cell adhesion and migration, Cancer Res. 69 (2009) 3713–3720, https://doi.org/10.1158/0008-5472.CAN-08-2515.

[79] G. Sagvolden, I. Giaver, E.O. Petterson, J. Feder, Cell adhesion force microscopy, Proc. Natl. Acad. Sci. 96 (1999) 471–476.

[80] C.P. Palmer, et al., Single cell adhesion measuring apparatus (SCAMA): application to cancer cell lines of different metastatic potential and voltage-gated Na+ channel expression, Eur. Biophys. J. 37 (2008) 359–368, https://doi.org/10.1007/s00249-007-0219-2.

[81] G. Weder, et al., Measuring cell adhesion forces during the cell cycle by force spectroscopy, Biointerphases 4 (2009) 27–34, https://doi.org/10.1116/1.3139962.

[82] C.D. Reyes, A.J. Garcia, A centrifugation cell adhesion assay for high-throughput screening of biomaterial surfaces, J. Biomed. Mater. Res. A 67 (2003) 328–333.

[83] A. Fuhrmann, A. Banisadr, P. Beri, T.D. Tlsty, A.J. Engler, Metastatic state of cancer cells may be indicated by adhesion strength, Biophys. J. 112 (2017) 736–745, https://doi.org/10.1016/j.bpj.2016.12.038.

[84] K.W. Kwon, et al., Label-free, microfluidic separation and enrichment of human breast cancer cells by adhesion difference, Lab Chip 7 (2007) 1461–1468, https://doi.org/10.1039/b710054j.

[85] A. Spencer, A.B. Baker, High throughput label free measurement of cancer cell adhesion kinetics under hemodynamic flow, Sci. Rep. 6 (2016) 19854, https://doi.org/10.1038/srep19854.

[86] A.J. Rice, et al., Matrix stiffness induces epithelial-mesenchymal transition and promotes chemoresistance in pancreatic cancer cells, Oncogenesis 6 (2017) e352, https://doi.org/10.1038/oncsis.2017.54.

[87] D. Lachowski, et al., Substrate rigidity controls activation and durotaxis in pancreatic stellate cells, Sci. Rep. 7 (2017) 2506, https://doi.org/10.1038/s41598-017-02689-x.

[88] A.W.M. Haining, et al., Mechanotransduction in talin through the interaction of the R8 domain with DLC1, PLoS Biol. 16 (2018) e2005599, https://doi.org/10.1371/journal.pbio.2005599.

[89] M.A. Wozniak, K. Modzelewska, L. Kwong, P.J. Keely, Focal adhesion regulation of cell behavior, Biochim. Biophys. Acta 1692 (2004) 103–119, https://doi.org/10.1016/j.bbamcr.2004.04.007.

[90] T.Q. Uyeda, Y. Iwadate, N. Umeki, A. Nagasaki, S. Yumura, Stretching actin filaments within cells enhances their affinity for the myosin II motor domain, PLoS One 6 (2011), https://doi.org/10.1371/journal.pone.0026200.

[91] J. Seo, J. Kim, Regulation of Hippo signaling by actin remodeling, BMB Rep. 51 (2018) 151–156, https://doi.org/10.5483/BMBRep.2018.51.3.012.

[92] J. Toyjanova, et al., 3D viscoelastic traction force microscopy, Soft Matter 10 (2014) 8095–8106, https://doi.org/10.1039/c4sm01271b.

[93] R.N. Palchesko, L. Zhang, Y. Sun, A.W. Feinberg, Development of polydimethylsiloxane substrates with tunable elastic modulus to study cell mechanobiology in muscle and nerve, PLoS One 7 (2012), https://doi.org/10.1371/journal.pone.0051499.

[94] L. Trichet, et al., Evidence of a large-scale mechanosensing mechanism for cellular adaptation to substrate stiffness, Proc. Natl. Acad. Sci. U. S. A 109 (2012) 6933–6938, https://doi.org/10.1073/pnas.1117810109.

[95] M.A. Wozniak, R. Desai, P.A. Solski, C.J. Der, P.J. Keely, ROCK-generated contractility regulates breast epithelial cell differentiation in response to the physical properties of a three-dimensional collagen matrix, J. Cell Biol. 163 (2003) 583–595, https://doi.org/10.1083/jcb.200305010.
[96] F.C. Chang, et al., PEG-chitosan hydrogel with tunable stiffness for study of drug response of breast cancer cells, Polymers (Basel) 8 (2016), https://doi.org/10.3390/polym8040112.
[97] B. Trappmann, C.S. Chen, How cells sense extracellular matrix stiffness: a material's perspective, Curr. Opin. Biotechnol. 24 (2013) 948–953, https://doi.org/10.1016/j.copbio.2013.03.020.
[98] M. Cavo, et al., Microenvironment complexity and matrix stiffness regulate breast cancer cell activity in a 3D in vitro model, Sci. Rep. 6 (2016) 35367, https://doi.org/10.1038/srep35367.
[99] E.L. Baker, R.T. Bonnecaze, M.H. Zaman, Extracellular matrix stiffness and architecture govern intracellular rheology in cancer, Biophys. J. 97 (2009) 1013–1021, https://doi.org/10.1016/j.bpj.2009.05.054.
[100] A.M. Throm Quinlan, L.N. Sierad, A.K. Capulli, L.E. Firstenberg, K.L. Billiar, Combining dynamic stretch and tunable stiffness to probe cell mechanobiology in vitro, PLoS One 6 (2011), https://doi.org/10.1371/journal.pone.0023272.
[101] D. Huh, et al., Reconstituting organ-level lung functions on a chip, Science 328 (2010) 1662–1668.
[102] R. Krishnan, et al., Reinforcement versus fluidization in cytoskeletal mechanoresponsiveness, PLoS One 4 (2009) e5486, https://doi.org/10.1371/journal.pone.0005486.
[103] J.M. Tse, et al., Mechanical compression drives cancer cells toward invasive phenotype, Proc. Natl. Acad. Sci. U. S. A 109 (2012) 911–916, https://doi.org/10.1073/pnas.1118910109.
[104] F. Montel, et al., Stress clamp experiments on multicellular tumor spheroids, Phys. Rev. Lett. 107 (2011).
[105] M. Delarue, et al., Compressive stress inhibits proliferation in tumor spheroids through a volume limitation, Biophys. J. 107 (2014) 1821–1828, https://doi.org/10.1016/j.bpj.2014.08.031.
[106] D. Irimia, M. Toner, Spontaneous migration of cancer cells under conditions of mechanical confinement, Integr. Biol. (Camb.) 1 (2009) 506–512, https://doi.org/10.1039/b908595e.
[107] A.C. Weitz, et al., Functional assay of cancer cell invasion potential based on mechanotransduction of focused ultrasound, Front. Oncol. 7 (2017) 161, https://doi.org/10.3389/fonc.2017.00161.
[108] M.J. Mitchell, M.R. King, Fluid shear stress sensitizes cancer cells to receptor-mediated apoptosis via trimeric death receptors, New J. Phys. 15 (2013) 015008, https://doi.org/10.1088/1367-2630/15/1/015008.
[109] H. Qazi, Z.D. Shi, J.M. Tarbell, Fluid shear stress regulates the invasive potential of glioma cells via modulation of migratory activity and matrix metalloproteinase expression, PLoS One 6 (2011), https://doi.org/10.1371/journal.pone.0020348.
[110] L.H. Ting, et al., Flow mechanotransduction regulates traction forces, intercellular forces, and adherens junctions, Am. J. Physiol. Heart Circ. Physiol. 302 (2012) H2220–H2229, https://doi.org/10.1152/ajpheart.00975.2011.
[111] M. Summers Moss, B. Sisken, S. Zimmer, K.W. Anderson, Adhesion of nonmetastatic and highly metastatic breast cancer cells to endothelial cells exposed to shear stress, Biorheology 36 (1999) 359–371.
[112] J.M. Barnes, J.T. Nauseef, M.D. Henry, Resistance to fluid shear stress is a conserved biophysical property of malignant cells, PLoS One 7 (2012), https://doi.org/10.1371/journal.pone.0050973.
[113] N.J. Sniadecki, et al., Magnetic microposts as an approach to apply forces to living cells, Proc. Natl. Acad. Sci. U. S. A 104 (2007) 14553–14558, https://doi.org/10.1073/pnas.0611613104.
[114] A. Elosegui-Artola, et al., Force triggers YAP nuclear entry by regulating transport across nuclear pores, Cell 171 (2017) 1397–1410. e1314, https://doi.org/10.1016/j.cell.2017.10.008.
[115] Z. Al-Rekabi, A.E. Pelling, Cross talk between matrix elasticity and mechanical force regulates myoblast traction dynamics, Phys. Biol. 10 (2013) 066003, https://doi.org/10.1088/1478-3975/10/6/066003.
[116] F. Falleroni, V. Torre, D. Cojoc, Cell mechanotransduction with piconewton forces applied by optical tweezers, Front. Cell. Neurosci. 12 (2018), https://doi.org/10.3389/fncel.2018.00130.

Conclusion

Sylvain Ladame
Department of Bioengineering, Imperial College London, London, United Kingdom

Despite a constant increase in the global burden of cancer worldwide (based on the combined numbers of new cases and deaths), a decline in cancer-related age-standardized mortality rates (ASMRs) by ~1% per year was reported over the last decade for Europe and the United States [1,2]. All cancers combined mortality rates are projected to fall by 15% in the United Kingdom between 2014 and 2035, to 280 deaths per 100,000 people by 2035 [3]. This was made possible at least in part by constant improvements in the ways cancer patients are diagnosed and treated, using new technologies, drugs, and devices that have been designed and developed by bioengineers. Working hand in hand with potential end users (GPs, nurses, surgeons, etc.) and through direct interactions with cancer patients and their families, bioengineers have the capacity to provide both clinicians and researchers new tools and technologies that address urgent, yet unmet, medical needs in a format that is suitable for their users and for the patients who will be on the receiving end. This textbook has presented an overview of the recent and most significant advances made by bioengineers to help diagnose cancer patients earlier, identify those who will benefit from treatment, and offer them with the best suited therapy. Diagnosis through public screening offers the best chances of detecting cancers early, before the appearance of the first clinical symptoms when patients are most likely to respond to treatment. This, however, requires not only biomarkers that are both highly specific and highly sensitive but also low-cost sensing technologies that are capable of detecting their presence in often complex biological samples (tissues, bodily fluids, etc.) and when possible at the point of need. One of the most promising route toward more cancer screening tests relies on the detection of cancer-specific molecular biomarkers from liquid biopsies (e.g., blood or urine) using minimally invasive and miniaturized sensing technologies. This is a fast expanding field in which bioengineers play a major role, as highlighted in Chapters 1.1–1.4.

Imaging technologies such as magnetic resonance imaging, optical imaging, and ultrasound imaging often provide complementary information that are typically not accessible from screening tests. This includes details on the size and location of the tumor that are both essential in guiding the clinician prior to or during targeted tissue biopsy or tumor excision. Technological advances in this area have led to significant improvements in imaging resolution, enabling the visualization of smaller tumors for early cancer detection. The physical principles behind each technique and their applications for cancer imaging were presented in Chapters 2.1–2.3.

Although early diagnosis is commonly associated with higher survival rate, reducing cancer mortality requires cancer patients to be offered the best suited treatment, whether it involves chemotherapy, immunotherapy, radiotherapy, surgical intervention, or any combination of them. Bioengineers play a front role in many of these areas by designing new drugs, improving their delivery specifically to the tumor site to minimize side effects, and providing surgeons with new robotic tools to improve precision. All these aspects were discussed in Chapters 3.1–3.3.

But all these technological advances also require a detailed understanding of how cancer cells develop, proliferate, and respond to treatment. This can be best assessed by providing the research community with carefully designed two- or three-dimensional in vitro models, as presented in Chapters 4.1 and 4.2.

In brief, this textbook intended to bridge the gap between two fast evolving disciplines: (molecular) bioengineering and cancer biology. Cancer research is a prime example of an area where new technologies are finding their way, through research to industry, radically altering the way this disease is being diagnosed, monitored, and treated. Although cancer survival has doubled in the last 40 years, progress has not advanced equally for all forms of the disease, and strong geographical inequalities remain that highlight the critical influence of accessibility to new technologies. Biomedical engineers hold the key to most current unmet needs in oncology. By providing a general overview of the impact bioengineers have had in this field, this book will hopefully inspire bioengineering students and pave the way to the next generation of researchers in both bioengineering and associated fields.

References

[1] C. La Vecchia, M. Rota, M. Malvezzi, E. Negri, Potential for improvement in cancer management: reducing mortality in the European Union, Oncologist 20 (2015) 495–498.

[2] R.L. Siegel, K.D. Miller, A. Jemal, Cancer statistics, 2015, CA Cancer J. Clin. 65 (2015) 5–29.

[3] C.R. Smittenaar, K.A. Petersen, K. Stewart, N. Moitt, Cancer incidence and mortality projections in the UK until 2035, Br. J. Cancer 115 (2016) 1147–1155.

Index

Note: Page numbers followed by *f* indicate figures and *t* indicate tables.

D

E

J

K

L

M

N

T

W

X

Y

Z

Made in the USA
Monee, IL
04 March 2026

45489303R00236